Process Modelling, Identification, and Control

Ján Mikleš · Miroslav Fikar

Process Modelling, Identification, and Control

With 187 Figures and 13 Tables

Springer

Ján Mikleš
Department of Information Engineering
and Process Control
Slovak University of Technology in Bratislava
Radlinského 9
812 37 Bratislava 1
Slovak Republic
jan.mikles@stuba.sk

Miroslav Fikar
Department of Information Engineering
and Process Control
Slovak University of Technology in Bratislava
Radlinského 9
812 37 Bratislava 1
Slovak Republic
miroslav.fikar@stuba.sk

ISBN 978-3-642-09112-4 e-ISBN 978-3-540-71970-0

Springer is a part of Springer Science+Business Media
springer.com

Cover design: Erich Kirchner, Heidelberg

To our wives, Eva and Zuzana, that showed so much
understanding during the preparation of the book

Preface

Control and automation in its broadest sense plays a fundamental role in process industries. Control assures stability of technologies, disturbance attenuation, safety of equipment and environment as well as optimal process operation from economic point of view. This book intends to present modern automatic control methods and their applications in process control in process industries. The processes studied mainly involve mass and heat transfer processes and chemical reactors.

It is assumed that the reader has already a basic knowledge about controlled processes and about differential and integral calculus as well as about matrix algebra. Automatic control problems involve mathematics more than it is usual in other engineering disciplines. The book treats problems in a similar way as it is in mathematics. The problem is formulated at first, then the theorem is stated. Only necessary conditions are usually proved and sufficiency is left aside as it follows from the physical nature of the problem solved. This helps to follow the engineering character of problems.

The intended audience of this book includes graduate students but can also be of interest to practising engineers or applied scientists.

Organisation of the book follows the requirement of presentation of the process control algorithms that take into account changes of static and dynamic properties of processes. There are several possibilities to capture these changes. As the control objects in process industries are mainly nonlinear and continuous-time, the first possibility is to use nonlinear mathematical models in the form of partial differential equations, or after some simplifications, nonlinear differential equations. Usage of these models in control is computationally very demanding. Therefore, it is suitable to use robust control methods based on linearised models with uncertainties. However, strong nonlinearities can decrease performance of robust control methods and adaptive control algorithms can possibly be used.

The book presents process control using self-tuning (ST) controllers based on linear models. There are several reasons for their use. ST controllers can take into account changes of operating regimes of nonlinear processes as well as changes in physical process parameters. These changes can be caused by ageing or wearing of materials, etc. Finally, ST controllers can eliminate unmeasured disturbances, process noise, and unmodelled dynamic properties of processes.

The last chapter of the book that discusses adaptive control makes use of all previous parts. Here, discrete and continuous-time ST control of singlevariable and multivariable processes is investigated. For its implementation it is necessary to have in disposition an algorithm of recursive identification that provides an up-to-date model of the controlled process. This model is then used in feedback process control based on either explicit or implicit pole placement methods.

A basic requirement for a successful control is to have a suitable model for control. It is known that the process design in process industries involves nonlinear models with states evolving in space and time. This corresponds to models described by partial differential equations. Automatic control based such models is very difficult as it requires additional capacity of control computers and can be used only in some special cases. Self-tuning control requires linear models only described by ordinary differential equations of low order. This is caused by the fact that higher order models can cause singularities in calculation of the control law.

The first chapter of the book explains basic concepts that can be encountered in process control.

The second chapter describes principles of process modelling. Some principal models as a packed absorption column, a distillation column, and a chemical reactor are developed. Mathematical model of a plug-flow reactor serves for explanation of transport time delay that together with dynamic time delay forms a basic phenomenon in continuous-time process models. Description of specific process models is accompanied by assumptions when the models in the form of ordinary differential equations can be valid. Finally, general process models are explained together with the method of linearisation and the notion of systems and processes.

The third chapter explains briefly the use of the Laplace transform in process modelling. State-space process models obtained in the previous chapter are analysed. A concept of state is defined and the Lyapunov stability of continuous-time processes is investigated. The physical interpretation of theoretical facts is explained on a model of a U-tube. Conditions of controllability and observability are applied to a continuously stirred tank reactor. The state-space process models obtained from material and heat balances are transformed into input-output models for the purposes of identification and control design.

The aim of the fourth chapter on dynamic behaviour of processes is to show responses of processes to deterministic and stochastic inputs. Computer simulations explain how Euler and Runge-Kutta methods for integration of differential equations can be used. Examples include MATLAB/Simulink, C, and BASIC implementations. Process examples in this chapter demonstrate differences between original nonlinear and linearised models. The part dealing with stochastic characteristics explains random variables and processes, the definition of the white noise and others needed in state observation and in stochastic control.

Direct digital process control that constitutes a part of discrete-time adaptive control needs discrete-time process models. Therefore, the fifth chapter deals with the Z-transform, conversion between continuous-time and sampled-data processes. Also discussed are stability, controllability, observability, and basic properties of discrete-time systems.

Both adaptive and non-adaptive control methods are usually based on a mathematical model of a process. The sixth chapter is divided into two parts. In the first one identification of process models is based on their response to the step change on input. The second part deals in more detail with recursive least-squares (RLS) methods. Although primarily developed for discrete-time models, it is shown how to use it also for continuous-time models. The RLS method described in the book has been in use for more than 20 years at the authors place. Here we describe its implementation in MATLAB/Simulink environment as a part of the IDTOOL Toolbox available for free in the Internet (see the web page of the book). IDTOOL includes both discrete-time and continuous-time process identification.

The aim of the seventh chapter is to show the basic feedback control configuration, open and closed-loop issues, steady state behaviour and control performance indices. The second part deals with the mostly used controllers in practise - PID controllers. Several problems solved heuristically with PID controllers are then more rigorously handled in next chapters dealing with optimal and predictive control.

The most important part of the book covers design of feedback controllers from the pole-placement point of view. Optimal control follows from the principle of minimum and from dynamic programming. At first, the problem of pole-placement is investigated in a detail based on state-space process models. The controller is then designed from a combination of a feedback and a state observer. This controller can then also be interpreted from input-output point of view and results in polynomial pole-placement controller design. The Youla-Kučera parametrisation is used to find all stabilising controllers for a given process. Also, its dual parametrisation is used to find all processes stabilised by a given controller. The Youla-Kučera parametrisation makes it possible to unify control design for both continuous-time and discrete-time systems using the polynomial approach. This is the reason why mostly continuous-time systems are studied. Dynamic programming is explained for both continuous-time and discrete-time systems. The part devoted to the Youla-Kučera parametrisation treats also finite time control – dead-beat (DB) that describes a discrete-time control of a continuous-time process that cannot be realised using a continuous-time controller. State-space quadratically (Q) optimal control of linear (L) process models, i.e. LQ control with observer and its polynomial equivalent are interpreted as the pole-placement problem. Similarly, LQ control with the Kalman filter (LQG) or H2 control can also be interpreted as the pole-placement control problem. An attention is paid to integral properties of controllers and it is shown that optimal controllers are often of the form of modified PID controllers. The examples in the chapter make a heavy use of the Polynomial Toolbox for MATLAB (`http://www.polyx.com`).

The chapter nine extends the optimal control design with predictive control approach. It is explained why this approach predetermines discrete-time process control. At the beginning, the basic principles and ingredients of predictive control are explained. The derivation of the predictors and controllers is then shown for both input-output and state-space process models. The chapter also deals with incorporation of constraints on input variables and discusses the issue of stability and stabilisability of predictive closed-loop system. Finally, the chapter concludes with both singlevariable and multivariable tuning and show examples of applications in process control.

The last chapter combines results of previous chapters in adaptive control. We prefer to use the so-called explicit, or self-tuning approach to adaptive control where process parameters are estimated on-line using recursive least-squares methods. Then, at each sampling instant, the available model is used in control design.

The book has an ambition to integrate interpretations of continuous-time and discrete-time control, deterministic and stochastic control with incorporation of sets of all stabilising controllers. Notations used in separate parts of the books have been adapted to this aim.

Some of the programs and figures of the examples in the book marked in the margin are freely available at the web page of the book: [www]
`http://www.kirp.chtf.stuba.sk/~fikar/books/mic/`.

The program sources are for MATLAB with toolboxes Simulink, Control System Toolbox, Polynomial Toolbox (`http://www.polyx.cz`), Multiparametric Toolbox (MPT, `http://control.ee.ethz.ch/~mpt`) and, IDTOOL (from the web site). The examples are verified in both simulation and real-time test cases using Real-time MATLAB toolbox and the industrial system SIMATIC STEP7.

Each chapter includes bibliographic references related to the text. Besides the classical material on process dynamics or feedback control, there are some parts of the book that follow from the journal publications of the authors.

Acknowledgements

The book has been in preparation in the form of study texts for almost 10 years and came into existence mainly at Department Information Engineering and Process Control established at Faculty of Chemical and Food Technology, Slovak University of Technology in 1960.

The authors wish to express their gratitude to all present and past employees of the department that helped to provide a creative environment and to all students, PhD students that commented the manuscript. Namely we would like to thank to our colleagues Ľ. Čirka, Š. Kožka, F. Jelenčiak, J. Dzivák, T. Hirmajer, K. Calík, L. Dermíšek, and to associated prof. M. Bakošová from Slovak University of Technology for comments to the manuscript that helped to find some errors and problems.

Authors also thank for inspiration and ideas to all with whom they had dialogues during numerous stays abroad at TU Darmstadt, Uni Birmingham, TU Lyngby, Uni Dortmund, and ENSIC Nancy.

We are very thankful for the constructive remarks in identification to prof. Prokop from T. Bata University in Zlín, in modelling and dynamic simulations of continuous-time processes to associated prof. Lavrin from Technical University of Košice and to prof. Alexík from University of Žilina. We thank for comments in parts of identification and optimal control to prof. Dostál from T. Bata University in Zlín and to prof. Roháľ from Faculty of Mechanical Engineering at Slovak University of Technology in Bratislava. The part on predictive control has greatly been enhanced by suggestions of M. Kvasnica and his work on MPT Toolbox and explicit predictive control during his stay at ETH Zürich.

We also thank to prof. Kučera and prof. Šebek from Czech Technical University in Prague for many inspiring ideas and long cooperation that has contributed to the conception of the book. We have been involved in the project EUROPOLY supported by European Union and have had the possibility to use the most recent versions of the Polynomial Toolbox for MATLAB.

We are thankful to Dr. Hippe from University of Nürnberg-Erlangen for comments to the manuscript.

Parts of the book were prepared during the stays of both authors at Ruhr University in Bochum that were supported by the Alexander von Humboldt Foundation. This support is very gratefully acknowledged. In this context we particularly thank to prof. Unbehauen from the Ruhr University in Bochum for his support, cooperation, and comments to the book.

Bratislava,
February 2007

Ján Mikleš
Miroslav Fikar

Contents

1 Introduction 1
1.1 Topics in Process Control 1
1.2 An Example of Process Control 3
1.2.1 Process 3
1.2.2 Steady-State 3
1.2.3 Process Control 4
1.2.4 Dynamical Properties of the Process 4
1.2.5 Feedback Process Control 5
1.2.6 Transient Performance of Feedback Control 6
1.2.7 Block Diagram 8
1.2.8 Feedforward Control 9
1.3 Development of Process Control 10
1.4 References 11

2 Mathematical Modelling of Processes 13
2.1 General Principles of Modelling 13
2.2 Examples of Dynamic Mathematical Models 16
2.2.1 Liquid Storage Systems 16
2.2.2 Heat Transfer Processes 19
2.2.3 Mass Transfer Processes 26
2.2.4 Chemical and Biochemical Reactors 31
2.3 General Process Models 33
2.4 Linearisation 39
2.5 Systems, Classification of Systems 45
2.6 References 46
2.7 Exercises 47

3 Analysis of Process Models 51
3.1 The Laplace Transform 51
3.1.1 Definition of the Laplace Transform 51
3.1.2 Laplace Transforms of Common Functions 53

3.1.3 Properties of the Laplace Transform ... 57
3.1.4 Inverse Laplace Transform ... 61
3.1.5 Solution of Linear Differential Equations by Laplace Transform Techniques ... 62
3.2 State-Space Process Models ... 66
3.2.1 Concept of State ... 67
3.2.2 Solution of State-Space Equations ... 67
3.2.3 Canonical Transformation ... 70
3.2.4 Stability, Controllability, and Observability of Continuous-Time Systems ... 71
3.2.5 Canonical Decomposition ... 83
3.3 Input-Output Process Models ... 84
3.3.1 SISO Continuous Systems with Constant Coefficients ... 84
3.3.2 Transfer Functions of Systems with Time Delays ... 93
3.3.3 Algebra of Transfer Functions for SISO Systems ... 96
3.3.4 Input Output Models of MIMO Systems - Matrix of Transfer Functions ... 100
3.3.5 BIBO Stability ... 103
3.3.6 Transformation of I/O Models into State-Space Models 104
3.3.7 I/O Models of MIMO Systems - Matrix Fraction Descriptions ... 107
3.4 References ... 112
3.5 Exercises ... 114

4 **Dynamical Behaviour of Processes** ... 117
4.1 Time Responses of Linear Systems to Unit Impulse and Unit Step ... 117
4.1.1 Unit Impulse Response ... 117
4.1.2 Unit Step Response ... 119
4.2 Computer Simulations ... 125
4.2.1 The Euler Method ... 126
4.2.2 The Runge-Kutta method ... 127
4.2.3 Runge-Kutta Method for a System of Differential Equations ... 129
4.2.4 Time Responses of Liquid Storage Systems ... 135
4.2.5 Time Responses of CSTR ... 137
4.3 Frequency Analysis ... 145
4.3.1 Response of the Heat Exchanger to Sinusoidal Input Signal ... 145
4.3.2 Definition of Frequency Responses ... 149
4.3.3 Frequency Characteristics of a First Order System ... 153
4.3.4 Frequency Characteristics of a Second Order System ... 156
4.3.5 Frequency Characteristics of an Integrator ... 158
4.3.6 Frequency Characteristics of Systems in a Series ... 159
4.4 Statistical Characteristics of Dynamic Systems ... 162

4.4.1 Fundamentals of Probability Theory ... 162
4.4.2 Random Variables ... 163
4.4.3 Stochastic Processes ... 169
4.4.4 White Noise ... 176
4.4.5 Response of a Linear System to Stochastic Input ... 178
4.4.6 Frequency Domain Analysis of a Linear System with Stochastic Input ... 183
4.5 References ... 185
4.6 Exercises ... 187

5 Discrete-Time Process Models ... 189
5.1 Computer Controlled and Sampled Data Systems ... 189
5.2 $\mathcal{Z}$ – Transform ... 195
5.3 Discrete-Time Transfer Functions ... 200
5.4 Input-Output Discrete-Time Models – Difference Equations ... 204
5.4.1 Direct Digital Control ... 205
5.5 State-Space Discrete-Time Models ... 207
5.6 Properties of Discrete-Time Systems ... 211
5.6.1 Stability ... 211
5.6.2 Controllability ... 212
5.6.3 Observability ... 212
5.6.4 Discrete-Time Feedback Systems – Control Performance 212
5.7 Examples of Discrete-Time Process Models ... 214
5.7.1 Discrete-Time Tank Model ... 214
5.7.2 Discrete-Time Model of Two Tanks in Series ... 215
5.7.3 Steady-State Discrete-Time Model of Heat Exchangers in Series ... 216
5.8 References ... 217
5.9 Exercises ... 218

6 Process Identification ... 221
6.1 Introduction ... 221
6.1.1 Models of Linear Dynamic Systems ... 223
6.2 Identification from Step Responses ... 225
6.2.1 First Order System ... 226
6.2.2 Underdamped Second Order System ... 227
6.2.3 System of a Higher Order ... 230
6.3 Least Squares Methods ... 233
6.3.1 Recursive Least Squares Method ... 235
6.3.2 Modifications of Recursive Least Squares ... 241
6.3.3 Identification of a Continuous-time Transfer Function ... 245
6.4 References ... 250
6.5 Exercises ... 251

7 The Control Problem and Design of Simple Controllers 253
7.1 Closed-Loop System 253
7.1.1 Feedback Control Problem Definition 255
7.2 Steady-State Behaviour 256
7.3 Control Performance Indices 257
7.3.1 Time Domain 257
7.3.2 Integral Criteria 259
7.3.3 Control Quality and Frequency Indices 260
7.3.4 Poles 263
7.4 PID Controller 266
7.4.1 Description of Components 266
7.4.2 PID Controller Structures 274
7.4.3 Setpoint Weighting 275
7.4.4 Simple Rules for Controller Selection 276
7.4.5 Practical Aspects 277
7.4.6 Controller Tuning 281
7.5 References 294
7.6 Exercises 294

8 Optimal Process Control 297
8.1 Problem of Optimal Control and Principle of Minimum 297
8.2 Feedback Optimal Control 305
8.3 Optimal Tracking, Servo Problem, and Disturbance Rejection . 317
8.3.1 Tracking Problem 318
8.3.2 Servo Problem 320
8.3.3 LQ Control with Integral Action 321
8.4 Dynamic Programming 322
8.4.1 Continuous-Time Systems 322
8.4.2 Dynamic Programming for Discrete-Time Systems 329
8.4.3 Optimal Feedback 331
8.5 Observers and State Estimation 335
8.5.1 State Observation 335
8.5.2 Kalman Filter 337
8.6 Analysis of State Feedback with Observer and Polynomial Pole Placement 340
8.6.1 Properties of State Feedback with Observer 340
8.6.2 Input-Output Interpretation of State Feedback with Observer 345
8.6.3 Diophantine Equations 351
8.6.4 Polynomial Pole Placement Control Design 354
8.6.5 Integrating Behaviour of Controller 355
8.6.6 Polynomial Pole Placement Design for Multivariable Systems 361
8.7 The Youla-Kučera Parametrisation 365
8.7.1 Fractional Representation 366

8.7.2 Parametrisation of Stabilising Controllers 368
8.7.3 Parametrised Controller in the State-Space Representation 370
8.7.4 Parametrisation of Transfer Functions of the Closed-Loop System 371
8.7.5 Dual Parametrisation 372
8.7.6 Parametrisation of Stabilising Controllers for Multivariable Systems 374
8.7.7 Parametrisation of Stabilising Controllers for Discrete-Time Systems 375
8.8 Observer LQ Control, State-Space and Polynomial Interpretations 380
8.8.1 Polynomial LQ Control Design with Observer for Singlevariable Systems 381
8.8.2 Polynomial LQ Design with State Estimation for Multivariable Systems 385
8.9 LQG Control, State-Space and Polynomial Interpretation 388
8.9.1 Singlevariable Polynomial LQG Control Design 388
8.9.2 Multivariable Polynomial LQG Control Design 392
8.10 H_2 Optimal Control 394
8.11 References 397
8.12 Exercises 400

9 Predictive Control 403
9.1 Introduction 403
9.2 Ingredients of MBPC 404
9.2.1 Models 404
9.2.2 Cost Function 406
9.3 Derivation and Implementation of Predictive Control 407
9.3.1 Derivation of the Predictor 407
9.3.2 Calculation of the Optimal Control 409
9.3.3 Closed-Loop Relations 411
9.3.4 Derivation of the Predictor from State-Space Models 412
9.3.5 Multivariable Input-Output Case 416
9.3.6 Implementation 416
9.3.7 Relation to Other Approaches 418
9.3.8 Continuous-Time Approaches 418
9.4 Constrained Control 419
9.5 Stability Results 420
9.5.1 Stability Results in GPC 421
9.5.2 Terminal Constraints 421
9.5.3 Infinite Horizons 425
9.5.4 Finite Terminal Penalty 425
9.6 Explicit Predictive Control 426
9.6.1 Quadratic Programming Definition 426

9.6.2 Explicit Solution ... 427
9.6.3 Multi-Parametric Toolbox ... 429
9.7 Tuning ... 431
9.7.1 Tuning based on the First Order Model ... 431
9.7.2 Multivariable Tuning based on the First Order Model ... 432
9.7.3 Output Horizon Tuning ... 433
9.7.4 λ Tuning ... 434
9.7.5 Tuning based on Model Following ... 434
9.7.6 The C polynomial ... 435
9.8 Examples ... 436
9.8.1 A Linear Process ... 436
9.8.2 Neural Network based GPC ... 438
9.8.3 pH Control ... 439
9.9 References ... 442
9.10 Exercises ... 444

10 Adaptive Control ... 445
10.1 Discrete-Time Self-Tuning Control ... 446
10.2 Continuous-Time Self-Tuning Control ... 447
10.3 Examples of Self-Tuning Control ... 447
10.3.1 Discrete-Time Adaptive Dead-Beat Control of a Second Order System ... 448
10.3.2 Continuous-Time Adaptive LQ Control of a Second Order System ... 451
10.3.3 Continuous-Time Adaptive MIMO Pole Placement Control ... 456
10.3.4 Adaptive Control of a Tubular Reactor ... 459
10.4 References ... 463

References ... 465

Index ... 475

List of Figures

1.1 A simple heat exchanger ... 3
1.2 Response of the process controlled with proportional feedback controller for a step change of disturbance variable ϑ_v ... 7
1.3 The scheme of the feedback control for the heat exchanger ... 8
1.4 The block scheme of the feedback control of the heat exchanger 9

2.1 A liquid storage system ... 16
2.2 An interacting tank-in-series process ... 17
2.3 Continuous stirred tank heated by steam in jacket ... 19
2.4 Series of heat exchangers ... 21
2.5 Double-pipe steam-heated exchanger and temperature profile along the exchanger length in steady-state ... 23
2.6 Temperature profile of ϑ in an exchanger element of length $d\sigma$ for time dt ... 23
2.7 A metal rod ... 24
2.8 A scheme of a packed countercurrent absorption column ... 26
2.9 Scheme of a continuous distillation column ... 28
2.10 Model representation of i-th tray ... 29
2.11 A nonisothermal CSTR ... 31
2.12 Classification of dynamical systems ... 46
2.13 A cone liquid storage process ... 48
2.14 Well mixed heat exchanger ... 48
2.15 A well mixed tank ... 49
2.16 Series of two CSTRs ... 50
2.17 A gas storage tank ... 50

3.1 A step function ... 53
3.2 An original and delayed function ... 59
3.3 A rectangular pulse function ... 60
3.4 A mixing process ... 68
3.5 A U-tube ... 74

3.6 Time response of the U-tube for initial conditions $(1,0)^T$ 76
3.7 Constant energy curves and state trajectory of the U-tube in the state plane 76
3.8 Canonical decomposition 84
3.9 Block scheme of a system with transfer function $G(s)$ 85
3.10 Two tanks in a series 86
3.11 Block scheme of two tanks in a series 88
3.12 Serial connection of n tanks 90
3.13 Block scheme of n tanks in a series 90
3.14 Simplified block scheme of n tanks in a series 91
3.15 Block scheme of a heat exchanger 92
3.16 Modified block scheme of a heat exchanger 92
3.17 Block scheme of a double-pipe heat exchanger 97
3.18 Serial connection 97
3.19 Parallel connection 97
3.20 Feedback connection 98
3.21 Moving of the branching point against the direction of signals .. 99
3.22 Moving of the branching point in the direction of signals 99
3.23 Moving of the summation point in the direction of signals 100
3.24 Moving of the summation point against the direction of signals . 100
3.25 Block scheme of controllable canonical form of a system 106
3.26 Block scheme of controllable canonical form of a second order system 107
3.27 Block scheme of observable canonical form of a system 108

4.1 Impulse response of the first order system 119
4.2 Step response of a first order system 120
4.3 Step responses of a first order system with time constants T_1, T_2, T_3 121
4.4 Step responses of the second order system for the various values of ζ 123
4.5 Step responses of the system with n equal time constants 124
4.6 Block scheme of the n-th order system connected in a series with time delay 124
4.7 Step response of the first order system with time delay 125
4.8 Step response of the second order system with the numerator $B(s) = b_1 s + 1$ 125
4.9 Simulink block scheme 133
4.10 Results from simulation 134
4.11 Simulink block scheme for the liquid storage system 137
4.12 Response of the tank to step change of q_0 138
4.13 Simulink block scheme for the nonlinear CSTR model 143
4.14 Responses of dimensionless deviation output concentration x_1 to step change of q_c 146

4.15 Responses of dimensionless deviation output temperature x_2 to step change of q_c . . . 146
4.16 Responses of dimensionless deviation cooling temperature x_3 to step change of q_c . . . 147
4.17 Ultimate response of the heat exchanger to sinusoidal input . . . 149
4.18 The Nyquist diagram for the heat exchanger . . . 153
4.19 The Bode diagram for the heat exchanger . . . 154
4.20 Asymptotes of the magnitude plot for a first order system . . . 155
4.21 Asymptotes of phase angle plot for a first order system . . . 156
4.22 Asymptotes of magnitude plot for a second order system . . . 157
4.23 Bode diagrams of an underdamped second order system ($Z_1 = 1$, $T_k = 1$) . . . 158
4.24 The Nyquist diagram of an integrator . . . 159
4.25 Bode diagram of an integrator . . . 160
4.26 The Nyquist diagram for the third order system . . . 161
4.27 Bode diagram for the third order system . . . 161
4.28 Graphical representation of the law of distribution of a random variable and of the associated distribution function . . . 163
4.29 Distribution function and corresponding probability density function of a continuous random variable . . . 166
4.30 Realisations of a stochastic process . . . 170
4.31 Power spectral density and auto-correlation function of white noise . . . 177
4.32 Power spectral density and auto-correlation function of the process given by (4.276) and (4.277) . . . 178
4.33 Block-scheme of a system with transfer function $G(s)$ . . . 183

5.1 A digital control system controlling a continuous-time process . . 189
5.2 Transformation of a continuous-time to a discrete-time signal . . 190
5.3 Continuous-time signal step reconstruction from its discrete-time counterpart . . . 191
5.4 A possible loss of information caused by sampling . . . 192
5.5 An ideal sampler . . . 192
5.6 Spectral density of a signal $y(t)$ . . . 193
5.7 Spectral density of a sampled signal $y^*(t)$ . . . 194
5.8 An ideal sampler and an impulse modulated signal . . . 195
5.9 Sample-and-hold device in series with a continuous-time system 201
5.10 Sampler and zero-order hold in series with a continuous-time system $G(s)$ . . . 202
5.11 Direct digital control . . . 205
5.12 Zero-order hold in series with a continuous-time system . . . 207
5.13 A feedback discrete-time system . . . 212
5.14 Series of heat exchangers . . . 216

6.1 Step response of the first order system . . . 226

6.2 Measured step response of a chemical reactor using the input change $\Delta u = 10$ (left), approximation of step response by the first order step response (right) 228
6.3 Step response of the second order underdamped system 228
6.4 Measured step response of an underdamped system (left). Normalised and approximated step response (right) 230
6.5 Step response of a higher order system 231
6.6 Normalised step response of the chemical reactor with lines drawn to estimate T_u, T_n (left), approximated step response (right) 233
6.7 Simulink schema for parameter estimation of the second order discrete system 244
6.8 Estimated parameter trajectories for the second order discrete system 246
6.9 Alternate Simulink schema for parameter estimation of the second order discrete-time system using blocks from library IDTOOL 246
6.10 Block scheme of identification of a continuous-time transfer function 247
6.11 Simulink schema for estimation of parameters of the second order continuous-time system 249
6.12 Estimated parameter trajectories for the second order continuous-time system 249
6.13 Alternate Simulink schema for parameter estimation of the second order continuous-time system using blocks from library IDTOOL 249

7.1 Detailed and simplified closed-loop system scheme 254
7.2 Open-loop system 255
7.3 A typical response of a controlled process to a step change of a setpoint variable 258
7.4 A typical response of the controlled process to a step change of disturbance 259
7.5 Gain and phase margins for a typical continuous system 261
7.6 Bandwidth, resonant peak, and resonance frequency in magnitude frequency response 262
7.7 Suitable region for pole locations of a closed-loop system 264
7.8 Curves with constant values of ζ, ω_0 ($\zeta_1 < \zeta_2 < \zeta_3, \omega_1 < \omega_2 < \omega_3$) 265
7.9 Response of a heat exchanger to a step change of the d with a P controller 269
7.10 Response of a heat exchanger to a step change of the w with a P controller 269
7.11 Proportional control to the setpoint value $w = 1$ with various gains Z_R of the controller 270
7.12 Heat exchanger – disturbance rejection with the PI controller . . 272

7.13 Heat exchanger – setpoint tracking with the PI controller 272
7.14 PI control with setpoint $w = 1$ and several values of T_I 273
7.15 Graphical representation of derivative controller effects 273
7.16 PD control with setpoint $w = 1$ and several values of T_D 274
7.17 Closed-loop system with a two degree-of-freedom controller 276
7.18 Windup phenomenon. (a) Problem, (b) Solution 278
7.19 Anti-windup controller 278
7.20 Step responses of selected standard transfer functions 282
7.21 The PID controller tuned according to Strejc 285
7.22 Heat exchanger control using pole placement design 287
7.23 Closed-Loop system with a relay 289
7.24 PID control tuned with the Ziegler-Nichols method 290
7.25 PID control using the Åström-Hägglund method 293

8.1 Optimal trajectories of input and state variables of the heat exchanger 306
8.2 Trajectories of $P(t)$ for various values of r 310
8.3 Optimal feedback control 311
8.4 Two heat exchangers in series 312
8.5 Simulink program to simulate feedback optimal control in Example 8.3 315
8.6 LQ optimal trajectories x_1, x_2, u_1 of heat exchangers 315
8.7 Optimal LQ output regulation 318
8.8 Optimal trajectory in n-dimensional state space 323
8.9 Optimal trajectory 324
8.10 Time scale with time intervals 330
8.11 Block diagram of the closed-loop system with a state observer . 335
8.12 Scheme of an observer 336
8.13 State feedback with observer 341
8.14 Block diagram of state estimation of a singlevariable system . . . 347
8.15 Polynomial state estimation 347
8.16 Interpretation of a pseudo-state 347
8.17 Realisation of a state feedback with an observer 348
8.18 State feedback control 349
8.19 Feedback control with one-degree-of-freedom controller 350
8.20 Closed-loop system with a pure integrator 355
8.21 State feedback control with state observation and reference input 362
8.22 Feedback control system 365
8.23 Block diagram of the parametrised controller 370
8.24 State feedback controller with observer 370
8.25 State parametrised controller 371
8.26 Closed-loop system response to the unit step disturbance on input 373
8.27 Parametrised controlled system in the closed-loop system 374
8.28 Discrete-time feedback closed-loop system 376

8.29 Simulink diagram for finite-time tracking of a continuous-time second order system . . . 380
8.30 Tracking in (a) finite (b) the least number of steps . . . 380
8.31 Scheme `optlq22s.mdl` in Simulink for multivariable continuous-time LQ control . . . 387
8.32 Trajectories of (a) output, and setpoint variables, (b) input variables of the multivariable continuous-time process controlled with LQ controller . . . 388
8.33 Block diagram of LQG control . . . 390
8.34 Standard control configuration . . . 395

9.1 Principle of MBPC . . . 404
9.2 Closed-loop response of the controlled system . . . 415
9.3 Controller regions and the optimal control for the double integrator example . . . 432
9.4 Increasing value of N_2 leads towards mean-level control . . . 437
9.5 Increasing value of λ - detuned dead-beat control . . . 437
9.6 Trajectories of the oxygen concentration (the process output) . . 439
9.7 Trajectories of the dilution rate (the manipulated variable) . . . 439
9.8 Neutralisation plant . . . 440
9.9 Setpoint tracking . . . 441
9.10 Disturbance rejection . . . 441

10.1 Block diagram of a self-tuning discrete-time controller . . . 446
10.2 Block diagram of a self-tuning continuous-time controller . . . 447
10.3 Scheme `ad031s.mdl` in Simulink for dead-beat tracking of a continuous-time second order system with a discrete-time adaptive controller . . . 449
10.4 Trajectories of (a) input, output, and setpoint variables, (b) identified parameters of the discrete-time process model controlled with a dead-beat controller . . . 451
10.5 Scheme `ad0313.mdl` in Simulink for LQ tracking of a continuous-time second order system with a continuous-time adaptive controller . . . 452
10.6 Trajectories of (a) input, output, and setpoint variables, (b) identified parameters of the continuous-time process model controlled with LQ controller . . . 455
10.7 Scheme `ad035s.mdl` in Simulink for tracking of a multivariable system with a continuous-time adaptive controller . . . 457
10.8 Trajectories of (a) output, and setpoint variables, (b) input variables, (c) identified parameters of the continuous-time process controlled with a pole placement controller . . . 459
10.9 Comparison of the dead-beat (db) and mean-level (ml) control strategy . . . 462
10.10 Effect of step disturbances in inlet concentration . . . 462

List of Tables

3.1 The Laplace transforms for common functions 56

4.1 Solution of the second order differential equation 134
4.2 The errors of the magnitude plot resulting from the use of asymptotes 155

5.1 Laplace and $\mathcal{Z}$-transforms of some common functions 198

7.1 Controller tuning using Strejc method 284
7.2 Ziegler-Nichols controller tuning based on permanent oscillations. PID_1 sets smaller overshoot, PID_2 is underdamped 289
7.3 The Ziegler-Nichols controller tuning based on the process step response 289
7.4 PI controller tuning based on step response for stable processes. Parameters are given as functions of τ of the form (7.90) 292
7.5 PID controller tuning based on step response for stable processes. Parameters are given as functions of τ of the form (7.90) 292
7.6 PI controller tuning based on step response for processes with pure integrator. Parameters are given as functions of τ of the form (7.90) 292
7.7 PID controller tuning based on step response for processes with pure integrator. Parameters are given as functions of τ of the form (7.90) 292
7.8 PI controller tuning based on frequency response for stable processes. Parameters are given as functions of κ of the form $a_0 \exp(a_1\kappa + a_2\kappa^2)$ 293
7.9 PID controller tuning based on frequency response for stable processes. Parameters are given as functions of κ of the form $a_0 \exp(a_1\kappa + a_2\kappa^2)$ 293

1

Introduction

This chapter serves as an introduction to process control. The aim is to show the necessity of process control and to emphasize its importance in industries and in design of modern technologies. Basic terms and problems of process control and modelling are explained on a simple example of heat exchanger control. Finally, a short history of development in process control is given.

1.1 Topics in Process Control

Continuous technologies consist of unit processes, that are rationally arranged and connected in such a way that the desired product is obtained effectively with certain inputs.

The most important technological requirement is safety. The technology must satisfy the desired quantity and quality of the final product, environmental claims, various technical and operational constraints, market requirements, etc. The operational conditions follow from minimum price and maximum profit.

Control system is the part of technology and in the framework of the whole technology which is a guarantee for satisfaction of the above given requirements. Control systems in the whole consist of technical devices and human factor. Control systems must satisfy

- disturbance attenuation,
- stability guarantee,
- optimal process operation.

Control is the purposeful influence on a controlled object (process) that ensures the fulfillment of the required objectives. In order to satisfy the safety and optimal operation of the technology and to meet product specifications, technical, and other constraints, tasks and problems of control must be divided into a hierarchy of subtasks and subproblems with control of unit processes at the lowest level.

The lowest control level may realise *continuous-time control* of some measured signals, for example to hold temperature at *constant value*. The second control level may perform static *optimisation* of the process so that optimal values of some signals (flows, temperatures) are calculated in certain time instants. These will be set and remain constant till the next optimisation instant. The optimisation may also be performed continuously. As the unit processes are connected, their operation is *coordinated* at the third level. The highest level is influenced by *market, resources*, etc.

The fundamental way of control on the lowest level is *feedback control*. Information about process output is used to calculate control (manipulated) signal, i.e. process output is fed back to process input.

There are several other methods of control, for example feed-forward. *Feedforward control* is a kind of control where the effect of control is not compared with the desired result. In this case we speak about *open-loop control*. If the feedback exists, *closed-loop system* results.

Process design of "modern" technologies is crucial for successful control. The design must be developed in such a way, that a "sufficiently large number of degrees of freedom" exists for the purpose of control. The control system must have the ability to operate the whole technology or the unit process in the required technology regime. The processes should be "well" controllable and the control system should have "good" information about the process, i.e. the design phase of the process should include a selection of suitable *measurements*. The use of computers in the process control enables to choose optimal structure of the technology based on claims formulated in advance. Projectants of "modern" technologies should be able to include all aspects of control in the design phase.

Experience from control praxis of "modern" technologies confirms the importance of assumptions about dynamical behaviour of processes and more complex control systems. The control centre of every "modern" technology is a place, where all information about operation is collected and where the operators have contact with technology (through keyboards and monitors of control computers) and are able to correct and interfere with technology. A good knowledge of technology and process control is a necessary assumption of qualified human influence of technology through control computers in order to achieve optimal performance.

All of our further considerations will be based upon *mathematical models of processes*. These models can be constructed from a physical and chemical nature of processes or can be abstract. The investigation of dynamical properties of processes as well as whole control systems gives rise to a need to look for effective means of differential and difference equation solutions. We will carefully examine dynamical properties of open and closed-loop systems. A fundamental part of each procedure for effective control design is the process *identification* as the real systems and their physical and chemical parameters are usually not known perfectly. We will give procedures for design of *control algorithms* that ensure effective and safe operation.

One of the ways to secure a high quality process control is to apply *adaptive control laws.* Adaptive control is characterised by gaining information about unknown process and by using the information about on-line changes to process control laws.

1.2 An Example of Process Control

We will now demonstrate problems of process dynamics and control on a simple example. The aim is to show some basic principles and problems connected with process control.

1.2.1 Process

Let us assume a heat exchanger shown in Fig. 1.1. Inflow to the exchanger is a liquid with a flow rate q and temperature ϑ_v. The task is to heat this liquid to a higher temperature ϑ_w. We assume that the heat flow from the heat source is independent from the liquid temperature and only dependent from the heat input ω. We further assume ideal mixing of the heated liquid and no heat loss. The accumulation ability of the exchanger walls is zero, the exchanger holdup, input and output flow rates, liquid density, and specific heat capacity of the liquid are constant. The temperature on the outlet of the exchanger ϑ is equal to the temperature inside the exchanger. The exchanger that is correctly designed has the temperature ϑ equal to ϑ_w.

The process of heat transfer realised in the heat exchanger is defined as our *controlled system.*

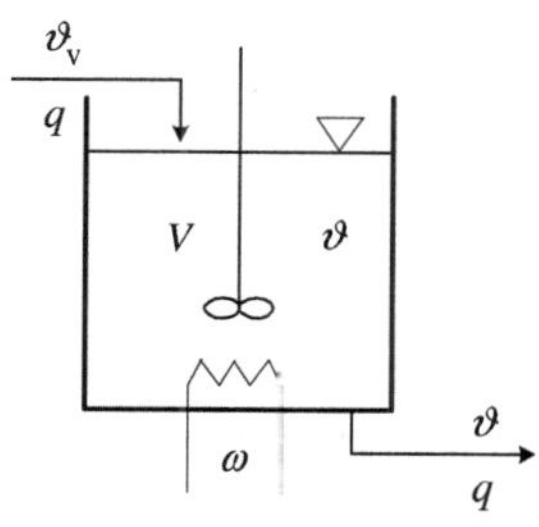

Fig. 1.1. A simple heat exchanger

1.2.2 Steady-State

The inlet temperature ϑ_v and the heat input ω are *input variables* of the process. The outlet temperature ϑ is process *output variable.* It is quite clear that every change of input variables ϑ_v, ω results in a change of output variable ϑ.

From this fact follows *direction of information transfer* of the process. The process is in the *steady state* if the input and output variables remain constant in time t.

The heat balance in the steady state is of the form

$$q\rho c_p(\vartheta^s - \vartheta_v^s) = \omega^s \tag{1.1}$$

where

ϑ^s is the output liquid temperature in the steady state,
ϑ_v^s is the input liquid temperature in the steady state,
ω^s is the heat input in the steady state,
q is volume flow rate of the liquid,
ρ is liquid density,
c_p is specific heat capacity of the liquid.

ϑ_v^s is the desired input temperature. For the suitable exchanger design, the output temperature in the steady state ϑ^s should be equal to the desired temperature ϑ_w. So the following equation follows

$$q\rho c_p(\vartheta_w - \vartheta_v^s) = \omega^s \tag{1.2}$$

It is clear, that if the input process variable ω^s is constant and if the process conditions change, the temperature ϑ would deviate from ϑ_w. The change of operational conditions means in our case the change in ϑ_v. The input temperature ϑ_v is then called *disturbance variable* and ϑ_w *setpoint variable.*

The heat exchanger should be designed in such a way that it can be possible to change the heat input so that the temperature ϑ would be equal to ϑ_w or be in its neighbourhood for all operational conditions of the process.

1.2.3 Process Control

Control of the heat transfer process in our case means to influence the process so that the output temperature ϑ will be kept close to ϑ_w. This influence is realised with changes in ω which is called *manipulated variable.* If there is a deviation ϑ from ϑ_w, it is necessary to adjust ω to achieve smaller deviation. This activity may be realised by a human operator and is based on the observation of the temperature ϑ. Therefore, a thermometer must be placed on the outlet of the exchanger. However, a human is not capable of high quality control. The task of the change of ω based on error between ϑ and ϑ_w can be realised automatically by some device. Such control method is called *automatic control.*

1.2.4 Dynamical Properties of the Process

In the case that the control is realised automatically then it is necessary to determine values of ω for each possible situation in advance. To make control decision in advance, the changes of ϑ as the result of changes in ω and ϑ_v must be known. The requirement of the knowledge about *process response* to changes of input variables is equivalent to knowledge about *dynamical properties of the process*, i.e. description of the process in *unsteady state*. The heat balance for the heat transfer process for a very short time Δt converging to zero is given by the equation

$$(q\rho c_p \vartheta_v \mathrm{d}t + \omega \mathrm{d}t) - (q\rho c_p \vartheta \mathrm{d}t) = (V\rho c_p \mathrm{d}\vartheta) \tag{1.3}$$

where V is the volume of the liquid in the exchanger. The equation (1.3) can be expressed in an abstract way as

(inlet heat) − (outlet heat) = (heat accumulation)

The dynamical properties of the heat exchanger given in Fig. 1.1 are given by the differential equation

$$V\rho c_p \frac{\mathrm{d}\vartheta}{\mathrm{d}t} + q\rho c_p \vartheta = q\rho c_p \vartheta_v + \omega \tag{1.4}$$

The heat balance in the steady state (1.1) may be derived from (1.4) in the case that $\frac{\mathrm{d}\vartheta}{\mathrm{d}t} = 0$. The use of (1.4) will be given later.

1.2.5 Feedback Process Control

As it was given above, process control may by realised either by human or automatically via *control device*. The control device performs the control actions practically in the same way as a human operator, but it is described exactly according to *control law*. The control device specified for the heat exchanger utilises information about the temperature ϑ and the desired temperature ϑ_w for the calculation of the heat input ω from formula formulated in advance. The difference between ϑ_w and ϑ is defined as *control error*. It is clear that we are trying to minimise the control error. The task is to determine the feedback control law to remove the control error optimally according to some criterion. The control law specifies the structure of the feedback controller as well as its properties if the structure is given.

The considerations above lead us to controller design that will change the heat input proportionally to the control error. This control law can be written as

$$\omega(t) = q\rho c_p(\vartheta_w - \vartheta_v^s) + Z_R(\vartheta_w - \vartheta(t)) \tag{1.5}$$

We speak about *proportional control* and *proportional controller*. Z_R is called the *proportional gain*. The proportional controller holds the heat input corresponding to the steady state as long as the temperature ϑ is equal to desired ϑ_w. The deviation between ϑ and ϑ_w results in nonzero control error and the controller changes the heat input proportionally to this error. If the control error has a plus sign, i.e. ϑ is greater as ϑ_w, the controller decreases heat input ω. In the opposite case, the heat input increases. This phenomenon is called *negative feedback*. The output signal of the process ϑ brings to the controller information about the process and is further transmitted via controller to the process input. Such kind of control is called *feedback control*. The quality of feedback control of the proportional controller may be influenced by the choice of controller gain Z_R. The equation (1.5) can be with the help of (1.2) written as

$$\omega(t) = \omega^s + Z_R(\vartheta_w - \vartheta(t)) \tag{1.6}$$

1.2.6 Transient Performance of Feedback Control

Putting the equation (1.6) into (1.4) we get

$$V\rho c_p \frac{\mathrm{d}\vartheta}{\mathrm{d}t} + (q\rho c_p + Z_R)\vartheta = q\rho c_p \vartheta_v + Z_R \vartheta_w + \omega^s \tag{1.7}$$

This equation can be arranged as

$$\frac{V}{q}\frac{\mathrm{d}\vartheta}{\mathrm{d}t} + \frac{q\rho c_p + Z_R}{q\rho c_p}\vartheta = \vartheta_v + \frac{Z_R}{q\rho c_p}\vartheta_w + \frac{1}{q\rho c_p}\omega^s \tag{1.8}$$

The variable $V/q = T_1$ has dimension of time and is called *time constant* of the heat exchanger. It is equal to time in which the exchanger is filled with liquid with flow rate q. We have assumed that the inlet temperature ϑ_v is a function of time t. For steady state ϑ_v^s is the input heat given as ω^s. We can determine the behaviour of $\vartheta(t)$ if ϑ_v, ϑ_w change. Let us assume that the process is controlled with feedback controller and is in the steady state given by values of $\vartheta_v^s, \omega^s, \vartheta^s$. In some time denoted by zero, we change the inlet temperature with the increment $\Delta\vartheta_v$. Idealised change of this temperature may by expressed mathematically as

$$\vartheta_v(t) = \begin{cases} \vartheta_v^s + \Delta\vartheta_v & t \geq 0 \\ \vartheta_v^s & t < 0 \end{cases} \tag{1.9}$$

To know the response of the process with the feedback proportional controller for the step change of the inlet temperature means to know the solution of the differential equation (1.8). The process is at $t = 0$ in the steady state and the initial condition is

$$\vartheta(0) = \vartheta_w \tag{1.10}$$

The solution of (1.8) if (1.9), (1.10) are valid is given as

$$\vartheta(t) = \vartheta_w + \Delta\vartheta_v \frac{q\rho c_p}{q\rho c_p + Z_R}\left(1 - \mathrm{e}^{-\frac{q\rho c_p + Z_R}{q\rho c_p}\frac{q}{V}t}\right) \tag{1.11}$$

The response of the heat transfer process controlled with the proportional controller for the step change of inlet temperature ϑ_v given by Eq. (1.9) is shown in Fig. 1.2 for several values of the controller gain Z_R. The investigation of the figure shows some important facts. The outlet temperature ϑ converges to some new steady state for $t \to \infty$. If the proportional controller is used, *steady state error* results. This means that there exists a difference between ϑ_w and ϑ at the time $t = \infty$. The steady state error is the largest if $Z_R = 0$. If the controller gain Z_R increases, steady state error decreases. If $Z_R = \infty$, then the steady state error is zero. Therefore our first intention would be to choose the largest possible Z_R. However, this would break some other closed-loop properties as will be shown later.

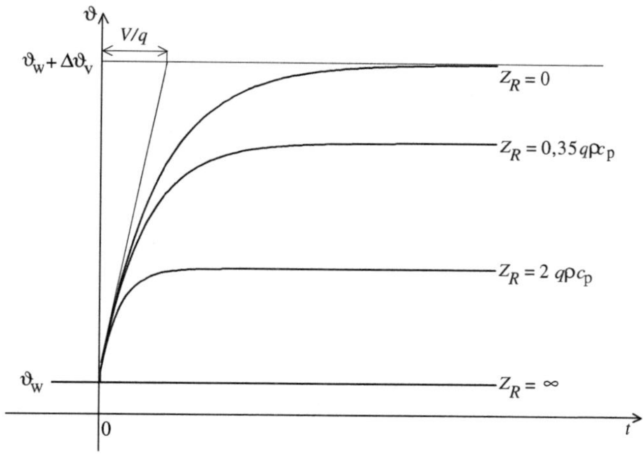

Fig. 1.2. Response of the process controlled with proportional feedback controller for a step change of disturbance variable ϑ_v

If the disturbance variable ϑ_v changes with time in the neighbourhood of its steady state value, the choice of large Z_R may cause large control deviations. However, it is in our interest that the control deviations are to be kept under some limits. Therefore, this kind of disturbance requires rather smaller values of controller gain Z_R and its choice is given as a compromise between these two requirements.

The situation may be improved if the controller consists of a proportional and integral part. Such a controller may remove the steady state error even with smaller gain.

It can be seen from (1.11) that $\vartheta(t)$ cannot grow beyond limits. We note however that the controlled system was described by the first order differential equation and was controlled with a proportional controller.

We can make the process model more realistic, for example, assuming the accumulation ability of its walls or dynamical properties of temperature measurement device. The model and the feedback control loop as well will then be described by a higher order differential equation. The solution of such a differential equation for similar conditions as in (1.11) can result in ϑ growing into infinity. This case represents *unstable response* of the closed loop system. The problem of stability is usually included into the general problem of *control quality*.

1.2.7 Block Diagram

In the previous sections the principal problems of feedback control were discussed. We have not dealt with technical issues of the feedback control implementation.

Consider again feedback control of the heat exchanger in Fig. 1.1. The necessary assumptions are i) to measure the outlet temperature ϑ and ii) the possibility of change of the heat input ω. We will assume that the heat input is realised by an electrical heater.

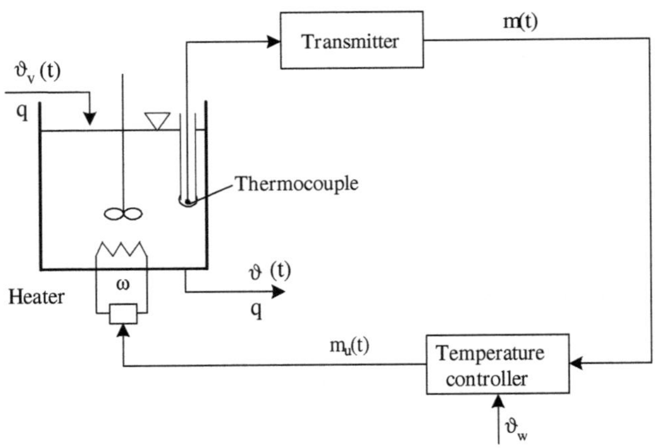

Fig. 1.3. The scheme of the feedback control for the heat exchanger

If the feedback control law is given then the feedback control of the heat exchanger may be realised as shown in Fig. 1.3. This scheme may be simplified for needs of analysis. Parts of the scheme will be depicted as blocks. The block scheme in Fig. 1.3 is shown in Fig. 1.4. The scheme gives physical interconnections and the information flow between the parts of the closed loop system. The signals represent physical variables as for example ϑ or instrumentation signals as for example m. Each block has its own input and output signal.

The outlet temperature is measured with a thermocouple. The thermocouple with its transmitter generates a voltage signal corresponding to the measured temperature. The dashed block represents the entire temperature controller and $m(t)$ is the input to the controller. The controller realises three activities:

1. the desired temperature ϑ_w is transformed into voltage signal m_w,
2. the control error is calculated as the difference between m_w and $m(t)$,
3. the control signal m_u is calculated from the control law.

All three activities are realised within the controller. The controller output $m_u(t)$ in volts is the input to the electric heater producing the corresponding heat input $\omega(t)$. The properties of each block in Fig. 1.4 are described by algebraic or differential equations.

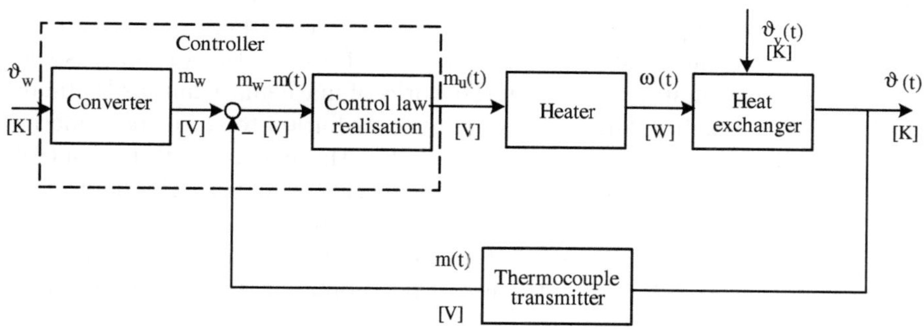

Fig. 1.4. The block scheme of the feedback control of the heat exchanger

Block schemes are usually simplified for the purpose of the investigation of control loops. The *simplified block scheme* consists of 2 blocks: control block and controlled object. Each block of the detailed block scheme must be included into one of these two blocks. Usually the simplified control block realizes the control law.

1.2.8 Feedforward Control

We can also consider another kind of the heat exchanger control when the disturbance variable ϑ_v is measured and used for the calculation of the heat input ω. This is called *feedforward control*. The effect of control is not compared with the expected result. In some cases of process control it is necessary and/or suitable to use a combination of feedforward and feedback control.

1.3 Development of Process Control

The history of automatic control began about 1788. At that time J. Watt developed a revolution controller for the steam engine. An analytic expression of the influence between controller and controlled object was presented by Maxwell in 1868. Correct mathematical interpretation of automatic control is given in the works of Stodola in 1893 and 1894. E. Routh in 1877 and Hurwitz in 1895 published works in which stability of automatic control and stability criteria were dealt with. An important contribution to the stability theory was presented by Nyquist (1932). The works of Oppelt (1939) and other authors showed that automatic control was established as an independent scientific branch.

Rapid development of discrete time control began in the time after the second world war. In continuous time control, the theory of transformation was used. The transformation of sequences defined as $\mathcal{Z}$-transform was introduced independently by Cypkin (1950), Ragazzini and Zadeh (1952).

A very important step in the development of automatic control was the state-space theory, first mentioned in the works of mathematicians as Bellman (1957) and Pontryagin (1962). An essential contribution to state-space methods belongs to Kalman (1960). He showed that the linear-quadratic control problem may be reduced to a solution of the Riccati equation. Paralel to the optimal control, the stochastic theory was being developed.

It was shown that automatic control problems have an algebraic character and the solutions were found by the use of polynomial methods (Rosenbrock, 1970).

In the fifties, the idea of *adaptive control* appeared in journals. The development of adaptive control was influenced by the theory of dual control (Feldbaum, 1965), parameter estimation (Eykhoff, 1974), and recursive algorithms for adaptive control (Cypkin, 1971).

The above given survey of development in automatic control also influenced development in process control. Before 1940, processes in the chemical industry and in industries with similar processes, were controlled practically only manually. If some controller were used, these were only very simple. The technologies were built with large tanks between processes in order to attenuate the influence of disturbances.

In the fifties, it was often uneconomical and sometimes also impossible to build technologies without automatic control as the capacities were larger and the demand of quality increased. The controllers used did not consider the complexity and dynamics of controlled processes.

In 1960-s the process control design began to take into considerations dynamical properties and bindings between processes. The process control used knowledge applied from astronautics and electrotechnics.

The seventies brought the demands on higher quality of control systems and integrated process and control design.

In the whole process control development, knowledge of processes and their *modelling* played an important role.

The development of process control was also influenced by the development of computers. The first ideas about the use of *digital computers* as a part of control system emerged in about 1950. However, computers were rather expensive and unreliable to use in process control. The first use was in *supervisory control.* The problem was to find the optimal operation conditions in the sense of static optimisation and the mathematical models of processes were developed to solve this task. In the sixties, the continuous control devices began to be replaced with digital equipment, the so called *direct digital process control.* The next step was an introduction of mini and microcomputers in the seventies as these were very cheap and also small applications could be equipped with them. Nowadays, the computer control is decisive for quality and effectivity of all modern technology.

1.4 References

Survey and development in automatic control are covered in:

K. Rörentrop. *Entwicklung der modernen Regelungstechnik.* Oldenbourg-Verlag, München, 1971.

H. Unbehauen. *Regelungstechnik I.* Vieweg, Braunschweig/Wiesbaden, 1986.

K. J. Åström and B. Wittenmark. *Computer Controlled Systems.* Prentice Hall, 1984.

A. Stodola. Über die Regulierung von Turbinen. *Schweizer Bauzeitung*, 22,23: 27 – 30, 17 – 18, 1893, 1894.

E. J. Routh. *A Treatise on the Stability of a Given State of Motion.* Mac Millan, London, 1877.

A. Hurwitz. Über die Bedinungen, unter welchen eine Gleichung nur Wurzeln mit negativen reellen Teilen besitzt. *Math. Annalen*, 46:273 – 284, 1895.

H. Nyquist. Regeneration theory. *Bell Syst. techn. J.*, 11:126 – 147, 1932.

W. Oppelt. Vergleichende Betrachtung verschiedener Regelaufgaben hinsichtlich der geeigneten Regelgesetzmäßigkeit. *Luftfahrtforschung*, 16: 447 – 472, 1939.

Y. Z. Cypkin. Theory of discontinuous control. *Automat. i Telemech.*, 3,5,5, 1949, 1949, 1950.

J. R. Ragazzini and L. A. Zadeh. The analysis of sampled-data control systems. *AIEE Trans.*, 75:141 – 151, 1952.

R. Bellman. *Dynamic Programming.* Princeton University Press, Princeton, New York, 1957.

L. S. Pontryagin, V. G. Boltyanskii, R. V. Gamkrelidze, and E. F. Mishchenko. *The Mathematical Theory of Optimal Processes.* Wiley, New York, 1962.

R. E. Kalman. On the general theory of control systems. In *Proc. First IFAC Congress, Moscow, Butterworths*, volume 1, pages 481 – 492, 1960a.

Some basic ideas about control and automatic control can be found in these books:

W. H. Ray. *Advanced Process Control.* McGraw-Hill, New York, 1981.

D. Chmúrny, J. Mikleš, P. Dostál, and J. Dvoran. *Modelling and Control of Processes and Systems in Chemical Technology.* Alfa, Bratislava, 1985. (in Slovak).

D. R. Coughanouwr and L. B. Koppel. *Process System Analysis and Control.* McGraw-Hill, New York, 1965.

G. Stephanopoulos. *Chemical Process Control, An Introduction to Theory and Practice.* Prentice Hall, Englewood Cliffs, New Jersey, 1984.

W. L. Luyben. *Process Modelling, Simulation and Control for Chemical Engineers.* McGraw Hill, Singapore, 2 edition, 1990.

C. J. Friedly. *Dynamic Behavior of Processes.* Prentice Hall, Inc., New Jersey, 1972.

J. M. Douglas. *Process Dynamics and Control.* Prentice Hall, Inc., New Jersey, 1972.

J. Mikleš. *Foundations of Technical Cybernetics.* ES SVŠT, Bratislava, 1973. (in Slovak).

W. Oppelt. *Kleines Handbuch technischer Regelvorgänge.* Verlag Chemie, Weinheim, 1972.

T. W. Weber. *An Introduction to Process Dynamics and Control.* Wiley, New York, 1973.

F. G. Shinskey. *Process Control Systems.* McGraw-Hill, New York, 1979.

2

Mathematical Modelling of Processes

This chapter explains general techniques that are used in the development of mathematical models of processes. It contains mathematical models of liquid storage systems, heat and mass transfer systems, chemical, and biochemical reactors. The remainder of the chapter explains the meaning of systems and their classification.

2.1 General Principles of Modelling

Schemes and block schemes of processes help to understand their qualitative behaviour. To express quantitative properties, mathematical descriptions are used. These descriptions are called *mathematical models.* Mathematical models are abstractions of real processes. They give a possibility to characterise behaviour of processes if their inputs are known. The validity range of models determines situations when models may be used. Models are used for control of continuous processes, investigation of process dynamical properties, optimal process design, or for the calculation of optimal process working conditions.

A process is always tied to an apparatus (heat exchangers, reactors, distillation columns, etc.) in which it takes place. Every process is determined with its physical and chemical nature that expresses its mass and energy bounds. Investigation of any typical process leads to the development of its mathematical model. This includes basic equations, variables and description of its static and dynamic behaviour. *Dynamical model* is important for control purposes. By the construction of *mathematical models of processes* it is necessary to know the problem of investigation and it is important to understand the investigated phenomenon thoroughly. If computer control is to be designed, a developed mathematical model should lead to the simplest control algorithm. If the basic use of a process model is to analyse the different process conditions including safe operation, a more complex and detailed model is needed. If a model is used in a computer simulation, it should at least include that part of the process that influences the process dynamics considerably.

Mathematical models can be divided into three groups, depending on how they are obtained:

Theoretical models developed using chemical and physical principles.
Empirical models obtained from mathematical analysis of process data.
Empirical-theoretical models obtained as a combination of theoretical and empirical approach to model design.

From the process operation point of view, processes can be divided into *continuous* and *batch*. It is clear that this fact must be considered in the design of mathematical models.

Theoretical models are derived from mass and energy balances. The balances in an unsteady-state are used to obtain dynamical models. Mass balances can be specified either in total mass of the system or in component balances. Variables expressing quantitative behaviour of processes are *natural state variables*. Changes of state variables are given by *state balance equations*. Dynamical mathematical models of processes are described by differential equations. Some processes are *processes with distributed parameters* and are described by partial differential equations (p.d.e). These usually contain first partial derivatives with respect to time and space variables and second partial derivatives with respect to space variables. However, the most important are dependencies of variables on one space variable. The first partial derivatives with respect to space variables show an existence of transport while the second derivatives follow from heat transfer, mass transfer resulting from molecular diffusion, etc. If *ideal mixing* is assumed, the modelled process does not contain changes of variables in space and its mathematical model is described by ordinary differential equations (o.d.e). Such models are referred to as *lumped parameter* type.

Mass balances for lumped parameter processes in an unsteady-state are given by the law of mass conservation and can be expressed as

$$\frac{\mathrm{d}(\rho V)}{\mathrm{d}t} = \sum_{i=1}^{m} \rho_i q_i - \sum_{j=1}^{r} \rho q_j \tag{2.1}$$

where

ρ, ρ_i - density,
V - volume,
q_i, q_j - volume flow rates,
m - number of inlet flows,
r - number of outlet flows.

Component mass balance of the k-th component can be expressed as

$$\frac{\mathrm{d}(c_k V)}{\mathrm{d}t} = \sum_{i=1}^{m} c_{ki} q_i - \sum_{j=1}^{r} c_k q_j + r_k V \tag{2.2}$$

where

c_k, c_{ki} - molar concentration,
V - volume,
q_i, q_j - volume flow rates,
m - number of inlet flows,
r - number of outlet flows,
r_k - rate of reaction per unit volume for k-th component.

Energy balances follow the general law of energy conservation and can be written as

$$\frac{\mathrm{d}(\rho V c_p \vartheta)}{\mathrm{d}t} = \sum_{i=1}^{m} \rho_i q_i c_{pi} \vartheta_i - \sum_{j=1}^{r} \rho q_j c_p \vartheta + \sum_{l=1}^{s} Q_l \tag{2.3}$$

where

ρ, ρ_i - density,
V - volume,
q_i, q_j - volume flow rates,
c_p, c_{pi} - specific heat capacities,
ϑ, ϑ_i - temperatures,
Q_l - heat per unit time,
m - number of inlet flows,
r - number of outlet flows,
s - number of heat sources and consumptions as well as heat brought in and taken away not in inlet and outlet streams.

To use a mathematical model for process simulation we must ensure that differential and algebraic equations describing the model give a unique relation among all inputs and outputs. This is equivalent to the requirement of unique solution of a set of algebraic equations. This means that the number of unknown variables must be equal to the number of independent model equations. In this connection, the term *degree of freedom* is introduced. Degree of freedom N_v is defined as the difference between the total number of unspecified inputs and outputs and the number of independent differential and algebraic equations. The model must be defined such that

$$N_v = 0 \tag{2.4}$$

Then the set of equations has a unique solution.

An approach to model design involves the finding of known constants and fixed parameters following from equipment dimensions, constant physical and chemical properties and so on. Next, it is necessary to specify the variables that will be obtained through a solution of the model differential and algebraic equations. Finally, it is necessary to specify the variables whose time behaviour is given by the process environment.

2.2 Examples of Dynamic Mathematical Models

In this section we present examples of mathematical models for liquid storage systems, heat and mass transfer systems, chemical, and biochemical reactors. Each example illustrates some typical properties of processes.

2.2.1 Liquid Storage Systems

Single-Tank Process

Let us examine a liquid storage system shown in Fig. 2.1. Input variable is the inlet volumetric flow rate q_0 and state variable the liquid height h. Mass balance for this process yields

$$\frac{\mathrm{d}(Fh\rho)}{\mathrm{d}t} = q_0\rho - q_1\rho \qquad (2.5)$$

where

t - time variable,
h - height of liquid in the tank,
q_0, q_1 - inlet and outlet volumetric flow rates,
F - cross-sectional area of the tank,
ρ - liquid density.

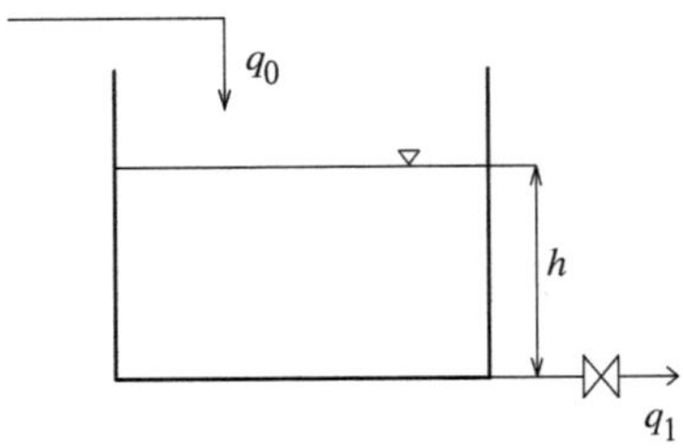

Fig. 2.1. A liquid storage system

Assume that liquid density and cross-sectional area are constant, then

$$F\frac{\mathrm{d}h}{\mathrm{d}t} = q_0 - q_1 \qquad (2.6)$$

q_0 is independent of the tank state and q_1 depends on the liquid height in the tank according to the relation

$$q_1 = k_1 f_1 \sqrt{2g}\sqrt{h} \qquad (2.7)$$

where

k_1 - constant,
f_1 - cross-sectional area of outflow opening,
g - acceleration gravity.

or

$$q_1 = k_{11}\sqrt{h} \tag{2.8}$$

Substituting q_1 from the equation (2.8) into (2.6) yields

$$\frac{\mathrm{d}h}{\mathrm{d}t} = \frac{q_0}{F} - \frac{k_{11}}{F}\sqrt{h} \tag{2.9}$$

Initial conditions can be arbitrary

$$h(0) = h_0 \tag{2.10}$$

The tank will be in a steady-state if

$$\frac{\mathrm{d}h}{\mathrm{d}t} = 0 \tag{2.11}$$

Let a steady-state be given by a constant flow rate q_0^s. The liquid height h^s then follows from Eq. (2.9) and (2.11) and is given as

$$h^s = \frac{(q_0^s)^2}{(k_{11})^2} \tag{2.12}$$

Interacting Tank-in-Series Process

Consider the interacting tank-in-series process shown in Fig. 2.2. The process input variable is the flow rate q_0.

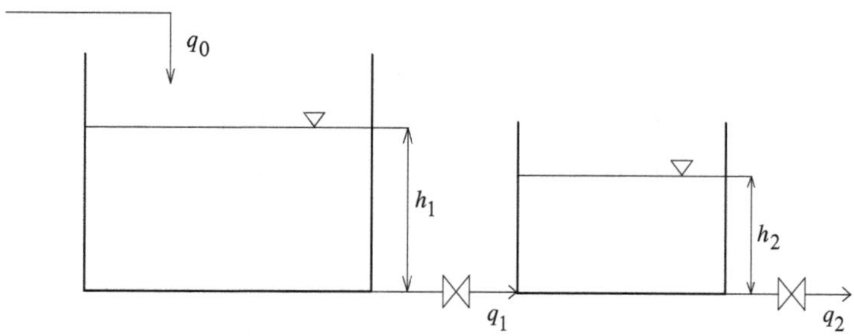

Fig. 2.2. An interacting tank-in-series process

The process state variables are heights of liquid in tanks h_1, h_2. Mass balance for the process yields

$$\frac{d(F_1h_1\rho)}{dt} = q_0\rho - q_1\rho \tag{2.13}$$

$$\frac{d(F_2h_2\rho)}{dt} = q_1\rho - q_2\rho \tag{2.14}$$

where

t - time variable,
h_1, h_2 - heights of liquid in the first and second tanks,
q_0 - inlet volumetric flow rate to the first tank,
q_1 - inlet volumetric flow rate to the second tank,
q_2 - outlet volumetric flow rate from the second tank,
F_1, F_2 - cross-sectional area of the tanks,
ρ - liquid density.

Assuming that ρ, F_1, F_2 are constant we can write

$$F_1\frac{dh_1}{dt} = q_0 - q_1 \tag{2.15}$$

$$F_2\frac{dh_2}{dt} = q_1 - q_2 \tag{2.16}$$

Inlet flow rate q_0 is independent of tank states whereas q_1 depends on the difference between liquid heights

$$q_1 = k_1 f_1 \sqrt{2g}\sqrt{h_1 - h_2} \tag{2.17}$$

where

k_1 - constant,
f_1 - cross-sectional area of the first tank outflow opening.

Outlet flow rate q_2 depends on liquid height in the second tank

$$q_2 = k_2 f_2 \sqrt{2g}\sqrt{h_2} \tag{2.18}$$

where

k_2 - constant,
f_2 - cross-sectional area of the second tank outflow opening.

Equations (2.17) and (2.18) can then be written as

$$q_1 = k_{11}\sqrt{h_1 - h_2} \tag{2.19}$$

$$q_2 = k_{22}\sqrt{h_2} \tag{2.20}$$

Substituting q_1 from Eq. (2.19) and q_2 from (2.20) into (2.15), (2.16) we get

$$\frac{dh_1}{dt} = \frac{q_0}{F_1} - \frac{k_{11}}{F_1}\sqrt{h_1 - h_2} \tag{2.21}$$

$$\frac{dh_2}{dt} = \frac{k_{11}}{F_1}\sqrt{h_1 - h_2} - \frac{k_{22}}{F_2}\sqrt{h_2} \tag{2.22}$$

with arbitrary initial conditions

$$h_1(0) = h_{10} \tag{2.23}$$
$$h_2(0) = h_{20} \tag{2.24}$$

The tanks will be in a steady-state if

$$\frac{\mathrm{d}h_1}{\mathrm{d}t} = 0 \tag{2.25}$$
$$\frac{\mathrm{d}h_2}{\mathrm{d}t} = 0 \tag{2.26}$$

Assume a steady-state flow rate q_0^s. The steady-state liquid levels in both tanks can be calculated from Eqs (2.21), (2.22), (2.25), (2.26) as

$$h_1^s = (q_0^s)^2 \left(\frac{1}{(k_{11})^2} + \frac{1}{(k_{22})^2} \right) \tag{2.27}$$
$$h_2^s = (q_0^s)^2 \frac{1}{(k_{22})^2} \tag{2.28}$$

2.2.2 Heat Transfer Processes

Heat Exchanger

Consider a heat exchanger for the heating of liquids shown in Fig. 2.3. The input variables are the temperatures ϑ_v, ϑ_p. The state variable is temperature ϑ.

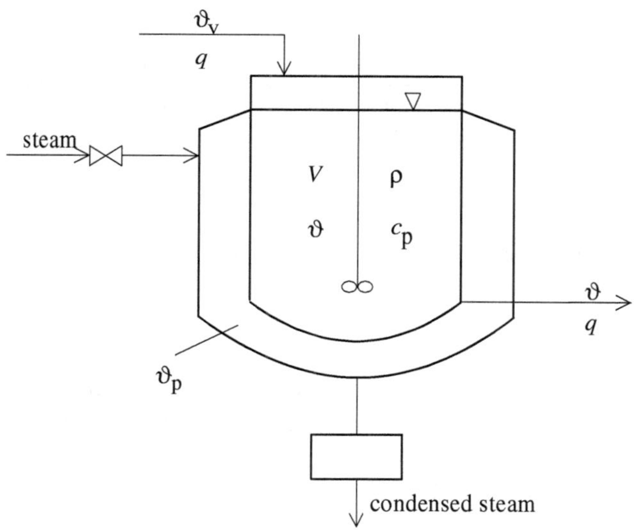

Fig. 2.3. Continuous stirred tank heated by steam in jacket

Assume that the wall accumulation ability is small compared to the liquid accumulation ability and can be neglected. Further assume spatially constant temperature inside of the tank as the heater is well mixed, constant liquid flow rate, density, and heat capacity. Then the heat balance equation becomes

$$V\rho c_p \frac{\mathrm{d}\vartheta}{\mathrm{d}t} = q\rho c_p \vartheta_v - q\rho c_p \vartheta + \alpha F(\vartheta_p - \vartheta) \tag{2.29}$$

where

t - time variable,
ϑ - temperature inside of the exchanger and in the outlet stream,
ϑ_v - temperature in the inlet stream,
ϑ_p - jacket temperature,
q - liquid volumetric flow rate,
ρ - liquid density,
V - volume of liquid in the tank,
c_p - liquid specific heat capacity,
F - heat transfer area of walls,
α - heat transfer coefficient.

Equation (2.29) can be rearranged as

$$\frac{V\rho c_p}{q\rho c_p + \alpha F} \frac{\mathrm{d}\vartheta}{\mathrm{d}t} = -\vartheta + \frac{\alpha F}{q\rho c_p + \alpha F}\vartheta_p + \frac{q\rho c_p}{q\rho c_p + \alpha F}\vartheta_v \tag{2.30}$$

or as

$$T_1 \frac{\mathrm{d}\vartheta}{\mathrm{d}t} = -\vartheta + Z_1\vartheta_p + Z_2\vartheta_v \tag{2.31}$$

where $T_1 = \dfrac{V\rho c_p}{q\rho c_p + \alpha F}$, $Z_1 = \dfrac{\alpha F}{q\rho c_p + \alpha F}$, $Z_2 = \dfrac{q\rho c_p}{q\rho c_p + \alpha F}$. The initial condition of Eq. (2.30) can be arbitrary

$$\vartheta(0) = \vartheta_0 \tag{2.32}$$

The heat exchanger will be in a steady-state if

$$\frac{\mathrm{d}\vartheta}{\mathrm{d}t} = 0 \tag{2.33}$$

Assume steady-state values of the input temperatures $\vartheta_p^s, \vartheta_v^s$. The steady-state outlet temperature ϑ^s can be calculated from Eqs. (2.30), (2.33) as

$$\vartheta^s = \frac{\alpha F}{q\rho c_p + \alpha F}\vartheta_p^s + \frac{q\rho c_p}{q\rho c_p + \alpha F}\vartheta_v^s \tag{2.34}$$

Series of Heat Exchangers

Consider a series of heat exchangers where a liquid is heated (Fig. 2.4). Assume that heat flows from heat sources into liquid are independent from liquid temperature. Further assume ideal liquid mixing and zero heat losses. We neglect accumulation ability of exchangers walls. Hold-ups of exchangers as well as flow rates, liquid specific heat capacity are constant.

Under these circumstances following heat balances result

$$\begin{aligned}
V_1\rho c_p\frac{\mathrm{d}\vartheta_1}{\mathrm{d}t} &= q\rho c_p\vartheta_0 - q\rho c_p\vartheta_1 + \omega_1 \\
V_2\rho c_p\frac{\mathrm{d}\vartheta_2}{\mathrm{d}t} &= q\rho c_p\vartheta_1 - q\rho c_p\vartheta_2 + \omega_2 \\
&\vdots \\
V_n\rho c_p\frac{\mathrm{d}\vartheta_n}{\mathrm{d}t} &= q\rho c_p\vartheta_{n-1} - q\rho c_p\vartheta_n + \omega_n
\end{aligned} \tag{2.35}$$

where

t - time variable,
$\vartheta_1,\ldots,\vartheta_n$ - temperature inside of the heat exchangers,
ϑ_0 - liquid temperature in the first tank inlet stream,
$\omega_1,\ldots,\omega_n$ - heat inputs,
q - liquid volumetric flow rate,
ρ - liquid density,
$V_1,\ldots,V_n$ - volumes of liquid in the tanks,
c_p - liquid specific heat capacity.

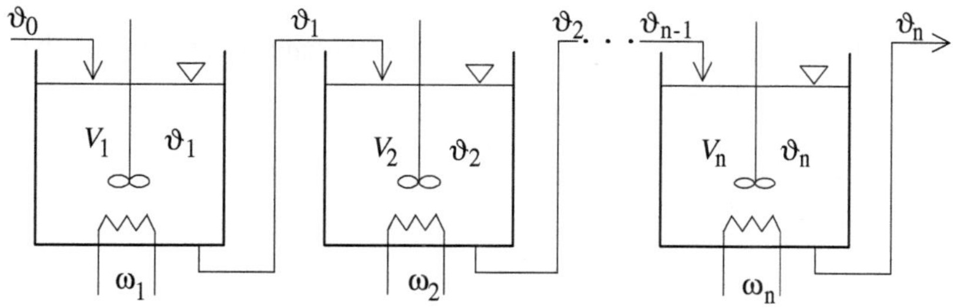

Fig. 2.4. Series of heat exchangers

The process input variables are heat inputs ω_i and inlet temperature ϑ_0. The process state variables are temperatures $\vartheta_1,\ldots,\vartheta_n$ and initial conditions are arbitrary

$$\vartheta_1(0) = \vartheta_{10}, \ldots, \vartheta_n(0) = \vartheta_{n0} \tag{2.36}$$

The process will be in a steady-state if

$$\frac{\mathrm{d}\vartheta_1}{\mathrm{d}t} = \frac{\mathrm{d}\vartheta_2}{\mathrm{d}t} = \cdots = \frac{\mathrm{d}\vartheta_n}{\mathrm{d}t} = 0 \tag{2.37}$$

Let the steady-state values of the process inputs ω_i, ϑ_0 be given. The steady-state temperatures inside the exchangers are

$$\begin{aligned}
\vartheta_1^s &= \vartheta_0^s + \frac{\omega_1^s}{q\rho c_p} \\
\vartheta_2^s &= \vartheta_1^s + \frac{\omega_2^s}{q\rho c_p} \\
&\vdots \\
\vartheta_n^s &= \vartheta_{n-1}^s + \frac{\omega_n^s}{q\rho c_p}
\end{aligned} \tag{2.38}$$

Double-Pipe Heat Exchanger

Figure 2.5 represents a single-pass, double-pipe steam-heated exchanger in which a liquid in the inner tube is heated by condensing steam. The process input variables are $\vartheta_p(t), \vartheta(0,t)$. The process state variable is the temperature $\vartheta(\sigma,t)$. We assume the steam temperature to be a function only of time, heat transfer only between inner and outer tube, plug flow of the liquid and zero heat capacity of the exchanger walls. We neglect heat conduction effects in the direction of liquid flow. It is further assumed that liquid flow, density, and specific heat capacity are constant.

Heat balance equation on the element of exchanger length $d\sigma$ can be derived according to Fig. 2.6

$$F_\sigma d\sigma \rho c_p \frac{\partial \vartheta}{\partial t} = q\rho c_p \vartheta - q\rho c_p \left(\vartheta + \frac{\partial \vartheta}{\partial \sigma} d\sigma\right) + \alpha F_d d\sigma(\vartheta_p - \vartheta) \tag{2.39}$$

where

t - time variable,
σ - space variable,
$\vartheta = \vartheta(\sigma, t)$ - liquid temperature in the inner tube,
$\vartheta_p = \vartheta_p(t)$ - liquid temperature in the outer tube,
q - liquid volumetric flow rate in the inner tube,
ρ - liquid density in the inner tube,
α - heat transfer coefficient,

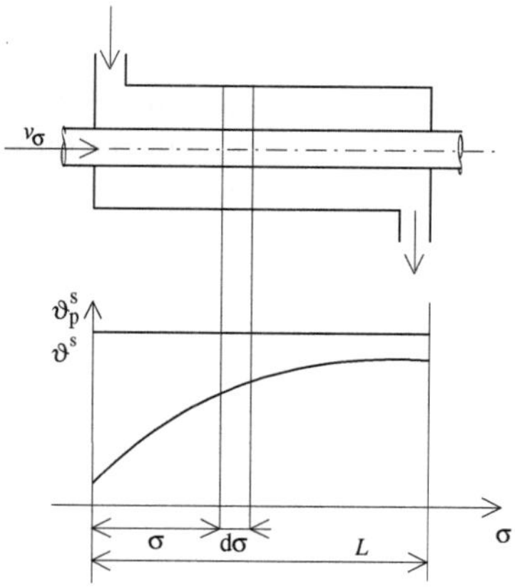

Fig. 2.5. Double-pipe steam-heated exchanger and temperature profile along the exchanger length in steady-state

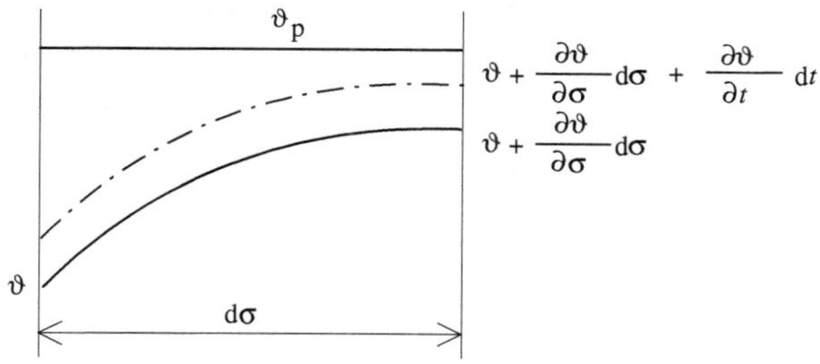

Fig. 2.6. Temperature profile of ϑ in an exchanger element of length $\mathrm{d}\sigma$ for time $\mathrm{d}t$

c_p - liquid specific heat capacity,
F_d - area of heat transfer per unit length,
F_σ - cross-sectional area of the inner tube.

The equation (2.39) can be rearranged to give

$$\frac{F_\sigma \rho c_p}{\alpha F_d}\frac{\partial\vartheta}{\partial t} = -\frac{q\rho c_p}{\alpha F_d}\frac{\partial\vartheta}{\partial\sigma} - \vartheta + \vartheta_p \tag{2.40}$$

or

$$T_1\frac{\partial\vartheta}{\partial t} = -v_\sigma T_1\frac{\partial\vartheta}{\partial\sigma} - \vartheta + \vartheta_p \tag{2.41}$$

where $T_1 = \dfrac{F_\sigma \rho c_p}{\alpha F_d}$, $v_\sigma = \dfrac{q}{F_\sigma}$.

Boundary condition of Eq. (2.41) is

$$\vartheta(0,t) = \vartheta^0(t) \tag{2.42}$$

and initial condition is

$$\vartheta(\sigma, 0) = \vartheta_0(\sigma) \tag{2.43}$$

Assume a steady-state inlet liquid temperature ϑ^{0s} and steam temperature ϑ_p^s. The temperature profile in the inner tube in the steady-state can be derived if

$$\frac{\partial \vartheta}{\partial t} = 0 \tag{2.44}$$

as

$$\vartheta^s(\sigma) = \vartheta_p^s - (\vartheta_p^s - \vartheta^{0s})\mathrm{e}^{-\dfrac{\sigma}{v_\sigma T_1}} \tag{2.45}$$

If $\alpha = 0$, Eq. (2.39) reads

$$\frac{\partial \vartheta}{\partial t} = -v_\sigma \frac{\partial \vartheta}{\partial \sigma} \tag{2.46}$$

while boundary and initial conditions remain the same. If the input variable is $\vartheta^0(t)$ and the output variable is $\vartheta(L, t)$, then Eq. (2.46) describes *pure time delay* with value

$$T_d = \frac{L}{v_\sigma} \tag{2.47}$$

Heat Conduction in a Solid Body

Consider a metal rod of length L in Fig. 2.7. Assume ideal insulation of the rod. Heat is brought in on the left side and withdrawn on the right side. Changes of densities of heat flows q_ω^0, q_L^0 influence the rod temperature $\vartheta(\sigma, t)$. Assume that heat conduction coefficient, density, and specific heat capacity of the rod are constant. We will derive unsteady heat flow through the rod. Heat balance on the rod element of length $\mathrm{d}\sigma$ for time $\mathrm{d}t$ can be derived from Fig. 2.7 as

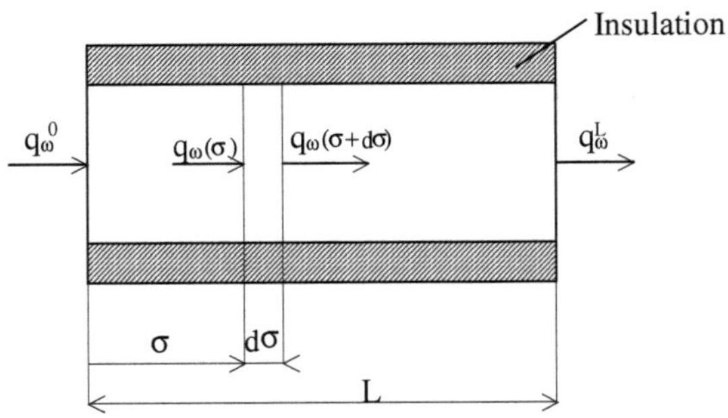

Fig. 2.7. A metal rod

$$F_\sigma d\sigma \rho c_p \frac{\partial \vartheta}{\partial t} = F_\sigma [q_\omega(\sigma) - q_\omega(\sigma + d\sigma)] \tag{2.48}$$

or

$$F_\sigma d\sigma \rho c_p \frac{\partial \vartheta}{\partial t} = -F_\sigma \frac{\partial q_\omega}{\partial \sigma} d\sigma \tag{2.49}$$

where

t - time variable,
σ - space variable,
$\vartheta = \vartheta(\sigma, t)$ - rod temperature,
ρ - rod density,
c_p - rod specific heat capacity,
F_σ - cross-sectional area of the rod,
$q_\omega(\sigma)$ - heat flow density (heat transfer velocity through unit area) at length σ,
$q_\omega(\sigma + d\sigma)$ - heat flow density at length $\sigma + d\sigma$.

From the Fourier law follows

$$q_\omega = -\lambda \frac{\partial \vartheta}{\partial \sigma} \tag{2.50}$$

where λ is the coefficient of thermal conductivity.

Substituting Eq. (2.50) into (2.49) yields

$$\frac{\partial \vartheta}{\partial t} = a \frac{\partial^2 \vartheta}{\partial \sigma^2} \tag{2.51}$$

where

$$a = \frac{\lambda}{\rho c_p} \tag{2.52}$$

is the *factor of heat conductivity.* The equation (2.51) requires boundary and initial conditions. The boundary conditions can be given with temperatures or temperature derivatives with respect to σ at the ends of the rod. For example

$$\vartheta(0, t) = \vartheta^0(t) \tag{2.53}$$
$$\vartheta(L, t) = \vartheta^L(t) \tag{2.54}$$

The initial condition for Eq. (2.51) is

$$\vartheta(\sigma, 0) = \vartheta_0(\sigma) \tag{2.55}$$

Consider the boundary conditions (2.53), (2.54). The process input variables are $\vartheta^0(t), \vartheta^L(t)$ and the state variable is $\vartheta(\sigma, t)$.

Assume steady-state temperatures $\vartheta^{0s}, \vartheta^{Ls}$. The temperature profile of the rod in the steady-state can be derived if

$$\frac{\partial \vartheta}{\partial t} = 0 \tag{2.56}$$

as

$$\vartheta^s(\sigma) = \vartheta^{0s} + \frac{\vartheta^{Ls} - \vartheta^{0s}}{L}\sigma \tag{2.57}$$

2.2.3 Mass Transfer Processes

Packed Absorption Column

A scheme of packed countercurrent absorption column is shown in Fig. 2.8 where

t - time variable,
σ - space variable,
L - height of column,
G - molar flow of gas phase,
$c_y = c_y(\sigma, t)$ - molar fraction concentration of a transferable component in gas phase,
Q - molar flow of liquid phase,
$c_x = c_x(\sigma, t)$ - molar fraction concentration of a transferable component in liquid phase.

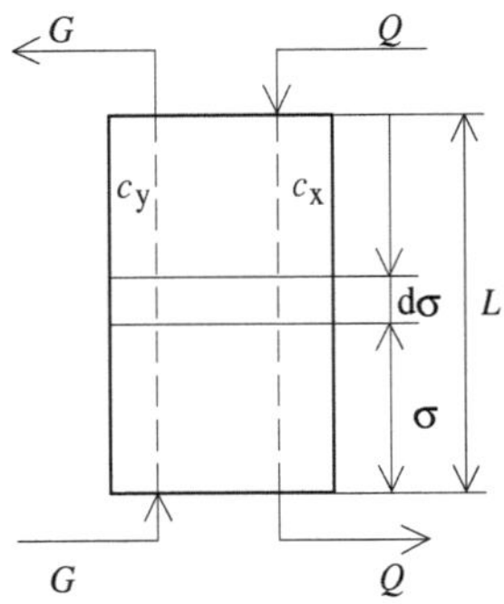

Fig. 2.8. A scheme of a packed countercurrent absorption column

Absorption represents a process of absorbing components of gaseous systems in liquids.

We assume ideal filling, plug flow of gas and liquid phases, negligible mixing and mass transfer in phase flow direction, uniform concentration profiles in both phases at cross surfaces, linear equilibrium curve, isothermal conditions, constant mass transfer coefficients, and constant flow rates G, Q.

Considering only the process state variables c_x, c_y and the above given simplifications and if only the physical process of absorption is considered then mass transfer is realised only in one direction. Then, the following equations result from general law of mass conservation.

For gas phase

$$-N = H_y \frac{\partial c_y}{\partial t} + G \frac{\partial c_y}{\partial \sigma} \tag{2.58}$$

H_y is gas molar hold-up in the column per unit length.

For liquid phase

$$N = H_x \frac{\partial c_x}{\partial t} - G \frac{\partial c_x}{\partial \sigma} \tag{2.59}$$

H_x is liquid molar hold-up in the column per unit length.

Under the above given conditions the following relation holds for mass transfer

$$N = K_G(c_y - c_y^*) \tag{2.60}$$

where

K_G - mass transfer coefficient [mol m^{-1} s^{-1}],
c_y^* - equilibrium concentration of liquid phase.

In the assumptions we stated that the equilibrium curve is linear, that is

$$c_y^* = K c_x \tag{2.61}$$

and K is some constant. Equations (2.58), (2.59) in conjunction with (2.60), (2.61) yield

$$H_y \frac{\partial c_y}{\partial t} + G \frac{\partial c_y}{\partial \sigma} = K_G(K c_x - c_y) \tag{2.62}$$

$$H_x \frac{\partial c_x}{\partial t} - G \frac{\partial c_x}{\partial \sigma} = K_G(c_y - K c_x) \tag{2.63}$$

In the case of the concurrent absorption column, the second term on the left side of Eq. (2.63) would have a positive sign, i.e. $+G(\partial c_x / \partial \sigma)$.

Boundary conditions of Eqs. (2.62), (2.63) are

$$c_y(0, t) = c_y^0(t) \tag{2.64}$$

$$c_x(L, t) = c_x^L(t) \tag{2.65}$$

and c_y^0, c_x^L are the process input variables.

Initial conditions of Eqs. (2.62), (2.63) are

$$c_y(\sigma, 0) = c_{y0}(\sigma) \tag{2.66}$$

$$c_x(\sigma, 0) = c_{x0}(\sigma) \tag{2.67}$$

Consider steady-state input concentration c_y^{0s}, c_x^{0s}. Profiles $c_y^s(\sigma), c_x^s(\sigma)$ can be calculated if

$$\frac{\partial c_y}{\partial t} = 0 \tag{2.68}$$

$$\frac{\partial c_x}{\partial t} = 0 \tag{2.69}$$

as solution of equations

$$G\frac{dc_y^s}{d\sigma} = K_G(Kc_x^s - c_y^s) \tag{2.70}$$

$$-Q\frac{dc_x^s}{d\sigma} = K_G(c_y^s - Kc_x^s) \tag{2.71}$$

with boundary conditions

$$c_y^s(0) = c_y^{0s} \tag{2.72}$$

$$c_x^s(L) = c_x^{Ls} \tag{2.73}$$

Binary Distillation Column

Distillation column represents a process of separation of liquids. A liquid stream is fed into the column, distillate is withdrawn from the condenser and the bottom product from the reboiler. Liquid flow falls down, it is collected in the reboiler where it is vaporised and as vapour flow gets back into the column. Vapour from the top tray condenses and is collected in the condenser. A part of the condensate is returned back to the column. The scheme of the distillation column is shown in Fig. 2.9.

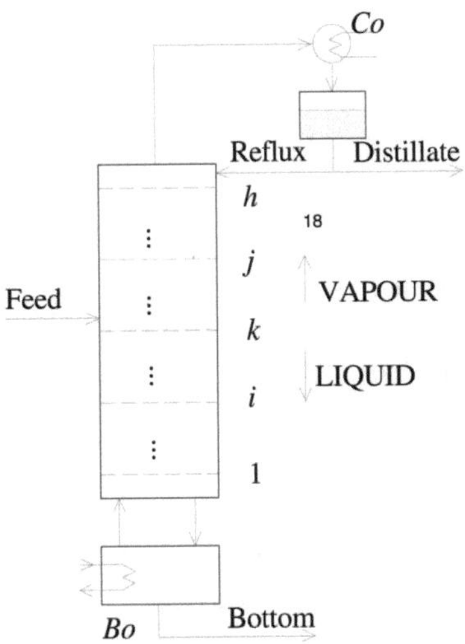

Fig. 2.9. Scheme of a distillation column, Co - condenser; Bo - reboiler; $1, \ldots, i, \ldots, k, \ldots, j, \ldots, h$ - tray number

We assume a binary system with constant relative volatility along the column with theoretical trays (100% efficiency - equilibrium between gas and liquid phases on trays). Vapour exiting the trays is in equilibrium with the tray liquid. Feed arriving on the feed tray boils. Vapour leaving the top tray is totally condensed in the condenser, the condenser is ideally mixed and the liquid within boils. We neglect the dynamics of the pipework. Liquid in the column reboiler is ideally mixed and boils. Liquid on every tray is well mixed and liquid hold-ups are constant in time. Vapour hold-up is negligible. We assume that the column is well insulated, heat losses are zero, and temperature changes along the column are small. We will not assume heat balances. We also consider constant liquid flow along the column and constant pressure.

Mathematical model of the column consists of mass balances of a more volatile component. Feed composition is usually considered as a disturbance and vapour flow as a manipulated variable.

Situation on i-th tray is represented in Fig. 2.10 where

G - vapour molar flow rate,
c_{yi}, c_{yi-1} - vapour molar fraction of a more volatile component,
R - reflux molar flow rate,
F - feed molar flow rate,
c_{xi}, c_{xi-1} - liquid molar fraction of a more volatile component,
H_{yi}, H_{xi} - vapour and liquid molar hold-ups on i-th tray.

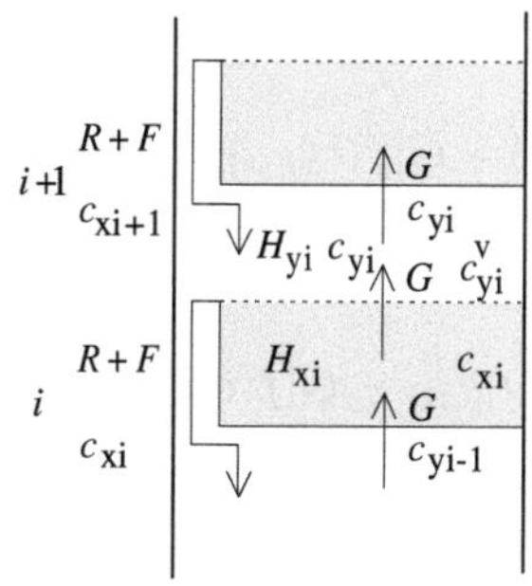

Fig. 2.10. Model representation of i-th tray

Mass balance of a more volatile component in the liquid phase on the i-th tray (stripping section) is given as

$$H_{xi}\frac{\mathrm{d}c_{xi}}{\mathrm{d}t} = (R+F)(c_{xi+1} - c_{xi}) + G(c_{yi-1} - c_{yi}^{v}) \qquad (2.74)$$

where t is time. Under the assumption of equilibrium on the tray follows

$$c_{yi}^{v} = c_{yi}^{*} = f(c_{xi}) \qquad (2.75)$$

Mass balance of a more volatile component in the vapour phase is

$$H_{yi}\frac{\mathrm{d}c_{yi}}{\mathrm{d}t} = G(c_{yi}^{v} - c_{yi}) \qquad (2.76)$$

We assume that vapour molar hold-up is small and the following simplification holds

$$c_{yi} \doteq c_{yi}^v \tag{2.77}$$

and the i-th tray is described by

$$H_{xi}\frac{\mathrm{d}c_{xi}}{\mathrm{d}t} = (R+F)(c_{xi+1} - c_{xi}) + G[f(c_{xi-1}) - f(c_{xi})] \tag{2.78}$$

Mass balance for k-th tray (feed tray) can be written as

$$H_{xk}\frac{\mathrm{d}c_{xk}}{\mathrm{d}t} = Rc_{xk+1} + Fc_{xF} - (R+F)c_{xk} + G[f(c_{xk-1}) - f(c_{xk})] \tag{2.79}$$

where c_{xF} is a molar fraction of a more volatile component in the feed stream.

Mass balances for other sections of the column are analogous:

- j-th tray (enriching section)

$$H_{xj}\frac{\mathrm{d}c_{xj}}{\mathrm{d}t} = R(c_{xj+1} - c_{xj}) + G[f(c_{xj-1}) - f(c_{xj})] \tag{2.80}$$

- h-th tray (top tray)

$$H_{xh}\frac{\mathrm{d}c_{xh}}{\mathrm{d}t} = R(c_{xD} - c_{xh}) + G[f(c_{xh-1}) - f(c_{xh})] \tag{2.81}$$

- Condenser

$$H_{xC}\frac{\mathrm{d}c_{xD}}{\mathrm{d}t} = -(R+D)c_{xD} + G[f(c_{xh})] \tag{2.82}$$

where

D - distillate molar flow,
c_{xD} - molar fraction of a more volatile component in condenser,
H_{xC} - liquid molar hold-up in condenser.

- first tray

$$H_{x1}\frac{\mathrm{d}c_{x1}}{\mathrm{d}t} = (R+F)(c_{x2} - c_{x1}) + G[f(c_{xW}) - f(c_{x1})] \tag{2.83}$$

where c_{xW} is molar fraction of a more volatile component in the bottom product.

- Reboiler

$$H_{xW}\frac{\mathrm{d}c_{xW}}{\mathrm{d}t} = (R+F)c_{x1} - Wc_{xW} - G[f(c_{xW})] \tag{2.84}$$

where W is a molar flow of the bottom product and H_{xW} is the reboiler molar hold-up.

The process state variables correspond to a liquid molar fraction of a more volatile component on trays, in the reboiler, and the condenser. Initial conditions of Eqs. (2.74)-(2.84) are

$$c_{xz}(0) = c_{xz0}, \quad z \in \{i, k, j, h, D, 1, W\} \tag{2.85}$$

The column is in a steady-state if all derivatives with respect to time in balance equations are zero. Steady-state is given by the choices of G^s and c^s_{xF} and is described by the following set of equations

$$0 = (R + F)(c^s_{xi+1} - c^s_{xi}) + G^s[f(c^s_{xi-1}) - f(c^s_{xi})] \tag{2.86}$$

$$0 = Rc^s_{xk+1} + Fc^s_{xF} - (R + F)c^s_{xk} + G^s[f(c^s_{xk-1}) - f(c^s_{xk})] \tag{2.87}$$

$$0 = R(c^s_{xj+1} - c^s_{xj}) + G^s[f(c^s_{xj-1}) - f(c^s_{xj})] \tag{2.88}$$

$$0 = R(c^s_{xD} - c^s_{xh}) + G^s[f(c^s_{xh-1}) - f(c^s_{xh})] \tag{2.89}$$

$$0 = -(R + D)c^s_{xD} + G^s[f(c^s_{xh})] \tag{2.90}$$

$$0 = (R + F)(c^s_{x2} - c^s_{x1}) + G^s[f(c^s_{xW}) - f(c^s_{x1})] \tag{2.91}$$

$$0 = (R + F)c^s_{x1} - (R + F - G^s)c^s_{xW} + G^s[f(c^s_{xW})] \tag{2.92}$$

2.2.4 Chemical and Biochemical Reactors

Continuous Stirred-Tank Reactor (CSTR)

Chemical reactors together with mass transfer processes constitute an important part of chemical technologies. From a control point of view, reactors belong to the most difficult processes. This is especially true for fast exothermal processes.

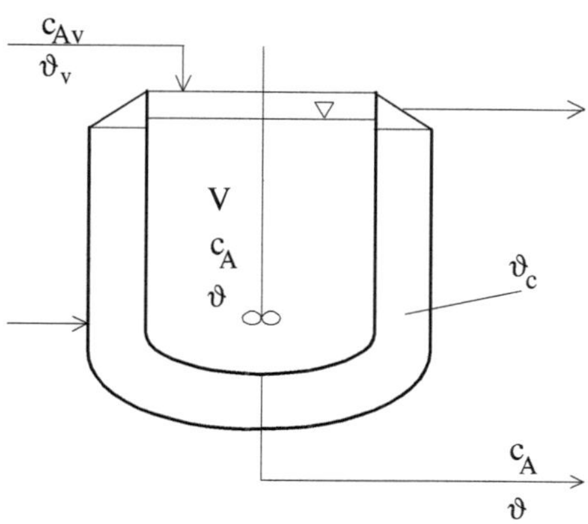

Fig. 2.11. A nonisothermal CSTR

We consider CSTR with a simple exothermal reaction $A \to B$ (Fig. 2.11). For the development of a mathematical model of the CSTR, the following assumptions are made: neglected heat capacity of inner walls of the reactor, constant density and specific heat capacity of liquid, constant reactor volume, constant overall heat transfer coefficient, and constant and equal input and output volumetric flow rates. As the reactor is well-mixed, the outlet stream concentration and temperature are identical with those in the tank.

Mass balance of component A can be expressed as

$$V\frac{\mathrm{d}c_A}{\mathrm{d}t} = qc_{Av} - qc_A - Vr(c_A, \vartheta) \tag{2.93}$$

where

t - time variable,
c_A - molar concentration of A (mole/volume) in the outlet stream,
c_{Av} - molar concentration of A (mole/volume) in the inlet stream,
V - reactor volume,
q - volumetric flow rate,
$r(c_A, \vartheta)$ - rate of reaction per unit volume,
ϑ - temperature of reaction mixture.

The rate of reaction is a strong function of concentration and temperature (Arrhenius law)

$$r(c_A, \vartheta) = kc_A = k_0 \mathrm{e}^{-\frac{E}{R\vartheta}} c_A \tag{2.94}$$

where k_0 is the frequency factor, E is the activation energy, and R is the gas constant.

Heat balance gives

$$V\rho c_p \frac{\mathrm{d}\vartheta}{\mathrm{d}t} = q\rho c_p \vartheta_v - q\rho c_p \vartheta - \alpha F(\vartheta - \vartheta_c) + V(-\Delta H) r(c_A, \vartheta) \tag{2.95}$$

where

ϑ_v - temperature in the inlet stream,
ϑ_c - cooling temperature,
ρ - liquid density,
c_p - liquid specific heat capacity,
α - overall heat transfer coefficient,
F - heat transfer area,
$(-\Delta H)$ - heat of reaction.

Initial conditions are

$$c_A(0) = c_{A0} \tag{2.96}$$

$$\vartheta(0) = \vartheta_0 \tag{2.97}$$

The process state variables are concentration c_A and temperature ϑ. The input variables are $\vartheta_c, c_{Av}, \vartheta_v$ and among them, the cooling temperature can be used as a manipulated variable.

The reactor is in the steady-state if derivatives with respect to time in equations (2.93), (2.95) are zero. Consider the steady-state input variables $\vartheta_c^s, c_{Av}^s, \vartheta_v^s$. The steady-state concentration and temperature can be calculated from the equations

$$0 = qc_{Av}^s - qc_A^s - Vr(c_A^s, \vartheta^s) \tag{2.98}$$

$$0 = q\rho c_p \vartheta_v^s - q\rho c_p \vartheta^s - \alpha F(\vartheta^s - \vartheta_c^s) + V(-\Delta H)r(c_A^s, \vartheta^s) \tag{2.99}$$

Bioreactor

Consider a typical *bioprocess* realised in a fed-batch stirred bioreactor. As an example of bioprocess, alcohol *fermentation* is assumed. Mathematical models of bioreactors usually include mass balances of *biomass, substrate* and *product.* Their concentrations in the reactor are process state variables. Assuming ideal mixing and other assumptions that are beyond the framework of this section, a mathematical model of alcohol fermentation is of the form

$$\frac{\mathrm{d}x}{\mathrm{d}t} = \mu x - Dx \tag{2.100}$$

$$\frac{\mathrm{d}s}{\mathrm{d}t} = -v_s x + D(s_f - s) \tag{2.101}$$

$$\frac{\mathrm{d}p}{\mathrm{d}t} = v_p x - Dp \tag{2.102}$$

where

x - biomass concentration,
s - substrate concentration,
p - product (alcohol) concentration,
s_f - inlet substrate concentration,
D - dilution rate,
μ - specific rate of biomass growth,
v_s - specific rate of substrate consumption,
v_p - specific rate of product creation.

The symbols x, s, p representing the process state variables are used in biochemical literature. The dilution rate can be used as a manipulated variable. The process kinetic properties are given by the relations

$$\mu = \text{function1}(x, s, p) \tag{2.103}$$

$$v_p = \text{function2}(x, s, p) \tag{2.104}$$

$$v_s = \text{function3}(x, s, p) \tag{2.105}$$

2.3 General Process Models

A general process model can be described by a set of ordinary differential and algebraic equations or in matrix-vector form. For control purposes, linearised mathematical models are used. In this section, deviation and dimensionless variables are explained. We show how to convert partial differential equations describing processes with distributed parameters into models with ordinary differential equations. Finally, we illustrate the use of these techniques on examples.

State Equations

As stated above, a suitable model for a large class of continuous technological processes is a set of ordinary differential equations of the form

$$\begin{aligned}
\frac{\mathrm{d}x_1(t)}{\mathrm{d}t} &= f_1(t, x_1(t), \ldots, x_n(t), u_1(t), \ldots, u_m(t), r_1(t), \ldots, r_s(t)) \\
\frac{\mathrm{d}x_2(t)}{\mathrm{d}t} &= f_2(t, x_1(t), \ldots, x_n(t), u_1(t), \ldots, u_m(t), r_1(t), \ldots, r_s(t)) \\
&\vdots \\
\frac{\mathrm{d}x_n(t)}{\mathrm{d}t} &= f_n(t, x_1(t), \ldots, x_n(t), u_1(t), \ldots, u_m(t), r_1(t), \ldots, r_s(t))
\end{aligned} \tag{2.106}$$

where

t - time variable,
$x_1, \ldots, x_n$ - state variables,
$u_1, \ldots, u_m$ - manipulated variables,
$r_1, \ldots, r_s$ - disturbance variables, nonmanipulable variables,
$f_1, \ldots, f_n$ - functions.

Typical technological processes can be described as complex systems. As processes are usually connected to other processes, the complexity of resulting systems increases. It is therefore necessary to investigate the problem of influence of processes and their contact to the environment which influences process with disturbances and manipulated variables. Process state variables are usually not completely measurable. A model of process measurement can be written as a set of algebraic equations

$$\begin{aligned}
y_1(t) &= g_1(t, x_1(t), \ldots, x_n(t), u_1(t), \ldots, u_m(t), r_{m1}(t), \ldots, r_{mt}(t)) \\
y_2(t) &= g_2(t, x_1(t), \ldots, x_n(t), u_1(t), \ldots, u_m(t), r_{m1}(t), \ldots, r_{mt}(t)) \\
&\vdots \\
y_r(t) &= g_r(t, x_1(t), \ldots, x_n(t), u_1(t), \ldots, u_m(t), r_{m1}(t), \ldots, r_{mt}(t))
\end{aligned} \tag{2.107}$$

where

$y_1, \ldots, y_r$ - measurable process output variables,
$r_{m1}, \ldots, r_{mt}$ - disturbance variables,
$g_1, \ldots, g_r$ - functions.

If the vectors of *state variables* $\boldsymbol{x}$, *manipulated variables* $\boldsymbol{u}$, *disturbance variables* $\boldsymbol{r}$, and vectors of functions $\boldsymbol{f}$ are defined as

$$\boldsymbol{x} = \begin{pmatrix} x_1 \\ \vdots \\ x_n \end{pmatrix}, \quad \boldsymbol{u} = \begin{pmatrix} u_1 \\ \vdots \\ u_m \end{pmatrix}, \quad \boldsymbol{r} = \begin{pmatrix} r_1 \\ \vdots \\ r_s \end{pmatrix}, \quad \boldsymbol{f} = \begin{pmatrix} f_1 \\ \vdots \\ f_n \end{pmatrix} \tag{2.108}$$

then the set of the equations (2.106) can be written more compactly

$$\frac{\mathrm{d}\boldsymbol{x}(t)}{\mathrm{d}t} = \boldsymbol{f}(t, \boldsymbol{x}(t), \boldsymbol{u}(t), \boldsymbol{r}(t)) \tag{2.109}$$

If the vectors of *output variables* $\boldsymbol{y}$, *disturbance variables* $\boldsymbol{r}_m$, and vectors of functions $\boldsymbol{g}$ are defined as

$$\boldsymbol{y} = \begin{pmatrix} y_1 \\ \vdots \\ y_r \end{pmatrix}, \quad \boldsymbol{r}_m = \begin{pmatrix} r_{m1} \\ \vdots \\ r_{mt} \end{pmatrix}, \quad \boldsymbol{g} = \begin{pmatrix} g_1 \\ \vdots \\ g_r \end{pmatrix} \tag{2.110}$$

then the set of the algebraic equations is rewritten as

$$\boldsymbol{y}(t) = \boldsymbol{g}(t, \boldsymbol{x}(t), \boldsymbol{u}(t), \boldsymbol{r}_m(t)) \tag{2.111}$$

There are two approaches of control design for processes with distributed parameters. The first approach called *late pass* to lumped parameter models uses model with partial differential equations (p.d.e.) for control design and the only exception is the final step - numerical solution. This approach preserves the nature of distributed systems which is advantageous, however it is also more demanding on the use of the advanced control theory of distributed systems.

The second approach called *early pass* to lumped parameter models is based on space discretisation of p.d.e's at the beginning of the control design problem.

Space discretisation means the division of a distributed process to a finite number of segments and it is assumed that each segment represents a lumped parameter system. The result of space discretisation is a process model described by a set of interconnected ordinary differential equations. The lumped parameter model can also be derived when partial derivatives with respect to space variables are replaced with corresponding differences. For the state variable $x(\sigma, t)$, $0 \le \sigma \le L$ holds

$$\left.\frac{\partial x(\sigma, t)}{\partial \sigma}\right|_k \doteq \frac{x(\sigma_k, t) - x(\sigma_{k-1}, t)}{\Delta\sigma} \tag{2.112}$$

$$\left.\frac{\partial^2 x(\sigma, t)}{\partial \sigma^2}\right|_k \doteq \frac{x(\sigma_{k+1}, t) - 2x(\sigma_k, t) + x(\sigma_{k-1}, t)}{(\Delta\sigma)^2} \tag{2.113}$$

where $\Delta\sigma = L/n$. L is the length of the process. The process is divided into n parts over the interval $[0, L]$, $k = 1, \ldots, n$. It can easily be shown that process models obtained directly from process segmentation and models derived by substitution of derivatives by differences are the same. There are many combinations of finite difference methods. When applied correctly, all are equivalent for $n \to \infty$.

An advantage of the early pass to lumped parameter models exists in fact that it is possible to use well developed control methods for lumped processes. However, a drawback can be found later as the controller derived does not necessarily satisfy all requirements laid on control quality. But in the majority of cases this approach produces satisfactory results. Space *discretisation* of processes with distributed parameters leads to models of type (2.109), (2.111).

The general state-variable equations (the general form of the state-space model) consist of the *state equations* (2.109) and the *output equations* (2.111).

For comparison of process properties in various situations it is advantageous to introduce *dimensionless variables.* These can be state, input, and output variables. Sometimes also dimensionless time and space variables are used.

Example 2.1: *Heat exchanger - state equation*

The heat exchanger shown in Fig. 2.3 is described by the differential equation

$$\frac{d\vartheta}{dt} = -\frac{1}{T_1}\vartheta + \frac{Z_1}{T_1}\vartheta_p + \frac{Z_2}{T_1}\vartheta_v$$

If

$$x_1 = \vartheta$$
$$u_1 = \vartheta_p$$
$$r_1 = \vartheta_v$$

then the state equation is

$$\frac{dx_1}{dt} = f_1(x_1, u_1, r_1)$$

where $f_1(x_1, u_1, r_1) = -\frac{1}{T_1}\vartheta + \frac{Z_1}{T_1}\vartheta_p + \frac{Z_2}{T_1}\vartheta_v$.

The output equation if temperature ϑ is measured is

$$y = x_1$$

Example 2.2: *CSTR - state equations*

Equations describing the dynamics of the CSTR shown in Fig. 2.11 are

$$\frac{dc_A}{dt} = \frac{q}{V}c_{Av} - \frac{q}{V}c_A - r(c_A, \vartheta)$$
$$\frac{d\vartheta}{dt} = \frac{q}{V}\vartheta_v - \frac{q}{V}\vartheta + \frac{\alpha F}{V\rho c_p}(\vartheta - \vartheta_c) + \frac{(-\Delta H)}{\rho c_p}r(c_A, \vartheta)$$

Introducing

$$
\begin{aligned}
x_1 &= c_A \\
x_2 &= \vartheta \\
u_1 &= \vartheta_c \\
r_1 &= c_{Av} \\
r_2 &= \vartheta_v
\end{aligned}
$$

and assuming that temperature measurement of ϑ is available, state and output equations are given as

$$
\begin{aligned}
\frac{\mathrm{d}\boldsymbol{x}}{\mathrm{d}t} &= \boldsymbol{f}(\boldsymbol{x}, \boldsymbol{u}, \boldsymbol{r}) \\
y_1 &= (0\ 1)\boldsymbol{x}
\end{aligned}
$$

where

$$
\boldsymbol{x} = \begin{pmatrix} x_1 \\ x_2 \end{pmatrix}, \quad \boldsymbol{u} = u_1, \quad \boldsymbol{r} = \begin{pmatrix} r_1 \\ r_2 \end{pmatrix}, \quad \boldsymbol{f} = \begin{pmatrix} f_1 \\ f_2 \end{pmatrix}
$$

$$
f_1 = \frac{q}{V} c_{Av} - \frac{q}{V} c_A - r(c_A, \vartheta)
$$

$$
f_2 = \frac{q}{V} \vartheta_v - \frac{q}{V} \vartheta + \frac{\alpha F}{V \rho c_p} (\vartheta - \vartheta_c) + \frac{(-\Delta H)}{\rho c_p} r(c_A, \vartheta)
$$

Example 2.3: *Double-pipe steam-heated exchanger - state equations*

Processes with distributed parameters are usually approximated by a series of well-mixed lumped parameter processes. This is also the case for the heat exchanger shown in Fig. 2.5 which is divided into n well-mixed heat exchangers. The space variable is divided into n equal lengths within the interval $[0, L]$. However, this division can also be realised differently. Mathematical model of the exchanger is of the form

$$
\frac{\partial \vartheta}{\partial t} = -v_\sigma \frac{\partial \vartheta}{\partial \sigma} - \frac{1}{T_1} \vartheta + \frac{1}{T_1} \vartheta_p
$$

Introduce

$$
\begin{aligned}
x_1(t) &= \vartheta\left(\frac{L}{n}, t\right) \\
x_2(t) &= \vartheta\left(\frac{2L}{n}, t\right) \\
&\ \vdots \\
x_n(t) &= \vartheta(L, t) \\
u_1(t) &= \vartheta_p(t) \\
r_1(t) &= \vartheta(0, t)
\end{aligned}
$$

and replace $\partial \vartheta / \partial \sigma$ with the corresponding difference. The resulting model consits of a set of ordinary differential equations

$$
\begin{aligned}
\frac{\mathrm{d}x_1}{\mathrm{d}t} &= -\frac{v_\sigma n}{L}(x_1 - r_1) - \frac{1}{T_1}x_1 + \frac{1}{T_1}u_1 \\
\frac{\mathrm{d}x_2}{\mathrm{d}t} &= -\frac{v_\sigma n}{L}(x_2 - x_1) - \frac{1}{T_1}x_2 + \frac{1}{T_1}u_1 \\
&\vdots \\
\frac{\mathrm{d}x_n}{\mathrm{d}t} &= -\frac{v_\sigma n}{L}(x_n - x_{n-1}) - \frac{1}{T_1}x_n + \frac{1}{T_1}u_1
\end{aligned}
$$

The state equation is given as

$$\frac{\mathrm{d}\boldsymbol{x}}{\mathrm{d}t} = \boldsymbol{A}\boldsymbol{x} + \boldsymbol{B}u_1 + \boldsymbol{H}r_1$$

where

$$
\begin{aligned}
\boldsymbol{x} &= (x_1 \ \dots \ x_n)^T \\
\boldsymbol{A} &= \begin{pmatrix}
-(\frac{v_\sigma n}{L} + \frac{1}{T_1}) & 0 & 0 \ \dots \ 0 & 0 \\
\frac{v_\sigma n}{L} & -(\frac{v_\sigma n}{L} + \frac{1}{T_1}) & 0 \ \dots \ 0 & 0 \\
\vdots & \vdots & \vdots \ \dots \ \vdots & \vdots \\
0 & 0 & 0 \ \dots \ \frac{v_\sigma n}{L} & -(\frac{v_\sigma n}{L} + \frac{1}{T_1})
\end{pmatrix} \\
\boldsymbol{B} &= \frac{1}{T_1}(1 \ \dots \ 1)^T \\
\boldsymbol{H} &= \left(\frac{v_\sigma n}{L} \ 0 \ \dots \ 0\right)^T
\end{aligned}
$$

Assume that temperature is measured at $\sigma_1 = 2L/n$ and/or $\sigma_2 = L$. Then the output equation is of the form

$$\boldsymbol{y} = \boldsymbol{C}\boldsymbol{x}$$

where

$$\boldsymbol{y} = \begin{pmatrix} y_1 \\ y_2 \end{pmatrix}, \quad \boldsymbol{C} = \begin{pmatrix} 0 \ 1 \ 0 \ \dots \ 0 \\ 0 \ 0 \ 0 \ \dots \ 1 \end{pmatrix}$$

or

$$y_1 = \boldsymbol{C}\boldsymbol{x}, \ \boldsymbol{C} = (0 \ 0 \ 0 \ \dots \ 1)$$

Example 2.4: *Heat exchanger - dimensionless variables*

The state equation for the heat exchanger shown in Fig. 2.3 is

$$\frac{\mathrm{d}\vartheta}{\mathrm{d}t'} = -\frac{1}{T_1}\vartheta + \frac{Z_1}{T_1}\vartheta_p + \frac{Z_2}{T_1}\vartheta_v$$

where t' is time variable. The exchanger is in a steady-state if $\mathrm{d}\vartheta/\mathrm{d}t' = 0$. Denote steady-state temperatures $\vartheta_p^s, \vartheta_v^s, \vartheta^s$. For the steady-state yields

$$\vartheta^s = Z_1\vartheta_p^s + Z_2\vartheta_v^s$$

Define dimensionless variables

$$x_1 = \frac{\vartheta}{\vartheta^s}$$

$$u_1 = \frac{\vartheta_p}{\vartheta_p^s}$$

$$r_1 = \frac{\vartheta_v}{\vartheta_v^s}$$

$$t = \frac{t'}{T_1}$$

then the state equation is given as

$$\frac{\mathrm{d}x_1}{\mathrm{d}t} = -x_1 + \frac{Z_1 \vartheta_p^s}{\vartheta^s} u_1 + \frac{Z_2 \vartheta_v^s}{\vartheta^s} r_1$$

with initial condition

$$x_1(0) = x_{10} = \frac{\vartheta(0)}{\vartheta^s}$$

2.4 Linearisation

Linearisation of nonlinear models plays an important role in practical control design. The principle of linearisation of nonlinear equations consists in supposition that process variables change very little and their *deviations from steady-state* remain small. Linear approximation can be obtained by using the Taylor series expansion and considering only linear terms. This approximation is then called *linearised model.* An advantage of linear models is their simplicity and their use can yield to analytical results.

Let us recall the *Taylor theorem*: Let a, x be different numbers, $k \geq 0$ and J is a closed interval with endpoints a, x. Let f be a function with continuous k-th derivative on J and $k+1$-th derivative within this interval. Then there exists a point ζ within J such that

$$f(x) = f(a) + \frac{\dot{f}(a)}{1!}(x-a) + \frac{\ddot{f}(a)}{2!}(x-a)^2 + \cdots + \frac{f^{(k)}(a)}{k!}(x-a)^k + R_k(x) \tag{2.114}$$

where $R_k(x) = \frac{f^{(k+1)}(\zeta)}{(k+1)!}(x-a)^{k+1}$ is the rest of the function f after the k-th term of the Taylor's polynomial.

Consider a process described by a set of equations

$$\frac{\mathrm{d}x_i'}{\mathrm{d}t} = f_i(\boldsymbol{x}', \boldsymbol{u}') = f_i(x_1', \ldots, x_n', u_1', \ldots, u_m'), \quad i = 1, \ldots, n \tag{2.115}$$

where

$\boldsymbol{x}'$ - vector of state variables $x'_1, \dots, x'_n$,
$\boldsymbol{u}'$ - vector of manipulated variables $u'_1, \dots, u'_m$.

Let the process state variables x'_i change in the neighbourhood of the steady-state x'^s_i under the influence of the manipulated variables u'_i. Then it is possible to approximate the process nonlinearities. The steady-state is given by the equation

$$0 = f_i(\boldsymbol{x}'^s, \boldsymbol{u}'^s) = f_i^s \tag{2.116}$$

We suppose that the solution of these equations is known for some $u'^s_j, j = 1, \dots, m$. The function $f_i(\bullet)$ is approximated by the Taylor series expansion truncated to only first order terms as

$$\begin{aligned} f_i(\boldsymbol{x}', \boldsymbol{u}') \doteq f_i(\boldsymbol{x}'^s, \boldsymbol{u}'^s) + & \\ + \left(\frac{\partial f_i}{\partial x'_1}\right)^s (x'_1 - x'^s_1) + \dots + \left(\frac{\partial f_i}{\partial x'_n}\right)^s (x'_n - x'^s_n) + & \\ + \left(\frac{\partial f_i}{\partial u'_1}\right)^s (u'_1 - u'^s_1) + \dots + \left(\frac{\partial f_i}{\partial u'_m}\right)^s (u'_m - u'^s_m) & \end{aligned} \tag{2.117}$$

$(\partial f_i/\partial x'_l)^s$, $l = 1, \dots, n$ and $(\partial f_i/\partial u'_j)^s$, $j = 1, \dots, m$ denote partial derivatives for $x'_l = x'^s_l$ and $u'_j = u'^s_j$, respectively. Therefore, these partial derivatives are constants

$$a_{il} = \left(\frac{\partial f_i}{\partial x'_l}\right)^s \qquad l = 1, \dots, n \tag{2.118}$$

$$b_{ij} = \left(\frac{\partial f_i}{\partial u'_j}\right)^s \qquad j = 1, \dots, m \tag{2.119}$$

From Eq. (2.116) follows that the first term on the right side of (2.117) is zero. Introducing state and manipulated deviation variables

$$x_i = x'_i - x'^s_i \tag{2.120}$$

$$u_j = u'_j - u'^s_j \tag{2.121}$$

gives

$$\begin{aligned} \frac{\mathrm{d}x'_1}{\mathrm{d}t} &= \frac{\mathrm{d}x_1}{\mathrm{d}t} = a_{11}x_1 + \dots + a_{1n}x_n + b_{11}u_1 + \dots + b_{1m}u_m \\ \frac{\mathrm{d}x'_2}{\mathrm{d}t} &= \frac{\mathrm{d}x_2}{\mathrm{d}t} = a_{21}x_1 + \dots + a_{2n}x_n + b_{21}u_1 + \dots + b_{2m}u_m \\ &\vdots \\ \frac{\mathrm{d}x'_n}{\mathrm{d}t} &= \frac{\mathrm{d}x_n}{\mathrm{d}t} = a_{n1}x_1 + \dots + a_{nn}x_n + b_{n1}u_1 + \dots + b_{nm}u_m \end{aligned} \tag{2.122}$$

We denote $\boldsymbol{x}$ the vector of *deviation state variables* and $\boldsymbol{u}$ the vector of *deviation manipulated variables*. Then (2.122) can be written as

$$\frac{\mathrm{d}\boldsymbol{x}}{\mathrm{d}t} = \boldsymbol{A}\boldsymbol{x} + \boldsymbol{B}\boldsymbol{u} \tag{2.123}$$

where

$$\boldsymbol{A} = \begin{pmatrix} a_{11} & a_{12} & \dots & a_{1n} \\ a_{21} & a_{22} & \dots & a_{2n} \\ \vdots & \vdots & \dots & \vdots \\ a_{n1} & a_{n2} & \dots & a_{nn} \end{pmatrix}$$

$$\boldsymbol{B} = \begin{pmatrix} b_{11} & b_{12} & \dots & b_{1m} \\ b_{21} & b_{22} & \dots & b_{2m} \\ \vdots & \vdots & \dots & \vdots \\ b_{n1} & b_{n2} & \dots & b_{nm} \end{pmatrix}$$

Equation (2.123) is a linearised differential equation. If initial state of (2.115) also represent steady-state of the modelled process then

$$\boldsymbol{x}(0) = \boldsymbol{0} \tag{2.124}$$

Equations (2.115), (2.123) describe dynamics of a process. The differences are as follows:

1. equation (2.123) is only an approximation,
2. equation (2.123) uses deviation variables,
3. equation (2.123) is linear with constant coefficients.

Linearisation of the process dynamics must be completed with linearisation of the output equation if this is nonlinear.

Consider the output equation of the form

$$y'_k = g_k(\boldsymbol{x}', \boldsymbol{u}'), \quad k = 1, \dots, r \tag{2.125}$$

where y'_k are the output variables. In the steady-state holds

$$y'^s_k = g_k(\boldsymbol{x}'^s, \boldsymbol{u}'^s) \tag{2.126}$$

Introducing *output deviation variables*

$$y_k = y'_k - y'^s_k \tag{2.127}$$

follows

$$y_k = g_k(\boldsymbol{x}'^s + \boldsymbol{x}, \boldsymbol{u}'^s + \boldsymbol{u}) - g_k(\boldsymbol{x}'^s, \boldsymbol{u}'^s) \tag{2.128}$$

Using the Taylor series expansion with only linear terms the following approximation holds

$$g_k(\boldsymbol{x}', \boldsymbol{u}') \doteq g_k(\boldsymbol{x}'^s, \boldsymbol{u}'^s) + \sum_{l=1}^{n} \left(\frac{\partial g_k}{\partial x'_l}\right)^s (x'_l - x'^s_l) + \sum_{j=1}^{m} \left(\frac{\partial g_k}{\partial u'_j}\right)^s (u'_j - u'^s_j) \tag{2.129}$$

and again the partial derivatives in (2.129) are constants

$$c_{kl} = \left(\frac{\partial g_k}{\partial x'_l}\right)^s \quad l = 1, \dots, n \tag{2.130}$$

$$d_{kj} = \left(\frac{\partial g_k}{\partial u'_j}\right)^s \quad j = 1, \dots, m \tag{2.131}$$

Output deviation variables are then of the form

$$\begin{aligned} y_1 &= c_{11}x_1 + \dots + c_{1n}x_n + d_{11}u_1 + \dots + d_{1m}u_m \\ y_2 &= c_{21}x_1 + \dots + c_{2n}x_n + d_{21}u_1 + \dots + d_{2m}u_m \\ &\vdots \\ y_r &= c_{r1}x_1 + \dots + c_{rn}x_n + d_{r1}u_1 + \dots + d_{rm}u_m \end{aligned} \tag{2.132}$$

If $\boldsymbol{y}$ denotes the vector of output deviation variables then the previous equation can more compactly be written as

$$\boldsymbol{y} = \boldsymbol{C}\boldsymbol{x} + \boldsymbol{D}\boldsymbol{u} \tag{2.133}$$

where

$$\boldsymbol{C} = \begin{pmatrix} c_{11} & c_{12} & \dots & c_{1n} \\ c_{21} & c_{22} & \dots & c_{2n} \\ \vdots & \vdots & \dots & \vdots \\ c_{r1} & c_{r2} & \dots & c_{rn} \end{pmatrix}$$

$$\boldsymbol{D} = \begin{pmatrix} d_{11} & d_{12} & \dots & d_{1m} \\ d_{21} & d_{22} & \dots & d_{2m} \\ \vdots & \vdots & \dots & \vdots \\ d_{r1} & d_{r2} & \dots & d_{rm} \end{pmatrix}$$

Equations (2.123) and (2.133) constitute together the *general linear state process model.* When it is obtained from the linearisation procedure, then it can only be used in the neighbourhood of the steady-state where linearisation was derived.

Example 2.5: *Liquid storage tank - linearisation*

Consider the liquid storage tank shown in Fig. 2.1. The state equation of this process is

$$\frac{\mathrm{d}h}{\mathrm{d}t} = f_1(h, q_0)$$

where

$$f_1(h, q_0) = -\frac{k_{11}}{F}\sqrt{h} + \frac{1}{F}q_0$$

The steady-state equation is

$$f_1(h^s, q_0^s) = -\frac{k_{11}}{F}\sqrt{h^s} + \frac{1}{F}q_0^s = 0$$

Linearised state equation for a neighbourhood of the steady-state given by q_0^s, h^s can be written as

$$\frac{dh}{dt} = \frac{d(h - h^s)}{dt} = -\frac{k_{11}}{2F\sqrt{h^s}}(h - h^s) + \frac{1}{F}(q_0 - q_0^s)$$

Introducing deviation variables

$$x_1 = h - h^s$$

$$u_1 = q_0 - q_0^s$$

and assuming that the level h is measured then the linearised model of the tank is of the form

$$\frac{dx_1}{dt} = a_{11}x_1 + b_{11}u_1$$

$$y_1 = x_1$$

where

$$a_{11} = -\frac{k_{11}}{2F\sqrt{h^s}}, \quad b_{11} = \frac{1}{F}$$

Example 2.6: *CSTR - linearisation*

Consider the CSTR shown in Fig. 2.11. The state equations for this reactor are

$$\frac{dc_A}{dt} = f_1(c_A, c_{Av}, \vartheta)$$

$$\frac{d\vartheta}{dt} = f_2(c_A, \vartheta, \vartheta_v, \vartheta_c)$$

where

$$f_1(c_A, c_{Av}, \vartheta) = \frac{q}{V}c_{Av} - \frac{q}{V}c_A - r(c_A, \vartheta)$$

$$f_2(c_A, \vartheta, \vartheta_v, \vartheta_c) = \frac{q}{V}\vartheta_v - \frac{q}{V}\vartheta - \frac{\alpha F}{V\rho c_p}(\vartheta - \vartheta_c) + \frac{(-\Delta H)}{\rho c_p}r(c_A, \vartheta)$$

Linearised dynamics equations for the neighbourhood of the steady-state given by steady-state input variables $\vartheta_c^s, c_{Av}^s, \vartheta_v^s$ and steady-state process state variables c_A^s, ϑ^s are of the form

$$\frac{dc_A}{dt} = \frac{d(c_A - c_A^s)}{dt} = \left(-\frac{q}{V} - \dot{r}_{c_A}(c_A^s, \vartheta^s)\right)(c_A - c_A^s)$$
$$+ (-\dot{r}_\vartheta(c_A^s, \vartheta^s))(\vartheta - \vartheta^s) + \frac{q}{V}(c_{Av} - c_{Av}^s)$$

$$\frac{d\vartheta}{dt} = \frac{d(\vartheta - \vartheta^s)}{dt} = \left(\frac{(-\Delta H)}{\rho c_p}\dot{r}_{c_A}(c_A^s, \vartheta^s)\right)(c_A - c_A^s)$$
$$+ \left(-\frac{q}{V} - \frac{\alpha F}{V\rho c_p} + \frac{(-\Delta H)}{\rho c_p}\dot{r}_\vartheta(c_A^s, \vartheta^s)\right)(\vartheta - \vartheta^s)$$
$$+ \frac{\alpha F}{V\rho c_p}(\vartheta_c - \vartheta_c^s) + \frac{q}{V}(\vartheta_v - \vartheta_v^s)$$

where

$$\dot{r}_{c_A}(c_A^s, \vartheta^s) = \left.\frac{\partial r(c_A, \vartheta)}{\partial c_A}\right|_{\substack{c_A = c_A^s \\ \vartheta = \vartheta^s}}$$

$$\dot{r}_{\vartheta}(c_A^s, \vartheta^s) = \left.\frac{\partial r(c_A, \vartheta)}{\partial \vartheta}\right|_{\substack{c_A = c_A^s \\ \vartheta = \vartheta^s}}$$

Introducing deviation variables

$$x_1 = c_A - c_A^s$$
$$x_2 = \vartheta - \vartheta^s$$
$$u_1 = \vartheta_c - \vartheta_c^s$$
$$r_1 = c_{Av} - c_{Av}^s$$
$$r_2 = \vartheta_v - \vartheta_v^s$$

and considering temperature measurements of ϑ then for the linearised process model follows

$$\frac{\mathrm{d}x_1}{\mathrm{d}t} = a_{11}x_1 + a_{12}x_2 + h_{11}r_1$$
$$\frac{\mathrm{d}x_2}{\mathrm{d}t} = a_{21}x_1 + a_{22}x_2 + b_{21}u_1 + h_{22}r_2$$
$$y_1 = x_2$$

where

$$a_{11} = -\frac{q}{V} - \dot{r}_{c_A}(c_A^s, \vartheta^s), \; a_{12} = -\dot{r}_{\vartheta}(c_A^s, \vartheta^s)$$

$$a_{21} = \frac{(-\Delta H)}{\rho c_p}\dot{r}_{c_A}(c_A^s, \vartheta^s), \; a_{22} = -\frac{q}{V} - \frac{\alpha F}{V\rho c_p} + \frac{(-\Delta H)}{\rho c_p}\dot{r}_{\vartheta}(c_A^s, \vartheta^s))$$

$$b_{21} = \frac{\alpha F}{V\rho c_p}$$

$$h_{11} = h_{22} = \frac{q}{V}$$

If the rate of reaction is given as (the first order reaction)

$$r(c_A, \vartheta) = c_A k_0 \mathrm{e}^{-\frac{E}{R\vartheta}}$$

then

$$\dot{r}_{c_A}(c_A^s, \vartheta^s) = k_0 \mathrm{e}^{-\dfrac{E}{R\vartheta^s}}$$

$$\dot{r}_{\vartheta}(c_A^s, \vartheta^s) = c_A^s k_0 \frac{E}{R(\vartheta^s)^2} \mathrm{e}^{-\dfrac{E}{R\vartheta^s}}$$

The deviation variables have the same meaning as before: x_1, x_2 are state deviation variables, u_1 is a manipulated deviation variable, and r_1, r_2 are disturbance deviation variables.

2.5 Systems, Classification of Systems

A *deterministic single-input single-output (SISO) system* is a physical device which has only one input $u(t)$ and the result of this influence is an observable output variable $y(t)$. The same initial conditions and the same function $u(t)$ lead to the same output function $y(t)$. This definition is easily extended to *deterministic multi-input multi-output (MIMO) systems* whose input variables are $u_1(t), \ldots, u_m(t)$ and output variables are $y_1(t), \ldots, y_r(t)$. The *concept of a system* is based on the relation between cause and consequence of input and output variables.

Continuous-time (CT) systems are systems with all variables defined for all time values.

Lumped parameter systems have influence between an input and output variables given by ordinary differential equations with derivatives with respect to time. Systems with distributed parameters are described by partial differential equations with derivatives with respect to time and space variables.

If the relation between an input and output variable for deterministic CT SISO system is given by ordinary differential equations with order greater than one, then it is necessary for determination of $y(t), t > t_0$ to know $u(t), t > t_0$ and output variable $y(t_0)$ with its derivatives at t_0 or some equivalent information. The necessity of knowledge about derivatives avoids introduction of *concept of state.*

Linear systems obey the law of superposition.

The systems described in Section 2.2 are examples of *physical systems.* The systems determined only by variables that define a relation between the system elements or between the system and its environment are called *abstract.* Every physical system has a corresponding abstract model but not vice versa. A notation of oriented systems can be introduced. This is every controlled system with declared input and output variables.

The relation between objects (processes) and systems can be explained as follows. If a process has defined some set of *typical important properties* significant for our investigations then we have defined a system on the process.

We note that we will not further pursue special details and differences between systems and mathematical relations describing their behaviour as it is not important for our purposes.

Analogously as continuous-time systems were defined, *discrete-time (DT) systems* have their variables defined only in certain time instants.

The process model examples were chosen to explain the procedure for simplification of models. Usually, two basic steps were performed. Models given by partial differential equations were transformed into ordinary differential equations and nonlinear models were linearised. Step-wise simplifications of process models led to models with linear differential equations. As computer control design is based on DT signals, the last transformation is toward DT systems.

A *Stochastic system* is characterised by variables known only with some probability.

Therefore, classification of *dynamical systems* can be clearly given as in Fig. 2.12.

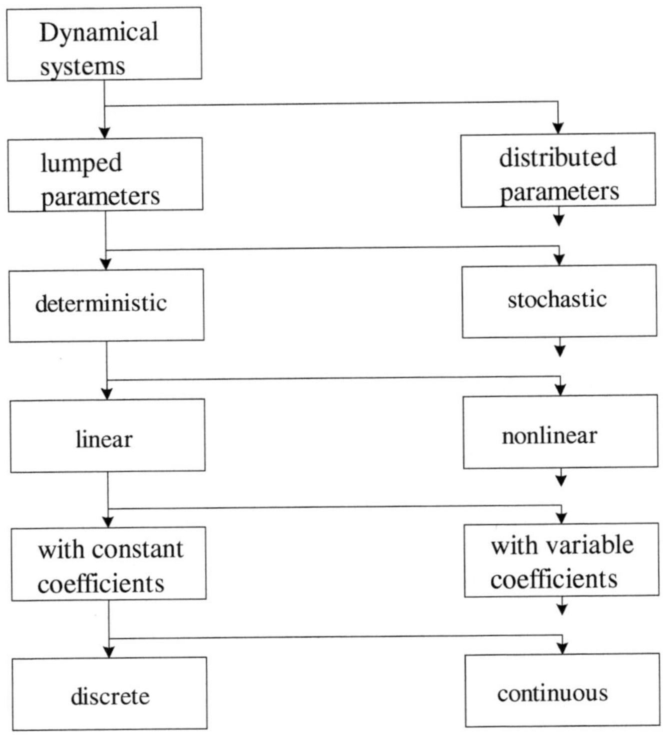

Fig. 2.12. Classification of dynamical systems

2.6 References

Mathematical models of processes are discussed in more detail in works:

J. Čermák, V. Peterka, and J. Závorka. *Dynamics of Controlled Systems in Thermal Energetics and Chemistry.* Academia, Praha, 1968. (in Czech).

C. J. Friedly. *Dynamic Behavior of Processes.* Prentice Hall, Inc., New Jersey, 1972.

L. N. Lipatov. *Typical Processes of Chemical Technology as Objects for Control.* Chimija, Moskva, 1973. (in Russian).

V. V. Kafarov, V. L Perov, and B. P. Mešalkin. *Principles of Mathematical Modelling of Systems in Chemical Technology.* Chimija, Moskva, 1974. (in Russian).

G. Stephanopoulos. *Chemical Process Control, An Introduction to Theory and Practice.* Prentice Hall, Englewood Cliffs, New Jersey, 1984.

D. Chmúrny, J. Mikleš, P. Dostál, and J. Dvoran. *Modelling and Control of Processes and Systems in Chemical Technology.* Alfa, Bratislava, 1985. (in Slovak).

W. L. Luyben. *Process Modelling, Simulation and Control for Chemical Engineers.* McGraw Hill, Singapore, 2 edition, 1990.

B. Wittenmark, J. K. Åström, and S. B. Jørgensen. *Process Control.* Lund Institute of Technology, Technical University of Denmark, Lyngby, 1992.

J. Ingham, I. J. Dunn, E. Henzle, and J. E. Přenosil. *Chemical Engineering Dynamics.* VCH Verlagsgesselschaft, Weinheim, 1994.

Mathematical models of unit processes are also given in many journal articles. Some of them are cited in books above. These are completed by articles dealing with bioprocesses:

J. Bhat, M. Chidambaram, and K. P. Madhavan. Robust control of a batch-fed fermentor. *Journal of Process Control,* 1:146 – 151, 1991.

B. Dahhou, M. Lakrori, I. Queinnec, E. Ferret, and A. Cherny. Control of continuous fermentation process. *Journal of Process Control,* 2:103 – 111, 1992.

M. Rauseier, P. Agrawal, and D. A. Melichamp. Non-linear adaptive control of fermentation processes utilizing a priori modelling knowledge. *Journal of Process Control,* 2:129 – 138, 1992.

Some works concerning definitions and properties of systems:

L. A. Zadeh and C. A. Desoer. *Linear System Theory - the State-space Approach.* McGraw-Hill, New York, 1963.

E. D. Gilles. *Systeme mit verteilten Parametern, Einführung in die Regelungstheorie.* Oldenbourg Verlag, München, 1973.

V. Strejc. *State-space Theory of Linear Control.* Academia, Praha, 1978. (in Czech).

A. A. Voronov. *Stability, Controllability, Observability.* Nauka, Moskva, 1979. (in Russian).

2.7 Exercises

Exercise 2.1:
Consider the liquid storage tank shown in Fig. 2.13. Assume constant liquid density and constant flow rate q_1. Flow rate q_2 can be expressed as

$$q_2 = k_{10}h + k_{11}\sqrt{h}$$

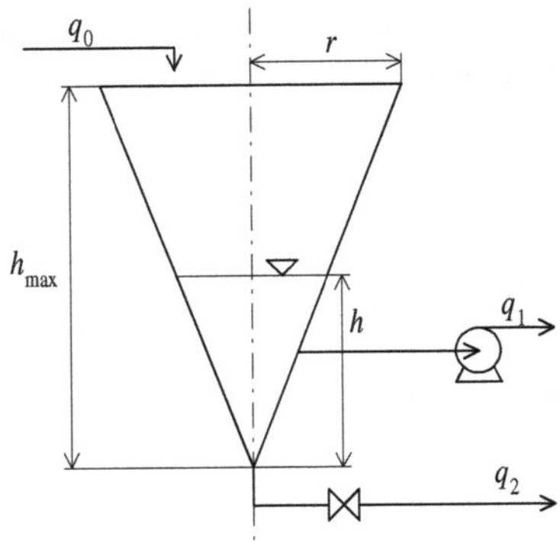

Fig. 2.13. A cone liquid storage process

Find:

1. state equation,
2. linearised process model.

Exercise 2.2:
A double vessel is used as a heat exchanger between two liquids separated by a wall (Fig. 2.14).

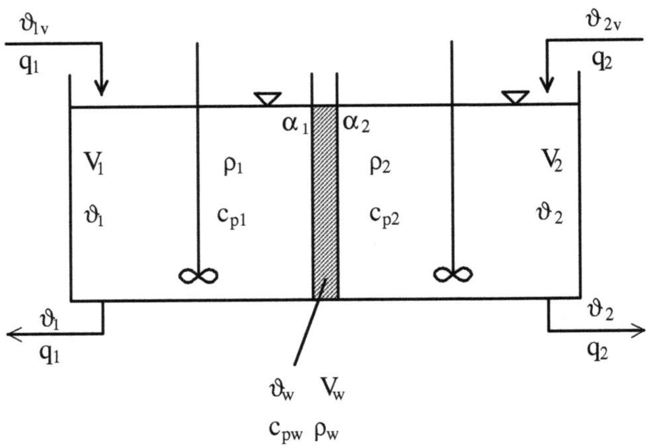

Fig. 2.14. Well mixed heat exchanger

Assume heating of a liquid with a constant volume V_2 with a liquid with a constant volume V_1. Heat transfer is considered only in direction vertical to the wall with temperature $\vartheta_w(t)$, volume V_w, density ρ_w, and specific heat capacity c_{pw}. Heat transfer from the process toits environment is neglected. Further, assume spatially constant temperatures ϑ_1 and ϑ_2, constant densities ρ_1, ρ_2, flow rates q_1, q_2, specific heat capacities c_{p1}, c_{p2}. α_1 is the heat transfer coefficient from liquid to wall and α_2 is the heat transfer coefficient from wall to liquid. The process state variables are $\vartheta_1, \vartheta_2, \vartheta_w$. The process input variables are $\vartheta_{1v}, \vartheta_{2v}$.

1. Find state equations,
2. introduce dimensionless variables and rewrite the state equations.

Exercise 2.3:
A tank is used for blending of liquids (Fig. 2.15). The tank is filled up with two pipelines with flow rates q_1, q_2. Both streams contain a component with constant concentrations c_0, c_1. The outlet stream has a flow rate q_2 and concentration c_2. Assume that the concentration within tank is c_2.

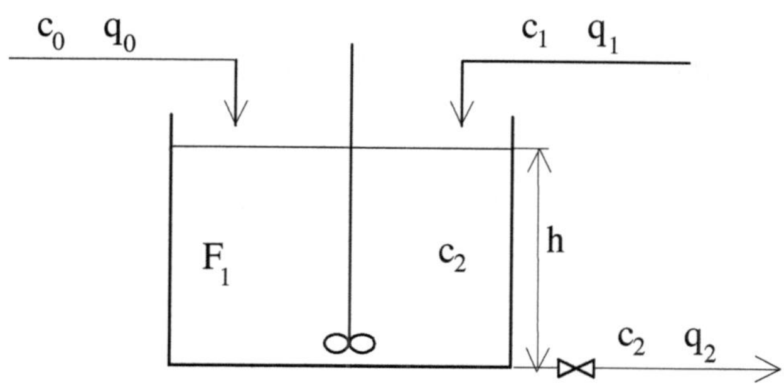

Fig. 2.15. A well mixed tank

Find:
1. state equations,
2. linearised process model.

Exercise 2.4:
An irreversible reaction $A \to B$ occurs in a series of CSTRs shown in Fig. 2.16. The assumptions are the same as for the reactor shown in Fig. 2.11.

1. Find state equations,
2. construct linearised process model.

Exercise 2.5:
Consider the gas tank shown in Fig. 2.17. A gas with pressure $p_0(t)$ flows through pneumatic resistance (capillary) R_1 to the tank with volume V. The pressure in the tank is $p_1(t)$. Molar flow rate G of the gas through resistance R_1 is

$$G = \frac{p_0 - p_1}{R_1}$$

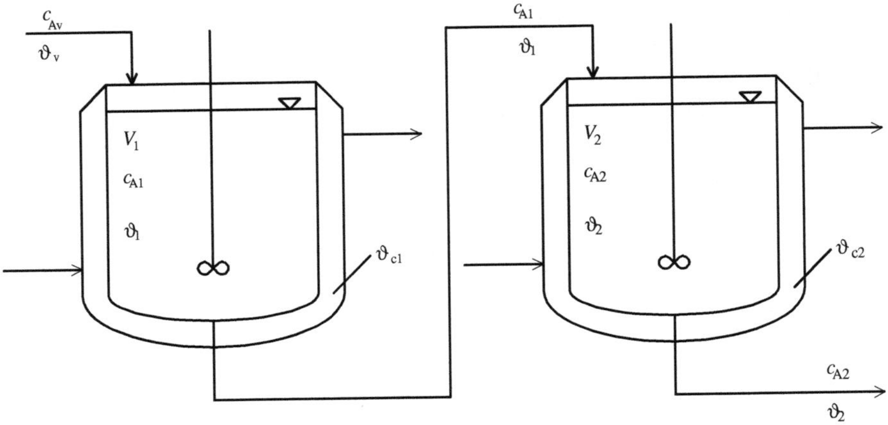

Fig. 2.16. Series of two CSTRs

Assume that the ideal gas law holds. Find state equation of the tank.

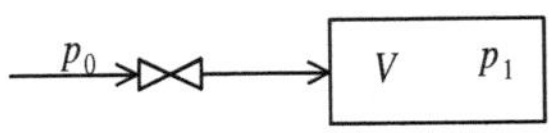

Fig. 2.17. A gas storage tank

3

Analysis of Process Models

Mathematical models describing behaviour of a large group of technological processes can under some simplifications be given by linear differential equations with constant coefficients. Similarly, other blocks of control loops can also be described by linear differential equations with constant coefficients. For investigation of the dynamical properties of processes it is necessary to solve differential equations with time as independent variable. Linear differential equations with constant coefficients can be very suitably solved with the help of the Laplace transform.

Analysis of dynamical systems is based on their state-space representation. The spate-space representation is closely tied to input-output representation of the systems that are described by input-output process models. In this chapter we will define the Laplace transform and show how to solve by means of it linear differential equations with constant coefficients. We introduce the definition of transfer function and transfer function matrix. Next, the concept of states and connection between state-space and input-output models will be given. We examine the problem of stability, controllability, and observability of continuous-time processes.

3.1 The Laplace Transform

The Laplace transform offers a very simple and elegant vehicle for the solution of differential equations with constant coefficients. It further enables to derive input-output models which are suitable for process identification and control. Moreover, it simplifies the qualitative analysis of process responses subject to various input signals.

3.1.1 Definition of the Laplace Transform

Consider a function $f(t)$. *The Laplace transform* is defined as

$$\mathcal{L}\{f(t)\} = \int_0^\infty f(t)\mathrm{e}^{-st}\mathrm{d}t \tag{3.1}$$

where $\mathcal{L}$ is an operator defined by the integral, $f(t)$ is some function of time. The Laplace transform is often written as

$$F(s) = \mathcal{L}\{f(t)\} \tag{3.2}$$

The function $f(t)$ given over an interval $0 \le t < \infty$ is called the time *original* and the function $F(s)$ its *Laplace transform.* The function $f(t)$ must satisfy some conditions. It must be piecewise continuous for all times from $t = 0$ to $t = \infty$. This requirement practically always holds for functions used in modelling and control. It follows from the definition integral that we transform the function from the time domain into s domain where s is a complex variable. Further it is clear that the Laplace transform of a function exists if the definition integral is bounded. This condition is fulfilled for all functions we will deal with.

The function $F(s)$ contains no information about $f(t)$ for $t < 0$. This is no real obstacle as t is the time variable usually defined as positive. Variables and systems are then usually defined such that

$$f(t) \equiv 0 \quad \text{for} \quad t < 0 \tag{3.3}$$

If the equation (3.3) is valid for the function $f(t)$, then this is uniquely given except at the points of incontinuities with the $\mathcal{L}$ transform

$$f(t) = \mathcal{L}^{-1}\{F(s)\} \tag{3.4}$$

This equation defines the *inverse Laplace transform.*

The Laplace transform is a *linear operator* and satisfies the *principle of superposition*

$$\mathcal{L}\{k_1 f_1(t) + k_2 f_2(t)\} = k_1\mathcal{L}\{f_1(t)\} + k_2\mathcal{L}\{f_2(t)\} \tag{3.5}$$

where k_1, k_2 are some constants. The proof follows from the definition integral

$$\begin{aligned}\mathcal{L}\{k_1 f_1(t) + k_2 f_2(t)\} &= \int_0^\infty [k_1 f_1(t) + k_2 f_2(t)]\mathrm{e}^{-st}\mathrm{d}t \\ &= k_1\int_0^\infty f_1(t)\mathrm{e}^{-st}\mathrm{d}t + k_2\int_0^\infty f_2(t)\mathrm{e}^{-st}\mathrm{d}t \\ &= k_1\mathcal{L}\{f_1(t)\} + k_2\mathcal{L}\{f_2(t)\}\end{aligned}$$

An important advantage of the Laplace transform stems from the fact that operations of derivation and integration are transformed into algebraic operations.

3.1.2 Laplace Transforms of Common Functions

Step Function

The Laplace transform of step function is very important as step functions and unit step functions are often used to investigate the process dynamical properties and in control applications.

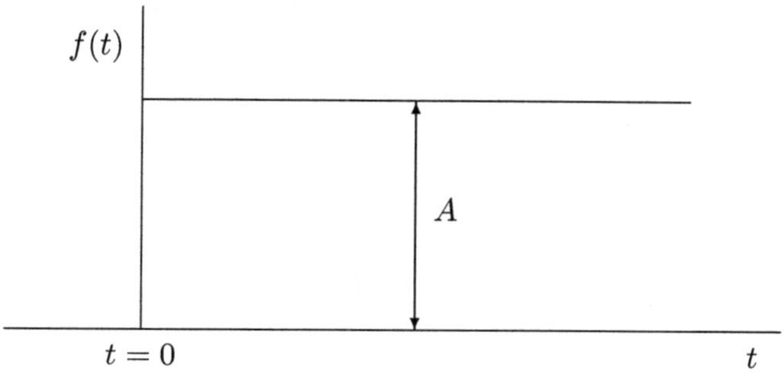

Fig. 3.1. A step function

The step function shown in Fig. 3.1 can be written as

$$f(t) = A1(t) \tag{3.6}$$

and $1(t)$ is *unit step function.* This is defined as

$$1(t) = \begin{cases} 1, t \geq 0 \\ 0, t < 0 \end{cases} \tag{3.7}$$

The Laplace transform of step function is

$$\mathcal{L}\{A1(t)\} = \frac{A}{s} \tag{3.8}$$

Proof:

$$\begin{aligned} \mathcal{L}\{A1(t)\} &= \int_0^\infty A1(t)\mathrm{e}^{-st}\mathrm{d}t = A\int_0^\infty \mathrm{e}^{-st}\mathrm{d}t \\ &= A\left[-\frac{1}{s}\mathrm{e}^{-st}\right]_0^\infty = A\frac{1}{(-s)}(\mathrm{e}^{-s\infty} - \mathrm{e}^{-s0}) \\ &= \frac{A}{s} \end{aligned}$$

The Laplace transform of the unit step functions is

$$\mathcal{L}\{1(t)\} = \frac{1}{s} \tag{3.9}$$

Exponential Function

The Laplace transform of an exponential function is of frequent use as exponential functions appear in the solution of linear differential equations. Consider an exponential function of the form

$$f(t) = \mathrm{e}^{-at}1(t) \tag{3.10}$$

hence $f(t) = \mathrm{e}^{-at}$ for $t \geq 0$ and $f(t) = 0$ for $t < 0$. The Laplace transform of this function is

$$\mathcal{L}\left\{\mathrm{e}^{-at}1(t)\right\} = \frac{1}{s+a} \tag{3.11}$$

Proof :

$$\begin{aligned}\mathcal{L}\left\{\mathrm{e}^{-at}1(t)\right\} &= \int_0^\infty \mathrm{e}^{-at}1(t)\mathrm{e}^{-st}\mathrm{d}t = \int_0^\infty \mathrm{e}^{-(s+a)t}\mathrm{d}t \\ &= -\frac{1}{s+a}\left[\mathrm{e}^{-(s+a)t}\right]_0^\infty \\ &= \frac{1}{s+a}\end{aligned}$$

From (3.10) follows that

$$\mathcal{L}\left\{\mathrm{e}^{at}1(t)\right\} = \frac{1}{s-a} \tag{3.12}$$

Ramp Function

Consider a ramp function of the form

$$f(t) = at1(t) \tag{3.13}$$

The Laplace transform of this function is

$$\mathcal{L}\left\{at1(t)\right\} = \frac{a}{s^2} \tag{3.14}$$

Proof :

$$\mathcal{L}\left\{at1(t)\right\} = \int_0^\infty at1(t)\mathrm{e}^{-st}\mathrm{d}t$$

Let us denote $u = at$ and $\dot{v} = \mathrm{e}^{-st}$ and use the rule of integrating by parts

$$(\dot{uv}) = u\dot{v} + \dot{u}v$$

$$\int u\dot{v}\mathrm{d}t = uv - \int \dot{u}v\mathrm{d}t$$

As $\dot{u} = a$ and $v = -\frac{1}{s}\mathrm{e}^{-st}$, the Laplace transform of the ramp function is

$$\begin{aligned}
\mathcal{L}\{at1(t)\} &= \left[at\frac{1}{(-s)}\mathrm{e}^{-st}\right]_0^\infty - a\int_0^\infty \frac{1}{(-s)}\mathrm{e}^{-st}\mathrm{d}t \\
&= (0-0) + \frac{a}{s}\left[at\frac{1}{(-s)}\mathrm{e}^{-st}\right]_0^\infty \\
&= \frac{a}{s^2}
\end{aligned}$$

Trigonometric Functions

Functions $\sin\omega t$ and $\cos\omega t$ are used in investigation of dynamical properties of processes and control systems. The process response to input variables of the form $\sin\omega t$ or $\cos\omega t$ is observed, where ω is the frequency in radians per time. The Laplace transform of these functions can be calculated using integration by parts or using the Euler identities

$$\begin{aligned}
\mathrm{e}^{\mathrm{j}\omega t} &= \cos\omega t + \mathrm{j}\sin\omega t \\
\mathrm{e}^{-\mathrm{j}\omega t} &= \cos\omega t - \mathrm{j}\sin\omega t \\
\mathrm{e}^{\mathrm{j}\omega t} + \mathrm{e}^{-\mathrm{j}\omega t} &= 2\cos\omega t \\
\mathrm{e}^{\mathrm{j}\omega t} - \mathrm{e}^{-\mathrm{j}\omega t} &= 2\mathrm{j}\sin\omega t
\end{aligned} \tag{3.15}$$

Consider a trigonometric function of the form

$$f(t) = (\sin\omega t)1(t) \tag{3.16}$$

The Laplace transform of this function is

$$\mathcal{L}\{(\sin\omega t)1(t)\} = \frac{\omega}{s^2+\omega^2} \tag{3.17}$$

Proof :

$$\begin{aligned}
\mathcal{L}\{(\sin\omega t)1(t)\} &= \int_0^\infty (\sin\omega t)1(t)\mathrm{e}^{-st}\mathrm{d}t = \int_0^\infty \frac{\mathrm{e}^{\mathrm{j}\omega t} - \mathrm{e}^{-\mathrm{j}\omega t}}{2\mathrm{j}}\mathrm{e}^{-st}\mathrm{d}t \\
&= \int_0^\infty \frac{1}{2\mathrm{j}}\mathrm{e}^{-(s-\mathrm{j}\omega)t}\mathrm{d}t - \int_0^\infty \frac{1}{2\mathrm{j}}\mathrm{e}^{-(s+\mathrm{j}\omega)t}\mathrm{d}t \\
&= \frac{1}{2\mathrm{j}}\left[\frac{\mathrm{e}^{-(s-\mathrm{j}\omega)t}}{-(s-\mathrm{j}\omega)}\right]_0^\infty + \frac{1}{2\mathrm{j}}\left[\frac{\mathrm{e}^{-(s+\mathrm{j}\omega)t}}{-(s+\mathrm{j}\omega)}\right]_0^\infty \\
&= \frac{1}{2\mathrm{j}}\left(\frac{1}{s-\mathrm{j}\omega}\right) - \frac{1}{2\mathrm{j}}\left(\frac{1}{s+\mathrm{j}\omega}\right) = \frac{\omega}{s^2+\omega^2}
\end{aligned}$$

The Laplace transform of other functions can be calculated in a similar manner. The list of the most commonly used functions together with their Laplace transforms is given in Table 3.1.

Table 3.1. The Laplace transforms for common functions

$f(t)$	$F(s)$
$\delta(t)$ - unit impulse function	1
$1(t)$ - unit step function	$\frac{1}{s}$
$1(t) - 1(t - T_v)$, T_v is a time constant	$\frac{1-\mathrm{e}^{-sT_v}}{s}$
$at1(t)$, a is a constant (ramp)	$\frac{a}{s^2}$
$at^{n-1}1(t)$, $n > 1$	$a\frac{(n-1)!}{s^n}$
$\mathrm{e}^{-at}1(t)$	$\frac{1}{s+a}$
$\frac{1}{T_1}\mathrm{e}^{-\frac{t}{T_1}}1(t)$	$\frac{1}{T_1 s+1}$
$(1 - \mathrm{e}^{-at})1(t)$	$\frac{a}{s(s+a)}$
$(1 - \mathrm{e}^{-\frac{t}{T_1}})1(t)$	$\frac{1}{s(T_1 s+1)}$
$\left(\frac{1}{a-b}(\mathrm{e}^{-bt} - \mathrm{e}^{-at})\right)1(t)$, a, b are constants	$\frac{1}{(s+a)(s+b)}$
$\left(\frac{c-a}{b-a}\mathrm{e}^{-at} + \frac{c-b}{a-b}\mathrm{e}^{-bt}\right)1(t)$, c is a constant	$\frac{s+c}{(s+a)(s+b)}$
$\frac{t^{n-1}\mathrm{e}^{-at}}{(n-1)!}1(t), n \geq 1$	$\frac{1}{(s+a)^n}$
$\left(\frac{1}{a}t - \frac{1-\mathrm{e}^{-at}}{a^2}\right)1(t)$	$\frac{1}{s^2(s+a)}$
$\left(\frac{1}{ab} + \frac{1}{a(a-b)}\mathrm{e}^{-at} + \frac{1}{b(b-a)}\mathrm{e}^{-bt}\right)1(t)$	$\frac{1}{s(s+a)(s+b)}$
$\left(\frac{c}{ab} + \frac{c-a}{a(a-b)}\mathrm{e}^{-at} + \frac{c-b}{b(b-a)}\mathrm{e}^{-bt}\right)1(t)$	$\frac{s+c}{s(s+a)(s+b)}$
$\sin\omega t\ 1(t)$, ω is a constant	$\frac{\omega}{s^2+\omega^2}$
$\cos\omega t\ 1(t)$	$\frac{s}{s^2+\omega^2}$
$\mathrm{e}^{-at}\sin\omega t\ 1(t)$	$\frac{\omega}{(s+a)^2+\omega^2}$
$\mathrm{e}^{-at}\cos\omega t\ 1(t)$	$\frac{s+a}{(s+a)^2+\omega^2}$
$\left\{1 - \mathrm{e}^{-\frac{\zeta t}{T_k}}\left[\cos\left(\sqrt{1-\zeta^2}\frac{t}{T_k}\right) + \frac{\zeta}{1-\zeta^2}\sin\left(\sqrt{1-\zeta^2}\frac{t}{T_k}\right)\right]\right\}1(t)$	$\frac{1}{s(T_k^2 s^2+2\zeta T_k s+1)}$
$0 \leq \lvert\zeta\rvert < 1$	
$\left[1 - \frac{1}{\sqrt{1-\zeta^2}}\mathrm{e}^{-\frac{\zeta t}{T_k}}\sin\left(\sqrt{1-\zeta^2}\frac{t}{T_k} + \varphi\right)\right]1(t)$	$\frac{1}{s(T_k^2 s^2+2\zeta T_k s+1)}$
$\varphi = \arctan\frac{\sqrt{1-\zeta^2}}{\zeta}, 0 \leq \lvert\zeta\rvert < 1$	

3.1.3 Properties of the Laplace Transform

Derivatives

The Laplace transform of derivatives are important as derivatives appear in linear differential equations. The transform of the first derivative of $f(t)$ is

$$\mathcal{L}\left\{\frac{\mathrm{d}f(t)}{\mathrm{d}t}\right\} = sF(s) - f(0) \tag{3.18}$$

Proof :

$$\begin{aligned}\mathcal{L}\left\{\frac{\mathrm{d}f(t)}{\mathrm{d}t}\right\} &= \int_0^\infty \dot{f}(t)\mathrm{e}^{-st}\mathrm{d}t \\ &= \left[f(t)\mathrm{e}^{-st}\right]_0^\infty - \int_0^\infty f(t)\mathrm{e}^{-st}(-s)\mathrm{d}t \\ &= sF(s) - f(0)\end{aligned}$$

The Laplace transform of the second derivative of $f(t)$ is

$$\mathcal{L}\left\{\frac{\mathrm{d}^2f(t)}{\mathrm{d}t^2}\right\} = s^2F(s) - sf(0) - \dot{f}(0) \tag{3.19}$$

Proof : Let us define a new function $\bar{f}(t) = \mathrm{d}f(t)/\mathrm{d}t$. Applying the equation (3.18) yields (3.19). Similarly for higher-order derivatives follows

$$\mathcal{L}\left\{\frac{\mathrm{d}^nf(t)}{\mathrm{d}t^n}\right\} = s^nF(s) - s^{n-1}f(0) - s^{n-2}\dot{f}(0) - \cdots - f^{(n-1)}(0) \tag{3.20}$$

Integral

The Laplace transform of the integral of $f(t)$ is

$$\mathcal{L}\left\{\int_0^t f(\tau)\mathrm{d}\tau\right\} = \frac{F(s)}{s} \tag{3.21}$$

Proof :

$$\mathcal{L}\left\{\int_0^t f(\tau)\mathrm{d}\tau\right\} = \int_0^\infty \left[\int_0^t f(\tau)\mathrm{d}\tau\right]\mathrm{e}^{-st}\mathrm{d}t$$

Let us denote $u = \int_0^t f(\tau)\mathrm{d}\tau, \dot{v} = \mathrm{e}^{-st}$ and use integration by parts. Because $\dot{u} = f(t), v = \frac{1}{(-s)}\mathrm{e}^{-st}$, the transform gives

$$\begin{aligned}\mathcal{L}\left\{\int_0^t f(\tau)\mathrm{d}\tau\right\} &= \left[\int_0^t f(\tau)\mathrm{d}\tau\frac{1}{(-s)}\mathrm{e}^{-st}\right]_0^\infty - \int_0^\infty f(t)\frac{1}{(-s)}\mathrm{e}^{-st}\mathrm{d}t \\ &= (0-0) + \frac{1}{s}\int_0^\infty f(t)\mathrm{e}^{-st}\mathrm{d}t \\ &= \frac{F(s)}{s}\end{aligned}$$

Convolution

The Laplace transform of convolution is important in situations when input variables of processes are general functions of time. Let functions $f_1(t)$ and $f_2(t)$ be transformed as $F_1(s)$ and $F_2(s)$ respectively. The convolution of the functions is defined as

$$f_1(t) \star f_2(t) = \int_0^t f_1(\tau) f_2(t-\tau) \mathrm{d}\tau \tag{3.22}$$

The Laplace transform of convolution is

$$\mathcal{L}\left\{\int_0^t f_1(\tau) f_2(t-\tau) \mathrm{d}\tau\right\} = \mathcal{L}\left\{\int_0^t f_1(t-\tau) f_2(\tau) \mathrm{d}\tau\right\} = F_1(s)F_2(s) \tag{3.23}$$

Proof.

$$\mathcal{L}\left\{\int_0^t f_1(\tau) f_2(t-\tau) \mathrm{d}\tau\right\} = \int_0^\infty \int_0^t f_1(\tau) f_2(t-\tau) \mathrm{d}\tau \mathrm{e}^{-st} \mathrm{d}t$$

Introduce a substitution $\eta = t - \tau, \mathrm{d}\eta = \mathrm{d}t$. Then

$$\begin{aligned}
\mathcal{L}\left\{\int_0^t f_1(\tau) f_2(t-\tau) \mathrm{d}\tau\right\} &= \int_{\eta=-\tau}^\infty \int_{\tau=0}^\infty f_1(\tau) f_2(\eta) \mathrm{e}^{-s(\eta+\tau)} \mathrm{d}\tau \mathrm{d}\eta \\
&= \int_0^\infty f_1(\tau) \mathrm{e}^{-s\tau} \mathrm{d}\tau \int_0^\infty f_2(\eta) \mathrm{e}^{-s\eta} \mathrm{d}\eta \\
&= F_1(s)F_2(s)
\end{aligned}$$

□

Final Value Theorem

The asymptotic value of $f(t), t \to \infty$ can be found (if $\lim_{t\to\infty} f(t)$ exists) as

$$f(\infty) = \lim_{t\to\infty} f(t) = \lim_{s=0} [sF(s)] \tag{3.24}$$

Proof. To prove the above equation we use the relation for the transform of a derivative (3.18)

$$\int_0^\infty \frac{\mathrm{d}f(t)}{\mathrm{d}t} \mathrm{e}^{-st} \mathrm{d}t = sF(s) - f(0)$$

and taking the limit as $s \to 0$

$$\begin{aligned}
\int_0^\infty \frac{\mathrm{d}f(t)}{\mathrm{d}t} \lim_{s\to 0} \mathrm{e}^{-st} \mathrm{d}t &= \lim_{s\to 0} [sF(s) - f(0)] \\
\lim_{t\to\infty} f(t) - f(0) &= \lim_{s\to 0} [sF(s)] - f(0) \\
\lim_{t\to\infty} f(t) &= \lim_{s\to 0} [sF(s)]
\end{aligned}$$

□

Initial Value Theorem

It can be proven that an initial value of a function can be calculated as

$$\lim_{t\to 0} f(t) = \lim_{s\to\infty} [sF(s)] \tag{3.25}$$

Time Delay

Time delays are phenomena commonly encountered in chemical and food processes and occur in mass transfer processes. Time delays exist implicitly in distributed parameter processes and explicitly in pure mass transport through piping. A typical example are some types of automatic gas analysers that are connected to a process via piping used for transport of analysed media. In this case, *time delay* is defined as time required for transport of analysed media from the process into the analyser.

Consider a function $f(t)$ given for $0 \le t < \infty$, $f(t) \equiv 0$ for $t < 0$. If the Laplace transform of this function is $F(s)$ then

$$\mathcal{L}\{f(t - T_d)\} = \mathrm{e}^{-T_d s} F(s) \tag{3.26}$$

where T_d is a time delay.

Proof : The relation between functions $f(t)$ and $f(t - T_d)$ is shown in Fig. 3.2.

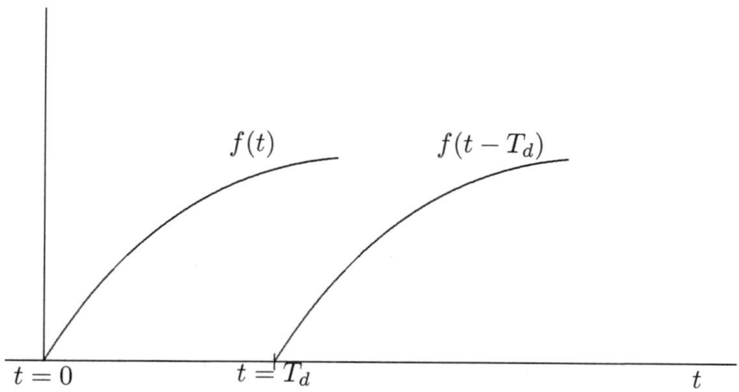

Fig. 3.2. An original and delayed function

Applying the definition integral to the function $f(t - T_d)$ yields

$$\begin{aligned}\mathcal{L}\{f(t - T_d)\} &= \int_0^\infty f(t - T_d)\mathrm{e}^{-st}\mathrm{d}t \\ &= \mathrm{e}^{-sT_d} \int_0^\infty f(t - T_d)\mathrm{e}^{-s(t-T_d)}\mathrm{d}(t - T_d)\end{aligned}$$

because $\mathrm{d}t = \mathrm{d}(t - T_d)$. Denoting $\tau = t - T_d$ follows

$$\begin{aligned}\mathcal{L}\{f(t-T_d)\} &= \mathrm{e}^{-sT_d}\int_0^\infty f(\tau)\mathrm{e}^{-s\tau}\mathrm{d}\tau \\ &= \mathrm{e}^{-sT_d}F(s)\end{aligned}$$

Unit Impulse Function

Unit impulse function plays a fundamental role in control analysis and synthesis. Although the derivation of its Laplace transform logically falls into the section dealing with elementary functions, it can be derived only with knowledge of the Laplace transform of delayed function.

Consider a function $f(t) = A1(t) - A1(t - T_d)$ illustrated in Fig. 3.3. The Laplace transform of this function is

$$\begin{aligned}\mathcal{L}\{A1(t) - A1(t-T_d)\} &= \frac{A}{s} - \frac{A\mathrm{e}^{-sT_d}}{s} \\ &= \frac{A(1-\mathrm{e}^{-sT_d})}{s}\end{aligned}$$

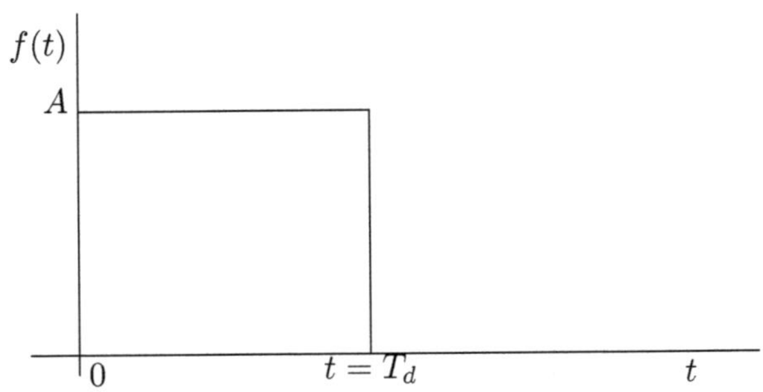

Fig. 3.3. A rectangular pulse function

If we substitute in the function $f(t)$ for $A = 1/T_d$ and take the limit case for T_d approaching zero, we obtain a function that is zero except for the point $t = 0$ where its value is infinity. The area of the pulse function in Fig. 3.3 is equal to one. This is also the way of defining a function usually denoted by $\delta(t)$ and for which follows

$$\int_{-\infty}^{\infty}\delta(t)\mathrm{d}t = 1 \tag{3.27}$$

It is called the *unit impulse function* or the *Dirac delta function.*

The Laplace transform of the unit impulse function is

$$\mathcal{L}\{\delta(t)\} = 1 \tag{3.28}$$

Proof :

$$\mathcal{L}\{\delta(t)\} = \lim_{T_d \to 0} \frac{1 - e^{-sT_d}}{T_d s}$$

The limit in the above equation can easily be found by application of L'Hospital's rule. Taking derivatives with respect to T_d of both numerator and denominator,

$$\mathcal{L}\{\delta(t)\} = \lim_{T_d \to 0} \frac{s e^{-sT_d}}{s} = 1$$

The unit impulse function is used as an idealised input variable in investigations of dynamical properties of processes.

3.1.4 Inverse Laplace Transform

When solving differential equations using the Laplace transform technique, the inverse Laplace transform can often be obtained from Table 3.1. However, a general function may not exactly match any of the entries in the table. Hence, a more general procedure is required. Every function can be factored as a sum of simpler functions whose Laplace transforms are in the table:

$$F(s) = F_1(s) + F_2(s) + \cdots + F_n(s) \tag{3.29}$$

Then the original solution can be found as

$$f(t) = f_1(t) + f_2(t) + \cdots + f_n(t) \tag{3.30}$$

where $f_i(t) = \mathcal{L}^{-1}\{F_i(s)\}, i = 1, \ldots, n$.

The function $F(s)$ is usually given as a rational function

$$F(s) = \frac{M(s)}{N(s)} \tag{3.31}$$

where

$M(s) = m_0 + m_1 s + \cdots + m_m s^m$ - numerator polynomial,
$N(s) = n_0 + n_1 s + \cdots + n_n s^n$ - denominator polynomial.

If $M(s)$ is a polynomial of a lower degree than $N(s)$, the function (3.31) is called *strictly proper rational function.* Otherwise, it is nonstrictly proper and can be written as a sum of some polynomial $T(s)$ and some strictly proper rational function of the form

$$\frac{M(s)}{N(s)} = T(s) + \frac{Z(s)}{N(s)} \tag{3.32}$$

Any strictly proper rational function can be written as a sum of strictly proper rational functions called *partial fractions* and the method of obtaining the partial fractions is called *partial fraction expansion.*

An intermediate step in partial fraction expansion is to find roots of the $N(s)$ polynomial. We can distinguish two cases when $N(s)$ has:

1. n different roots,
2. multiple roots.

Different Roots

If the denominator of (3.31) has the roots $s_1, \dots, s_n$, the the function $F(s)$ can be written as

$$F(s) = \frac{M(s)}{n_n(s - s_1)(s - s_2)\dots(s - s_n)} \tag{3.33}$$

Expansion of $F(s)$ into partial fractions yields

$$F(s) = \frac{K_1}{s - s_1} + \frac{K_2}{s - s_2} + \dots + \frac{K_n}{s - s_n} \tag{3.34}$$

and the original $f(t)$ is

$$f(t) = K_1 \mathrm{e}^{s_1 t} + K_2 \mathrm{e}^{s_2 t} + \dots + K_n \mathrm{e}^{s_n t} \tag{3.35}$$

Note that if $N(s)$ has complex roots $s_{1,2} = a \pm \mathrm{j}b$, then for $F(s)$ follows

$$F(s) = \frac{K_1}{s - (a + \mathrm{j}b)} + \frac{K_2}{s - (a - \mathrm{j}b)} + \cdots \tag{3.36}$$

$$= \frac{\beta_0 + \beta_1 s}{\alpha_0 + \alpha_1 s + s^2} + \cdots \tag{3.37}$$

The original function corresponding to this term can be found by an inverse Laplace transform using the combination of trigonometric entries in Table 3.1 (see example 3.3b).

Multiple Roots

If a root s_1 of the polynomial $N(s)$ occurs k-times, then the function $F(s)$ must be factored as

$$F(s) = \frac{K_1}{s - s_1} + \frac{K_2}{(s - s_1)^2} + \dots + \frac{K_k}{(s - s_1)^k} + \cdots \tag{3.38}$$

and the corresponding original $f(t)$ can be found from Table 3.1.

3.1.5 Solution of Linear Differential Equations by Laplace Transform Techniques

Linear differential equations are solved by the means of the Laplace transform very simply with the following procedure:

1. Take Laplace transform of the differential equation,

2. solve the resulting algebraic equation,
3. find the inverse of the transformed output variable.

Example 3.1: *Solution of the 1st order ODE with zero initial condition*
Consider the heat exchanger shown in Fig. 2.3. The state equation is of the form

$$\frac{\mathrm{d}\vartheta(t)}{\mathrm{d}t} = -\frac{1}{T_1}\vartheta(t) + \frac{Z_1}{T_1}\vartheta_p(t) + \frac{Z_2}{T_1}\vartheta_v(t)$$

The exchanger is in a steady-state if $\mathrm{d}\vartheta(t)/\mathrm{d}t = 0$. Let the steady-state temperatures be given as $\vartheta_p^s, \vartheta_v^s, \vartheta^s$. Introduce deviation variables

$$x_1(t) = \vartheta(t) - \vartheta^s$$
$$u_1(t) = \vartheta_p(t) - \vartheta_p^s$$
$$r_1(t) = \vartheta_v(t) - \vartheta_v^s$$

then the state equation is

$$\frac{\mathrm{d}x_1(t)}{\mathrm{d}t} = -\frac{1}{T_1}x_1(t) + \frac{Z_1}{T_1}u_1(t) + \frac{Z_2}{T_1}r_1(t)$$

The output equation if temperature ϑ is measured is

$$y_1(t) = x_1(t)$$

so the differential equation describing the heat exchanger is

$$\frac{\mathrm{d}y_1(t)}{\mathrm{d}t} = -\frac{1}{T_1}y_1(t) + \frac{Z_1}{T_1}u_1(t) + \frac{Z_2}{T_1}r_1(t)$$

Let us assume that the exchanger is up to time t in the steady-state, hence

$$y_1(0) = 0, u_1(0) = 0, r_1(0) = 0 \quad \text{for} \quad t < 0$$

Let us assume that at time $t = 0$ begins the input $u_1(t)$ to change as a function of time $u_1(t) = Z_u \mathrm{e}^{-t/T_u}$. The question is the behaviour of $y_1(t), t \geq 0$. From a pure mathematical point of view this is equivalent to the solution of a differential equation

$$T_1\frac{\mathrm{d}y_1(t)}{\mathrm{d}t} + y_1(t) = Z_1 Z_u \mathrm{e}^{-t/T_u}$$

with initial condition $y_1(0) = 0$. The first step is the Laplace transform of this equation which yields

$$T_1\mathcal{L}\left\{\frac{\mathrm{d}y_1(t)}{\mathrm{d}t}\right\} + \mathcal{L}\{y_1(t)\} = Z_1 Z_u \mathcal{L}\left\{\mathrm{e}^{-t/T_u}\right\}$$

$$T_1 s Y_1(s) + Y_1(s) = Z_1 Z_u \frac{T_u}{T_u s + 1}$$

Solution of this equation for $Y_1(s)$ is

$$Y_1(s) = \frac{Z_1 Z_u T_u}{(T_1 s + 1)(T_u s + 1)}$$

The right hand side of these equations can be factored as

$$Y_1(s) = \frac{A}{T_1 s + 1} + \frac{B}{T_u s + 1} = \frac{Z_1 Z_u T_u}{T_1 - T_u}\left(\frac{T_1}{T_1 s + 1} - \frac{T_u}{T_u s + 1}\right)$$

The inverse Laplace transform can be calculated using Table 3.1 and is given as

$$y_1(t) = \frac{Z_1 Z_u T_u}{T_1 - T_u}\left(\mathrm{e}^{-\frac{t}{T_1}} - \mathrm{e}^{-\frac{t}{T_u}}\right)$$

Example 3.2: *Solution of the 1st order ODE with a nonzero initial condition*
Consider the previous example but with the conditions $y_1(0) = y_{10}$ and $u_1(t) = 0$ for $t \geq 0$. This is mathematically equivalent to the differential equation

$$T_1 \frac{\mathrm{d}y_1(t)}{\mathrm{d}t} + y_1(t) = 0, \quad y_1(0) = y_{10}$$

Taking the Laplace transform, term by term using Table 3.1 :

$$T_1 \mathcal{L}\left\{\frac{\mathrm{d}y_1(t)}{\mathrm{d}t}\right\} + \mathcal{L}\left\{y_1(t)\right\} = 0$$

$$T_1[sY_1(s) - y_1(0)] + Y_1(s) = 0$$

Rearranging and factoring out $Y_1(s)$, we obtain

$$Y_1(s) = \frac{y_{10}}{T_1 s + 1}$$

Now we can take the inverse Laplace transform and obtain

$$y_1(t) = \frac{y_{10}}{T_1}\mathrm{e}^{-\frac{t}{T_1}}.$$

Example 3.3: *Solution of the 2nd order ODE*
a) Consider a second order differential equation

$$\ddot{y}(t) + 3\dot{y}(t) + 2y(t) = 2u(t)$$

and assume zero initial conditions $y(0) = \dot{y}(0) = 0$. This case frequently occurs for process models with deviation variables that are up to time $t = 0$ in a steady-state. Let us find the solution of this differential equation for unit step function $u(t) = 1(t)$.
After taking the Laplace transform, the differential equation gives

$$(s^2 + 3s + 2)Y(s) = 2\frac{1}{s}$$
$$Y(s) = \frac{2}{s(s^2 + 3s + 2)}$$
$$Y(s) = \frac{2}{s(s+1)(s+2)}$$

The denominator roots are all different and partial fraction expansion is of the form

$$\frac{2}{s(s+1)(s+2)} = \frac{K_1}{s} + \frac{K_2}{s+1} + \frac{K_3}{s+2}$$

The coefficients K_1, K_2, K_3 can be calculated by multiplying both sides of this equation with the denominator and equating the coefficients of each power of s:

$$\left.\begin{array}{l} s^2 : K_1 + K_2 + K_3 = 0 \\ s^1 : 3K_1 + 2K_2 + K_3 = 0 \\ s^0 : 2K_1 = 2 \end{array}\right\} K_1 = 1, K_2 = -2, K_3 = 1$$

The solution of the differential equation can now be read from Table 3.1:

$$y(t) = 1 - 2\mathrm{e}^{-t} + \mathrm{e}^{-2t}$$

b) Consider a second order differential equation

$$\ddot{y}(t) + 2\dot{y}(t) + 5y(t) = 2u(t)$$

and assume zero initial conditions $y(0) = \dot{y}(0) = 0$. Find the solution of this differential equation for unit step function $u(t) = 1(t)$.
Take the Laplace transform

$$(s^2 + 2s + 5)Y(s) = 2\frac{1}{s}$$
$$Y(s) = \frac{2}{s(s^2 + 2s + 5)}$$

The denominator has one real root and two complex conjugate roots, hence the partial fraction expansion is of the form

$$\frac{2}{s(s^2 + 2s + 5)} = \frac{K_1}{s} + \frac{K_2 s + K_3}{s^2 + 2s + 5} = \frac{2}{5}\left(\frac{1}{s} - \frac{2 + s}{s^2 + 2s + 5}\right)$$

where the coefficients K_1, K_2, K_3 have been found as in the previous example. The second term on the right side of the previous equation is not in Table 3.1 but can be manipulated to obtain a sum of trigonometric terms. Firstly, the denominator is rearranged by completing the squares to $(s + 1)^2 + 4$ and the numerator is then rewritten to match numerators of trigonometric expressions. Hence

$$\frac{1+2s}{s^2+2s+5} = \frac{2+s}{(s+1)^2+4} = \frac{(s+1)-\frac{1}{2}2}{(s+1)^2+4}$$
$$= \frac{s+1}{(s+1)^2+4} - \frac{1}{2}\frac{2}{(s+1)^2+4}$$

and $Y(s)$ can be written as

$$Y(s) = \frac{2}{5}\left(\frac{1}{s} - \frac{s+1}{(s+1)^2+4} - \frac{1}{2}\frac{2}{(s+1)^2+4}\right)$$

Taking the inverse Laplace transform, term by term, yields

$$Y(s) = \frac{2}{5}\left(1 - e^{-t}\cos 2t - \frac{1}{2}e^{-t}\sin 2t\right)$$

c) Consider a second order differential equation

$$\ddot{y}(t) + 2\dot{y}(t) + 1y(t) = 2, \quad y(0) = \dot{y}(0) = 0$$

Take the Laplace transform

$$(s^2+2s+1)Y(s) = 2\frac{1}{s}$$
$$Y(s) = \frac{2}{s(s^2+2s+1)}$$
$$Y(s) = \frac{2}{s(s+1)^2}$$

The denominator has one single root $s_1 = 0$ and one double root $s_{2,3} = -1$. The partial fraction expansion is of the form

$$\frac{2}{s(s+1)^2} = \frac{K_1}{s} + \frac{K_2}{s+1} + \frac{K_3}{(s+1)^2}$$

and the solution from Table 3.1 reads

$$y(t) = 2 - 2(1-t)e^{-t}$$

3.2 State-Space Process Models

Investigation of processes as dynamical systems is based on *theoretical state-space balance equations*. State-space variables may generally be abstract. If a model of a process is described by state-space equations, we speak about *state-space representation*. This representation includes a description of linear as well as nonlinear models. In this section we introduce the concept of state, solution of state-space equations, canonical representations and transformations, and some properties of systems.

3.2.1 Concept of State

Consider a continuous-time MIMO system with m input variables and r output variables. The relation between input and output variables can be expressed as (see also Section 2.3)

$$\frac{\mathrm{d}\boldsymbol{x}(t)}{\mathrm{d}t} = \boldsymbol{f}(\boldsymbol{x}(t), \boldsymbol{u}(t)) \tag{3.39}$$

$$\boldsymbol{y}(t) = \boldsymbol{g}(\boldsymbol{x}(t), \boldsymbol{u}(t)) \tag{3.40}$$

where $\boldsymbol{x}(t)$ is a vector of state-space variables, $\boldsymbol{u}(t)$ is a vector of input variables, and $\boldsymbol{y}(t)$ is a vector of output variables.

The *state of a system* at time t_0 is a minimum amount of information which (in the absence of external excitation) is sufficient to determine uniquely the evolution of the system for $t \geq t_0$.

If the vector $\boldsymbol{x}(t_0)$ and the vector of input variables $\boldsymbol{u}(t)$ for $t > t_0$ are known then this knowledge suffices to determine $\boldsymbol{y}(t)$, $t > t_0$, thus

$$\boldsymbol{y}(t_0, t] = \boldsymbol{y}\{\boldsymbol{x}(t_0), \boldsymbol{u}(t_0, t]\} \tag{3.41}$$

where $\boldsymbol{u}(t_0, t], \boldsymbol{y}(t_0, t]$ are vectors of input and output variables over the interval $(t_0, t]$ respectively.

The above equation is equivalent to

$$\boldsymbol{x}(t_0, t] = \boldsymbol{x}\{\boldsymbol{x}(t_0), \boldsymbol{u}(t_0, t]\} \tag{3.42}$$

Therefore, the knowledge about the states at $t = t_0$ removes the necessity to know the past behavior of the system in order to forecast its future and the future evolution of states is dependent only on its present state and future inputs.

This definition of state will be clearer when we introduce a solution of state-space equation for the general functions of input variables.

3.2.2 Solution of State-Space Equations

Solution of state-space equations will be specified only for linear systems with constant coefficients with the aid of Laplace transform techniques. Firstly, a simple example will be given and then it will be generalised.

Example 3.4: *Mixing process - solution of state-space equations*

Consider a process of mixing shown in Fig. 3.4 with mathematical model described by the equation

$$V\frac{\mathrm{d}c_1}{\mathrm{d}t} = qc_0 - qc_1$$

where c_0, c_1 are concentrations with dimensions mass/volume, V is a constant volume of the vessel, and q is a constant volumetric flow rate.

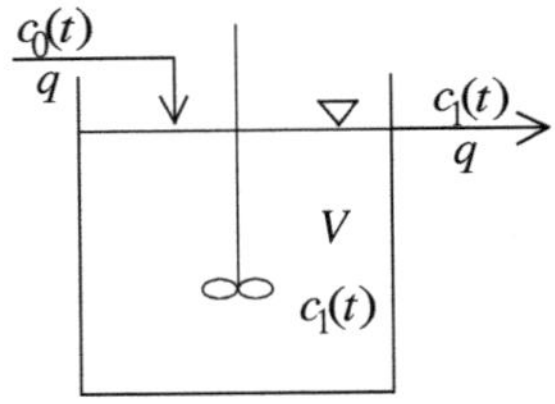

Fig. 3.4. A mixing process

In the steady-state holds

$$qc_0^s - qc_1^s = 0$$

Introduce deviation variables

$$x = c_1 - c_1^s$$
$$u = c_0 - c_0^s$$

and define the process output variable $y = x$. Then the process state-space equations are of the form

$$\frac{\mathrm{d}x}{\mathrm{d}t} = ax + bu$$
$$y = cx$$

where $a = -1/T_1, b = 1/T_1, c = 1$. $T_1 = V/q$ is the process time constant. Assume that the system is at $t_0 = 0$ in the state $x(0) = x_0$. Then the time solution can be calculated by applying the Laplace transform:

$$sX(s) - x(0) = aX(s) + bU(s)$$

$$X(s) = \frac{1}{s-a}x(0) + \frac{b}{s-a}U(s)$$

The time domain description $x(t)$ can be read from Table 3.1 for the first term and from the convolution transformation for the second term and is given as

$$x(t) = \mathrm{e}^{at}x(0) + \int_0^t \mathrm{e}^{a(t-\tau)}bu(\tau)\mathrm{d}\tau$$

and for $y(t)$

$$y(t) = c\mathrm{e}^{at}x(0) + c\int_0^t \mathrm{e}^{a(t-\tau)}bu(\tau)\mathrm{d}\tau$$

After substituting for the constants yields for $y(t)$

$$y(t) = \mathrm{e}^{-\frac{q}{V}t}x(0) + \frac{V}{q}\int_0^t \mathrm{e}^{-\frac{q}{V}(t-\tau)}u(\tau)\mathrm{d}\tau$$

Solution of State-Space Equations for the Multivariable Case

The solution for the multivariable case is analogous as in the previous example. Each state equation is transformed with the Laplace transform applied and transformed back into the time domain. The procedure is simplified if we use matrix notation.

Consider state-space equations

$$\frac{\mathrm{d}\boldsymbol{x}(t)}{\mathrm{d}t} = \boldsymbol{A}\boldsymbol{x}(t) + \boldsymbol{B}\boldsymbol{u}(t), \quad \boldsymbol{x}(0) = \boldsymbol{x}_0 \tag{3.43}$$

$$\boldsymbol{y}(t) = \boldsymbol{C}\boldsymbol{x}(t) \tag{3.44}$$

Taking the Laplace transform yields

$$s\boldsymbol{X}(s) - \boldsymbol{x}_0 = \boldsymbol{A}\boldsymbol{X}(s) + \boldsymbol{B}\boldsymbol{U}(s) \tag{3.45}$$

$$\boldsymbol{X}(s) = (s\boldsymbol{I} - \boldsymbol{A})^{-1}\boldsymbol{x}_0 + (s\boldsymbol{I} - \boldsymbol{A})^{-1}\boldsymbol{B}\boldsymbol{U}(s) \tag{3.46}$$

and after the inverse transformation for $\boldsymbol{x}(t), \boldsymbol{y}(t)$ hold

$$\boldsymbol{x}(t) = \mathrm{e}^{\boldsymbol{A}t}\boldsymbol{x}(0) + \int_0^t \mathrm{e}^{\boldsymbol{A}(t-\tau)}\boldsymbol{B}\boldsymbol{U}(\tau)\mathrm{d}\tau \tag{3.47}$$

$$\boldsymbol{y}(t) = \boldsymbol{C}\mathrm{e}^{\boldsymbol{A}t}\boldsymbol{x}(0) + \boldsymbol{C}\int_0^t \mathrm{e}^{\boldsymbol{A}(t-\tau)}\boldsymbol{B}\boldsymbol{U}(\tau)\mathrm{d}\tau \tag{3.48}$$

$$\mathrm{e}^{\boldsymbol{A}t} = \mathcal{L}^{-1}\left\{(s\boldsymbol{I} - \boldsymbol{A})^{-1}\right\} \tag{3.49}$$

The equation (3.48) shows some important properties and features. Its solution consists of two parts: initial conditions term (zero-input response) and input term dependent on $\boldsymbol{u}(t)$ (zero-state response).

The solution of (3.43) for *free system* ($\boldsymbol{u}(t) = \boldsymbol{0}$) is

$$\boldsymbol{x}(t) = \mathrm{e}^{\boldsymbol{A}t}\boldsymbol{x}(0) \tag{3.50}$$

and the exponential term is defined as

$$\mathrm{e}^{\boldsymbol{A}t} = \sum_{i=1}^{\infty} \boldsymbol{A}^i \frac{t^i}{i!} \tag{3.51}$$

The matrix

$$\boldsymbol{\Phi}(t) = \mathrm{e}^{\boldsymbol{A}t} = \mathcal{L}^{-1}\left\{(s\boldsymbol{I} - \boldsymbol{A})^{-1}\right\} \tag{3.52}$$

is called the *state transition matrix*, (*fundamental matrix, matrix exponential*). The solution of (3.43) for $u(t)$ is then

$$\boldsymbol{x}(t) = \boldsymbol{\Phi}(t - t_0)\boldsymbol{x}(t_0) \tag{3.53}$$

The matrix exponential satisfies the following identities:

$$x(t_0) = \Phi(t_0 - t_0)x(t_0) \quad \Rightarrow \quad \Phi(0) = I \tag{3.54}$$

$$x(t_2) = \Phi(t_2 - t_1)x(t_1) \tag{3.55}$$

$$x(t_2) = \Phi(t_2 - t_1)\Phi(t_1 - t_0)x(t_0) \tag{3.56}$$

The equation (3.52) shows that the *system matrix* $\boldsymbol{A}$ plays a crucial role in the solution of state-space equations. Elements of this matrix depend on coefficients of mass and heat transfer, activation energies, flow rates, etc. Solution of the state-space equations is therefore influenced by physical and chemical properties of processes.

The solution of state-space equations depends on roots of the *characteristic equation*

$$\det(s\boldsymbol{I} - \boldsymbol{A}) = 0 \tag{3.57}$$

This will be clarified from the next example

Example 3.5: *Calculation of matrix exponential*

Consider a matrix

$$\boldsymbol{A} = \begin{pmatrix} -1 & -1 \\ 0 & -2 \end{pmatrix}$$

The matrix exponential corresponding to $\boldsymbol{A}$ is defined in equation (3.52) as

$$\begin{aligned}
\Phi(t) &= \mathcal{L}^{-1}\left\{\left[s\begin{pmatrix} 1 & 0 \\ 0 & 1 \end{pmatrix} - \begin{pmatrix} -1 & -1 \\ 0 & -2 \end{pmatrix}\right]^{-1}\right\} \\
&= \mathcal{L}^{-1}\left\{\begin{bmatrix} s+1 & 1 \\ 0 & s+2 \end{bmatrix}^{-1}\right\} \\
&= \mathcal{L}^{-1}\left\{\frac{1}{\det\begin{pmatrix} s+1 & 1 \\ 0 & s+2 \end{pmatrix}}\begin{pmatrix} s+2 & -1 \\ 0 & s+1 \end{pmatrix}\right\} \\
&= \mathcal{L}^{-1}\left\{\frac{1}{(s+1)(s+2)}\begin{pmatrix} s+2 & -1 \\ 0 & s+1 \end{pmatrix}\right\} \\
&= \mathcal{L}^{-1}\left\{\begin{pmatrix} \frac{1}{s+1} & \frac{-1}{(s+1)(s+2)} \\ 0 & \frac{1}{s+2} \end{pmatrix}\right\}
\end{aligned}$$

The elements of $\Phi(t)$ are found from Table 3.1 as

$$\Phi(t) = \begin{pmatrix} e^{-t} & e^{-2t} - e^{-t} \\ 0 & e^{-2t} \end{pmatrix}$$

3.2.3 Canonical Transformation

Eigenvalues of $\boldsymbol{A}$, $\lambda_1, \ldots, \lambda_n$ are given as solutions of the equation

$$\det(\boldsymbol{A} - \lambda \boldsymbol{I}) = 0 \tag{3.58}$$

If the eigenvalues of $\boldsymbol{A}$ are distinct, then a nonsingular matrix $\boldsymbol{T}$ exists, such that

$$\boldsymbol{\Lambda} = \boldsymbol{T}^{-1}\boldsymbol{A}\boldsymbol{T} \tag{3.59}$$

is an diagonal matrix of the form

$$\boldsymbol{\Lambda} = \begin{pmatrix} \lambda_1 & 0 & \dots & 0 \\ 0 & \lambda_2 & \dots & 0 \\ \vdots & & & \vdots \\ 0 & 0 & \dots & \lambda_n \end{pmatrix} \tag{3.60}$$

The *canonical transformation* (3.59) can be used for direct calculation of $\mathrm{e}^{-\boldsymbol{A}t}$. Substituting $\boldsymbol{A}$ from (3.59) into the equation

$$\frac{\mathrm{d}\boldsymbol{x}(t)}{\mathrm{d}t} = \boldsymbol{A}\boldsymbol{x}(t), \quad \boldsymbol{x}(0) = \boldsymbol{I} \tag{3.61}$$

gives

$$\frac{\mathrm{d}(\boldsymbol{T}^{-1}\boldsymbol{x})}{\mathrm{d}t} = \boldsymbol{\Lambda}\boldsymbol{T}^{-1}\boldsymbol{x}, \quad \boldsymbol{T}^{-1}\boldsymbol{x}(0) = \boldsymbol{T}^{-1} \tag{3.62}$$

Solution of the above equation is

$$\boldsymbol{T}^{-1}\boldsymbol{x} = \mathrm{e}^{-\boldsymbol{\Lambda}t}\boldsymbol{T}^{-1} \tag{3.63}$$

or

$$\boldsymbol{x} = \boldsymbol{T}\mathrm{e}^{-\boldsymbol{\Lambda}t}\boldsymbol{T}^{-1} \tag{3.64}$$

and therefore

$$\boldsymbol{\Phi}(t) = \boldsymbol{T}\mathrm{e}^{-\boldsymbol{\Lambda}t}\boldsymbol{T}^{-1} \tag{3.65}$$

where

$$\mathrm{e}^{\boldsymbol{\Lambda}t} = \begin{pmatrix} \mathrm{e}^{\lambda_1 t} & 0 & \dots & 0 \\ 0 & \mathrm{e}^{\lambda_2 t} & \dots & 0 \\ \vdots & & & \vdots \\ 0 & 0 & \dots & \mathrm{e}^{\lambda_n t} \end{pmatrix} \tag{3.66}$$

3.2.4 Stability, Controllability, and Observability of Continuous-Time Systems

Stability, controllability, and observability are basic properties of systems closely related to state-space models. These properties can be utilised for system analysis and synthesis.

Stability of Continuous-Time Systems

An important aspect of system behaviour is *stability*. System can be defined as stable if its response to bounded inputs is also bounded. The concept of stability is of great practical interest as nonstable control systems are unacceptable. Stability can also be determined without an analytical solution of process equations which is important for nonlinear systems.

Consider a system

$$\frac{\mathrm{d}\boldsymbol{x}(t)}{\mathrm{d}t} = \boldsymbol{f}(\boldsymbol{x}(t), \boldsymbol{u}(t), t), \quad \boldsymbol{x}(t_0) = \boldsymbol{x}_0 \tag{3.67}$$

Such a system is called *forced* as the vector of input variables $\boldsymbol{u}(t)$ appears on the right hand side of the equation. However, stability can be studied on *free (zero-input)* systems given by the equation

$$\frac{\mathrm{d}\boldsymbol{x}(t)}{\mathrm{d}t} = \boldsymbol{f}(\boldsymbol{x}(t), t), \quad \boldsymbol{x}(t_0) = \boldsymbol{x}_0 \tag{3.68}$$

$\boldsymbol{u}(t)$ does not appear in the previous equation, which is equivalent to processes with constant inputs. If time t appears explicitly as an argument in process dynamics equations we speak about *nonautonomous* system, otherwise about *autonomous* system.

In our discussion about stability of (3.68) we will consider stability of motion of $\boldsymbol{x}^s(t)$ that corresponds to constant values of input variables. Let us for this purpose investigate any solution (motion) of the forced system $\boldsymbol{x}(t)$ that is at $t = 0$ in the neighbourhood of $\boldsymbol{x}^s(t)$. The problem of stability is closely connected to the question if for $t \geq 0$ remains $\boldsymbol{x}(t)$ in the neighbourhood of $\boldsymbol{x}^s(t)$. Let us define deviation

$$\tilde{\boldsymbol{x}}(t) = \boldsymbol{x}(t) - \boldsymbol{x}^s(t) \tag{3.69}$$

then,

$$\begin{aligned}
\frac{\mathrm{d}\tilde{\boldsymbol{x}}(t)}{\mathrm{d}t} + \frac{\mathrm{d}\boldsymbol{x}^s(t)}{\mathrm{d}t} &= \boldsymbol{f}(\tilde{\boldsymbol{x}}(t) + \boldsymbol{x}^s(t), \boldsymbol{u}(t), t) \\
\frac{\mathrm{d}\tilde{\boldsymbol{x}}(t)}{\mathrm{d}t} &= \boldsymbol{f}(\tilde{\boldsymbol{x}}(t) + \boldsymbol{x}^s(t), \boldsymbol{u}(t), t) - \boldsymbol{f}(\boldsymbol{x}^s(t), t) \\
\frac{\mathrm{d}\tilde{\boldsymbol{x}}(t)}{\mathrm{d}t} &= \tilde{\boldsymbol{f}}(\tilde{\boldsymbol{x}}(t), \boldsymbol{u}(t), t)
\end{aligned} \tag{3.70}$$

The solution $\boldsymbol{x}^s(t)$ in (3.70) corresponds for all $t > 0$ to relation $\tilde{\boldsymbol{x}}(t) = \mathbf{0}$ and $\dot{\tilde{\boldsymbol{x}}}(t) = \mathbf{0}$. Therefore the state $\tilde{\boldsymbol{x}}(t) = \mathbf{0}$ is called *equilibrium state* of the system described by (3.70). This equation can always be constructed and stability of equilibrium point can be interpreted as stability in the beginning of the state-space.

Stability theorems given below are valid for nonautonomous systems. However, such systems are very rare in common processes. In connection to the above ideas about equilibrium point we will restrict our discussion to systems given by

$$\frac{\mathrm{d}\boldsymbol{x}(t)}{\mathrm{d}t} = \boldsymbol{f}(\boldsymbol{x}(t)), \quad \boldsymbol{x}(t_0) = \boldsymbol{x}_0 \tag{3.71}$$

The equilibrium state $\boldsymbol{x}^e = \mathbf{0}$ of this system obeys the relation

$$\boldsymbol{f}(\mathbf{0}) = \mathbf{0} \tag{3.72}$$

as $\mathrm{d}\boldsymbol{x}/\mathrm{d}t = \mathbf{0}$

We assume that the solution of the equation (3.71) exists and is unique.

Stability can be intuitively defined as follows: If $\boldsymbol{x}^e = \mathbf{0}$ is the equilibrium point of the system (3.71), then we may say that $\boldsymbol{x}^e = \mathbf{0}$ is the stable equilibrium point if the solution of (3.71) $\boldsymbol{x}(t) = \boldsymbol{x}[\boldsymbol{x}(t_0), t]$ that begins in some state $\boldsymbol{x}(t_0)$ "close" to the equilibrium point $\boldsymbol{x}^e = \mathbf{0}$ remains in the neighbourhood of $\boldsymbol{x}^e = \mathbf{0}$ or the solution approaches this state.

The equilibrium state $\boldsymbol{x}^e = \mathbf{0}$ is unstable if the solution $\boldsymbol{x}(t) = \boldsymbol{x}[\boldsymbol{x}(t_0), t]$ that begins in some state $\boldsymbol{x}(t_0)$ diverges from the neighbourhood of $\boldsymbol{x}^e = \mathbf{0}$.

Next, we state the definitions of stability from Lyapunov asymptotic stability and asymptotic stability in large.

Lyapunov stability: The system (3.71) is stable in the equilibrium state $\boldsymbol{x}^e = \mathbf{0}$ if for any given $\varepsilon > 0$, there exists $\delta(\varepsilon) > 0$ such that for all $\boldsymbol{x}(t_0)$ such that $\|\boldsymbol{x}(t_0)\| \leq \delta$ implies $\|\boldsymbol{x}[\boldsymbol{x}(t_0), t]\| \leq \varepsilon$ for all $t \geq 0$.

Asymptotic (internal) stability: The system (3.71) is asymptotically stable in the equilibrium state $\boldsymbol{x}^e = \mathbf{0}$ if it is Lyapunov stable and if all $\boldsymbol{x}(t) = \boldsymbol{x}[\boldsymbol{x}(t_0), t]$ that begin sufficiently close to the equilibrium state $\boldsymbol{x}^e = \mathbf{0}$ satisfy the condition $\lim_{t\to\infty} \|\boldsymbol{x}(t)\| = 0$.

Asymptotic stability in large: The system (3.71) is asymptotically stable in large in the equilibrium state $\boldsymbol{x}^e = \mathbf{0}$ if it is asymptotic stable for all initial states $\boldsymbol{x}(t_0)$.

In the above definitions, the notation $\|\boldsymbol{x}\|$ has been used for the Euclidean norm of a vector $\boldsymbol{x}(t)$ that is defined as the distance of the point given by the coordinates of $\boldsymbol{x}$ from equilibrium point $\boldsymbol{x}^e = \mathbf{0}$ and given as $\|\boldsymbol{x}\| = (\boldsymbol{x}^T\boldsymbol{x})^{1/2}$.

Note 3.1. Norm of a vector is some function transforming any vector $\boldsymbol{x} \in R^n$ to some real number $\|\boldsymbol{x}\|$ with the following properties:

1. $\|\boldsymbol{x}\| \geq 0$,
2. $\|\boldsymbol{x}\| = 0$ iff $\boldsymbol{x} = \mathbf{0}$,
3. $\|k\boldsymbol{x}\| = |k| \, \|\boldsymbol{x}\|$ for any k,
4. $\|\boldsymbol{x} + \boldsymbol{y}\| \leq \|\boldsymbol{x}\| + \|\boldsymbol{y}\|$.

Some examples of norms are $\|\boldsymbol{x}\| = (\boldsymbol{x}^T\boldsymbol{x})^{1/2}$, $\|\boldsymbol{x}\| = \sum_{i=1}^{n} |x_i|$, $\|\boldsymbol{x}\| = \max |x_i|$. It can be proven that all these norms satisfy properties 1-4.

Example 3.6: *Physical interpretation – U-tube*

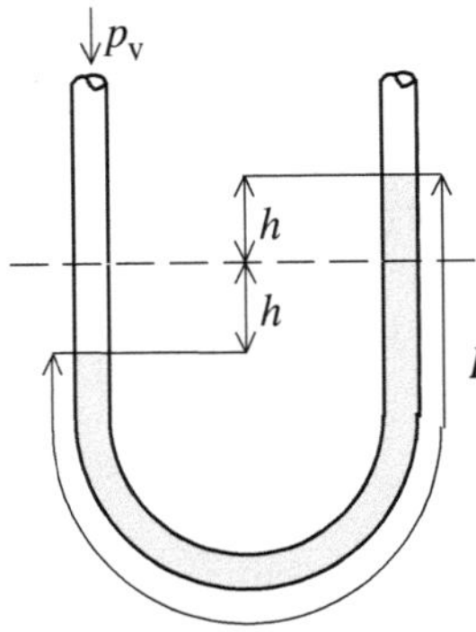

Fig. 3.5. A U-tube

Consider a U-tube as an example of the second order system. Mathematical model of this system can be derived from Fig. 3.5 considering the equilibrium of forces.

We assume that if specific pressure changes, the force with which the liquid flow is inhibited, is proportional to the speed of the liquid. Furthermore, we assume that the second Newton law is applicable. The following equation holds for the equilibrium of forces

$$Fp_v = 2Fg\rho h + kF\frac{\mathrm{d}h}{\mathrm{d}t} + FL\rho\frac{\mathrm{d}^2h}{\mathrm{d}t^2}$$

or

$$\frac{\mathrm{d}^2h}{\mathrm{d}t^2} + \frac{k}{L\rho}\frac{\mathrm{d}h}{\mathrm{d}t} + \frac{2g}{L}h = \frac{1}{L\rho}p_v$$

where

F - inner cross-sectional area of tube,
k - coefficient,
p_v - specific pressure,
g - acceleration of gravity,
ρ - density of liquid.

If the input is zero then the mathematical model is of the form

$$\frac{\mathrm{d}^2x_1}{\mathrm{d}t^2} + a_1\frac{\mathrm{d}x_1}{\mathrm{d}t} + a_0x_1 = 0$$

where $x_1 = h - h^s, a_0 = 2g/L, a_1 = k/L\rho$. The speed of liquid flow will be denoted by $x_2 = \mathrm{d}x_1/\mathrm{d}t$. If x_1, x_2 are elements of state vector $\boldsymbol{x}$ then the dynamics of the U-tube is given as

$$\frac{\mathrm{d}x_1}{\mathrm{d}t} = x_2$$

$$\frac{\mathrm{d}x_2}{\mathrm{d}t} = -a_0x_1 - a_1x_2$$

If we consider $a_0 = 1, a_1 = 1, \boldsymbol{x}(0) = (1,0)^T$ then the solution of the differential equations is shown in Fig. 3.6. At any time instant the total system energy is given as a sum of kinetic and potential energies of liquid

$$V(x_1, x_2) = FL\rho\frac{x_2^2}{2} + \int_0^{x_1} 2Fg\rho x \mathrm{d}x$$

Energy V satisfies the following conditions: $V(\boldsymbol{x}) > 0, \boldsymbol{x} \neq \boldsymbol{0}$ and $V(\boldsymbol{0}) = 0$.
These conditions show that the sum of kinetic and potential energies is positive with the exception when liquid is in the equilibrium state $\boldsymbol{x}^e = \boldsymbol{0}$ when $\mathrm{d}x_1/\mathrm{d}t = \mathrm{d}x_2/\mathrm{d}t = 0$.
The change of V in time is given as

$$\frac{\mathrm{d}V}{\mathrm{d}t} = \frac{\partial V}{\partial x_1}\frac{\mathrm{d}x_1}{\mathrm{d}t} + \frac{\partial V}{\partial x_2}\frac{\mathrm{d}x_2}{\mathrm{d}t}$$

$$\frac{\mathrm{d}V}{\mathrm{d}t} = 2Fg\rho x_1 x_2 + FL\rho x_2\left(-\frac{2g}{L}x_1 - \frac{k}{L\rho}x_2\right)$$

$$\frac{\mathrm{d}V}{\mathrm{d}t} = -Fkx_2^2$$

As $k > 0$, time derivative of V is always negative except if $x_2 = 0$ when $\mathrm{d}V/\mathrm{d}t = 0$ and hence V cannot increase. If $x_2 = 0$ the dynamics of the tube shows that

$$\frac{\mathrm{d}x_2}{\mathrm{d}t} = -\frac{2g}{L}x_1$$

is nonzero (except $\boldsymbol{x}^e = \boldsymbol{0}$). The system cannot remain in a nonequilibrium state for which $x_2 = 0$ and always reaches the equilibrium state which is stable. The sum of the energies V is given as

$$V(x_1, x_2) = 2Fg\rho\frac{x_1^2}{2} + FL\rho\frac{x_2^2}{2}$$

$$V(x_1, x_2) = \frac{F\rho}{2}(2gx_1^2 + Lx_2^2)$$

Fig. 3.7 shows the state plane with curves of constant energy levels $V_1 < V_2 < V_3$ and state trajectory corresponding to Fig. 3.6 where x_1, x_2 are plotted as function of parameter t.

Conclusions about system behaviour and about state trajectory in the state plane can be generalised by general state-space. It is clear that some results about system properties can also be derived without analytical solution of state-space equations.

Stability theory of Lyapunov assumes the existence of the Lyapunov function $V(\boldsymbol{x})$. The continuous function $V(\boldsymbol{x})$ with continuous derivatives is called *positive definite* in some neighbourhood Δ of state origin if

$$V(\boldsymbol{0}) = 0 \tag{3.73}$$

and

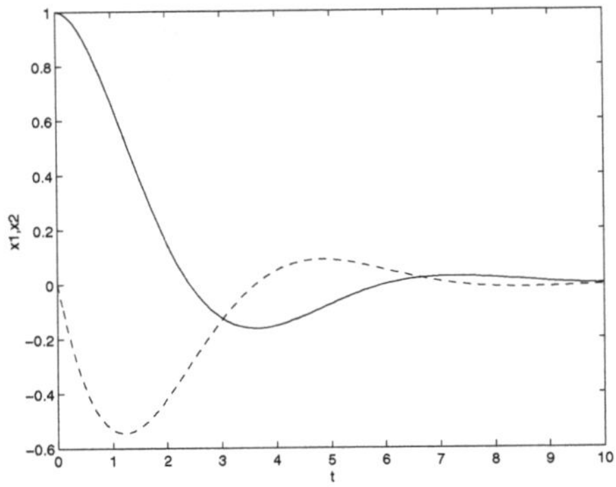

Fig. 3.6. Time response of the U-tube for initial conditions $(1,0)^T$

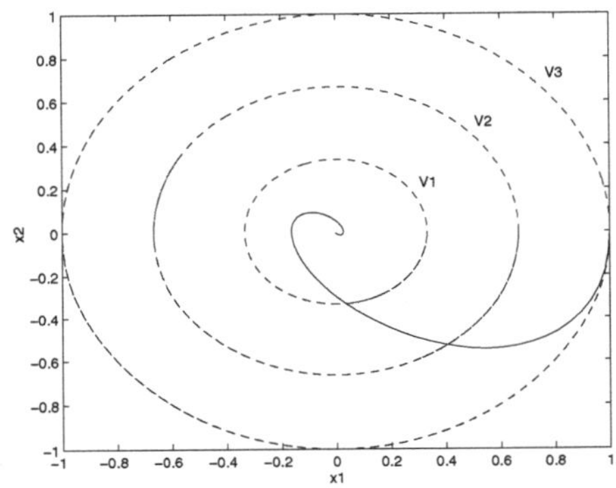

Fig. 3.7. Constant energy curves and state trajectory of the U-tube in the state plane

$$V(\boldsymbol{x}) > 0 \tag{3.74}$$

for all $\boldsymbol{x} \neq \mathbf{0}$ within Δ. If (3.74) is replaced by

$$V(\boldsymbol{x}) \geq 0 \tag{3.75}$$

for all $\boldsymbol{x} \in \Delta$ then $V(\boldsymbol{x})$ is *positive semidefinite*. Definitions of *negative definite* and *negative semidefinite* functions follow analogously.

Various definitions of stability for the system $\mathrm{d}\boldsymbol{x}(t)/\mathrm{d}t = \boldsymbol{f}(\boldsymbol{x}), \boldsymbol{f}(\mathbf{0}) = \mathbf{0}$ lead to the following theorems:

Stability in Lyapunov sense: If a positive definite function $V(\boldsymbol{x})$ can be chosen such that

$$\frac{\mathrm{d}V}{\mathrm{d}t} = \left(\frac{\partial V}{\partial \boldsymbol{x}}\right)^T \boldsymbol{f}(\boldsymbol{x}) \leq 0 \tag{3.76}$$

then the system is stable in origin in the Lyapunov sense.

The function $V(\boldsymbol{x})$ satisfying this theorem is called the *Lyapunov function*.

Asymptotic stability: If a positive definite function $V(\boldsymbol{x})$ can be chosen such that

$$\frac{\mathrm{d}V}{\mathrm{d}t} = \left(\frac{\partial V}{\partial \boldsymbol{x}}\right)^T \boldsymbol{f}(\boldsymbol{x}) < 0, \quad \boldsymbol{x} \neq \boldsymbol{0} \tag{3.77}$$

then the system is asymptotically stable in origin.

Asymptotic stability in large: If the conditions of asymptotic stability are satisfied for all $\boldsymbol{x}$ and if $V(\boldsymbol{x}) \to \infty$ for $\|\boldsymbol{x}\| \to \infty$ then the system is asymptotically stable by large in origin.

There is no general procedure for the construction of the Lyapunov function. If such a function exists then it is not unique. Often it is chosen in the form

$$V(\boldsymbol{x}) = \sum_{k=1}^{n}\sum_{r=1}^{n} K_{rk} x_k x_r \tag{3.78}$$

K_{rk} are real constants, $K_{rk} = K_{kr}$ so (3.78) can be written as

$$V(\boldsymbol{x}) = \boldsymbol{x}^T \boldsymbol{K} \boldsymbol{x} \tag{3.79}$$

and $\boldsymbol{K}$ is symmetric matrix. $V(\boldsymbol{x})$ is positive definite if and only if the determinants

$$K_{11}, \begin{vmatrix} K_{11}, K_{12} \\ K_{21}, K_{22} \end{vmatrix}, \begin{vmatrix} K_{11}, K_{12}, K_{13} \\ K_{21}, K_{22}, K_{23} \\ K_{31}, K_{32}, K_{33} \end{vmatrix}, \ldots \tag{3.80}$$

are greater than zero.

Asymptotic stability of linear systems: Linear system

$$\frac{\mathrm{d}\boldsymbol{x}(t)}{\mathrm{d}t} = \boldsymbol{A}\boldsymbol{x}(t) \tag{3.81}$$

is *asymptotically stable* (in large) if and only if one of the following properties is valid:

1. *Lyapunov equation*

   $$\boldsymbol{A}^T \boldsymbol{K} + \boldsymbol{K}\boldsymbol{A} = -\boldsymbol{\mu} \tag{3.82}$$

   where $\boldsymbol{\mu}$ is any symmetric positive definite matrix, has a unique positive definite symmetric solution $\boldsymbol{K}$.
2. all eigenvalues of *system matrix* $\boldsymbol{A}$, i.e. all roots of *characteristic polynomial* $\det(s\boldsymbol{I} - \boldsymbol{A})$ have negative real parts.

Proof: We prove only the sufficient part of 1. Consider the Lyapunov function of the form

$$V(\boldsymbol{x}) = \boldsymbol{x}^T \boldsymbol{K} \boldsymbol{x} \tag{3.83}$$

if $\boldsymbol{K}$ is a positive definite then

$$V(\boldsymbol{x}) > 0 \ , \quad \boldsymbol{x} \neq \boldsymbol{0} \tag{3.84}$$

$$V(\boldsymbol{0}) = 0 \tag{3.85}$$

and for $\mathrm{d}V/\mathrm{d}t$ holds

$$\frac{\mathrm{d}V(\boldsymbol{x})}{\mathrm{d}t} = \left(\frac{\mathrm{d}\boldsymbol{x}}{\mathrm{d}t}\right)^T \boldsymbol{K}\boldsymbol{x} + \boldsymbol{x}^T\boldsymbol{K}\frac{\mathrm{d}\boldsymbol{x}}{\mathrm{d}t} \tag{3.86}$$

Substituting $\mathrm{d}\boldsymbol{x}/\mathrm{d}t$ from Eq. (3.81) yields

$$\frac{\mathrm{d}V(\boldsymbol{x})}{\mathrm{d}t} = \boldsymbol{x}^T\boldsymbol{A}^T\boldsymbol{K}\boldsymbol{x} + \boldsymbol{x}^T\boldsymbol{K}\boldsymbol{A}\boldsymbol{x} \tag{3.87}$$

$$\frac{\mathrm{d}V(\boldsymbol{x})}{\mathrm{d}t} = \boldsymbol{x}^T(\boldsymbol{A}^T\boldsymbol{K} + \boldsymbol{K}\boldsymbol{A})\boldsymbol{x} \tag{3.88}$$

Applying (3.82) we get

$$\frac{\mathrm{d}V(\boldsymbol{x})}{\mathrm{d}t} = -\boldsymbol{x}^T\boldsymbol{\mu}\boldsymbol{x} \tag{3.89}$$

and because $\boldsymbol{\mu}$ is a positive definite matrix then

$$\frac{\mathrm{d}V(\boldsymbol{x})}{\mathrm{d}t} < 0 \tag{3.90}$$

for all $\boldsymbol{x} \neq \boldsymbol{0}$ and the system is asymptotically stable in origin. As the Lyapunov function can be written as

$$V(\boldsymbol{x}) = \|\boldsymbol{x}\|^2 \tag{3.91}$$

and therefore

$$V(\boldsymbol{x}) \rightarrow \infty \quad \text{for } \|\boldsymbol{x}\| \rightarrow \infty \tag{3.92}$$

The corresponding norm is defined as $(\boldsymbol{x}^T\boldsymbol{K}\boldsymbol{x})^{1/2}$. It can easily be shown that $\boldsymbol{K}$ exists and all conditions of the theorem on asymptotic stability by large in origin are fulfilled. The second part of the proof - necessity - is much harder to prove.

The choice of $\boldsymbol{\mu}$ for computations is usually

$$\boldsymbol{\mu} = \boldsymbol{I} \tag{3.93}$$

Controllability of Continuous-Time Systems

The concept of controllability together with observability is of fundamental importance in theory of automatic control.

Definition of controllability of linear system

$$\frac{d\boldsymbol{x}(t)}{dt} = \boldsymbol{A}(t)\boldsymbol{x}(t) + \boldsymbol{B}(t)\boldsymbol{u}(t) \tag{3.94}$$

is as follows: A state $\boldsymbol{x}(t_0) \neq \boldsymbol{0}$ of the system (3.94) is *controllable* if the system can be driven from this state to state $\boldsymbol{x}(t_1) = \boldsymbol{0}$ by applying suitable $\boldsymbol{u}(t)$ within finite time $t_1 - t_0, t \in [t_0, t_1]$.

If every state is controllable then the system is *completely controllable.*

Definition of reachable of linear systems: A state $\boldsymbol{x}(t_1)$ of the system (3.94) is *reachable* if the system can be driven from the state $\boldsymbol{x}(t_0) = \boldsymbol{0}$ to $\boldsymbol{x}(t_1)$ by applying suitable $\boldsymbol{u}(t)$ within finite time $t_1 - t_0, t \in [t_0, t_1]$.

If every state is reachable then the system is *completely reachable.*

For linear systems with constant coefficients (linear time invariant systems) are all reachable states controllable and it is sufficient to speak about controllability. Often the definitions are simplified and we can speak that the system is completely controllable (shortly controllable) if there exists such $\boldsymbol{u}(t)$ that drives the system from the arbitrary initial state $\boldsymbol{x}(t_0)$ to the final state $\boldsymbol{x}(t_1)$ within a finite time $t_1 - t_0, t \in [t_0, t_1]$.

Theorem (Controllability of linear continuous systems with constant coefficients): The system

$$\frac{d\boldsymbol{x}(t)}{dt} = \boldsymbol{A}\boldsymbol{x}(t) + \boldsymbol{B}\boldsymbol{u}(t) \tag{3.95}$$

$$\boldsymbol{y}(t) = \boldsymbol{C}\boldsymbol{x}(t) \tag{3.96}$$

is completely controllable if and only if rank of *controllability matrix* $\boldsymbol{Q}_c$ is equal to n. $\boldsymbol{Q}_c[n \times nm]$ is defined as

$$\boldsymbol{Q}_c = (\boldsymbol{B}\ \boldsymbol{A}\boldsymbol{B}\ \boldsymbol{A}^2\boldsymbol{B} \ldots \boldsymbol{A}^{n-1}\boldsymbol{B}) \tag{3.97}$$

where n is the dimension of the vector $\boldsymbol{x}$ and m is the dimension of the vector $\boldsymbol{u}$.

Proof: We prove only the "if" part. Solution of the Eq. (3.95) with initial condition $\boldsymbol{x}(t_0)$ is

$$\boldsymbol{x}(t) = e^{\boldsymbol{A}t}\boldsymbol{x}(t_0) + \int_0^t e^{\boldsymbol{A}(t-\tau)}\boldsymbol{B}\boldsymbol{u}(\tau)d\tau \tag{3.98}$$

For $t = t_1$ follows

$$\boldsymbol{x}(t_1) = e^{\boldsymbol{A}t_1}\boldsymbol{x}(t_0) + e^{\boldsymbol{A}t_1}\int_0^{t_1} e^{-\boldsymbol{A}\tau}\boldsymbol{B}\boldsymbol{u}(\tau)d\tau \tag{3.99}$$

The function $e^{-\boldsymbol{A}\tau}$ can be rewritten with the aid of the Cayley-Hamilton theorem as

$$e^{-\boldsymbol{A}\tau} = k_0(\tau)\boldsymbol{I} + k_1(\tau)\boldsymbol{A} + k_2(\tau)\boldsymbol{A}^2 + \cdots + k_{n-1}(\tau)\boldsymbol{A}^{n-1} \tag{3.100}$$

Substituting for $e^{-\boldsymbol{A}\tau}$ from (3.100) into (3.99) yields

$$\begin{aligned}\boldsymbol{x}(t_1) = \mathrm{e}^{\boldsymbol{A}t_1}\boldsymbol{x}(t_0) + \mathrm{e}^{\boldsymbol{A}t_1}\int_0^{t_1} (k_0(\tau)\boldsymbol{B} + k_1(\tau)\boldsymbol{AB} + \\ +k_2(\tau)\boldsymbol{A}^2\boldsymbol{B} + \cdots + k_{n-1}(\tau)\boldsymbol{A}^{n-1}\boldsymbol{B})\boldsymbol{u}(\tau)\mathrm{d}\tau\end{aligned} \tag{3.101}$$

or

$$\begin{aligned}\boldsymbol{x}(t_1) &= \mathrm{e}^{\boldsymbol{A}t_1}\boldsymbol{x}(t_0) + \\ &+\mathrm{e}^{\boldsymbol{A}t_1}\int_0^{t_1} (\boldsymbol{B}\ \boldsymbol{AB}\ \boldsymbol{A}^2\boldsymbol{B}\ldots\boldsymbol{A}^{n-1}\boldsymbol{B}) \times \\ &\times \begin{pmatrix} k_0(\tau)\boldsymbol{u}(\tau) \\ k_1(\tau)\boldsymbol{u}(\tau) \\ k_2(\tau)\boldsymbol{u}(\tau) \\ \vdots \\ k_{n-1}(\tau)\boldsymbol{u}(\tau) \end{pmatrix} \mathrm{d}\tau\end{aligned} \tag{3.102}$$

Complete controllability means that for all $\boldsymbol{x}(t_0) \neq \boldsymbol{0}$ there exists a finite time $t_1 - t_0$ and suitable $\boldsymbol{u}(t)$ such that

$$-\boldsymbol{x}(t_0) = (\boldsymbol{B}\ \boldsymbol{AB}\ \boldsymbol{A}^2\boldsymbol{B}\ldots\boldsymbol{A}^{n-1}\boldsymbol{B})\int_0^{t_1} \begin{pmatrix} k_0(\tau)\boldsymbol{u}(\tau) \\ k_1(\tau)\boldsymbol{u}(\tau) \\ k_2(\tau)\boldsymbol{u}(\tau) \\ \vdots \\ k_{n-1}(\tau)\boldsymbol{u}(\tau) \end{pmatrix} \mathrm{d}\tau \tag{3.103}$$

From this equation follows that any vector $-\boldsymbol{x}(t_0)$ can be expressed as a linear combination of the columns of $\boldsymbol{Q}_c$. The system is controllable if the integrand in (3.102) allows the influence of $\boldsymbol{u}$ to reach all the states $\boldsymbol{x}$. Hence complete controllability is equivalent to the condition of rank of $\boldsymbol{Q}_c$ being equal to n. The controllability theorem enables a simple check of system controllability with regard to $\boldsymbol{x}$. The test with regard to $\boldsymbol{y}$ can be derived analogously and is given below.

Theorem (Output controllability of linear systems with constant coefficients): The system output $\boldsymbol{y}$ of (3.95), (3.96) is completely controllable if and only if the rank of *controllability matrix* $\boldsymbol{Q}_c^y[r \times nm]$ is equal to r (with r being dimension of the output vector) where

$$\boldsymbol{Q}_c^y = (\boldsymbol{CB}\ \boldsymbol{CAB}\ \boldsymbol{CA}^2\boldsymbol{B}\ldots\boldsymbol{CA}^{n-1}\boldsymbol{B}) \tag{3.104}$$

We note that the controllability conditions are also valid for linear systems with time-varying coefficients if $\boldsymbol{A}(t), \boldsymbol{B}(t)$ are known functions of time. The conditions for nonlinear systems are derived only for some special cases. Fortunately, in the majority of practical cases, controllability of nonlinear systems is satisfied if the corresponding linearised system is controllable.

Example 3.7: *CSTR - controllability*

Linearised state-space model of CSTR (see Example 2.6) is of the form

$$\frac{\mathrm{d}x_1(t)}{\mathrm{d}t} = a_{11}x_1(t) + a_{12}x_2(t)$$
$$\frac{\mathrm{d}x_2(t)}{\mathrm{d}t} = a_{21}x_1(t) + a_{22}x_2(t) + b_{21}u_1(t)$$

or

$$\frac{\mathrm{d}\boldsymbol{x}(t)}{\mathrm{d}t} = \boldsymbol{A}\boldsymbol{x}(t) + \boldsymbol{B}u_1(t)$$

where

$$\boldsymbol{A} = \begin{pmatrix} a_{11} & a_{12} \\ a_{21} & a_{22} \end{pmatrix}, \quad \boldsymbol{B} = \begin{pmatrix} 0 \\ b_{21} \end{pmatrix}$$

The controllability matrix $\boldsymbol{Q}_c$ is

$$\boldsymbol{Q}_c = (\boldsymbol{B}|\boldsymbol{A}\boldsymbol{B}) = \begin{pmatrix} 0 & a_{12}b_{21} \\ b_{21} & a_{22}b_{21} \end{pmatrix}$$

and has rank equal to 2 and the system is completely controllable. It is clear that this is valid for all steady-states and hence the corresponding nonlinear model of the reactor is controllable.

Observability

States of a system are in the majority of cases measurable only partially or they are nonmeasurable. Therefore it is not possible to realise a control that assumes knowledge of state variables. In this connection a question arises whether it is possible to determine state vector from output measurements. We speak about observability and reconstructibility. To investigate observability, only a free system can be considered.

Definition of observability: A state $\boldsymbol{x}(t_0)$ of the system

$$\frac{\mathrm{d}\boldsymbol{x}(t)}{\mathrm{d}t} = \boldsymbol{A}(t)\boldsymbol{x}(t) \tag{3.105}$$
$$\boldsymbol{y}(t) = \boldsymbol{C}(t)\boldsymbol{x}(t) \tag{3.106}$$

is observable if it can be determined from knowledge about $\boldsymbol{y}(t)$ within a finite time $t \in [t_0, t_1]$. If every state $\boldsymbol{x}(t_0)$ can be determined from the output vector $\boldsymbol{y}(t)$ within arbitrary finite interval $t \in [t_0, t_1]$ then the system is completely observable.

Definition of reconstructibility : A state of system $\boldsymbol{x}(t_0)$ is reconstructible if it can be determined from knowledge about $\boldsymbol{y}(t)$ within a finite time $t \in [t_{00}, t_0]$. If every state $\boldsymbol{x}(t_0)$ can be determined from the output vector $\boldsymbol{y}(t)$ within arbitrary finite interval $t \in [t_{00}, t_0]$ then the system is completely reconstructible.

Similarly as in the case of controllability and reachability, the terms observability of a system and reconstructibility of a system are used for simplicity. For linear time-invariant systems, both terms are interchangeable.

Theorem: Observability of linear continuous systems with constant coefficients: The system

$$\frac{d\boldsymbol{x}(t)}{dt} = \boldsymbol{A}\boldsymbol{x}(t) \tag{3.107}$$

$$\boldsymbol{y}(t) = \boldsymbol{C}\boldsymbol{x}(t) \tag{3.108}$$

is completely observable if and only if rank of *observability matrix* $\boldsymbol{Q}_o$ is equal to n. The matrix $\boldsymbol{Q}_o[nr \times n]$ is given as

$$\boldsymbol{Q_o} = \begin{pmatrix} \boldsymbol{C} \\ \boldsymbol{CA} \\ \boldsymbol{CA}^2 \\ \vdots \\ \boldsymbol{CA}^{n-1} \end{pmatrix} \tag{3.109}$$

Proof: We prove only the "if" part. Solution of the Eq. (3.107) is

$$\boldsymbol{x}(t) = e^{\boldsymbol{A}t}\boldsymbol{x}(t_0) \tag{3.110}$$

According to the Cayley-Hamilton theorem, the function $e^{-\boldsymbol{A}t}$ can be written as

$$e^{-\boldsymbol{A}t} = k_0(t)\boldsymbol{I} + k_1(t)\boldsymbol{A} + k_2(t)\boldsymbol{A}^2 + \cdots + k_{n-1}(t)\boldsymbol{A}^{n-1} \tag{3.111}$$

Substituting Eq. (3.111) into (3.110) yields

$$\boldsymbol{x}(t) = [k_0(t)\boldsymbol{I} + k_1(t)\boldsymbol{A} + k_2(t)\boldsymbol{A}^2 + \cdots + k_{n-1}(t)\boldsymbol{A}^{n-1}]\boldsymbol{x}(t_0) \tag{3.112}$$

Equation (3.108) now gives

$$\boldsymbol{y}(t) = [k_0(t)\boldsymbol{C}+k_1(t)\boldsymbol{CA}+k_2(t)\boldsymbol{CA}^2+\cdots+k_{n-1}(t)\boldsymbol{CA}^{n-1}]\boldsymbol{x}(t_0) \tag{3.113}$$

or

$$\boldsymbol{x}(t_0) = \left[\int_{t_0}^{t_1} (\boldsymbol{k}(t)\boldsymbol{Q}_o)^T(\boldsymbol{k}(t)\boldsymbol{Q}_o)dt\right]^{-1} \int_{t_0}^{t_1} (\boldsymbol{k}(t)\boldsymbol{Q}_o)^T\boldsymbol{y}(t)dt \tag{3.114}$$

where $\boldsymbol{k}(t) = [k_0(t), k_1(t), \dots, k_{n-1}(t)]$.

If the system is observable, it must be possible to determine $\boldsymbol{x}(t_0)$ from (3.114). Hence the inverse of $\int_{t_0}^{t_1}(\boldsymbol{k}(t)\boldsymbol{Q}_o)^T(\boldsymbol{k}(t)\boldsymbol{Q}_o)dt$ must exist and the matrix

$$\int_{t_0}^{t_1} (\boldsymbol{k}(t)\boldsymbol{Q}_o)^T(\boldsymbol{k}(t)\boldsymbol{Q}_o)dt = \boldsymbol{Q_o}^T \int_{t_0}^{t_1} (\boldsymbol{k}^T(t)\boldsymbol{k}(t)dt\boldsymbol{Q_o} \tag{3.115}$$

must be nonsingular. It can be shown that the matrix $\boldsymbol{k}^T(t)\boldsymbol{k}(t)$ is nonsingular and observability is satisfied if and only if $\text{rank}(\boldsymbol{Q_o}) = n$.

We note that observability and reconstructibility conditions for linear continuous systems with constant coefficients are the same.

Example 3.8: *CSTR - observability*

Consider the linearised model of CSTR from Example 2.6

$$\frac{\mathrm{d}x_1(t)}{\mathrm{d}t} = a_{11}x_1(t) + a_{12}x_2(t)$$

$$\frac{\mathrm{d}x_2(t)}{\mathrm{d}t} = a_{21}x_1(t) + a_{22}x_2(t)$$

$$y_1(t) = x_2(t)$$

The matrices $\boldsymbol{A}, \boldsymbol{C}$ are

$$\boldsymbol{A} = \begin{pmatrix} a_{11} & a_{12} \\ a_{21} & a_{22} \end{pmatrix}, \quad \boldsymbol{C} = (0, 1)$$

and for $\boldsymbol{Q}_o$ yields

$$\boldsymbol{Q}_o = \begin{pmatrix} 0 & 1 \\ a_{21} & a_{22} \end{pmatrix}$$

Rank of $\boldsymbol{Q}_o$ is 2 and the system is observable.
(Recall that $a_{21} = (-\Delta H)\dot{r}_{c_A}(c_a^s, \vartheta^s)/\rho c_p$)

3.2.5 Canonical Decomposition

Any continuous linear system with constant coefficients can be transformed into a special state-space form such that four separated subsystems result:

(A)controllable and observable subsystem,
(B)controllable and nonobservable subsystem,
(C)noncontrollable and observable subsystem,
(D)noncontrollable and nonobservable subsystem.

This division is called *canonical decomposition* and is shown in Fig. 3.8. Only subsystem A can be calculated from input and output relations.

The system eigenvalues can be also divided into 4 groups:

(A)controllable and observable modes,
(B)controllable and nonobservable modes,
(C)noncontrollable and observable modes,
(D)noncontrollable and nonobservable modes.

State-space model of continuous linear systems with constant coefficients is said to be minimal if it is controllable and observable.

State-space models of processes are more general than I/O models as they can also contain noncontrollable and nonobservable parts that are cancelled in I/O models.

Sometimes the notation *detectability* and *stabilisability* is used. A system is said to be detectable if all nonobservable eigenvalues are asymptotically stable and it is stabilisable if all nonstable eigenvalues are controllable.

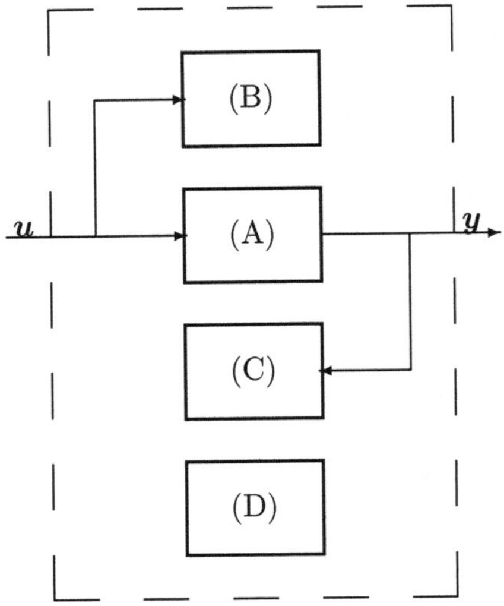

Fig. 3.8. Canonical decomposition

3.3 Input-Output Process Models

In this section we focus our attention to transfer properties of processes. We show the relations between state-space and I/O models.

3.3.1 SISO Continuous Systems with Constant Coefficients

Linear continuous SISO (single input, single output) systems with constant coefficients with input $u(t)$ and output $y(t)$ can be described by a differential equation in the form

$$a_n \frac{\mathrm{d}^n y(t)}{\mathrm{d}t^n} + a_{n-1} \frac{\mathrm{d}^{n-1} y(t)}{\mathrm{d}t^{n-1}} + \cdots + a_0 y(t) = b_m \frac{\mathrm{d}^m u(t)}{\mathrm{d}t^m} + \cdots + b_0 u(t) \quad (3.116)$$

where we suppose that $u(t)$ and $y(t)$ are deviation variables. After taking the Laplace transform and assuming zero initial conditions we get

$$(a_n s^n + a_{n-1} s^{n-1} + \cdots + a_0) Y(s) = (b_m s^m + \cdots + b_0) U(s) \quad (3.117)$$

or

$$G(s) = \frac{Y(s)}{U(s)} = \frac{B(s)}{A(s)} \quad (3.118)$$

where

$$B(s) = b_m s^m + b_{m-1} s^{m-1} + \cdots + b_0$$
$$A(s) = a_n s^n + a_{n-1} s^{n-1} + \cdots + a_0$$

$G(s)$ is called a *transfer function* of the system and is defined as the ratio between the Laplace transforms of output and input with zero initial conditions.

Note 3.2. Transfer functions use the variable s of the Laplace transform. Introducing the *derivation operator* $p = \mathrm{d}/\mathrm{d}t$ then the relation

$$G(p) = \frac{Y(p)}{U(p)} = \frac{B(p)}{A(p)} \tag{3.119}$$

is only another way of writing Eq. (3.116).

A transfer function $G(s)$ corresponds to physical reality if

$$n \geq m \tag{3.120}$$

Consider the case when this condition is not fulfilled, when $n = 1, m = 0$

$$a_0 y = b_1 \frac{\mathrm{d}u}{\mathrm{d}t} + b_0 u \tag{3.121}$$

If $u(t) = 1(t)$ (step change) then the system response is given as a sum of two functions. The first function is an impulse function and the second is a step function. As any real process cannot show on output impulse behaviour, the case $n < m$ does not occur in real systems and the relation (3.120) is called the condition of *physical realisability.*

The relation

$$Y(s) = G(s)U(s) \tag{3.122}$$

can be illustrated by the block scheme shown in Fig. 3.9 where the block corresponds to $G(s)$. The scheme shows that if input to system is $U(s)$ then output is the function $G(s)U(s)$.

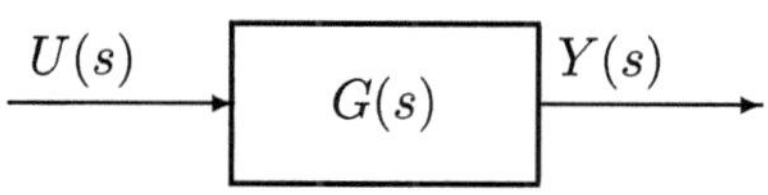

Fig. 3.9. Block scheme of a system with transfer function $G(s)$

Example 3.9: *Transfer function of a liquid storage system*

Consider the tank shown in Fig. 2.1. State-space equations for this system are of the form (see 2.5)

$$\frac{\mathrm{d}x_1}{\mathrm{d}t} = a_{11}x_1 + b_{11}u$$
$$y_1 = x_1$$

where $x_1 = h - h^s, u = q_0 - q_0^s$. After taking the Laplace transform and considering the fact that $x_1(0) = 0$ follows

$$sX_1(s) = a_{11}X_1(s) + b_{11}U(s)$$
$$Y_1(s) = X_1(s)$$

and

$$(s - a_{11})Y_1(s) = b_{11}U(s)$$

Hence, the transfer function of this process is

$$G_1(s) = \frac{b_0}{a_1 s + 1} = \frac{Z_1}{T_1 s + 1}$$

where $a_1 = T_1 = (2F\sqrt{h^s})/k_{11}$, $b_0 = Z_1 = (2\sqrt{h^s})/k_{11}$. T_1 is time constant and Z_1 gain of the first order system.

Example 3.10: *Two tanks in a series - transfer function*
Consider two tanks shown in Fig. 3.10. The level h_1 is not influenced by the level h_2.

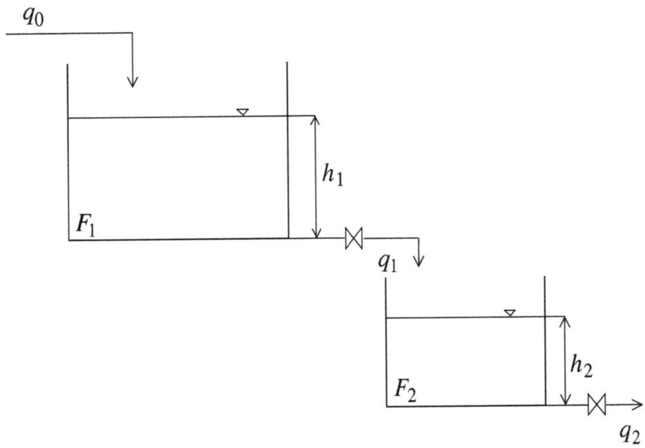

Fig. 3.10. Two tanks in a series

The dynamical properties of the first tank can be described as

$$F_1 \frac{dh_1}{dt} = q_0 - q_1$$

and the output equation is of the form

$$q_1 = k_{11}\sqrt{h_1}$$

The dynamical properties can also be written as

$$\frac{dh_1}{dt} = f_1(h_1, q_0)$$

where

$$f_1(h_1, q_0) = -\frac{k_{11}}{F_1}\sqrt{h_1} + \frac{1}{F_1}q_0$$

For the steady-state follows

$$0 = -\frac{k_{11}}{F_1}\sqrt{h_1^s} + \frac{1}{F_1}q_0^s$$
$$q_1^s = k_{11}\sqrt{h_1^s}$$

Linearised dynamical properties in the neighbourhood of the steady-state are of the form

$$\frac{dh_1}{dt} = \frac{d(h_1 - h_1^s)}{dt} = -\frac{k_{11}}{2F_1\sqrt{h_1^s}}(h_1 - h_1^s) + \frac{1}{F_1}(q_0 - q_0^s)$$

and linearised output equation is

$$q_1 - q_1^s = \frac{k_{11}}{2\sqrt{h_1^s}}(h_1 - h_1^s)$$

Let us introduce deviation variables

$$x_1 = h_1 - h_1^s$$
$$u = q_0 - q_0^s$$
$$y_1 = q_1 - q_1^s$$

Linear state-space model of the first tank is

$$\frac{dx_1}{dt} = a_{11}x_1 + b_{11}u$$
$$y_1 = c_{11}x_1$$

where

$$a_{11} = -\frac{k_{11}}{2F_1\sqrt{h_1^s}}, \quad b_{11} = \frac{1}{F_1}, \quad c_{11} = \frac{k_{11}}{2\sqrt{h_1^s}}.$$

After applying the Laplace transform to these equations and using the fact that initial conditions are zero we obtain

$$sX_1(s) = a_{11}X_1(s) + b_{11}U(s)$$
$$Y(s) = c_{11}X_1(s)$$

or

$$(s - a_{11})Y(s) = c_{11}b_{11}U(s)$$

The first tank transfer function $G_1(s)$ is

$$G_1(s) = \frac{Y_1(s)}{U(s)} = \frac{1}{a_1 s + 1} = \frac{1}{T_1 s + 1}$$

where $a_1 = T_1 = (2F\sqrt{h_1^s})/k_{11}$ and the gain is equal to one.

The transfer function of the second tank $G_2(s)$ can be derived when considering deviation variables $x_2 = y = h_2 - h_2^s$ and the relation $q_2 = k_{22}\sqrt{h_2}$ and is given as

$$G_2(s) = \frac{Y(s)}{Y_1(s)} = \frac{Z_2}{T_2 s + 1}$$

where $T_2 = (2F_2\sqrt{h_2^s})/k_{22}$, $Z_2 = (2\sqrt{h_2^s})/k_{22}$. The output $Y(s)$ can be written as

$$Y(s) = \frac{Z_2}{T_2 s + 1} Y_1(s)$$

$$Y(s) = \frac{Z_2}{T_2 s + 1} \frac{1}{T_1 s + 1} U(s)$$

The overall transfer function of both tanks in a series is then

$$G(s) = \frac{Y(s)}{U(s)} = \frac{Z_2}{T_2 s + 1} \frac{1}{T_1 s + 1}$$

Block scheme of this system is shown in Fig. 3.11.

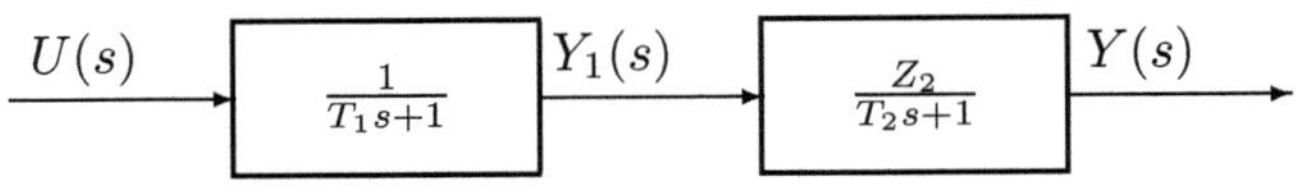

Fig. 3.11. Block scheme of two tanks in a series

Note 3.3. The example given above shows serial connection of two systems where the second system does not influence the behaviour of the first system. We can speak about "one-way" effect.

When the systems influence each other, the overall transfer function cannot be obtained as a product of transfer functions of subsystems. This is shown in the next example dealing with the interacting two tanks in a series (See Fig. 2.2). Mathematical model of this system described by equations (2.21) and (2.22) can be linearised in the neighbourhood of the steady-state given by flow rate q_0^s and levels h_1^s, h_2^s as

$$\frac{\mathrm{d}h_1}{\mathrm{d}t} = \frac{\mathrm{d}(h_1 - h_1^s)}{\mathrm{d}t} = \frac{1}{F_1}(q_0 - q_0^s) - \frac{k_{11}}{2F_1\sqrt{h_1^s - h_2^s}}[(h_1 - h_1^s) - (h_2 - h_2^s)]$$

$$\frac{\mathrm{d}h_2}{\mathrm{d}t} = \frac{\mathrm{d}(h_2 - h_2^s)}{\mathrm{d}t} = \frac{k_{11}}{2F_2\sqrt{h_1^s - h_2^s}}[(h_1 - h_1^s) - (h_2 - h_2^s)] - \frac{k_{22}}{2F_2\sqrt{h_2^s}}(h_2 - h_2^s)$$

Introducing deviation variables

$$x_1 = h_1 - h_1^s, \quad u = q_0 - q_0^s, \quad y = x_2 = h_2 - h_2^s$$

yields the linear model

$$\frac{\mathrm{d}x_1}{\mathrm{d}t} = a_{11}x_1 + a_{12}x_2 + b_{11}u$$
$$\frac{\mathrm{d}x_2}{\mathrm{d}t} = a_{21}x_1 + a_{22}x_2$$
$$y = x_2$$

where

$$a_{11} = -\frac{k_{11}}{2F_1\sqrt{h_1^s - h_2^s}}, \ a_{12} = -a_{11}, \qquad b_{11} = \frac{1}{F_1}$$
$$a_{21} = -\frac{k_{11}}{2F_2\sqrt{h_1^s - h_2^s}}, \ a_{22} = -a_{21} - \frac{k_{22}}{2F_2\sqrt{h_2^s}},$$

Taking the Laplace transform yields

$$sX_1(s) = a_{11}X_1(s) + a_{12}X_2(s) + b_{11}U(s)$$
$$sX_2(s) = a_{21}X_1(s) + a_{22}X_2(s)$$
$$Y(s) = X_2(s)$$

or

$$(s^2 - (a_{11} + a_{22})s + (a_{11}a_{22} - a_{12}a_{21}))Y(s) = a_{21}b_{11}U(s)$$

and hence the transfer function is given as

$$G(s) = \frac{Y(s)}{U(s)} = \frac{b_0}{a_2s^2 + a_1s + 1}$$

where

$$b_0 = \frac{a_{21}b_{11}}{a_{11}a_{22} - a_{12}a_{21}}$$
$$a_2 = \frac{1}{a_{11}a_{22} - a_{12}a_{21}}$$
$$a_1 = -\frac{a_{11} + a_{22}}{a_{11}a_{22} - a_{12}a_{21}}$$

Example 3.11: *n tanks in a series - transfer function*

Assume n tanks in a series as shown in Fig. 3.12 and the corresponding block scheme in Fig. 3.13. The variable $U(s)$ denotes the Laplace transform of $u(t) = q_0(t) - q_0^s$, $Y_i(s)$ are the Laplace transforms of $y_i(t) = q_i(t) - q_0^s$, $i = 1 \dots n-1$, $Y(s)$ is the Laplace transform of $y(t) = h_n(t) - h_n^s$. $T_1, T_2, \dots, T_n$ are time constants and Z_n is gain.

Similarly as in the case of the two tanks without interaction, the partial input and output variables are tied up with the following relations

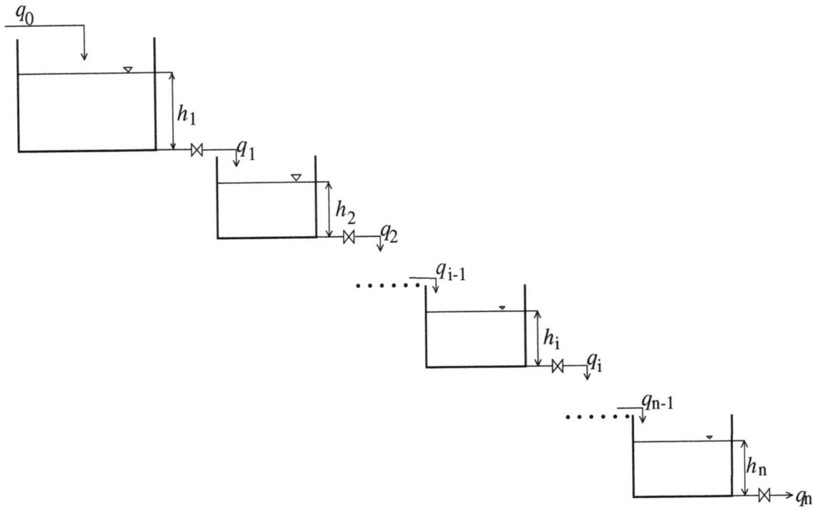

Fig. 3.12. Serial connection of n tanks

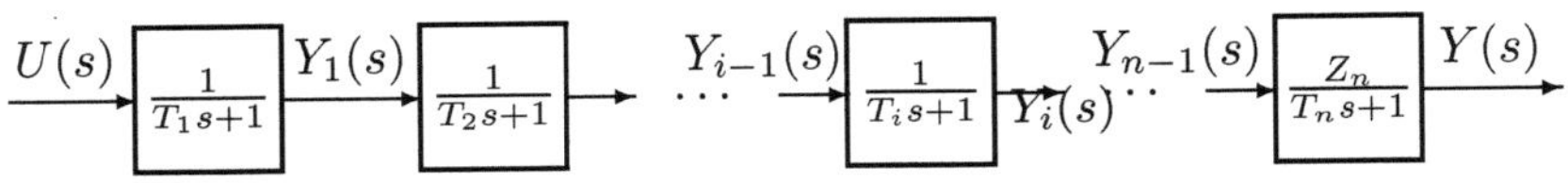

Fig. 3.13. Block scheme of n tanks in a series

$$
\begin{aligned}
Y_1(s) &= \frac{1}{T_1 s+1} U(s) \\
Y_2(s) &= \frac{1}{T_2 s+1} Y_1(s) \\
&\vdots \\
Y_i(s) &= \frac{1}{T_i s+1} Y_{i-1}(s) \\
&\vdots \\
Y(s) &= \frac{Z_n}{T_n s+1} Y_{n-1}(s)
\end{aligned}
$$

The overall input variable is $U(s)$ and the overall output variable is $Y(s)$. The overall transfer function is then

$$
G(s) = \frac{Y(s)}{U(s)} = \frac{Z_n}{\prod_{i=1}^{n}(T_i s+1)}
$$

Simplified block scheme of this system is shown in Fig. 3.14.

Example 3.12: *U-tube : transfer function*
Mathematical model of the U-tube shown in Fig. 3.5 is of the form

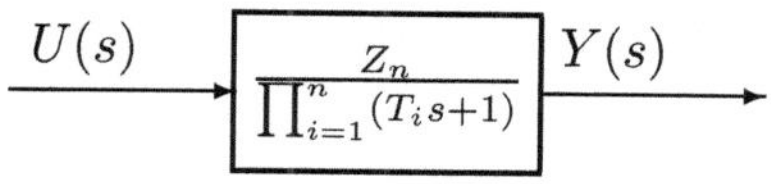

Fig. 3.14. Simplified block scheme of n tanks in a series

$$\frac{L}{2g}\frac{\mathrm{d}^2h}{\mathrm{d}t^2} + \frac{k}{2g\rho}\frac{\mathrm{d}h}{\mathrm{d}t} + h = \frac{1}{2g\rho}p_v$$

Steady-state is determined by level of liquid $h = h^s = 0$. We denote the output deviation variable as $y = h$ and the input deviation variable as $u = h_v = p_v/2g\rho$. Further let us introduce $1/\omega_k^2 = L/2g, 2\zeta/\omega_k = k/2g\rho$ where ω_k is a critical frequency and ζ is a damping coefficient. The terms critical frequency and damping coefficient will become clear from analysis of the solution of the differential equation describing dynamical properties of the U-tube. Mathematical model can be then rewritten as

$$\frac{1}{\omega_k^2}\frac{\mathrm{d}^2y}{\mathrm{d}t^2} + 2\frac{\zeta}{\omega_k}\frac{\mathrm{d}y}{\mathrm{d}t} + y = u$$

and the corresponding transfer function as

$$G(s) = \frac{Y(s)}{U(s)} = \frac{1}{T_k^2s^2 + 2\zeta T_k s + 1}$$

where $T_k = 1/\omega_k$.

Note 3.4. Mathematical model of the U-tube shows that step function on input can result in an oscillatory response. Therefore, U-tube is able of to produce its own oscillations. This is in contrast to other systems of the second order that can be decomposed into two systems of the first order and cannot produce the oscillations.

Example 3.13: *Heat exchanger - transfer function*

Mathematical model of a heat exchanger was developed in the Section 2.2 and was shown to be in the form

$$T_1\frac{\mathrm{d}y_1}{\mathrm{d}t} + y_1 = Z_1u_1 + Z_2r_1$$

where $y_1 = \vartheta = \vartheta^s$, $u_1 = \vartheta_p - \vartheta_p^s$, $r_1 = \vartheta_v - \vartheta_v^s$ and T_1, T_2, Z_2 are constants. The output variable is the natural state variable $y_1 = x_1$.

To determine the heat exchanger response to the change of inlet temperature ϑ_v it is necessary to set $u_1 = 0$ and analogously if response of the process to the jacket temperature change is desired then $r_1 = 0$. The variable u_1 is usually assumed to be a manipulated variable and r_1 acts as a disturbance.

Taking the Laplace transform and considering zero initial conditions yields

$$(T_1 s + 1)Y_1(s) = Z_1 U_1(s) + Z_2 R_1(s)$$

if $R_1(s) = 0$ then

$$G_1(s) = \frac{Y_1(s)}{U_1(s)} = \frac{Z_1}{T_1 s + 1}$$

if $U_1(s) = 0$ then

$$G_2(s) = \frac{Y_1(s)}{R_1(s)} = \frac{Z_2}{T_1 s + 1}$$

$Y_1(s)$ can be written as

$$Y_1(s) = \frac{Z_1}{T_1 s + 1} U_1(s) + \frac{Z_2}{T_1 s + 1} R_1(s)$$

$$Y_1(s) = G_1(s) U_1(s) + G_2(s) R_1(s)$$

Block scheme of this process is shown in Fig. 3.15 or in Fig. 3.16 where r_1 is moved from output to input to the system. This has an importance in design of control systems because modified block scheme simplifies some considerations.

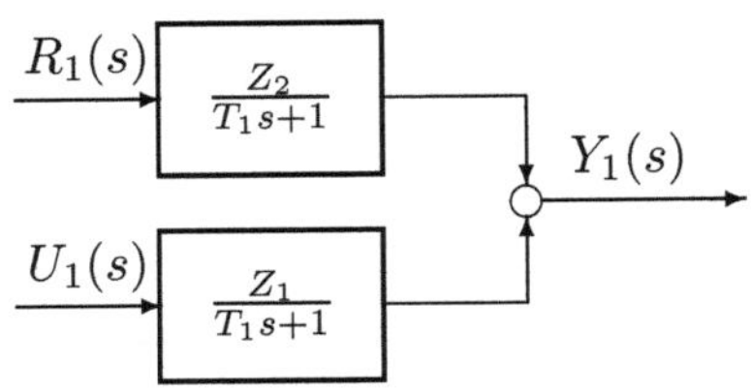

Fig. 3.15. Block scheme of a heat exchanger

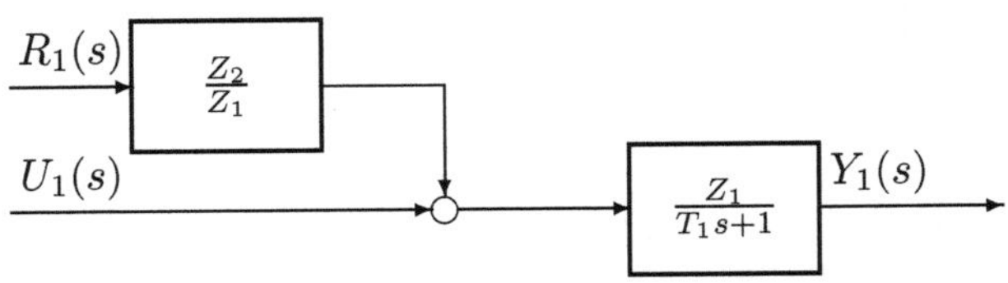

Fig. 3.16. Modified block scheme of a heat exchanger

Example 3.14: *CSTR - transfer function*

Consider the CSTR shown in Fig. 2.11. Let us introduce deviation variables

$$y_1 = x_1 = c_A - c_A^s$$

$$y_2 = x_2 = \vartheta - \vartheta^s$$

$$u_1 = c_{Av} - c_{Av}^s$$

$$u_2 = \vartheta_c - \vartheta_c^s$$

The linearised mathematical model is then of the form

$$\frac{dx_1}{dt} = a_{11}x_1 + a_{12}x_2 + b_{11}u_1$$

$$\frac{dx_2}{dt} = a_{21}x_1 + a_{22}x_2 + b_{22}u_2$$

Compared to the Example 2.6, $b_{11} = h_{11}, b_{22} = b_{21}$ and inlet temperature is assumed to be constant. We define the following transfer functions

$$G_{11}(s) = \frac{Y_1(s)}{U_1(s)} \quad G_{12}(s) = \frac{Y_1(s)}{U_2(s)}$$

$$G_{21}(s) = \frac{Y_2(s)}{U_1(s)} \quad G_{22}(s) = \frac{Y_2(s)}{U_2(s)}$$

Taking the Laplace transform of linearised mathematical model follows

$$sX_1(s) = a_{11}X_1(s) + a_{12}X_2(s) + b_{11}U_1(s)$$

$$sX_2(s) = a_{21}X_1(s) + a_{22}X_2(s) + b_{22}U_2(s)$$

The transfer function $G_{11}(s)$ can be derived if $U_2(s) = 0$. Analogously, other transfer functions can also be obtained.

$$G_{11}(s) = \frac{b_{11}s - a_{22}b_{11}}{s^2 - (a_{11} + a_{22})s + (a_{11}a_{22} - a_{12}a_{21})}$$

$$G_{12}(s) = \frac{a_{12}b_{22}}{s^2 - (a_{11} + a_{22})s + (a_{11}a_{22} - a_{12}a_{21})}$$

$$G_{21}(s) = \frac{a_{21}b_{11}}{s^2 - (a_{11} + a_{22})s + (a_{11}a_{22} - a_{12}a_{21})}$$

$$G_{22}(s) = \frac{b_{22}s - a_{11}b_{22}}{s^2 - (a_{11} + a_{22})s + (a_{11}a_{22} - a_{12}a_{21})}$$

3.3.2 Transfer Functions of Systems with Time Delays

Consider a process described by the differential equation

$$\frac{\partial x_1(\sigma,t)}{\partial t} + v_\sigma \frac{\partial x_1(\sigma,t)}{\partial \sigma} = 0, \; x_1(\sigma,0) = 0 \tag{3.123}$$

This equation is a description of a mathematical model of the double pipe heat exchanger shown in Fig. 2.5 with $\alpha = 0$ and x_1 being deviation temperature in the inner pipe. The process input variable is

$$u(t) = x_1(0,t) = x_1^0(t) \tag{3.124}$$

and the output variable is defined as

$$y(t) = x_1(L,t) \tag{3.125}$$

The system defined with the above equations is called *pure time delay*.

After taking the Laplace transform with argument t we get

$$v_\sigma \frac{\partial X_1(\sigma,s)}{\partial \sigma} + sX_1(\sigma,s) = 0 \tag{3.126}$$

where

$$X_1(\sigma, s) = \int_0^\infty x_1(\sigma, t)\mathrm{e}^{-st}\mathrm{d}t \tag{3.127}$$

Applying the Laplace transform with argument σ yields

$$v_\sigma q\bar{X}_1(q, s) - v_\sigma U(s) + s\bar{X}_1(q, s) = 0 \tag{3.128}$$

where

$$\bar{X}_1(q, s) = \int_0^\infty X_1(\sigma, s)\mathrm{e}^{-q\sigma}\mathrm{d}\sigma \tag{3.129}$$

$$U(s) = X_1(0, s) \tag{3.130}$$

From Eq. (3.128) follows

$$\bar{X}_1(q, s) = \frac{1}{q + \frac{s}{v_\sigma}} U(s) \tag{3.131}$$

This equation can be transformed back into σ domain

$$X_1(\sigma, s) = \mathrm{e}^{-\frac{\sigma}{v_\sigma}s} U(s) \tag{3.132}$$

The corresponding transfer function of pure time delay for any $\sigma \in [0, L]$ is

$$G_{d\sigma} = \frac{X_1(\sigma, s)}{U(s)} = \mathrm{e}^{-\frac{\sigma}{v_\sigma}s} \tag{3.133}$$

and for $\sigma = L$

$$G_d = \frac{Y(s)}{U(s)} = \mathrm{e}^{-T_d s} \tag{3.134}$$

where $T_d = L/v_\sigma$.

Let us now consider only part of the process of length $\Delta\sigma$ that is perfectly mixed. The equation (3.123) can be approximated as

$$\frac{\mathrm{d}x_1(\Delta\sigma, t)}{\mathrm{d}t} = v_\sigma \frac{-x_1(\Delta\sigma, t) + u(t)}{\Delta\sigma} \tag{3.135}$$

This equation after taking the Laplace transform is of the form

$$\frac{X_1(\Delta\sigma, t)}{U(s)} = \frac{1}{1 + \frac{\Delta\sigma}{v_\sigma}s} \tag{3.136}$$

Because the term $\mathrm{e}^{\frac{\Delta\sigma}{v_\sigma}s}$ can be written as

$$\frac{1}{\mathrm{e}^{\frac{\Delta\sigma}{v_\sigma}s}} = \frac{1}{1 + \frac{\Delta\sigma}{v_\sigma}s + \frac{1}{2}\frac{\Delta\sigma^2}{v_\sigma^2}s^2 + \cdots} \tag{3.137}$$

then the right hand side term of (3.136) can be viewed as the first order approximation of time delay term.

Let us further assume mass balance of the process of length $\Delta\sigma$ of the form

$$\frac{\mathrm{d}}{\mathrm{d}t}\left(\frac{u(t)+x_1(\Delta\sigma,t)}{2}\right) = v_\sigma \frac{-x_1(\Delta\sigma,t)+u(t)}{\Delta\sigma} \tag{3.138}$$

Taking the Laplace transform becomes

$$\frac{X_1(\Delta\sigma,t)}{U(s)} = \frac{1-\frac{1}{2}\frac{\Delta\sigma}{v_\sigma}s}{1+\frac{1}{2}\frac{\Delta\sigma}{v_\sigma}s} \tag{3.139}$$

This equation can be understood as the first order Pade approximation of time delay term. Similarly for the second order Pade approximation yields

$$\frac{X_1(\Delta\sigma,t)}{U(s)} = \frac{1-\frac{1}{2}\frac{\Delta\sigma}{v_\sigma}s+\frac{1}{12}\frac{\Delta\sigma^2}{v_\sigma^2}s^2}{1+\frac{1}{2}\frac{\Delta\sigma}{v_\sigma}s+\frac{1}{12}\frac{\Delta\sigma^2}{v_\sigma^2}s^2} \tag{3.140}$$

Example 3.15: *Double-pipe heat exchanger - transfer functions*

Consider the heat exchanger shown in Fig. 2.5. It can be described by the following differential equation

$$T_1\frac{\partial\vartheta(\sigma,t)}{\partial t} + v_\sigma T_1\frac{\partial\vartheta(\sigma,t)}{\partial\sigma} + \vartheta(\sigma,t) = \vartheta_p(t)$$

Assume boundary and initial conditions of the form

$$\begin{aligned}
\vartheta(0,t) &= \vartheta^0(t)\\
\vartheta(\sigma,0) &= \vartheta^s(\sigma) = \vartheta_p^s - (\vartheta_p^s - \vartheta_p^{0s})\mathrm{e}^{-\frac{\sigma}{v_\sigma T_1}}\\
\vartheta_p(t) &= \vartheta_p^s,\ t<0\\
\vartheta^0(t) &= \vartheta^{0s},\ t<0
\end{aligned}$$

and deviation variables

$$\begin{aligned}
x_1(\sigma,t) &= \vartheta(\sigma,t) - \vartheta^s(\sigma)\\
u_1(t) &= \vartheta_p(t) - \vartheta_p^s\\
u_2(t) &= \vartheta^0(t) - \vartheta^{0s}
\end{aligned}$$

The differential equation of the heat exchanger then becomes

$$T_1\frac{\partial x_1(\sigma,t)}{\partial t} + v_\sigma T_1\frac{\partial x_1(\sigma,t)}{\partial\sigma} + x_1(\sigma,t) = u_1(t)$$

with boundary and initial conditions

$$\begin{aligned}
x_1(0,t) &= u_2(t)\\
x_1(\sigma,0) &= 0\\
u_1(t) &= 0,\ t<0\\
u_2(t) &= 0,\ t<0
\end{aligned}$$

Taking the Laplace transform with an argument t yields

$$(T_1 s + 1)X_1(\sigma, s) + v_\sigma T_1 \frac{\partial X_1(\sigma, s)}{\partial \sigma} = U_1(s)$$

where

$$X_1(\sigma, s) = \int_0^\infty x_1(\sigma, t)\mathrm{e}^{-st}\mathrm{d}t$$

The second Laplace transform gives

$$(T_1 s + v_\sigma T_1 q + 1)\bar{X}_1(q, s) = \frac{1}{q}U_1(s) + v_\sigma T_1 U_2(s)$$

where

$$\bar{X}_1(q, s) = \int_0^\infty X_1(\sigma, s)\mathrm{e}^{-q\sigma}\mathrm{d}\sigma$$
$$U_2(s) = X_1(0, s)$$

$\bar{X}_1(q, s)$ can be written as

$$\bar{X}_1(q, s) = \frac{1}{T_1 s + 1}\frac{a}{q(q + a)}U_1(s) + \frac{1}{q + a}U_2(s)$$

where $a = (T_1 s + 1)/v_\sigma T_1$.

The inverse Laplace transform according to σ gives

$$X_1(\sigma, s) = \frac{1}{T_1 s + 1}\left(1 - \mathrm{e}^{-\frac{T_1 s + 1}{v_\sigma T_1}\sigma}\right)U_1(s) + \mathrm{e}^{-\frac{T_1 s + 1}{v_\sigma T_1}\sigma}U_2(s)$$

which shows that the transfer functions are of the form

$$G_{1k} = \frac{Y_k(s)}{U_1(s)} = \frac{1}{T_1 s + 1}\left(1 - \mathrm{e}^{-\frac{\sigma_k}{v_\sigma T_1}}\mathrm{e}^{-\frac{\sigma_k}{v_\sigma}s}\right)$$
$$G_{2k} = \frac{Y_k(s)}{U_2(s)} = \mathrm{e}^{-\frac{\sigma_k}{v_\sigma T_1}}\mathrm{e}^{-\frac{\sigma_k}{v_\sigma}s}$$

where

$$Y_k(s) = X_1(\sigma_k, s),\ k = 1, 2, \ldots, r;\ 0 \le \sigma_k \le L$$

Block scheme of the double-pipe heat exchanger is shown in Fig. 3.17.

3.3.3 Algebra of Transfer Functions for SISO Systems

Investigation of block schemes reveals the fact that all schemes can be decomposed into 3 basic connections: serial, parallel, and feedback. The rules that enable to determine the overall transfer function of a system composed from basic blocks are called *algebra of transfer functions.*

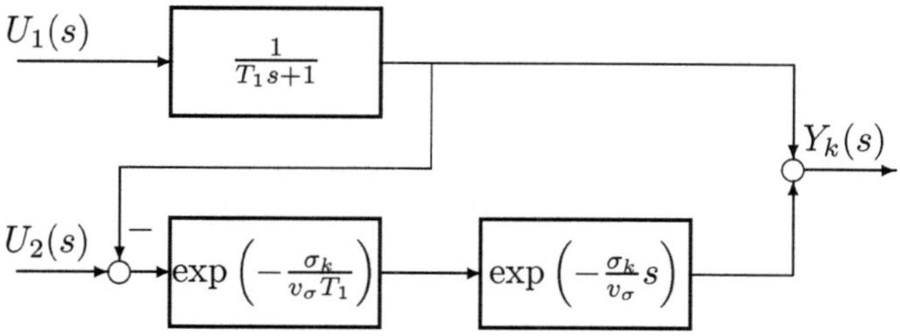

Fig. 3.17. Block scheme of a double-pipe heat exchanger

Serial Connection

Serial connection results in a situation when the output variable of the first block is the input variable of the second block (Fig. 3.18). The overall transfer function can be written as

$$G(s) = G_1(s)G_2(s) \tag{3.141}$$

Fig. 3.18. Serial connection

Generally when n blocks are connected in series, the transfer function is given as a product of partial transfer functions:

$$G(s) = G_1(s)G_2(s)\dots G_n(s) \tag{3.142}$$

Parallel Connection

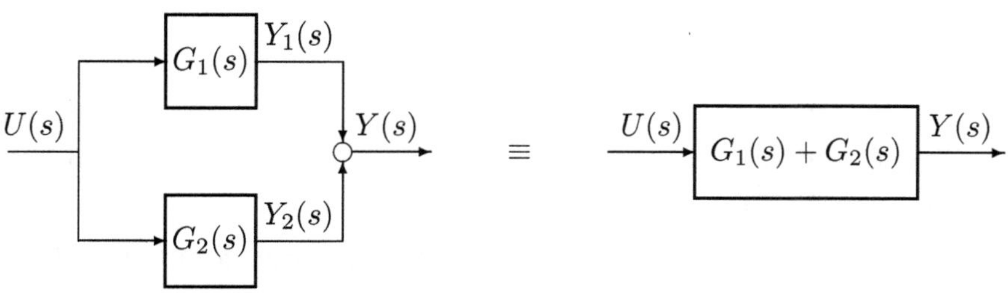

Fig. 3.19. Parallel connection

Parallel connection is characterised by one input variable for all systems. Output variable is given as the sum of partial outputs (Fig. 3.19). Parallel connection is characterised by the equations

$$Y_1(s) = G_1(s)U(s) \tag{3.143}$$

$$Y_2(s) = G_2(s)U(s) \tag{3.144}$$

$$Y(s) = Y_1(s) + Y_2(s) \tag{3.145}$$

Substituting $Y_1(s)$ from (3.143) and $Y_2(s)$ from (3.144) into (3.145) yields

$$Y(s) = [G_1(s) + G_2(s)]U(s) = G(s)U(s) \tag{3.146}$$

and

$$G(s) = G_1(s) + G_2(s) \tag{3.147}$$

In general, the overall transfer functions is given as the sum of partial transfer functions

$$G(s) = \sum_{i=1}^{n} G_i(s) \tag{3.148}$$

Feedback Connection

Feedback connection of two blocks results when output variables of each block are fed back as the input of the other block (Fig. 3.20).

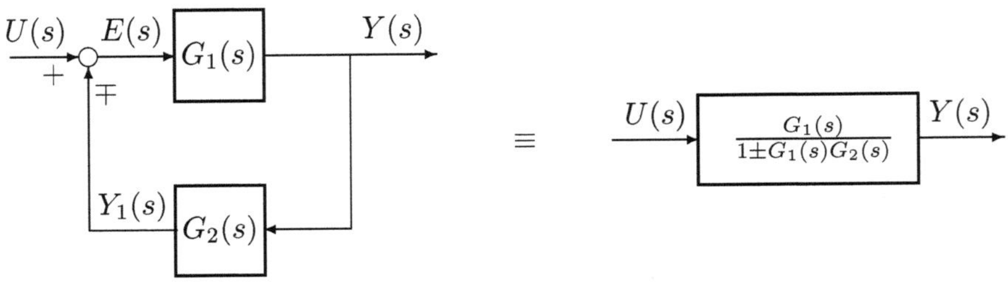

Fig. 3.20. Feedback connection

For the feedback connection holds

$$Y(s) = G_1(s)E(s) \tag{3.149}$$

$$Y_1(s) = G_2(s)Y(s) \tag{3.150}$$

$$E(s) = U(s) \mp Y_1(s) \tag{3.151}$$

The minus sign in the Eq. (3.151) corresponds to negative feedback and the plus sign to positive feedback. From these equations follow

$$Y_1(s) = G_1(s)G_2(s)E(s) \tag{3.152}$$

$$E(s) = \frac{1}{1 \pm G_1(s)G_2(s)}U(s) \tag{3.153}$$

$$Y(s) = \frac{G_1(s)}{1 \pm G_1(s)G_2(s)}U(s) = G(s)U(s) \tag{3.154}$$

The overall transfer function is then given as

$$G(s) = \frac{G_1(s)}{1 \pm G_1(s)G_2(s)} \tag{3.155}$$

The feedback transfer function is a ratio with the numerator given as the transfer function between the input and output signals and with the denominator given as a sum (negative feedback) or difference (positive feedback) of 1 and transfer function of the corresponding open-loop system.

Rule for Moving of the Branching Point

When the branching point is moved against the direction of the previous signal then the moved branch must contain all blocks which are between the original and new branching point (Fig. 3.21).

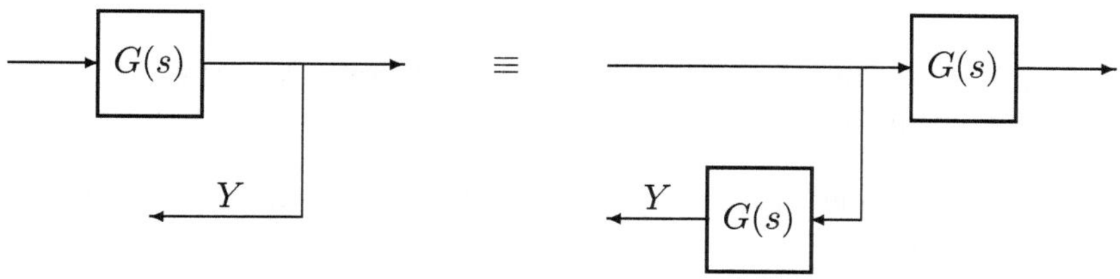

Fig. 3.21. Moving of the branching point against the direction of signals

The opposite situation is when the branching point is moved in the direction of the signal flow. In this case the moved branch contains inverses of the relevant blocks (Fig. 3.22).

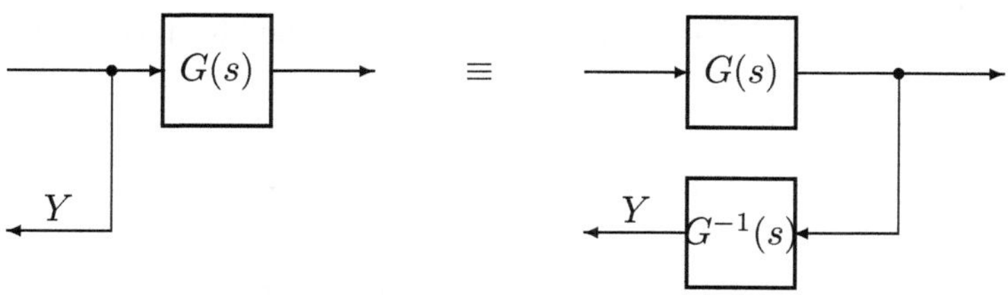

Fig. 3.22. Moving of the branching point in the direction of signals

Rule for Moving of the Summation Point

Moving of the summation point is an inverse action to moving of the branching point. The rules are shown in Figs. 3.23 and 3.24.

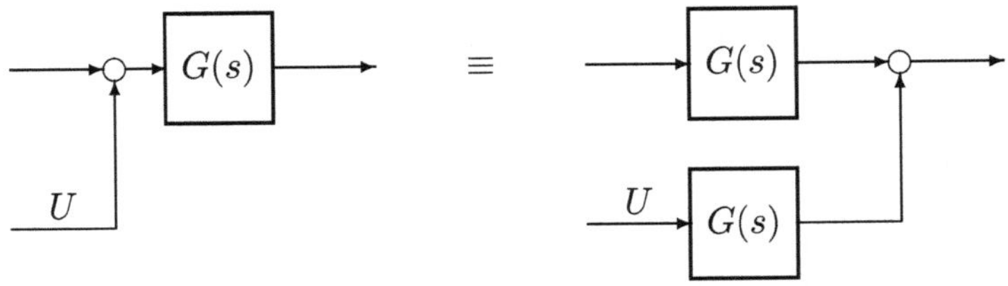

Fig. 3.23. Moving of the summation point in the direction of signals

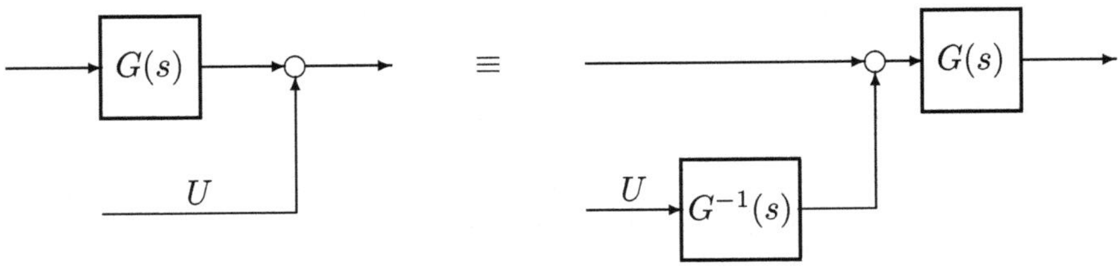

Fig. 3.24. Moving of the summation point against the direction of signals

3.3.4 Input Output Models of MIMO Systems - Matrix of Transfer Functions

The standard and natural description of MIMO systems is in the form of state-space equations

$$\frac{\mathrm{d}\boldsymbol{x}(t)}{\mathrm{d}t} = \boldsymbol{A}\boldsymbol{x}(t) + \boldsymbol{B}\boldsymbol{u}(t) \tag{3.156}$$

$$\boldsymbol{y}(t) = \boldsymbol{C}\boldsymbol{x}(t) + \boldsymbol{D}\boldsymbol{u}(t) \tag{3.157}$$

where $\boldsymbol{x}[n \times 1]$ is the vector of state variables, $\boldsymbol{u}[m \times 1]$ is the vector of input variables, $\boldsymbol{y}[r \times 1]$ is the vector of output variables, and $\boldsymbol{A}, \boldsymbol{B}, \boldsymbol{C}, \boldsymbol{D}$ are constant matrices of appropriate dimensions.

When all variables are deviation variables and $\boldsymbol{x}(0) = \boldsymbol{0}$ then the input-output (I/O) properties of this system can be determined from the convolution multiplication

$$\boldsymbol{y}(t) = \int_0^t \boldsymbol{g}(t-\tau)\boldsymbol{u}(\tau)\mathrm{d}\tau \tag{3.158}$$

where $\boldsymbol{g}(t)$ is *matrix of impulse responses* of the $[r \times m]$ system and is given as

$$\boldsymbol{g}(t) = \begin{cases} \boldsymbol{0} & t < 0 \\ \boldsymbol{C}\mathrm{e}^{\boldsymbol{A}t}\boldsymbol{B} + \boldsymbol{D}\delta(t) & t \geq 0 \end{cases} \tag{3.159}$$

and $\delta(t)$ is the Dirac delta function.

Consider now the system given by Eqs. (3.156), (3.157). Taking the Laplace transform yields

$$\boldsymbol{Y}(s) = \left(\boldsymbol{C}(s\boldsymbol{I} - \boldsymbol{A})^{-1}\boldsymbol{B} + \boldsymbol{D}\right)\boldsymbol{U}(s) \tag{3.160}$$

or

$$\boldsymbol{Y}(s) = \boldsymbol{G}(s)\boldsymbol{U}(s) \tag{3.161}$$

where

$$\boldsymbol{G}(s) = \boldsymbol{C}(s\boldsymbol{I} - \boldsymbol{A})^{-1}\boldsymbol{B} + \boldsymbol{D} \tag{3.162}$$

is $[r \times m]$ *transfer function matrix* of linear continuous system with constant coefficients. This matrix is the Laplace transform of matrix $\boldsymbol{g}(t)$

$$\boldsymbol{G}(s) = \int_0^\infty \boldsymbol{g}(t)\mathrm{e}^{-st}\mathrm{d}t \tag{3.163}$$

The matrix $\boldsymbol{G}(s)$ derived from the original state-space model is the same as the transfer function matrix of controllable and observable part of this system. The noncontrollable and nonobservable modes of the system are cancelled in the process of transformation from state-space models into I/O models. Often, there are tasks of the inverse transformation from I/O to state-space (SS) model. It must be emphasised that one I/O model corresponds to an infinite number of state-space models. We then speak about a *state-space realisation of I/O model. Minimum realisation* fulfils the properties of controllability and observability. Hence an unambiguous relation between $\boldsymbol{G}(s)$ and its state-space realisation exists only if the state-space model is minimal.

If we deal with SISO systems we can write

$$G(s) = \boldsymbol{C}(s\boldsymbol{I} - \boldsymbol{A})^{-1}\boldsymbol{B} + \boldsymbol{D} = \frac{B(s)}{A(s)} \tag{3.164}$$

If the state-space model is the minimal realisation of $G(s)$ then

$$\det(s\boldsymbol{I} - \boldsymbol{A}) = A(s) \tag{3.165}$$

The degree of characteristic polynomial $A(s)$ is equal to n where n is the number of states of state-space model. We call n as *system order.*

Any transfer function can be one of the following forms:

1.

$$G(s) = \frac{b_m s^m + b_{m-1}s^{m-1} + \cdots + b_0}{a_n s^n + a_{n-1}s^{n-1} + \cdots + a_0}, \quad n \geq m \tag{3.166}$$

2.

$$G(s) = \frac{b_m(s - s_{N1})(s - s_{N2})\dots(s - s_{Nm})}{a_n(s - s_1)(s - s_2)\dots(s - s_n)} \tag{3.167}$$

Roots of the characteristic polynomial $s_1, \dots, s_n$ are *system poles.* Roots of numerator polynomial $s_{N1}, \dots, s_{Nn}$ are *system zeros.*

3.

$$G(s) = Z_s \frac{(T_{N1}s+1)(T_{N2}s+1)\dots(T_{Nm}s+1)}{(T_1 s+1)(T_2 s+1)\dots(T_n s+1)} \tag{3.168}$$

where $T_1, \dots, T_n, T_{N1}, \dots, T_{Nm}$ are time constants and $Z_s = b_0/a_0$ is the system gain. This expression for transfer function can only be written if all poles and zeros are real. Time constants correspond to negative inverses of poles and zeros.

Transfer function matrix $\boldsymbol{G}(s)$ has dimensions $[r \times m]$. An element of this matrix $G_{kj}(s)$ is the transfer function corresponding to input u_j and output y_k.

$$G_{kj}(s) = \frac{Y_k(s)}{U_j(s)} \tag{3.169}$$

The matrix $\boldsymbol{G}(s)$ can also be written as

$$\boldsymbol{G}(s) = \frac{\boldsymbol{C}\mathrm{adj}(s\boldsymbol{I}-\boldsymbol{A})\boldsymbol{B} + \boldsymbol{D}d(s)}{d(s)} \tag{3.170}$$

where $d(s) = |s\boldsymbol{I} - \boldsymbol{A}|$.

As all elements of $\mathrm{adj}(s\boldsymbol{I}-\boldsymbol{A})$ are polynomials with a degree less than or equal to $n-1$ and polynomial $d(s)$ is of degree n then all transfer functions $G_{kj}(s)$ have a degree of numerator less than or equal to the degree of the denominator. $\boldsymbol{G}(s)$ is *proper rational function matrix*. When $\boldsymbol{D} = \boldsymbol{0}$ then all numerator degrees are less than the denominator degrees and $\boldsymbol{G}(s)$ is *strictly proper rational function matrix.*

Definition of proper and strictly proper transfer function matrix $\boldsymbol{G}(s)$: A rational matrix $\boldsymbol{G}(s)[r \times m]$ is proper if all its transfer functions satisfy $\lim_{|s|\to\infty} G_{kj}(s) < \infty$. A rational matrix $\boldsymbol{G}(s)$ is strictly proper if for all its elements hold $\lim_{|s|\to\infty} G_{kj}(s) = 0$. The numerator degree of a *proper SISO system* is smaller or equal to the denominator degree. The numerator degree of a *strictly proper SISO system* is smaller as the denominator degree.

Roots of polynomial $d(s)$ are poles of $\boldsymbol{G}(s)$. If no cancellation of roots occurs during the calculation of $\boldsymbol{G}(s)$ then the matrix poles are the same as *system poles.*

If all poles of $\boldsymbol{G}(s)$ are located in the left half plane of the complex plane then the *frequency transfer function matrix* that is defined as *Fourier transformation* of $\boldsymbol{g}(t)$ exists and can be obtained by the substitution $s = \mathrm{j}\omega$, i.e.

$$\boldsymbol{G}(\mathrm{j}\omega) = \boldsymbol{C}(\mathrm{j}\omega\boldsymbol{I} - \boldsymbol{A})^{-1}\boldsymbol{B} + \boldsymbol{D} \tag{3.171}$$

The Fourier transform is defined as

$$\boldsymbol{F}(\mathrm{j}\omega) \equiv \int_{-\infty}^{\infty} \boldsymbol{f}(t)\mathrm{e}^{-\mathrm{j}\omega t}\mathrm{d}t \tag{3.172}$$

$\boldsymbol{G}(\mathrm{j}\omega)$ is called the *frequency transfer function matrix*. The values of $\boldsymbol{G}(\mathrm{j}\omega)$ are for any real ω given as values of $\boldsymbol{G}(s)$ for $s = \mathrm{j}\omega$.

$$\boldsymbol{G}(\mathrm{j}\omega) = \boldsymbol{G}(s)|_{s=\mathrm{j}\omega} \tag{3.173}$$

This function can be introduced not only for stable but for arbitrary transfer functions. However, if $\boldsymbol{G}(s)$ has a root on imaginary axis $s_i = \mathrm{j}\beta_i$ then $\boldsymbol{G}(\mathrm{j}\omega)$ has an infinite value for $\omega = \beta_i$.

Example 3.16: *CSTR - transfer function matrix*

Consider the CSTR shown in Fig. 2.11 and assume the same notation as in the Example 3.14. The state-space model matrices are

$$\boldsymbol{A} = \begin{pmatrix} a_{11} & a_{12} \\ a_{21} & a_{22} \end{pmatrix}, \boldsymbol{B} = \begin{pmatrix} b_{11} & 0 \\ 0 & b_{22} \end{pmatrix}, \boldsymbol{C} = I_2$$

From (3.162) follows

$$\begin{aligned} \boldsymbol{G}(s) &= \boldsymbol{I}_2 \left[s\boldsymbol{I}_2 - \begin{pmatrix} a_{11} & a_{12} \\ a_{21} & a_{22} \end{pmatrix} \right]^{-1} \begin{pmatrix} b_{11} & 0 \\ 0 & b_{22} \end{pmatrix} \\ &= \begin{pmatrix} s - a_{11} & -a_{12} \\ -a_{21} & s - a_{22} \end{pmatrix}^{-1} \begin{pmatrix} b_{11} & 0 \\ 0 & b_{22} \end{pmatrix} \\ &= \frac{1}{(s - a_{11})(s - a_{22}) - a_{12}a_{21}} \begin{pmatrix} s - a_{22} & a_{12} \\ a_{21} & s - a_{11} \end{pmatrix} \begin{pmatrix} b_{11} & 0 \\ 0 & b_{22} \end{pmatrix} \\ &= \frac{1}{(s - a_{11})(s - a_{22}) - a_{12}a_{21}} \begin{pmatrix} b_{11}s - a_{22}b_{11} & a_{12}b_{22} \\ a_{21}b_{11} & b_{22}s - a_{11}b_{22} \end{pmatrix} \end{aligned}$$

The partial transfer functions of $\boldsymbol{G}(s)$ are the same as in the example 3.14.

3.3.5 BIBO Stability

BIBO stability plays an important role among different definitions of stability. The abbreviation stands for *Bounded Input, Bounded Output.* Roughly said, a system is BIBO stable if any bounded input gives a bounded output. This is also the reason why we sometimes speak about BIBO stability as of *external stability.*

Definition of BIBO stability: A linear continuous system with constant coefficients (3.156), (3.157) with zero initial state $\boldsymbol{x}(t_0)$ is BIBO stable if for all t_0 and for all inputs $\boldsymbol{u}(t)$ that are finite on $[t_0, \infty)$ is output $\boldsymbol{y}(t)$ also finite on $[t_0, \infty)$.

Theorem: BIBO stability. A linear continuous system with constant coefficients (3.156), (3.157) is BIBO stable if and only if

$$\int_0^\infty \|\boldsymbol{g}(\tau)\| \, \mathrm{d}\tau < \infty \tag{3.174}$$

where the norm is induced by the norm on u.

An alternate theorem about BIBO stability states: A linear continuous system with constant coefficients (3.156), (3.157) is BIBO stable if and only if all poles of transfer function matrix $\boldsymbol{G}(s)$ lie in the open left half plane of the complex plane.

Asymptotic stability of linear continuous systems with constant coefficients implies BIBO stability but the opposite case need not be true.

3.3.6 Transformation of I/O Models into State-Space Models

In the previous sections we found out that an input-output model can be transformed into infinitely many state-space models. In this section we show the procedures of this transformation that lead to controllable and observable canonical forms for SISO systems.

Controllable Canonical Form

Consider a system with transfer function in the form

$$G(s) = \frac{b_m s^m + b_{m-1} s^{m-1} + \cdots + b_1 s + b_0}{s^n + a_{n-1} s^{n-1} + \cdots + a_1 s + a_0}, \quad n \geq m \tag{3.175}$$

Let us introduce an auxiliary variable $z(t)$ and its Laplace transform $Z(s)$ such that

$$\frac{Y(s)}{Z(s)} = b_m s^m + b_{m-1} s^{m-1} + \cdots + b_1 s + b_0 \tag{3.176}$$

$$\frac{Z(s)}{U(s)} = \frac{1}{s^n + a_{n-1} s^{n-1} + \cdots + a_1 s + a_0} \tag{3.177}$$

Equation (3.177) corresponds to the following differential equation

$$\frac{\mathrm{d}^n z(t)}{\mathrm{d}t^n} + a_{n-1} \frac{\mathrm{d}^{n-1} z(t)}{\mathrm{d}t^{n-1}} + \cdots + a_1 \dot{z}(t) + a_0 z(t) = u(t) \tag{3.178}$$

Now let us define state variables by the following relations

$$\frac{\mathrm{d}^i z(t)}{\mathrm{d}t^i} = x_{i+1}(t), \quad i = 0, 1, \ldots, n-1 \tag{3.179}$$

$$\frac{\mathrm{d}x_i(t)}{\mathrm{d}t} = x_{i+1}(t), \quad i = 1, 2, \ldots, n-1 \tag{3.180}$$

$$\tag{3.181}$$

Equation (3.178) can now be transformed into n first order differential equations

$$
\begin{aligned}
\frac{\mathrm{d}x_1(t)}{\mathrm{d}t} &= x_2(t) \\
\frac{\mathrm{d}x_2(t)}{\mathrm{d}t} &= x_3(t) \\
&\vdots \\
\frac{\mathrm{d}x_{n-1}(t)}{\mathrm{d}t} &= x_n(t) \\
\frac{\mathrm{d}x_n(t)}{\mathrm{d}t} &= \frac{\mathrm{d}^n z(t)}{\mathrm{d}t^n} = -a_{n-1}x_n - \cdots - a_1 x_2(t) - a_0 x_1(t) + u(t)
\end{aligned}
\tag{3.182}
$$

When $n = m$ then Eq. (3.176) corresponds to

$$y(t) = b_0 x_1(t) + b_1 x_2(t) + \cdots + b_{n-1} x_n(t) + b_n \dot{x}_n(t) \tag{3.183}$$

$\dot{x}_n(t)$ from this equation can be obtained from Eq. (3.182) and yields

$$
\begin{aligned}
y(t) = {} & (b_0 - a_0 b_n) x_1(t) + (b_1 - a_1 b_n) x_2(t) + \cdots + \\
& + (b_{n-1} - a_{n-1} b_n) x_n(t) + b_n u(t)
\end{aligned}
\tag{3.184}
$$

Equations (3.182) and (3.184) form a general state-space model of the form

$$\frac{\mathrm{d}\boldsymbol{x}(t)}{\mathrm{d}t} = \boldsymbol{A}_c \boldsymbol{x}(t) + \boldsymbol{B}_c \boldsymbol{u}(t) \tag{3.185}$$

$$\boldsymbol{y}(t) = \boldsymbol{C}_c \boldsymbol{x}(t) + \boldsymbol{D}_c \boldsymbol{u}(t) \tag{3.186}$$

where $\boldsymbol{x} = (x_1, x_2, \ldots, x_n)^T$,

$$
\boldsymbol{A}_c = \begin{pmatrix}
0 & 1 & 0 & \ldots & 0 \\
0 & 0 & 1 & \ldots & 0 \\
\vdots & & & \ddots & \vdots \\
0 & 0 & 0 & \ldots & 1 \\
-a_0 & -a_1 & -a_2 & \ldots & -a_{n-1}
\end{pmatrix}, \quad
\boldsymbol{B}_c = \begin{pmatrix} 0 \\ 0 \\ \vdots \\ 0 \\ 1 \end{pmatrix}
$$

$$
\boldsymbol{C}_c = \begin{pmatrix} b_0 - a_0 b_n & b_1 - a_1 b_n & \ldots & b_{n-1} - a_{n-1} b_n \end{pmatrix}, \quad \boldsymbol{D}_c = b_n
$$

We see that if $m < n$ then $\boldsymbol{D} = \boldsymbol{0}$. This system of state-space equations can be shown to be always controllable but it need not be observable. We speak about *controllable canonical form* of a system. The corresponding block scheme is shown in Fig. 3.25.

Example 3.17: *Controllable canonical form of a second order system*

Consider a system described by the following differential equation

$$\ddot{y}(t) + a_1 \dot{y}(t) + a_0 y(t) = b_1 \dot{u}(t) + b_0 u(t)$$

and corresponding transfer function

$$\frac{Y(s)}{U(s)} = \frac{b_1 s + b_0}{s^2 + a_1 s + a_0}$$

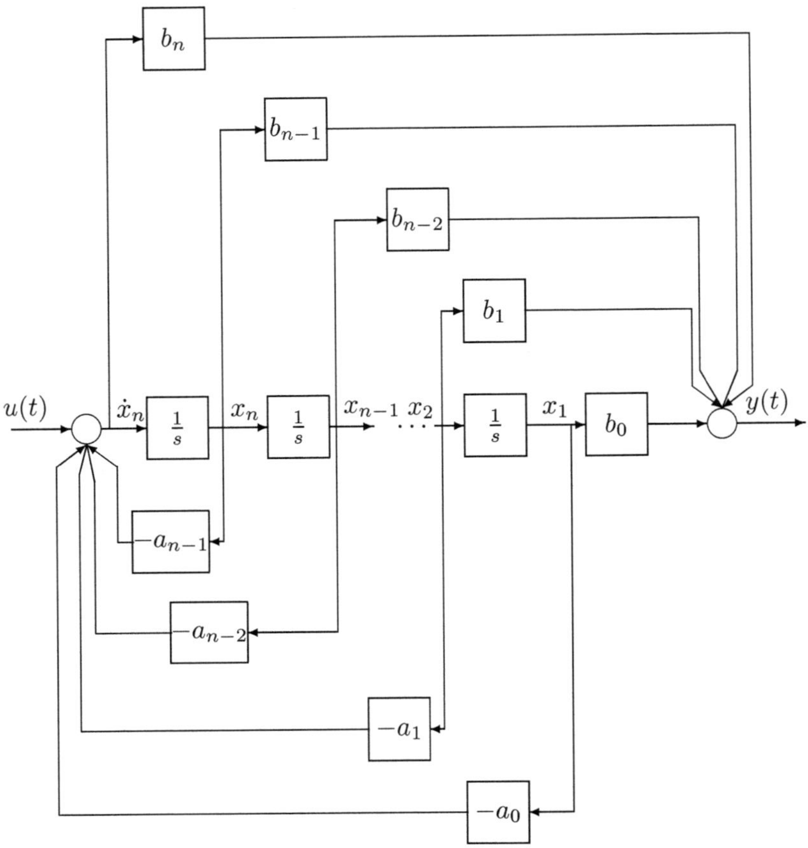

Fig. 3.25. Block scheme of controllable canonical form of a system

We introduce $Z(s)$ such the following equations hold

$$\frac{Y(s)}{Z(s)} = b_1 s + b_0$$

$$\frac{Z(s)}{U(s)} = \frac{1}{s^2 + a_1 s + a_0}$$

State-space equations can be written as

$$\frac{\mathrm{d}x_1(t)}{\mathrm{d}t} = x_2(t)$$

$$\frac{\mathrm{d}x_2 t)}{\mathrm{d}t} = -a_0 x_1(t) - a_1 x_2(t) + u(t)$$

$$y(t) = b_0 x_1(t) + b_1 x_2(t)$$

and the corresponding block scheme is shown in Fig. 3.26.

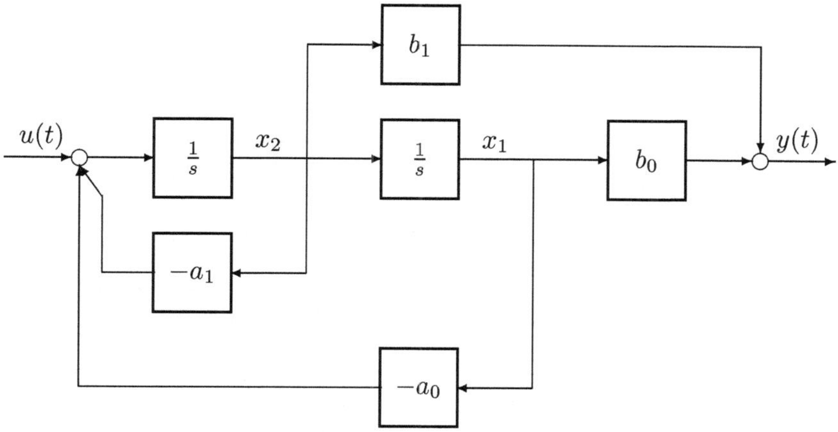

Fig. 3.26. Block scheme of controllable canonical form of a second order system

Observable Canonical Form

Consider again the system with transfer function given by (3.175) and assume $n = m$. *Observable canonical form* of this system is given by

$$\frac{\mathrm{d}\boldsymbol{x}(t)}{\mathrm{d}t} = \boldsymbol{A}_o\boldsymbol{x}(t) + \boldsymbol{B}_o\boldsymbol{u}(t) \tag{3.187}$$

$$\boldsymbol{y}(t) = \boldsymbol{C}_o\boldsymbol{x}(t) + \boldsymbol{D}_o\boldsymbol{u}(t) \tag{3.188}$$

where $\boldsymbol{x} = (x_1, x_2, \ldots, x_n)^T$,

$$\boldsymbol{A}_o = \begin{pmatrix} -a_{n-1} & 1 & 0 & \ldots & 0 \\ -a_{n-2} & 0 & 1 & \ldots & 0 \\ \vdots & & & \ddots & \vdots \\ -a_1 & 0 & 0 & \ldots & 1 \\ -a_0 & 0 & 0 & \ldots & 0 \end{pmatrix}, \quad \boldsymbol{B}_o = \begin{pmatrix} b_{n-1} - a_{n-1}b_n \\ b_{n-2} - a_{n-2}b_n \\ \vdots \\ b_1 - a_1 b_n \\ b_0 - a_0 b_n \end{pmatrix}$$

$$\boldsymbol{C}_o = \begin{pmatrix} 1 & 0 & \ldots & 0 \end{pmatrix}, \boldsymbol{D}_o = b_n$$

The corresponding block scheme is shown in Fig. 3.27.

3.3.7 I/O Models of MIMO Systems - Matrix Fraction Descriptions

Transfer function matrices provide a way to describe I/O (*external*) models of MIMO linear continuous systems. An alternate way is to give descriptions of such systems in polynomial matrix fractions that provide a natural generalisation of the singlevariable concept. A fundamental element of such descriptions is the polynomial matrix or matrix polynomial.

Real polynomial matrix is a matrix

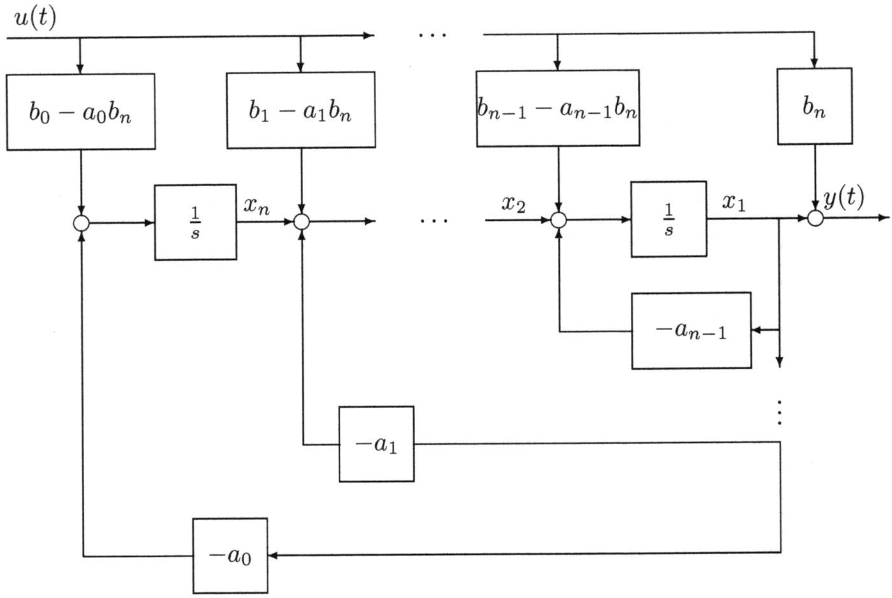

Fig. 3.27. Block scheme of observable canonical form of a system

$$\boldsymbol{P}(s) = \begin{pmatrix} p_{11}(s) & p_{12}(s) & \dots & p_{1m}(s) \\ p_{21}(s) & p_{22}(s) & \dots & p_{2m}(s) \\ \vdots & & & \vdots \\ p_{n1}(s) & p_{n2}(s) & \dots & p_{nm}(s) \end{pmatrix} \tag{3.189}$$

with elements being polynomials

$$p_{ij}(s) = p_{ij0} + p_{ij1}s + \dots + p_{ijd_{ij}} s^{d_{ij}} \tag{3.190}$$

where p_{ijk} are real numbers and $i = 1 \dots n$, $j = 1 \dots m$, $k = 0 \dots d_{ij}$.

An element $p_{ij}(s)$ identically equal to zero has according to definition a degree equal to minus one.

Row degree of i-th row of $\boldsymbol{P}$ is denoted by r_i and it is the maximum degree of all polynomials in the i-th row ($r_i = \max_j d_{ij}$). Analogously are defined column degrees c_j as $\max_i d_{ij}$.

Degree of polynomial matrix $\boldsymbol{P}$ is denoted by $\deg \boldsymbol{P}$ and is defined as the maximum degree of all polynomials of $\boldsymbol{P}$ ($\max_{ij} d_{ij}$).

An alternate way of writing (3.189) is as *matrix polynomial*

$$\boldsymbol{P}(s) = \boldsymbol{P}_0 + \boldsymbol{P}_1 s + \dots + \boldsymbol{P}_d s^d \tag{3.191}$$

where

$$\boldsymbol{P}_k = \begin{pmatrix} p_{11k} & p_{12k} & \dots & p_{1mk} \\ p_{21k} & p_{22k} & \dots & p_{2mk} \\ \vdots & & & \vdots \\ p_{n1k} & p_{n2k} & \dots & p_{nmk} \end{pmatrix} \qquad k = 0, \dots, d$$

$$d = \deg \boldsymbol{P}$$

A square polynomial matrix $\boldsymbol{P}(s)$ is nonsingular if $\det \boldsymbol{P}(s)$ is not identically equal to zero.

Roots of a square polynomial matrix are roots of the determinant of the polynomial matrix. A polynomial matrix is *stable* if all its roots lie in the open left half plane of the complex plane.

An *unimodular* polynomial matrix is a square polynomial matrix with a determinant equal to a nonzero constant. $\boldsymbol{P}(s)$ is unimodular if its inverse is also unimodular.

Rank of polynomial matrix is the highest order of a nonzero minor of this matrix.

For any $\boldsymbol{P}(s)$ there exist unimodular matrices $\boldsymbol{U}(s)$ and $\boldsymbol{V}(s)$ such that

$$\boldsymbol{U}(s)\boldsymbol{P}(s)\boldsymbol{V}(s) = \boldsymbol{\Lambda}(s) \tag{3.192}$$

where

$$\boldsymbol{\Lambda}(s) = \begin{pmatrix} \lambda_1(s) & 0 & \dots & 0 & 0 \\ 0 & \lambda_2(s) & \dots & 0 & 0 \\ \vdots & & & \vdots & \vdots \\ 0 & 0 & \dots & \lambda_r(s) & 0 \\ 0 & 0 & \dots & 0 & 0 \end{pmatrix}$$

λ_i are unique monic polynomials (i.e. polynomials with a unit first coefficient) for which holds λ_i divides λ_j for $i < j \leq r$ and r is equal to rank $\boldsymbol{P}(s)$. The matrix $\boldsymbol{\Lambda}(s)[n \times m]$ is called the Smith form of polynomial matrix.

Elementary row (column) operations on $\boldsymbol{P}(s)$ are:

1. interchange of any two rows (columns) of $\boldsymbol{P}(s)$,
2. multiplication of any row (column) of $\boldsymbol{P}(s)$ by any nonzero number,
3. addition to any row (column) of a polynomial multiple of any other row (column).

For any polynomial matrix we can find elementary row and column operations transforming the matrix into the Smith form.

A $[r \times m]$ transfer function matrix $\boldsymbol{G}(s)$ can be of the form

$$\boldsymbol{G}(s) = \frac{\boldsymbol{M}(s)}{d(s)} \tag{3.193}$$

where $d(s)$ is the least common multiplier of denominators of elements of $\boldsymbol{G}(s)$ and $\boldsymbol{M}(s)$ is a polynomial matrix. The matrix $\boldsymbol{G}(s)$ from Eq. (3.193) can also be written as

$$\boldsymbol{G}(s) = \boldsymbol{B}_R(s)\boldsymbol{A}_R^{-1}(s) \tag{3.194}$$

where $\boldsymbol{B}_R(s) = \boldsymbol{M}(s)$, $\boldsymbol{A}_R(s) = d(s)\boldsymbol{I}_m$. It is also possible to write

$$\boldsymbol{G}(s) = \boldsymbol{A}_L^{-1}(s)\boldsymbol{B}_L(s) \tag{3.195}$$

where $\boldsymbol{B}_L(s) = \boldsymbol{M}(s)$, $\boldsymbol{A}_L(s) = d(s)\boldsymbol{I}_r$.

The description of $\boldsymbol{G}(s)$ given by (3.194) is called *right matrix fraction description* (RMFD). The degree of determinant of $\boldsymbol{A}_R(s)$ is

$$\deg\det \boldsymbol{A}_R(s) = d_d m \tag{3.196}$$

where d_d is a degree of $d(s)$ and m is a dimension of vector $\boldsymbol{u}$.

Analogously, the description of (3.194) is called *left matrix fraction description* (LMFD). The degree of determinant of $\boldsymbol{A}_L(s)$ is

$$\deg\det \boldsymbol{A}_L(s) = d_d r \tag{3.197}$$

where r is the dimension of vector $\boldsymbol{y}$.

Given $\boldsymbol{G}(s)$ there are infinitely many LMFD's and RMFD's.

Minimum degree of determinant of any LMFD (RMFD) of $\boldsymbol{G}(s)$ is equal to the minimum order of some realisation of $\boldsymbol{G}(s)$.

If some RMFD of $\boldsymbol{G}(s)$ is of the form

$$\boldsymbol{G}(s) = \boldsymbol{B}_{R1}(s)\boldsymbol{A}_{R1}^{-1}(s) \tag{3.198}$$

and some other RMFD of the form

$$\boldsymbol{G}(s) = \boldsymbol{B}_R(s)\boldsymbol{A}_R^{-1}(s) \tag{3.199}$$

then $\boldsymbol{B}_{R1}(s) = \boldsymbol{B}_R(s)\boldsymbol{W}(s)$, $\boldsymbol{A}_{R1}(s) = \boldsymbol{A}_R(s)\boldsymbol{W}(s)$, and $\boldsymbol{W}(s)$ is some polynomial matrix and it is common right divisor of $\boldsymbol{B}_{R1}(s)$, $\boldsymbol{A}_{R1}(s)$. Analogously, the common left divisor can be defined.

Definition of relatively right (left) prime (RRP-RLP) polynomial matrices: Polynomial matrices $\boldsymbol{B}(s)$, $\boldsymbol{A}(s)$ with the same number of columns (rows) are RRP (RLP) if their right (left) common divisors are unimodular matrices.

Matrix fraction description of $\boldsymbol{G}(s)$ given by $\boldsymbol{A}(s), \boldsymbol{B}(s)$ is *right (left) irreducible* if $\boldsymbol{A}(s), \boldsymbol{B}(s)$ are RRP (RLP).

The process of obtaining irreducible MFD is related to greatest common divisors.

Greatest right (left) common divisor (GRCD-GLCD) of polynomial matrices $\boldsymbol{A}(s), \boldsymbol{B}(s)$ with the same number of columns (rows) is a polynomial matrix $\boldsymbol{R}(s)$ that satisfies the following conditions:

- $\boldsymbol{R}(s)$ is common right (left) divisor of $\boldsymbol{A}(s), \boldsymbol{B}(s)$,
- if $\boldsymbol{R}_1(s)$ is any common right (left) divisor of $\boldsymbol{A}(s), \boldsymbol{B}(s)$ then $\boldsymbol{R}_1(s)$ is right (left) divisor of $\boldsymbol{R}(s)$.

Lemma: relatively prime polynomial matrices: Polynomial matrices $\boldsymbol{B}_R(s)$, $\boldsymbol{A}_R(s)$ are RRP if and only if there exist polynomial matrices $\boldsymbol{X}_L(s)$, $\boldsymbol{Y}_L(s)$ such that the following Bezout identity is satisfied

$$\boldsymbol{Y}_L(s)\boldsymbol{B}_R(s) + \boldsymbol{X}_L(s)\boldsymbol{A}_R(s) = \boldsymbol{I}_m \tag{3.200}$$

Polynomial matrices $\boldsymbol{B}_L(s), \boldsymbol{A}_L(s)$ are RLP if and only if there exist polynomial matrices $\boldsymbol{X}_R(s)$, $\boldsymbol{Y}_R(s)$ such that the following Bezout identity is satisfied

$$\boldsymbol{B}_L(s)\boldsymbol{Y}_R(s) + \boldsymbol{A}_L(s)\boldsymbol{X}_R(s) = \boldsymbol{I}_r \tag{3.201}$$

For any polynomial matrices $\boldsymbol{B}_R(s)[r \times m]$ and $\boldsymbol{A}_R(s)[m \times m]$ an unimodular matrix $\boldsymbol{V}(s)$ exists such that

$$\boldsymbol{V}(s) = \begin{pmatrix} \boldsymbol{V}_{11}(s) & \boldsymbol{V}_{12}(s) \\ \boldsymbol{V}_{21}(s) & \boldsymbol{V}_{22}(s) \end{pmatrix}, \quad \begin{matrix} \boldsymbol{V}_{11}(s) \in [m \times r] & \boldsymbol{V}_{12}(s) \in [m \times m] \\ \boldsymbol{V}_{21}(s) \in [r \times r] & \boldsymbol{V}_{22}(s) \in [r \times m] \end{matrix} \tag{3.202}$$

and

$$\boldsymbol{V}(s) \begin{pmatrix} \boldsymbol{B}_R(s) \\ \boldsymbol{A}_R(s) \end{pmatrix} = \begin{pmatrix} \boldsymbol{R}(s) \\ \boldsymbol{0} \end{pmatrix} \tag{3.203}$$

$\boldsymbol{R}(s)[m \times m]$ is GRCD$(\boldsymbol{B}_R(s), \boldsymbol{A}_R(s))$. The couples $\boldsymbol{V}_{11}, \boldsymbol{V}_{12}$ and $\boldsymbol{V}_{21}, \boldsymbol{V}_{22}$ are RLP. An analogous property holds for LMFD: For any polynomial matrices $\boldsymbol{B}_L(s)[r \times m]$ and $\boldsymbol{A}_L(s)[r \times r]$ an unimodular matrix $\boldsymbol{U}(s)$ exists such that

$$\boldsymbol{U}(s) = \begin{pmatrix} \boldsymbol{U}_{11}(s) & \boldsymbol{U}_{12}(s) \\ \boldsymbol{U}_{21}(s) & \boldsymbol{U}_{22}(s) \end{pmatrix}, \quad \begin{matrix} \boldsymbol{U}_{11}(s) \in [r \times r] & \boldsymbol{U}_{12}(s) \in [r \times m] \\ \boldsymbol{U}_{21}(s) \in [m \times r] & \boldsymbol{U}_{22}(s) \in [m \times m] \end{matrix} \tag{3.204}$$

and

$$(\boldsymbol{A}_L(s) \ \boldsymbol{B}_L(s)) \ \boldsymbol{U}(s) = (\boldsymbol{L}(s) \ \boldsymbol{0}) \tag{3.205}$$

$\boldsymbol{L}(s)[r \times r]$ is GLCD$(\boldsymbol{B}_L(s), \boldsymbol{A}_L(s))$. The couples $\boldsymbol{U}_{11}, \boldsymbol{U}_{21}$ and $\boldsymbol{U}_{12}, \boldsymbol{U}_{22}$ are RRP.

Equations (3.203), (3.205) can be used to obtain irreducible MFD of $\boldsymbol{G}(s)$. When assuming RMFD the $\boldsymbol{G}(s)$ is given as

$$\boldsymbol{G}(s) = \boldsymbol{B}_R(s)\boldsymbol{A}_R^{-1}(s) \tag{3.206}$$

where $\boldsymbol{B}_R(s) = -\boldsymbol{V}_{12}(s)$ and $\boldsymbol{A}_R(s) = \boldsymbol{V}_{22}(s)$.

Lemma: division algorithm: Let $\boldsymbol{A}(s)[m \times m]$ be a nonsingular polynomial matrix. Then for any $\boldsymbol{B}(s)[r \times m]$ there exist unique polynomial matrices $\boldsymbol{Q}(s)$, $\boldsymbol{R}(s)$ such that

$$\boldsymbol{B}(s) = \boldsymbol{Q}(s)\boldsymbol{A}(s) + \boldsymbol{R}(s) \tag{3.207}$$

and $\boldsymbol{R}(s)\boldsymbol{A}^{-1}(s)$ is strictly proper.

The previous lemma deals with right division algorithm. Analogously, the left division can be defined.

This lemma can be used in the process of finding of a strictly proper part of the given R(L)MFD.

Lemma: minimal realisation of MFD: A MFD realisation with a degree equal to the denominator determinant degree is minimal if and only if the MFD is irreducible.

Lemma: BIBO stability: If the matrix transfer function $\boldsymbol{G}(s)$ is given by Eq. (3.194) then it is BIBO stable if and only if all roots of $\det \boldsymbol{A}_R(s)$ lie in the open left half plane of the complex plane. (analogously for LMFD).

Spectral factorisation: Consider a real polynomial matrix $\boldsymbol{B}(s)[m \times m]$ such that

$$\boldsymbol{B}^T(-s) = \boldsymbol{B}(s) \tag{3.208}$$

$$\boldsymbol{B}(\mathrm{j}\omega) > 0 \qquad \forall \omega \in R \tag{3.209}$$

Right spectral factor of $\boldsymbol{B}(s)$ is some stable polynomial matrix $\boldsymbol{A}(s)[m \times m]$ that satisfies the following relation

$$\boldsymbol{B}(s) = \boldsymbol{A}^T(-s)\boldsymbol{A}(s) \tag{3.210}$$

3.4 References

The use of the Laplace transform in theory of automatic control has been treated in large number of textbooks; for instance,

B. K. Čemodanov et al. *Mathematical Foundations of Automatic Control I.* Vyššaja škola, Moskva, 1977. (in Russian).

K. Reinisch. *Kybernetische Grundlagen und Beschreibung kontinuericher Systeme.* VEB Verlag Technik, Berlin, 1974.

H. Unbehauen. *Regelungstechnik I.* Vieweg, Braunschweig/Wiesbaden, 1986.

J. Mikleš and V. Hutla. *Theory of Automatic Control.* Alfa, Bratislava, 1986. (in Slovak).

State-space models and their analysis are discussed in

C. J. Friedly. *Dynamic Behavior of Processes.* Prentice Hall, Inc., New Jersey, 1972.

J. Mikleš and V. Hutla. *Theory of Automatic Control.* Alfa, Bratislava, 1986. (in Slovak).

H. Unbehauen. *Regelungstechnik II.* Vieweg, Braunschweig/Wiesbaden, 1987.

L. B. Koppel. *Introduction to Control Theory with Application to Process Control.* Prentice Hall, Englewood Cliffs, New Jersey, 1968.

L. A. Zadeh and C. A. Desoer. *Linear System Theory - the State-space Approach.* McGraw-Hill, New York, 1963.

A. A. Voronov. *Stability, Controllability, Observability.* Nauka, Moskva, 1979. (in Russian).

H. P. Geering. *Mess- und Regelungstechnik.* Springer Verlag, Berlin, 1990.

H. Kwakernaak and R. Sivan. *Linear Optimal Control Systems.* Wiley, New York, 1972.

W. H. Ray. *Advanced Process Control.* McGraw-Hill, New York, 1981.

M. Athans and P. L. Falb. *Optimal Control.* McGraw-Hill, New York, 1966.

R. E. Kalman, Y. C. Ho, and K. S. Narendra. Controllability of linear dynamical systems in contributions to differential equations. *Interscience Publishers*, V1(4):189 – 213, 1963.

P. L. Kalman, R. E. Falb and M. Arib. *Examples of Mathematical Systems Theory.* Mir, Moskva, 1971. (in Russian).

E. D. Gilles. *Systeme mit verteilten Parametern, Einführung in die Regelungstheorie.* Oldenbourg Verlag, München, 1973.

D. Chmúrny, J. Mikleš, P. Dostál, and J. Dvoran. *Modelling and Control of Processes and Systems in Chemical Technology.* Alfa, Bratislava, 1985. (in Slovak).

A. Mészáros and J. Mikleš. On observability and controllability of a tubular chemical reactor. *Chem. prumysl*, 33(2):57 – 60, 1983. (in Slovak).

P. Dostál, J. Mikleš, and A. Mészáros. *Theory of Automatic Control. Exercises II.* ES SVŠT, Bratislava, 1983. (in Slovak).

A. M. Lyapunov. *General Problem of Stability of Motion.* Fizmatgiz, Moskva, 1959. (in Russian).

R. E. Kalman and J. E. Bertram. Control system analysis and design via the second method of Lyapunov. *J. Basic Engineering*, 82:371 – 399, 1960.

J. Mikleš. *Theory of Automatic Control of Processes in Chemical Technology, Part II.* ES SVŠT, Bratislava, 1980. (in Slovak).

The subject of input-output models is considered as a classical theme in textbooks on automatic control; for example,

G. Stephanopoulos. *Chemical Process Control, An Introduction to Theory and Practice.* Prentice Hall, Englewood Cliffs, New Jersey, 1984.

Y. Z. Cypkin. *Foundations of Theory of Automatic Systems.* Nauka, Moskva, 1977. (in Russian).

C. A. Desoer and M. Vidyasagar. *Feedback Systems: Input-Output Properties.* Academic Press, New York, 1975.

C. T. Chen. *Linear System Theory and Design.* Holt, Rinehart and Winston, New York, 1984.

Further information about input-output models as well as about matrix fraction descriptions can be found in

W. A. Wolovich. *Linear Multivariable Systems.* Springer Verlag, New York, 1974.

H. H. Rosenbrock. *State-space and Multivariable Theory.* Nelson, London, 1970.

T. Kailaith. *Linear Systems.* Prentice Hall, Englewood Cliffs, New Jersey, 1980.

V. Kučera. *Discrete Linear Control: The Polynomial Equation Approach.* Wiley, Chichester, 1979.

J. Ježek and V. Kučera. Efficient algorithm for matrix spectral factorization. *Automatica*, 21:663 – 669, 1985.

J. Ježek. Symmetric matrix polynomial equations. *Kybernetika*, 22:19 – 30, 1986.

3.5 Exercises

Exercise 3.1:
Consider the two tanks shown in Fig. 3.10. A linearised mathematical model of this process is of the form

$$\frac{\mathrm{d}x_1}{\mathrm{d}t} = a_{11}x_1 + b_{11}u$$
$$\frac{\mathrm{d}x_2}{\mathrm{d}t} = a_{21}x_1 + a_{22}x_2$$
$$y = x_2$$

where

$$a_{11} = -\frac{k_{11}}{2F_1\sqrt{h_1^s}}, \; a_{21} = \frac{k_{11}}{2F_2\sqrt{h_1^s}}$$
$$a_{22} = -\frac{k_{22}}{2F_2\sqrt{h_2^s}}, \; b_{11} = \frac{1}{F_1}$$

Find:

1. state transition matrix of this system,
2. if $x_1(0) = x_2(0) = 0$ give expressions for functions
$$x_1(t) = f_1(u(t))$$
$$x_2(t) = f_2(u(t))$$
$$y(t) = f_3(u(t))$$

Exercise 3.2:
Consider CSTR shown in Fig. 2.11 and examine its stability. The rate of reaction is given as (see example 2.6)

$$r(c_A, \vartheta) = kc_A = k_0 \mathrm{e}^{-\frac{E}{R\vartheta}} c_A$$

Suppose that concentration c_{Av} and temperatures ϑ_v, ϑ_c are constant. Perform the following tasks:

1. define steady-state of the reactor and find the model in this steady-state so that $\mathrm{d}c_A/\mathrm{d}t = \mathrm{d}\vartheta/\mathrm{d}t = 0$,
2. define deviation variables for reactor concentration and temperature and find a nonlinear model of the reactor with deviation variables,
3. perform linearisation and determine state-space description,
4. determine stability conditions according to the Lyapunov equation (3.82). We assume that if the reactor is asymptotically stable in large in origin then it is asymptotically stable in origin.

Exercise 3.3:
Consider the mixing process shown in Fig. 2.15. The task is to linearise the process for the input variables q_0, q_1 and output variables h, c_2 and to determine its transfer function matrix.

Exercise 3.4:
Consider a SISO system described by the following differential equation

$$\ddot{y}(t) + a_1\dot{y}(t) + a_0 y(t) = b_1\dot{u}(t) + b_0 u(t)$$

Find an observable canonical form of this system and its block scheme.

Exercise 3.5:
Assume 2I/2O system with transfer function matrix given as LMFD (3.195) where

$$\boldsymbol{A}_L(s) = \begin{pmatrix} 1 + a_1 s & a_2 s \\ a_3 s & 1 + a_4 s \end{pmatrix}$$

$$\boldsymbol{B}_L(s) = \begin{pmatrix} b_1 & b_2 \\ b_3 & b_4 \end{pmatrix}$$

By using the method of comparing coefficients, find the corresponding RMFD (3.194) where

$$\boldsymbol{A}_R(s) = \begin{pmatrix} a_{1R} + a_{2R}s & a_{3R} + a_{4R}s \\ a_{5R} + a_{6R}s & a_{7R} + a_{8R}s \end{pmatrix}$$

$$\boldsymbol{B}_R(s) = \begin{pmatrix} b_{1R} & b_{2R} \\ b_{3R} & b_{4R} \end{pmatrix}$$

Elements of matrix

$$\boldsymbol{A}_{R0} = \begin{pmatrix} a_{1R} & a_{3R} \\ a_{5R} & a_{7R} \end{pmatrix}$$

can be chosen freely, but $\boldsymbol{A}_{R0}$ must be nonsingular.

4

Dynamical Behaviour of Processes

Process responses to various simple types of input variables are valuable for process control design. In this chapter three basic process responses are studied: impulse, step, and frequency responses. These characteristics are usually investigated by means of computer simulations. In this connection we show and explain computer codes that numerically solve systems of differential equations in the programming languages BASIC, C, and MATLAB.

The end of this chapter deals with process responses for the case of stochastic input variables.

4.1 Time Responses of Linear Systems to Unit Impulse and Unit Step

4.1.1 Unit Impulse Response

Consider a system described by a transfer function $G(s)$ and for which holds

$$Y(s) = G(s)U(s) \tag{4.1}$$

If the system input variable $u(t)$ is the unit impulse $\delta(t)$ then

$$U(s) = \mathcal{L}\{\delta(t)\} = 1 \tag{4.2}$$

and the system response is given as

$$y(t) = g(t) \tag{4.3}$$

where $g(t) = \mathcal{L}^{-1}\{G(s)\}$ is *system response to the unit impulse* if the system initial conditions are zero, $g(t)$ is called *impulse response* or *weighting function.*

If we start from the solution of state-space equations (3.47)

$$\boldsymbol{x}(t) = \mathrm{e}^{\boldsymbol{A}t}\boldsymbol{x}(0) + \int_0^t \mathrm{e}^{\boldsymbol{A}(t-\tau)}\boldsymbol{B}u(\tau)\mathrm{d}\tau \tag{4.4}$$

$$y(t) = \boldsymbol{C}\boldsymbol{x}(t) + Du(t) \tag{4.5}$$

and replace $u(t)$ with $\delta(t)$ we get

$$\boldsymbol{x}(t) = \mathrm{e}^{\boldsymbol{A}t}\boldsymbol{x}(0) + \mathrm{e}^{\boldsymbol{A}t}\boldsymbol{B} \tag{4.6}$$

$$y(t) = \boldsymbol{C}\mathrm{e}^{\boldsymbol{A}t}\boldsymbol{x}(0) + \boldsymbol{C}\mathrm{e}^{\boldsymbol{A}t}\boldsymbol{B} + D\delta(t) \tag{4.7}$$

For $\boldsymbol{x}(0) = 0$ then follows

$$y(t) = \boldsymbol{C}\mathrm{e}^{\boldsymbol{A}t}\boldsymbol{B} + D\delta(t) = g(t) \tag{4.8}$$

Consider the transfer function $G(s)$ of the form

$$G(s) = \frac{b_n s^n + b_{n-1}s^{n-1} + \cdots + b_0}{a_n s^n + a_{n-1}s^{n-1} + \cdots + a_0} \tag{4.9}$$

The initial value theorem gives

$$g(0) = \lim_{s\rightarrow\infty} sG(s) = \begin{cases} \infty, & \text{if } b_n \neq 0 \\ \frac{b_{n-1}}{a_n}, & \text{if } b_n = 0 \\ 0, & \text{if } b_n = b_{n-1} = 0 \end{cases} \tag{4.10}$$

and $g(t) = 0$ for $t < 0$.

If for the impulse response holds $g(t) = 0$ for $t < 0$ then we speak about *causal* system.

From the Duhamel integral

$$y(t) = \int_0^t g(t-\tau)u(\tau)\mathrm{d}\tau \tag{4.11}$$

follows that if the condition

$$\int_0^t |g(t)|\mathrm{d}t < \infty \tag{4.12}$$

holds then any bounded input to the system results in bounded system output.

WWW **Example 4.1:** *Impulse response of the first order system*

Assume a system with transfer function

$$G(s) = \frac{1}{T_1 s + 1}$$

then the corresponding weighting function is the inverse Laplace transform of $G(s)$

$$g(t) = \frac{1}{T_1}\mathrm{e}^{-\frac{t}{T_1}}$$

The graphical representation of this function is shown in Fig. 4.1.

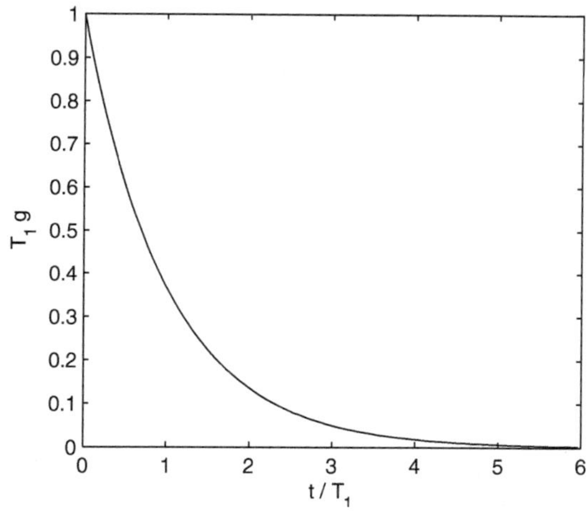

Fig. 4.1. Impulse response of the first order system

4.1.2 Unit Step Response

Step response is a response of a system with zero initial conditions to the unit step function $1(t)$. Consider a system with transfer function $G(s)$ for which holds

$$Y(s) = G(s)U(s) \tag{4.13}$$

If the system input variable $u(t)$ is the unit step function

$$u(t) = 1(t) \tag{4.14}$$

then the system response (for zero initial conditions) is

$$y(t) = \mathcal{L}^{-1}\left\{G(s)\frac{1}{s}\right\} \tag{4.15}$$

From this equation it is clear that step response is a time counterpart of the term $G(s)/s$ or equivalently $G(s)/s$ is the Laplace transform of step response. The impulse response is the time derivative of the step response.

Consider again the state-space approach. For $u(t) = 1(t)$ we get from (3.47)

$$\boldsymbol{x}(t) = \mathrm{e}^{\boldsymbol{A}t}\boldsymbol{x}(0) + \int_0^t \mathrm{e}^{\boldsymbol{A}(t-\tau)}\boldsymbol{B}u(t)\mathrm{d}\tau \tag{4.16}$$

$$\boldsymbol{x}(t) = \mathrm{e}^{\boldsymbol{A}t}\boldsymbol{x}(0) + \mathrm{e}^{\boldsymbol{A}t}(-\boldsymbol{A}^{-1})(\mathrm{e}^{-\boldsymbol{A}t} - \boldsymbol{I})\boldsymbol{B} \tag{4.17}$$

$$\boldsymbol{x}(t) = \mathrm{e}^{\boldsymbol{A}t}\boldsymbol{x}(0) + (\mathrm{e}^{\boldsymbol{A}t} - \boldsymbol{I})\boldsymbol{A}^{-1}\boldsymbol{B} \tag{4.18}$$

$$y(t) = \boldsymbol{C}\mathrm{e}^{\boldsymbol{A}t}\boldsymbol{x}(0) + \boldsymbol{C}(\mathrm{e}^{\boldsymbol{A}t} - \boldsymbol{I})\boldsymbol{A}^{-1}\boldsymbol{B} + D \tag{4.19}$$

For $\boldsymbol{x}(0) = \boldsymbol{0}$ holds

$$y(t) = \boldsymbol{C}(\mathrm{e}^{\boldsymbol{A}t} - \boldsymbol{I})\boldsymbol{A}^{-1}\boldsymbol{B} + D \tag{4.20}$$

If all eigenvalues of $\boldsymbol{A}$ have negative real parts, the steady-state value of step response is equal to $G(0)$. This follows from the Final value theorem (see page 58)

$$\lim_{t\to\infty} y(t) = \lim_{s=0} G(s) = -\boldsymbol{C}\boldsymbol{A}^{-1}\boldsymbol{B} + D = \frac{b_0}{a_0} \tag{4.21}$$

The term b_0/a_0 is called *(steady-state) gain* of the system.

www

Example 4.2: *Step response of first order system*

Assume a process that can be described as

$$T_1 \frac{\mathrm{d}y}{\mathrm{d}t} + y = Z_1 u$$

This is an example of the first order system with the transfer function

$$G(s) = \frac{Z_1}{T_1 s + 1}$$

The corresponding step response is given as

$$y(t) = Z_1(1 - \mathrm{e}^{-\frac{t}{T_1}})$$

Z_1 the gain and T_1 time constant of this system. Step response of this system is shown in Fig 4.2.

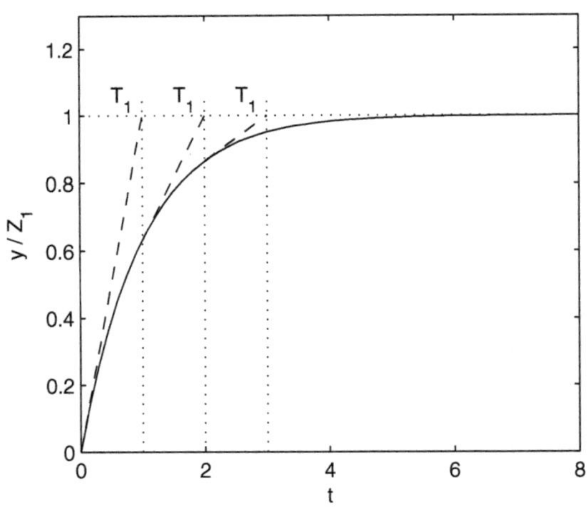

Fig. 4.2. Step response of a first order system

Step responses of the first order system with various time constants are shown in Fig 4.3. The relation between time constants is $T_1 < T_2 < T_3$.

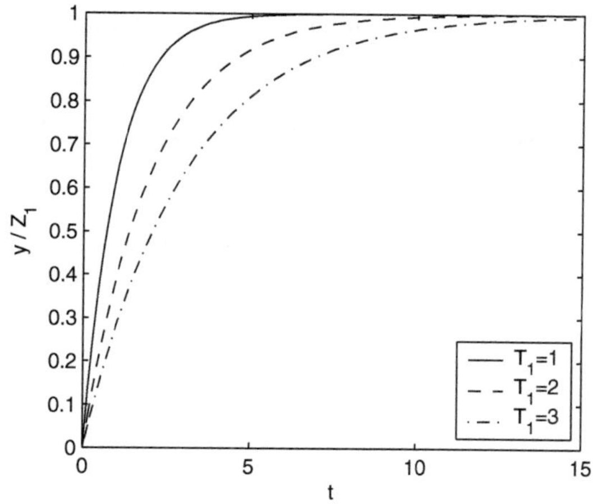

Fig. 4.3. Step responses of a first order system with time constants T_1, T_2, T_3

W **Example 4.3:** *Step responses of higher order systems*
Consider two systems with transfer functions of the form

$$G_1(s) = \frac{Z_1}{T_1 s + 1}, \qquad G_2(s) = \frac{Z_2}{T_2 s + 1}$$

connected in series. The overall transfer function is given as their product

$$\frac{Y(s)}{U(s)} = \frac{Z_1 Z_2}{(T_1 s + 1)(T_2 s + 1)}$$

The corresponding step response function can be calculated as

$$y(t) = Z_1 Z_2 \left[1 - \frac{T_1}{T_1 - T_2} e^{-\frac{t}{T_1}} + \frac{T_2}{T_1 - T_2} e^{-\frac{t}{T_2}}\right]$$

or

$$y(t) = Z_1 Z_2 \left[1 - \frac{T_1 T_2}{T_1 - T_2} \left(\frac{1}{T_2} e^{-\frac{t}{T_1}} - \frac{1}{T_1} e^{-\frac{t}{T_2}}\right)\right]$$

Consider now a second order system with the transfer function given by

$$G(s) = \frac{Y(s)}{U(s)} = \frac{Z_s}{T_k^2 s^2 + 2\zeta T_k s + 1}$$

As it was shown in the Example 3.12, such transfer function can result from the mathematical model of a U-tube.
The characteristic form of the step response depends on the roots of the characteristic equation

$$T_k^2 s^2 + 2\zeta T_k s + 1 = 0$$

If T_k represents the time constant then the dumping factor ζ plays a crucial role in the properties of the step response. In the following analysis the case $\zeta < 0$ will be automatically excluded as that corresponding to an unstable system. We will focus on the following cases of roots:
Case a: $\zeta > 1$ - two different real roots,
Case b: $\zeta = 1$ - double real root,
Case c: $0 < \zeta < 1$ - two complex conjugate roots.
Case a: If $\zeta > 1$ then the characteristic equation can be factorised as follows

$$T_k^2 s^2 + 2\zeta T_k s + 1 = (T_1 s + 1)(T_2 s + 1)$$

where

$$\begin{aligned} T_k^2 &= T_1 T_2 \\ 2\zeta T_k &= T_1 + T_2 \end{aligned} \quad \text{or} \quad \begin{aligned} T_k &= \sqrt{T_1 T_2} \\ \zeta &= \frac{T_1 + T_2}{2\sqrt{T_1 T_2}} \end{aligned}$$

Another possibility how to factorise the characteristic equation is

$$T_k^2 s^2 + 2\zeta T_k s + 1 = \left(\frac{T_k}{\zeta - \sqrt{\zeta^2 - 1}} s + 1\right)\left(\frac{T_k}{\zeta + \sqrt{\zeta^2 - 1}} s + 1\right)$$

Now the constants T_1, T_2 are of the form

$$T_1 = \frac{T_k}{\zeta - \sqrt{\zeta^2 - 1}}, \quad T_2 = \frac{T_k}{\zeta + \sqrt{\zeta^2 - 1}}$$

Case b: If $\zeta = 1$ then $T_1 = T_k$, $T_2 = T_k$.
Case c: If $0 < \zeta < 1$ then the transfer function can be rewritten as

$$G(s) = \frac{Z_s}{T_k^2 \left(s^2 + \frac{2\zeta}{T_k} s + \frac{1}{T_k^2}\right)}$$

and the solution of the characteristic equation is given by

$$s_{1,2} = \frac{-\frac{2\zeta}{T_k} \pm \sqrt{4\frac{\zeta^2}{T_k^2} - 4\frac{1}{T_k^2}}}{2}$$

$$s_{1,2} = \frac{-\zeta \pm \sqrt{\zeta^2 - 1}}{T_k}$$

The corresponding transfer functions are found from the inverse Laplace transform and are of the form
Case a:

$$y(t) = Z_s \left(1 - \frac{T_1 e^{-t/T_1} - T_2 e^{-t/T_2}}{T_1 - T_2}\right)$$

Case b:

$$y(t) = Z_s \left[1 - \left(1 - \frac{t}{T_k}\right) e^{-t/T_k}\right]$$

Case c:

$$y(t) = Z_s \left[1 - \frac{1}{\sqrt{1-\zeta^2}} e^{-\frac{\zeta}{T_k}t} \sin\left(\frac{\sqrt{1-\zeta^2}}{T_k} t + \arctan\frac{\sqrt{1-\zeta^2}}{\zeta}\right)\right]$$

For the sake of completeness, if $\zeta = 0$ the step response contains only a sinus function. Step responses for various values of ζ are shown in Fig. 4.4.

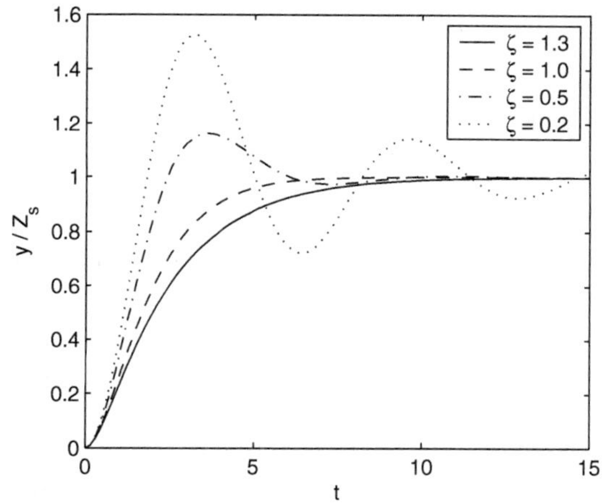

Fig. 4.4. Step responses of the second order system for the various values of ζ

Consider now the system consisting of two first order systems connected in a series. The worst case concerning system inertia occurs if $T_1 = T_2$. In this case the tangent in the inflex point has the largest value. If $T_2 \ll T_1$ then the overall characteristic of the system approaches a first order system with a time constant T_1.

A generalisation of this phenomenon shows that if the system consists of n-in-series connected systems, then the system inertia is the largest if all time constants are equal.

If i-th subsystem is of the form

$$G_i(s) = \frac{Z_i}{T_i s + 1}$$

then for the overall transfer function yields

$$G(s) = \frac{Y(s)}{U(s)} = \frac{\prod_{i=1}^{n} Z_i}{\prod_{i=1}^{n} (T_i s + 1)}$$

If $T_s = T_1 = T_2 = \cdots = T_n$ then the system residence time will be the largest. Consider unit step function on input and $Z_s = Z_1 Z_2 \cdots Z_n$. The step responses for various n are given in Fig. 4.5.

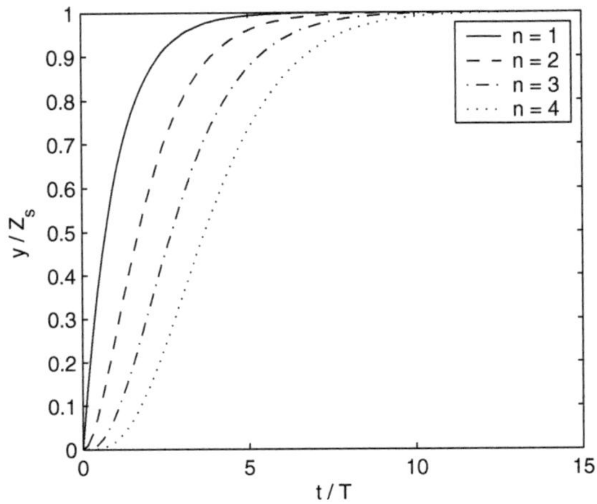

Fig. 4.5. Step responses of the system with n equal time constants

WWW **Example 4.4:** *Step response of the n-th order system connected in a series with time delay*

Fig. 4.6 shows the block scheme of a system composed of n-th order system and pure time delay connected in a series. The corresponding step response is shown in Fig. 4.7 where it is considered that $n = 1$.

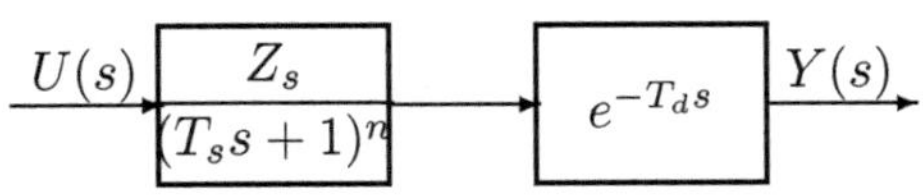

Fig. 4.6. Block scheme of the n-th order system connected in a series with time delay

WWW **Example 4.5:** *Step response of 2nd order system with the numerator* $B(s) = b_0 + b_1 s$

As it was shown in Example 3.14, some transfer function numerators of the CSTR can contain first order polynomials. Investigation of such step responses is therefore of practical importance. An especially interesting case is when the numerator polynomial has a positive root.

Consider for example the following system

$$G(s) = \frac{b_1 s + 1}{s^2 + 2.6s + 1}$$

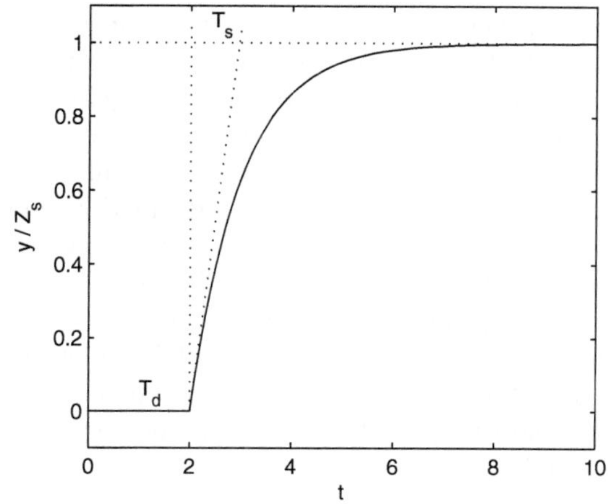

Fig. 4.7. Step response of the first order system with time delay

The corresponding step response is illustrated in Fig. 4.8.

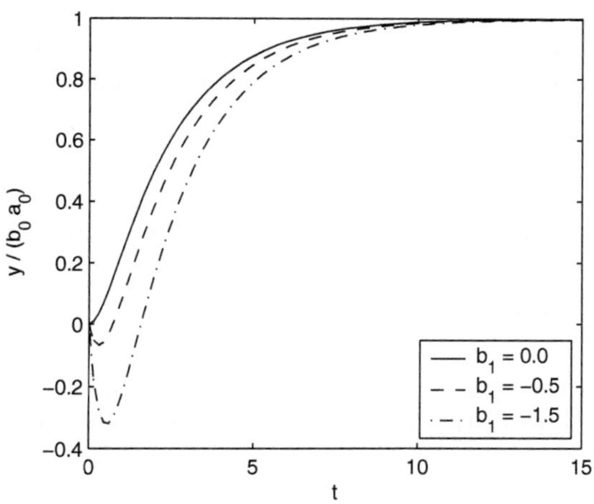

Fig. 4.8. Step response of the second order system with the numerator $B(s) = b_1s+1$

4.2 Computer Simulations

As it was shown in the previous pages, the investigation of process behaviour requires a solution of differential equations. Analytical solutions can only be found for processes described by linear differential equations with constant coefficients. If the differential equations that describe dynamical behaviour of a process are nonlinear, then it is either very difficult or impossible to find the analytical solution. In such cases it is necessary to utilise numerical methods. These procedures transform the original differential equations into difference equations that can be solved iteratively on a computer. The drawback of this type of solution is a loss of generality as numerical values of the initial conditions, coefficients of the model, and its input functions must be specified. However, in the majority of cases there does not exist any other approach as a numerical solution of differential equations. The use of numerical methods for the determination of process responses is called *simulation*. There is a large number of simulation methods. We will explain Euler and Runge-Kutta methods. The Euler method will be used for the explanation of principles of numerical methods. The Runge-Kutta method is the most versatile approach that is extensively used.

4.2.1 The Euler Method

Consider a process model in the form

$$\frac{\mathrm{d}x(t)}{\mathrm{d}t} = f(t, x(t)), \qquad x(t_0) = x_0 \tag{4.22}$$

At first we transform this equation into its difference equation counterpart. We start from the definition of a derivative of a function

$$\frac{\mathrm{d}x}{\mathrm{d}t} = \lim_{\Delta t \to 0} \frac{x(t+\Delta t) - x(t)}{\Delta t} \tag{4.23}$$

if Δt is sufficiently small, the derivative can be approximated as

$$\frac{\mathrm{d}x}{\mathrm{d}t} \doteq \frac{x(t+\Delta t) - x(t)}{\Delta t} \tag{4.24}$$

Now, suppose that the right hand side of (4.22) is constant over some interval $(t, t+\Delta t)$ and substitute the left hand side derivative from (4.24). Then we can write

$$\frac{x(t+\Delta t) - x(t)}{\Delta t} = f(t, x(t)) \tag{4.25}$$

or

$$x(t+\Delta t) = x(t) + \Delta t f(t, x(t)) \tag{4.26}$$

The assumptions that led to Eq. (4.26) are only justified if Δt is sufficiently small. At time $t = t_0$ we can write

$$x(t_0+\Delta t) = x(t_0) + \Delta t f(t_0, x(t_0)) \tag{4.27}$$

and at time $t_1 = t_0 + \Delta t$

$$x(t_1 + \Delta t) = x(t_1) + \Delta t f(t_1, x(t_1)) \tag{4.28}$$

In general, for $t = t_k$, $t_{k+1} = t_k + \Delta t$ Eq. (4.26) yields

$$x(t_{k+1}) = x(t_k) + \Delta t f(t_k, x(t_k)) \tag{4.29}$$

Consider now the following differential equation

$$\frac{\mathrm{d}x(t)}{\mathrm{d}t} = f(t, x(t), u(t)), \qquad x(t_0) = x_0 \tag{4.30}$$

We assume again that the right hand side is constant over the interval (t_k, t_{k+1}) and is equal to $f(t_k, x(t_k), u(t_k))$. Applying the approximation (4.24) yields

$$x(t_{k+1}) = x(t_k) + \Delta t f(t_k, x(t_k), u(t_k)) \tag{4.31}$$

In this equation we can notice that the continuous-time variables $x(t), u(t)$ have been replaced by discrete variables $x(t_k), u(t_k)$. Let us denote

$$x(t_k) \equiv x_k \tag{4.32}$$

$$u(t_k) \equiv u_k \tag{4.33}$$

and we obtain a recursive relation called *difference equation*

$$x_{k+1} = x_k + \Delta t f(t_k, x_k, u_k) \tag{4.34}$$

that can be solved recursively for $k = 0, 1, 2, \ldots$ for the given initial value x_0.

Equation (4.34) constitutes *the Euler method* of solving the differential equation (4.30) and it is easily programmable on a computer. The difference $h = t_{k+1} - t_k$ is usually called *integration step.*

As the basic Euler method is only very crude and inaccurate, the following modification of *modified Euler method* was introduced

$$x_{k+1} = x_k + \frac{h}{2}(f_k + f_{k+1}) \tag{4.35}$$

where

$$t_k = t_0 + kh, \; k = 0, 1, 2, \ldots$$
$$f_k = f(t_k, x(t_k), u(t_k))$$
$$f_{k+1} = f[t_{k+1}, x(t_k) + hf(t_k, x(t_k), u(t_k)), u(t_{k+1})]$$

4.2.2 The Runge-Kutta method

This method is based on the Taylor expansion of a function. The Taylor expansion helps to express the solution $x(t + \Delta t)$ of a differential equation in relation to $x(t)$ and its time derivatives as follows

$$x(t + \Delta t) = x(t) + \Delta t\dot{x}(t) + \frac{1}{2}(\Delta t)^2\ddot{x}(t) + \cdots \tag{4.36}$$

If the solved differential equation is of the form

$$\frac{\mathrm{d}x(t)}{\mathrm{d}t} = f(t, x(t)) \tag{4.37}$$

then the time derivatives can be expressed as

$$\begin{aligned} \dot{x}(t) &= f(t, x(t)) \\ \ddot{x}(t) &= \frac{\mathrm{d}f}{\mathrm{d}t} = \frac{\partial f}{\partial t} + \frac{\partial f}{\partial x}\frac{\mathrm{d}x}{\mathrm{d}t} = \frac{\partial f}{\partial t} + \frac{\partial f}{\partial x}f \\ &\vdots \end{aligned} \tag{4.38}$$

Substituting (4.38) into (4.36) yields

$$x(t + \Delta t) = x(t) + \Delta t f + \frac{1}{2}(\Delta t)^2(f_t + f_x f) + \cdots \tag{4.39}$$

where $f = f(t, x(t))$, $f_t = \partial f/\partial t$, $f_x = \partial f/\partial x$.

The solution $x(t+\Delta t)$ in Eq. (4.39) depends on the knowledge of derivatives of the function f. However, the higher order derivatives of f are difficult to obtain. Therefore only some first terms of (4.39) are assumed to be significant and others are neglected. The Taylor expansion is truncated and forms the basis for *Runge-Kutta methods*. The number of terms determines order of the Runge-Kutta method.

Assume that the integration step is given as

$$t_{k+1} = t_k + h \tag{4.40}$$

The second order Runge-Kutta method is based on the difference equation

$$x(t_{k+1}) = x(t_k) + h\dot{x}(t_k) + \frac{1}{2}h^2\ddot{x}(t_k) \tag{4.41}$$

or

$$x_{k+1} = x_k + hf_k + \frac{1}{2}h^2(f_t + f_x f)_k \tag{4.42}$$

The recursive relation suitable for numerical solution is then given by

$$x_{k+1} = x_k + \gamma_1 k_1 + \gamma_2 k_2 \tag{4.43}$$

where γ_1, γ_2 are weighting constants and

$$k_1 = hf(t_k, x_k) \tag{4.44}$$

$$k_2 = hf(t_k + \alpha_1 h, x_k + \beta_1 k_1) \tag{4.45}$$

and α_1, α_2 are some constants. The proof that (4.43) is a recursive solution following from the second order Runge-Kutta method can be shown as follows. Allow at first perform the Taylor expansion for k_2

$$k_2 = h[f_k + (f_t)_k \alpha_1 h + (f_x)_k \beta_1 h f_k + \cdots] \tag{4.46}$$

and neglect all terms not explicitly given. Substituting k_1 from (4.44) and k_2 from (4.46) into (4.43) gives

$$x_{k+1} = x_k + h(\gamma_1 f_k + \gamma_2 f_k) + h^2[\gamma_2 \alpha_1 (f_t)_k + \gamma_2 \beta_1 (f_x)_k f_k] \tag{4.47}$$

Comparison of (4.41) and (4.47) gives

$$\gamma_1 + \gamma_2 = 1 \tag{4.48}$$

$$\gamma_2 \alpha_1 = \frac{1}{2} \tag{4.49}$$

$$\gamma_2 \beta_1 = \frac{1}{2} \tag{4.50}$$

This showed that (4.43) is a recursive solution that follows from the second order Runge-Kutta method.

The best known recursive equations suitable for a numerical solution of differential equations is the fourth order Runge-Kutta method that is of the form

$$x_{k+1} = x_k + \frac{1}{6}k_1 + \frac{1}{3}k_2 + \frac{1}{3}k_3 + \frac{1}{6}k_4 \tag{4.51}$$

where

$$\begin{aligned} k_1 &= hf(t_k, x_k) \\ k_2 &= hf(t_k + \tfrac{1}{2}h, x_k + \tfrac{1}{2}k_1) \\ k_3 &= hf(t_k + \tfrac{1}{2}h, x_k + \tfrac{1}{2}k_2) \\ k_4 &= hf(t_k + h, x_k + k_3) \end{aligned} \tag{4.52}$$

4.2.3 Runge-Kutta Method for a System of Differential Equations

The Runge-Kutta method can be used for the solution of a system of differential equations

$$\frac{\mathrm{d}\boldsymbol{x}(t)}{\mathrm{d}t} = \boldsymbol{f}(t, \boldsymbol{x}(t)), \qquad \boldsymbol{x}(t_0) = \boldsymbol{x}_0 \tag{4.53}$$

where $\boldsymbol{x} = (x_1, \ldots, x_n)^T$.

Vectorised equivalents of equations (4.41), (4.43), (4.44), (4.45) are as follows

$$\boldsymbol{x}(t_{k+1}) = \boldsymbol{x}(t_k) + h\dot{\boldsymbol{x}}(t_k) + \frac{1}{2}h^2 \ddot{\boldsymbol{x}}(t_k) \tag{4.54}$$

$$\boldsymbol{x}_{k+1} = \boldsymbol{x}_k + \gamma_1 \boldsymbol{k}_1 + \gamma_2 \boldsymbol{k}_2 \tag{4.55}$$

$$\boldsymbol{k}_1 = h\boldsymbol{f}(t_k, \boldsymbol{x}_k) \tag{4.56}$$

$$\boldsymbol{k}_2 = h\boldsymbol{f}(t_k + \alpha_1 h, \boldsymbol{x}_k + \beta_1 \boldsymbol{k}_1) \tag{4.57}$$

The programs for implementation of the fourth order Runge-Kutta method in various computer languages are given in the next example.

www

Example 4.6: *Programs for the solution of state-space equations*
We will explain the use of the fourth order Runge-Kutta method applied to the following second order differential equation

$$T_1 T_2 \frac{\mathrm{d}^2 y_2}{\mathrm{d}t^2} + (T_1 + T_2)\frac{\mathrm{d}y_2}{\mathrm{d}t} + y_2 = Z_1 u$$

with zero initial conditions and for $u(t) = 1(t)$. T_1, T_2 are time constants and Z_1 gain of this system. At first we transform this differential equation into state-space so that two differential equations of the first order result

$$\frac{\mathrm{d}x_1}{\mathrm{d}t} = \frac{Z_1 u - x_2 - (T_1 + T_2)x_1}{T_1 T_2}$$

$$\frac{\mathrm{d}x_2}{\mathrm{d}t} = x_1$$

The program written in GW-BASIC is given in Program 4.1. The state-space differential equations are defined on lines 550, 560. The solution $y_1(t) = x_1(t)$, $y_2(t) = x_2(t)$ calculated with this program is given in Table 4.1. The values of variable $y_2(t)$ represent the step response of the system with transfer function

$$G(s) = \frac{Y(s)}{U(s)} = \frac{Z_1}{(T_1 s + 1)(T_2 s + 1)}$$

Program 4.2 is written in C. The example of the solution in the simulation environment MATLAB/Simulink is given in Program 4.3. This represents `m-file` that can be introduced as S-function into Simulink block scheme shown in Fig. 4.9. The graphical solution is then shown in Fig. 4.10 and it is the same as in Tab. 4.1.

Program 4.1 (Simulation program in GW-BASIC)

```
5 REM ruku4.bas
10 REM solution of the ODE system
20 REM  n  number of equations
30 REM  h  integration step
50 REM y(1),y(2),...,y(n)  initial conditions
55 DATA 2: READ n
58 DIM y(n), x(n), f(n), k(4, n)
60 DATA  .5,        0, 0
70 READ h
80 FOR i = 1 TO n: READ y(i): NEXT i
140 PRINT "t", "y1", "y2"
160 PRINT t, y(1), y(2)
200 FOR k = 0 TO 19
205    FOR i = 1 TO n: x(i) = y(i): NEXT i: GOSUB 470
```

```
240   FOR i = 1 TO n
242     k(1, i) = h * f(i) : x(i) = y(i) + k(1, i) / 2
244   NEXT i: GOSUB 470
290   FOR i = 1 TO n
292     k(2, i) = h * f(i): x(i) = y(i) + k(2, i) / 2
294   NEXT i: GOSUB 470
340   FOR i = 1 TO n
342     k(3, i) = h * f(i): x(i) = y(i) + k(3, i)
344   NEXT i: GOSUB 470
390   FOR i = 1 TO n
392     k(4, i) = h * f(i)
410     y(i) = y(i)+(k(1,i)+2*k(2,i)+2*k(3,i)+k(4,i)) /6
420   NEXT i
430   t = t + h
440   PRINT t, y(1), y(2)
450 NEXT k
460 END
470 REM assignments
480 z1 = 1: te1 = 1: te2 = 2
510 u = 1
520 x1 = x(1): x2 = x(2)
540 REM funkcie
550 f(1) = (z1 * u - x2 - (te1 + te2) * x1) / (te1 * te2)
560 f(2) = x1
570 RETURN
```

Program 4.2 (Simulation program in C)

```
#include <stdio.h>

void rk45 (rouble *u,rouble *y, rouble *f, rouble dt);
void fun(rouble y[], rouble f[], rouble u[]);

#define N 2 /* number of ODEs */

int main(void)
{
  rouble t=0, tend=10, dt=0.5;
  rouble y[N], u[1];
  rouble f[N];

  u[0]=1;y[0]=0;y[1]=0;
  printf("%f  %f   %f\n",t, y[0],y[1]);
  do{
    rk45 (u, y, f, dt);
    t+=dt;
```

```
    printf("%f  %f   %f\n",t, y[0],y[1]);
  }while (t<tend);
  return 0;
}

void fun(rouble y[], rouble f[], rouble u[])
{
  static rouble te1=1, te2=2, z=1;
  f[0]=(z*u[0]-y[1]-(te1+te2)*y[0])/(te1*te2);
  f[1]=y[0];
}

void rk45 (rouble *u, rouble *y, rouble *f, rouble dt)
{
  int i,j;
  rouble yold[N], fh[4*N];

  for (i=0 ; i<N ; i++)
    yold[i]=y[i];
  for(i=0; i<4; i++){
    fun(y, f, u);
    for(j=0;j<N; j++){
      fh[i*N+j]=dt*f[j];
      if(i<2) y[j]=yold[j]+0.5*fh[i*N+j];
      if(i==2) y[j]=yold[j]+fh[i*N+j];
    }
  }
  for(i=0; i<N; i++)
    y[i]=yold[i]+(fh[i]+2.0*(fh[N+i]+fh[2*N+i])+fh[3*N+i])/6.0;

}
```

Program 4.3 (Source code in MATLAB)

```
function [sys,x0,str,ts] = simss(t,x,u,flag)
  z1 = 1;
  te1 = 1;
  te2 = 2;
  den = 1/(te1 * te2);
  A = [-(te1+te2)/den  -1/den
           1           0];

  B = [z1/den; 0];

  C = [1 0; 0 1];
```

```
  switch flag,
   case 0
    [sys,x0,str,ts] = mdlInitializeSizes;
   case 1,
    sys = mdlDerivatives(t,x,u,A,B);
   case 3,
    sys = mdlOutputs(t,x,u,C);
   case 9
    sys = [];
  end

function [sys,x0,str,ts] = mdlInitializeSizes()
  sizes = simsizes;

  sizes.NumContStates  = 2;
  sizes.NumDiscStates  = 0;
  sizes.NumOutputs     = 2;
  sizes.NumInputs      = 1;
  sizes.DirFeedthrough = 0;
  sizes.NumSampleTimes = 1;

  sys = simsizes(sizes);
  x0  = [0; 0];
  str = [];
  ts  = [0 0];

function sys = mdlDerivatives(t,x,u,A,B)
  sys = A*x + B*u;

function sys = mdlOutputs(t,x,u,C)
  sys = C*x;
```

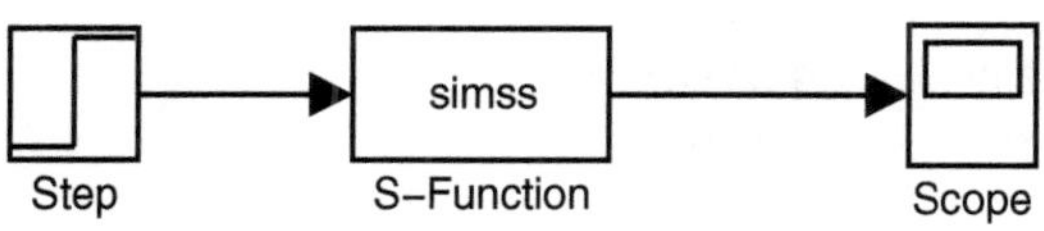

Fig. 4.9. Simulink block scheme

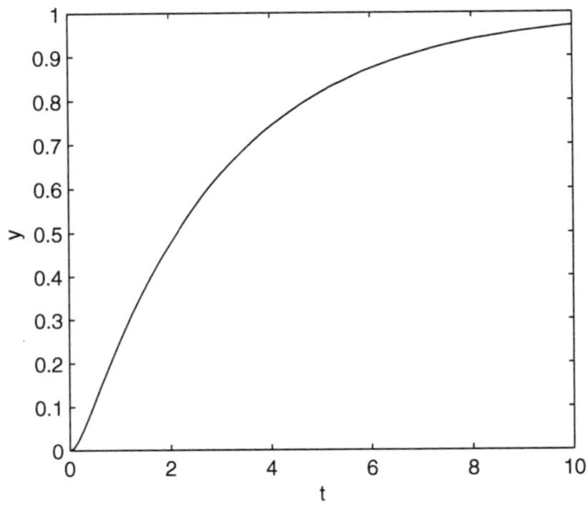

Fig. 4.10. Results from simulation

Table 4.1. Solution of the second order differential equation

t	y1	y2
0	0	0
.5	.1720378	.0491536
1	.238372	.1550852
1.5	.2489854	.2786338
2	.2323444	.3997614
2.5	.2042715	.5092093
3	.1732381	.6036183
3.5	.1435049	.6827089
4	.1169724	.7476815
4.5	.0926008	.8003312
5	.07532854	.8425783
5.5	.05983029	.8762348
6	.04730249	.9029045
6.5	.03726806	.9239527
7	.02928466	.9405137
7.5	.02296489	.9535139
8	.01798097	.9637005
8.5	.01406181	.9716715
9	.01098670	.9779023
9.5	.00857792	.9827687
10	.00669353	.9865671

4.2.4 Time Responses of Liquid Storage Systems

Consider the liquid storage system shown in Fig. 2.1. Assume its mathematical model in the form

$$F_1\frac{\mathrm{d}h_1}{\mathrm{d}t} + c_1 h_1 + c_2\sqrt{h_1} = q_0 \tag{4.58}$$

where c_1 and c_2 are constants obtained from measurements on a real process. The steady-state is given by the following equation

$$c_1 h_1^s + g(h_1^s) = q_0^s, \quad \text{where} \quad g(h_1^s) = c_2\sqrt{h_1^s} \tag{4.59}$$

Let us introduce the deviation variables

$$\begin{aligned} x_1 &= h_1 - h_1^s \\ u_1 &= q_0 - q_0^s \end{aligned} \tag{4.60}$$

The mathematical model can then be rewritten as

$$F_1\frac{\mathrm{d}x_1}{\mathrm{d}t} + c_1 x_1 + c_1 h_1^s + c_2\sqrt{x_1 + h_1^s} = u_1 + q_0^s \tag{4.61}$$

Substracting (4.58) from (4.61) yields

$$F_1\frac{\mathrm{d}x_1}{\mathrm{d}t} + c_1 x_1 + c_2\sqrt{x_1 + h_1^s} - c_2\sqrt{h_1^s} = u_1 \tag{4.62}$$

Let us introduce the function

$$G(x_1) = g(x_1 + h_1^s) - g(h_1^s) \tag{4.63}$$

then the mathematical model is finally given with deviation variables as

$$F_1\frac{\mathrm{d}x_1}{\mathrm{d}t} + c_1 x_1 + G(x_1) = u_1 \tag{4.64}$$

The Simulink block scheme that uses the MATLAB file `hs11m.m` (Program 4.4) as the S-function for solution of Eq. (4.64) is shown in Fig.4.11. www

Program 4.4 (MATLAB file hs11m.m)

```
function [sys,x0,str,ts] = hs11m(t,x,u,flag)
% Deviation model of the level tank;

% F1*(dx1/dt)+c1*x1+c2*(x1+h1s)^(1/2)-c2*(h1s)^(1/2)=u1 ;
% h1s =1.5 dm, q0s= 0.006359 dm^3/s     ;
% c1= 0.00153322 dm^2/s ,  c2 = 0.00331442 (dm^5/2)/s  ;

% F1 = 1.44dm^2,  Step change q00s = new value for t>=0;
% u1 is constrained as <-0.006359, 0.004161 dm^3/s>;
```

```
  switch flag,
   case 0
    [sys,x0,str,ts] = mdlInitializeSizes;
   case 1,
    sys = mdlDerivatives(t,x,u);
   case 3,
    sys = mdlOutputs(t,x,u);
   case 9
    sys = [];
  end

function [sys,x0,str,ts] = mdlInitializeSizes()
  sizes = simsizes;

  sizes.NumContStates  = 1;
  sizes.NumDiscStates  = 0;
  sizes.NumOutputs     = 1;
  sizes.NumInputs      = 1;
  sizes.DirFeedthrough = 0;
  sizes.NumSampleTimes = 1;

  sys = simsizes(sizes);
  x0  = [0];
  str = [];
  ts  = [0 0];

function sys = mdlDerivatives(t,x,u)
        c1 = 0.00153322;
        f1 = 1.44;
        a1 = -(c1/f1);
        c2 = 0.00331442;
        a2 = -(c2/f1);
        b1 = 1/f1;
        h1s = 1.5;
% Right hand sides;
        x1 = x(1)+h1s;
        if x1 < 0
          x1 = 0;
        end
        sys(1) = a1*x(1) + a2*sqrt(x1) - a2*sqrt(h1s) + b1*u(1);

function sys = mdlOutputs(t,x,u)
  sys = x;
```

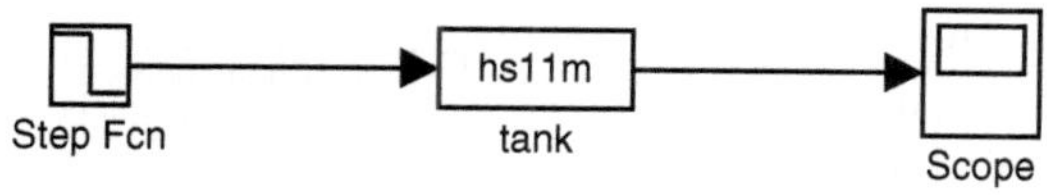

Fig. 4.11. Simulink block scheme for the liquid storage system

Linearised mathematical model in the neighbourhood of the steady-state $x_1 = 0$ is

$$\frac{\mathrm{d}x_1}{\mathrm{d}t} = -\frac{c_1}{F_1}x_1 - \frac{1}{F_1}\frac{\partial G(0)}{\partial x_1}x_1 + \frac{1}{F_1}u_1 \tag{4.65}$$

where

$$\frac{\partial G}{\partial x_1} = \frac{c_2}{2}\frac{1}{\sqrt{x_1 + h_1^s}} \quad , \quad \frac{\partial G(0)}{\partial x_1} = \frac{c_2}{2}\frac{1}{\sqrt{h_1^s}}$$

and finally

$$\frac{\mathrm{d}x_1}{\mathrm{d}t} = \left(-\frac{c_1}{F_1} - \frac{c_2}{2F_1\sqrt{h_1^s}}\right)x_1 + \frac{1}{F_1}u_1 \tag{4.66}$$

Fig. 4.12 shows the response of the tank to step change of the flow q_0 equal to $-0.0043\,\mathrm{dm^3\,s^{-1}}$. The steady-state before the change was characterised by the flow $q_0^s = 0.006359\,\mathrm{dm^3\,s^{-1}}$ and the tank level $h_1^s = 1.5\,\mathrm{dm}$. The coefficients c_1, c_2 and the crossover area F_1 corresponding to the real liquid storage tank are

$$\begin{aligned} c_1 &= 1.53322.10^{-3}\mathrm{dm^2\,s^{-1}} \\ c_2 &= 3.31142.10^{-3}\mathrm{dm^{2.5}\,s^{-1}} \\ F_1 &= 1.44\mathrm{dm^2} \end{aligned}$$

Fig. 4.12 also shows the step response of the linearised model (4.66). Both curves can serve for analysis of the error resulting from linearisation of the mathematical model of the tank.

4.2.5 Time Responses of CSTR

Consider a CSTR with a cooling jacket. In the reactor, an irreversible exothermic reaction takes place. We assume that its mathematical model is in the form

$$\frac{\mathrm{d}c_A}{\mathrm{d}t'} = \frac{q}{V}c_{Av} - c_A(\frac{q}{V} + k) \tag{4.67}$$

$$\frac{\mathrm{d}\vartheta}{\mathrm{d}t'} = \frac{q}{V}\vartheta_v - \frac{q}{V}\vartheta - \frac{\alpha F}{V\rho c_p}(\vartheta - \vartheta_c) + \frac{kH_r}{\rho c_p}c_A \tag{4.68}$$

$$\frac{\mathrm{d}\vartheta_c}{\mathrm{d}t'} = \frac{q_c}{V_c}\vartheta_{cv} - \frac{q_c}{V_c}\vartheta_c + \frac{\alpha F}{V_c\rho_c c_{pc}}(\vartheta - \vartheta_c) \tag{4.69}$$

$$k = k_0 \exp\left(-\frac{g}{\vartheta}\right) \tag{4.70}$$

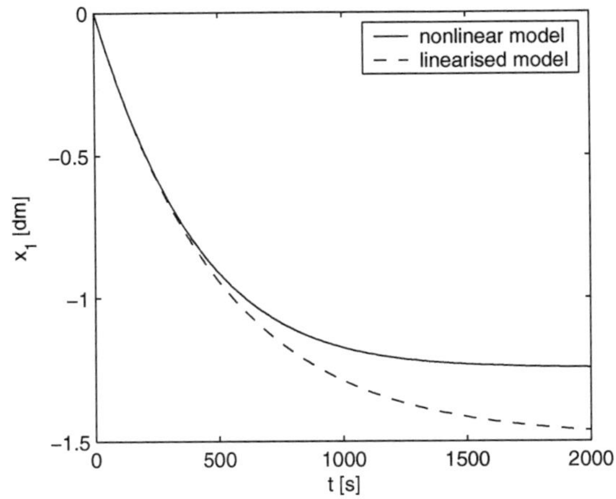

Fig. 4.12. Response of the tank to step change of q_0

where

t' - time variable,
c_A - molar concentration of A (mole/volume) in the outlet stream,
c_{Av} - molar concentration of A (mole/volume) in the inlet stream,
V - reactor volume,
q - volumetric flow rate,
ϑ - temperature of reaction mixture,
ϑ_v - temperature in the inlet stream,
F - heat transfer area,
α - overall heat transfer coefficient,
ρ - liquid density,
c_p - liquid specific heat capacity,
H_r - heat of reaction,
ϑ_c - cooling temperature,
ϑ_{cv} - cooling temperature in the inlet cooling stream,
q_c - volumetric flow rate of coolant,
V_c - coolant volume in the jacket,
ρ_c - coolant density,
c_{pc} - coolant specific heat capacity,
k_0 - frequency factor,
g - activation energy divided by the gas constant.

Compared to the CSTR model from page 31 we included the equation (4.69) describing the behaviour of cooling temperature ϑ_c. The state-space variables are $c_A, \vartheta, \vartheta_c$, the input variable is q_c, and the output variable is ϑ. We assume that the variables $c_{Av}, \vartheta_v, \vartheta_{cv}$ are held constant and the other parameters of the reactor are also constant.

The reactor model can be rewritten as

$$\frac{dx'_1}{dt} = 1 - x'_1 - f'_1(x'_1, x'_2) \tag{4.71}$$
$$\frac{dx'_2}{dt} = b'_1 - x'_2 - f'_2(x'_2) + f'_3(x'_3) + f'_1(x'_1, x'_2) \tag{4.72}$$
$$\frac{dx'_3}{dt} = f'_4(x'_2) - f'_5(x'_3) - f'_6(u', x'_3) + f'_7(u') \tag{4.73}$$
$$k = k_0 \exp\left(-\frac{g\rho c_p}{c^s_{Av} H_r x'_2}\right) \tag{4.74}$$

where

$$\begin{aligned}
x'_1 &= \frac{c_A}{c^s_{Av}} , & x'_2 &= \frac{\rho c_p \vartheta}{c^s_{Av} H_r} \\
x'_3 &= \frac{\rho_c c_{pc} \vartheta_c}{c^s_{Av} H_r} , & t &= \frac{q}{V} t' \\
u' &= \frac{q_c}{q^s_c} , & b'_1 &= \frac{\rho c_p \vartheta^s_v}{c^s_{Av} H_r} \\
f'_1(x'_1, x'_2) &= \frac{V}{q} x'_1 k_0 \exp\left(-\frac{g\rho c_p}{c^s_{Av} H_r x'_2}\right) , & f'_2(x'_2) &= \frac{\alpha F}{V \rho c_p} x'_2 \\
f'_3(x'_3) &= \frac{\alpha F}{V \rho_c c_{pc}} x'_3 , & f'_4(x'_2) &= \frac{\alpha F V}{q V_c \rho c_p} x'_2 \\
f'_5(x'_3) &= \frac{\alpha F V}{q V_c \rho_c c_p} x'_3 , & f'_6(u', x'_3) &= \frac{V q^s_c}{q V_c} u' x'_3 \\
f'_7(u') &= \frac{\rho_c c_{pc} V q^s_c \vartheta^s_{cv}}{c^s_{Av} H_r q V_c}
\end{aligned}$$

and variables with superscript ^{s} denote steady-state values.

In the steady-state for state variables x'_1, x'_2, x'_3 holds

$$0 = 1 - x'^s_1 - f'_1(x'^s_1, x'^s_2) \tag{4.75}$$
$$0 = b'_1 - x'^s_2 - f'_2(x'^s_2) + f'_3(x'^s_3) + f'_1(x'^s_1, x'^s_2) \tag{4.76}$$
$$0 = f'_4(x'^s_2) - f'_5(x'^s_3) - f'_6(u'^s, x'^s_3) + f'_7(u'^s) \tag{4.77}$$

We define deviation variables and functions

$$x_1 = x'_1 - x'^s_1 \tag{4.78}$$
$$x_2 = x'_2 - x'^s_2 \tag{4.79}$$
$$x_3 = x'_3 - x'^s_3 \tag{4.80}$$
$$u = u' - u'^s \tag{4.81}$$
$$f_1(x_1, x_2) = f'_1(x_1 + x'^s_1, x_2 + x'^s_2) - f'_1(x'^s_1, x'^s_2) \tag{4.82}$$
$$f_2(x_2) = f'_2(x_2 + x'^s_2) - f'_2(x'^s_2) \tag{4.83}$$
$$f_3(x_3) = f'_3(x_3 + x'^s_3) - f'_3(x'^s_3) \tag{4.84}$$
$$f_4(x_2) = f'_4(x_2 + x'^s_2) - f'_4(x'^s_2) \tag{4.85}$$
$$f_5(x_3) = f'_5(x_3 + x'^s_3) - f'_5(x'^s_3) \tag{4.86}$$
$$f_6(u, x_3) = f'_6(u + u'^s_1, x_3 + x'^s_3) - f'_6(u'^s, x'^s_3) \tag{4.87}$$
$$f_7(u) = f'_7(u + u'^s) - f'_7(u'^s) \tag{4.88}$$

The original mathematical model of the reactor given by (4.67) - (4.69) can finally be rewritten as

$$\frac{\mathrm{d}x_1}{\mathrm{d}t} = -x_1 - f_1(x_1, x_2) \tag{4.89}$$

$$\frac{\mathrm{d}x_2}{\mathrm{d}t} = -x_2 - f_2(x_2) + f_3(x_3) + f_1(x_1, x_2) \tag{4.90}$$

$$\frac{\mathrm{d}x_3}{\mathrm{d}t} = f_4(x_2) - f_5(x_3) - f_6(u, x_3) + f_7(u) \tag{4.91}$$

www Figure 4.13 shows the Simulink block-scheme that uses the program `rea7m1.m` (Program 4.5) as its S-function for the solution of the differential equations (4.89) - (4.91).

Program 4.5 (MATLAB file rea7m1.m)

```
function [sys,x0,str,ts] = rea7m1(t,x,u,flag)
% Non-linear deviation model of the CSTR;
% 3 equations, Reaction of the type A ----> B;

%    k0 = a1           [1/min];
%    g = a2            [K];
%    ro = a3           [kg/m3];
%    cp = a4           [kJ/kg.K];
%    V = a5            [m3];
%    Hr = a6           [kJ/kmol];
%    F = a7            [m2];
%    al = a8           [kJ/m2.min.K];
%    roc = a9          [kg/m3];
%    cpc = a10         [kJ/kg.K];
%    Vc = a11          [m3];
%    qs = a12          [m3/min];
%    Thvs = a13        [K];
%    cavs = a14        [kmol/m3];
%    Thcvs = a15       [K];
%    qcs = a16         [m3/min];
%    cas = a17         [kmol/m3];
%    Ths = a18         [K];
%    Thcs = a19        [K];

%    x(1) - concentration of A, dimensionless, deviation;
%    x(2) - temperature of the mixture, dimensionless, deviation;
%    x(3) - jacket temperature, dimensionless, deviation;
%    u(1) - cooling flowrate, dimensionless, deviation;
%    u=qc/qcs ;
%    u(1)=<-0.2,   0.2> = cca<-0.001m3/min,0.001m3/min>;
```

```
  switch flag,
   case 0
    [sys,x0,str,ts] = mdlInitializeSizes;
   case 1,
    sys = mdlDerivatives(t,x,u);
   case 3,
    sys = mdlOutputs(t,x,u);
   case 9
    sys = [];
  end

function [sys,x0,str,ts] = mdlInitializeSizes()
  sizes = simsizes;

  sizes.NumContStates  = 3;
  sizes.NumDiscStates  = 0;
  sizes.NumOutputs     = 6;
  sizes.NumInputs      = 1;
  sizes.DirFeedthrough = 0;
  sizes.NumSampleTimes = 1;

  sys = simsizes(sizes);
  x0  = [0; 0; 0];
  str = [];
  ts  = [0 0];

function sys = mdlDerivatives(t,x,u)
% Calculation of f1c, f2c,...,f7c in the steady-state;
  a1=10000000000000;
  a2=11078;
  a3=970;
  a4=4.05;
  a5=1.8;
  a6=149280;
  a7=5.04;
  a8=130;
  a9=998;
  a10=4.18;
  a11=0.7;
  a12=0.25;
  a13=370;
  a14=0.88;
  a15=285;
  a16=0.05;
  a17=0.0345;
```

```
  a18=385.877;
  a19=361.51753;


  x1cs = a17/a14;
  x2cs = (a3*a4*a18)/(a14*a6);
  x3cs = (a9*a10*a19)/(a14*a6);
  u1cs = a16/a16;
  f1cs = ((a5/a12)*x1cs*a1)*(exp((-a2*a3*a4)/(a14*a6*x2cs)));
  f2cs = ((a7*a8)/(a12*a3*a4))*x2cs;
  f3cs = ((a7*a8)/(a12*a9*a10))*x3cs;
  f4cs = ((a7*a8*a5)/(a12*a11*a3*a4))*x2cs;
  f5cs = ((a5*a7*a8)/(a12*a11*a9*a10))*x3cs;
  f6cs = ((a5*a16)/(a12*a11))*u1cs*x3cs;
  f7cs = ((a9*a10*a5*a16*a15)/(a14*a6*a12*a11))*u1cs;

  x1c = x(1) + x1cs;
  x2c = x(2) + x2cs;
  x3c = x(3) + x3cs;
  u1c = u(1) + u1cs;
  f1c = ((a5/a12)*x1c*a1)*(exp((-a2*a3*a4)/(a14*a6*x2c)));
  f2c = ((a7*a8)/(a12*a3*a4))*x2c;
  f3c = ((a7*a8)/(a12*a9*a10))*x3c;
  f4c = ((a7*a8*a5)/(a12*a11*a3*a4))*x2c;
  f5c = ((a5*a7*a8)/(a12*a11*a9*a10))*x3c;
  f6c = ((a5*a16)/(a12*a11))*u1c*x3c;
  f7c = ((a9*a10*a5*a16*a15)/(a14*a6*a12*a11))*u1c;
  f1 = f1c - f1cs;
  f2 = f2c - f2cs;
  f3 = f3c - f3cs;
  f4 = f4c - f4cs;
  f5 = f5c - f5cs;
  f6 = f6c - f6cs;
  f7 = f7c - f7cs;

% Right hand sides of ODEs;
  sys(1)=-x(1) - f1;
  sys(2)=-x(2) - f2 + f3 + f1;
  sys(3)=f4 - f5 - f6 + f7;

function sys = mdlOutputs(t,x,u)
  a3=970;
  a4=4.05;
  a6=149280;
  a9=998;
```

```
a10=4.18;
a14=0.88;

sys(1)=x(1);
sys(4)=a14*x(1);
sys(2)=x(2);
sys(5)=(a14*a6*x(2))/(a3*a4);
sys(3)=x(3);
sys(6)=(a14*a6*x(3))/(a9*a10);
```

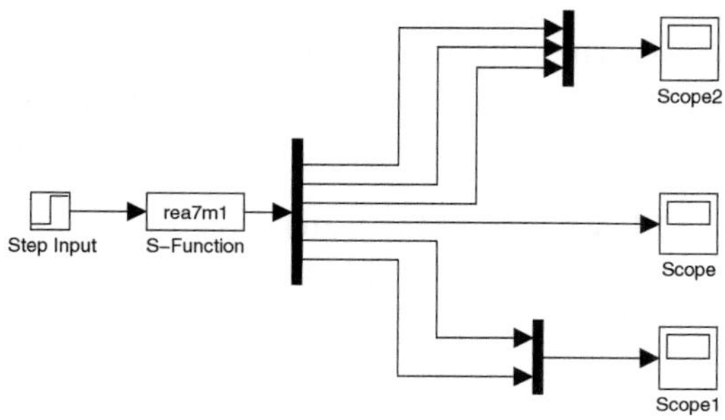

Fig. 4.13. Simulink block scheme for the nonlinear CSTR model

Linearised mathematical model of the CSTR with steady-state $x_1 = x_2 = x_3 = 0$ is of the form

$$\frac{\mathrm{d}x_1}{\mathrm{d}t} = -x_1 - \frac{\partial f_1(0,0)}{\partial x_1}x_1 - \frac{\partial f_1(0,0)}{\partial x_2}x_2 \tag{4.92}$$

$$\frac{\mathrm{d}x_2}{\mathrm{d}t} = -x_2 - \frac{\partial f_2(0)}{\partial x_2}x_2 + \frac{\partial f_3(0)}{\partial x_3}x_3 + \frac{\partial f_1(0,0)}{\partial x_1}x_1 + \frac{\partial f_1(0,0)}{\partial x_2}x_2 \tag{4.93}$$

$$\begin{aligned}\frac{\mathrm{d}x_3}{\mathrm{d}t} &= \frac{\partial f_4(0)}{\partial x_2}x_2 - \frac{\partial f_5(0)}{\partial x_3}x_3 \\ &\quad - \frac{\partial f_6(0,0)}{\partial x_3}x_3 - \frac{\partial f_6(0,0)}{\partial u}u + \frac{\partial f_7(0)}{\partial u}u\end{aligned} \tag{4.94}$$

or

$$\frac{\mathrm{d}\boldsymbol{x}}{\mathrm{d}t} = \boldsymbol{A}\boldsymbol{x} + \boldsymbol{B}u \tag{4.95}$$

where

$$\boldsymbol{x} = (x_1, x_2, x_3)^T$$

$$\boldsymbol{A} = \begin{pmatrix} a_{s11} & a_{s12} & a_{s13} \\ a_{s21} & a_{s22} & a_{s23} \\ a_{s31} & a_{s32} & a_{s33} \end{pmatrix}, \quad \boldsymbol{B} = \begin{pmatrix} 0 \\ 0 \\ b_{s31} \end{pmatrix}$$

$$a_{s11} = -1 - \frac{\partial f_1(0,0)}{\partial x_1}, \qquad a_{s12} = -\frac{\partial f_1(0,0)}{\partial x_2}$$

$$a_{s13} = 0, \qquad a_{s21} = \frac{\partial f_1(0,0)}{\partial x_1}$$

$$a_{s22} = -1 - \frac{\partial f_2(0)}{\partial x_2} + \frac{\partial f_1(0,0)}{\partial x_2}, \qquad a_{s23} = \frac{\partial f_3(0)}{\partial x_3}$$

$$a_{s31} = 0, \qquad a_{s32} = \frac{\partial f_4(0)}{\partial x_2}$$

$$a_{s33} = -\frac{\partial f_5(0)}{\partial x_3} + \frac{\partial f_6(0,0)}{\partial x_3}, \qquad b_{s31} = \frac{\partial f_7(0)}{\partial u} - \frac{\partial f_6(0,0)}{\partial u}$$

Figures 4.14, 4.15, 4.16 show responses of the CSTR to step change of q_c ($\Delta u = 10$). Numerical values of all variables are given in Program 4.6.

File `kolire8.m` (Program 4.6) calculates for the given steady-state of the CSTR parameters of the corresponding linearised mathematical model.

Program 4.6 (MATLAB file kolire8.m)

```
% Linearised model of the CSTR, 3 equations;
% state-space equations, Reaction A ----> B;

%   k0 = a1         [1/min];
%   g = a2          [K];
%   ro = a3         [kg/m3];
%   cp = a4         [kJ/kg.K];
%   V = a5          [m3];
%   Hr = a6         [kJ/kmol];
%   F = a7          [m2];
%   al = a8         [kJ/m2.min.K];
%   roc = a9        [kg/m3];
%   cpc = a10       [kJ/kg.K];
%   Vc = a11        [m3];
%   qs = a12        [m3/min];
%   Thvs = a13      [K];
%   cavs = a14      [kmol/m3];
%   Thcvs = a15     [K];
%   qcs = a16       [m3/min];
%   cas = a17       [kmol/m3];
%   Ths = a18       [K];
%   Thcs = a19      [K];

%   x(1) - concentration of A, dimensionless, deviation;
%   x(2) - temperature of the mixture, dimensionless, deviation;
%   x(3) - jacket temperature, dimensionless, deviation;
%   u(1) - cooling flowrate, dimensionless, deviation;
%   u=qc/qcs ;
```

```
%   u(1)=<-0.2,    0.2> = cca<-0.001m3/min,0.001m3/min>;

a1=10000000000000; a2=11078;
a3=970; a4=4.05;
a5=1.8; a6=149280;
a7=5.04; a8=130;
a9=998; a10=4.18;
a11=0.7; a12=0.25;
a13=370; a14=0.88;
a15=285; a16=0.05;
a17=0.0345; a18=385.877;
a19=361.51753;

% Calculation of the state-space coefficients;
as11=-1-((a5*a1)/a12)*(exp(-a2/a18));
as12=-((a5*a1*a6*a2*a17)/(a12*a3*a4*((a18)^2)))*(exp(-a2/a18));
as13=0;
as21=((a5*a1)/a12)*(exp(-a2/a18));
as22=-1-((a7*a8)/(a12*a3*a4));
as221=((a5*a1*a6*a2*a17)/(a12*a3*a4*((a18)^2)))*(exp(-a2/a18));
as22=as22 + as221;
as23=(a7*a8)/(a12*a9*a10);
as31=0;
as32=(a7*a8*a5)/(a12*a11*a3*a4);
as33=-(a7*a8*a5)/(a12*a11*a9*a10)-(a5*a16)/(a12*a11);
bs11=0;
bs21=0;
bs31=-(a5*a16*a9*a10*a19)/(a12*a11*a14*a6);
bs31=bs31+(a9*a10*a5*a16*a15)/(a14*a6*a12*a11);
```

4.3 Frequency Analysis

4.3.1 Response of the Heat Exchanger to Sinusoidal Input Signal

Consider a jacketed heat exchanger described by the differential equation (see Example 3.13)

$$T_1 \frac{\mathrm{d}y}{\mathrm{d}t} + y = Z_1 u \tag{4.96}$$

where $y = \vartheta - \vartheta^s$ is the deviation of the liquid temperature from its steady state value, $u = \vartheta_p - \vartheta_p^s$ is the deviation of the jacket temperature from its

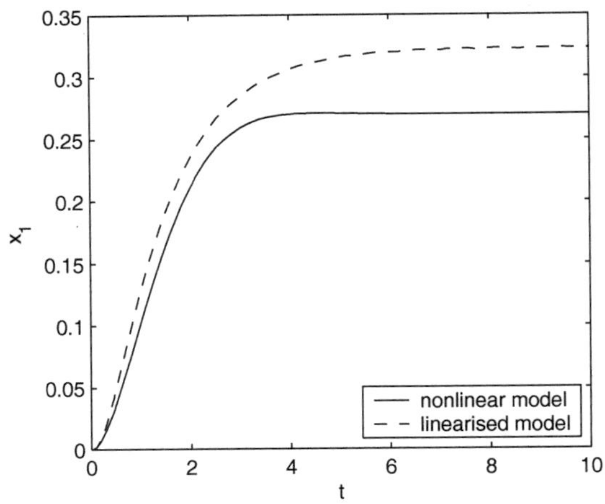

Fig. 4.14. Responses of dimensionless deviation output concentration x_1 to step change of q_c

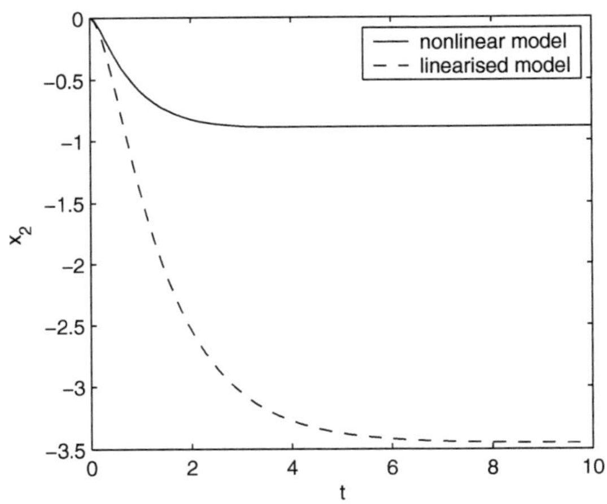

Fig. 4.15. Responses of dimensionless deviation output temperature x_2 to step change of q_c

steady state value, T_1 is the process time constant, and Z_1 is the process steady-state gain.

The process transfer function of (4.96) is given as

$$G(s) = \frac{Y(s)}{U(s)} = \frac{Z_1}{T_1 s + 1} \tag{4.97}$$

Let the following sinusoidal signal with amplitude A_1 and frequency ω influence the heat exchanger:

$$u(t) = A_1 \sin \omega t \; 1(t) \tag{4.98}$$

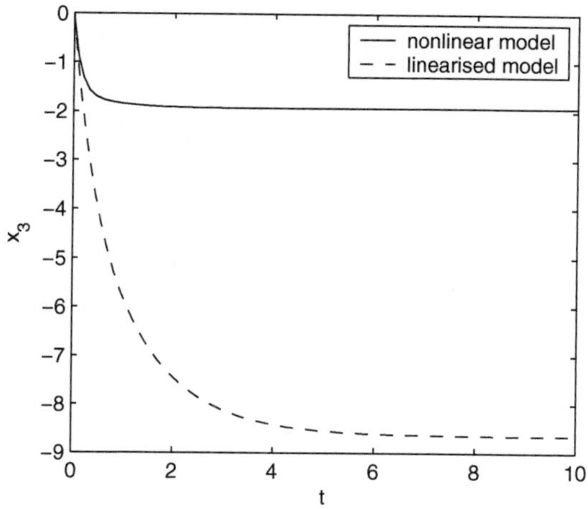

Fig. 4.16. Responses of dimensionless deviation cooling temperature x_3 to step change of q_c

Then the Laplace transform of the process output can be written as

$$Y(s) = \frac{Z_1}{T_1 s + 1} \frac{A_1 \omega}{s^2 + \omega^2} \tag{4.99}$$

$$Y(s) = \frac{Z_1 A_1 \omega}{T_1 s^3 + s^2 + T_1 \omega^2 s + \omega^2} \tag{4.100}$$

$$Y(s) = \frac{\frac{Z_1 A_1 \omega}{T_1}}{s^3 + \frac{1}{T_1} s^2 + \omega^2 s + \frac{\omega^2}{T_1}} \tag{4.101}$$

The denominator roots are:

$$s_1 = -\mathrm{j}\omega, \; s_2 = \mathrm{j}\omega, \; s_3 = -1/T_1$$

After taking the inverse Laplace transform the output of the process is described by

$$\begin{aligned} y(t) = & \frac{\frac{Z_1 A_1}{T_1}\omega(-\mathrm{j}\omega)}{-\mathrm{j}\omega[3(-\mathrm{j}\omega)^2 + \frac{2}{T_1}(-\mathrm{j}\omega) + \omega^2]} \mathrm{e}^{-\mathrm{j}\omega t} + \\ & + \frac{\frac{Z_1 A_1}{T_1}\omega(\mathrm{j}\omega)}{\mathrm{j}\omega[3(\mathrm{j}\omega)^2 + \frac{2}{T_1}(\mathrm{j}\omega) + \omega^2]} \mathrm{e}^{\mathrm{j}\omega t} + \\ & + \frac{\frac{Z_1 A_1}{T_1}\omega(-\frac{1}{T_1})}{-\frac{1}{T_1}[2(-\frac{1}{T_1})^2 + \frac{2}{T_1}(-\frac{1}{T_1}) + \omega^2]} \mathrm{e}^{-\frac{t}{T_1}} \end{aligned} \tag{4.102}$$

$$y(t) = \frac{Z_1 A_1}{-2\omega T_1 - 2\mathrm{j}} \mathrm{e}^{-\mathrm{j}\omega t} + \frac{Z_1 A_1}{-2\omega T_1 + 2\mathrm{j}} \mathrm{e}^{\mathrm{j}\omega t} + K \mathrm{e}^{-\frac{t}{T_1}} \tag{4.103}$$

where $K = (2\omega T_1 A_1)/(2T_1 - 2 + \omega^2 T_1^2)$.

If the sine wave is continued for a long time, the exponential term disappears and the remaining terms can be further manipulated to yield

$$y(t) = \frac{Z_1 A_1}{-2\omega T_1 - 2\mathrm{j}} \frac{-2\omega T_1 + 2\mathrm{j}}{-2\omega T_1 + 2\mathrm{j}} \mathrm{e}^{-\mathrm{j}\omega t} + \frac{Z_1 A_1}{-2\omega T_1 + 2\mathrm{j}} \frac{-2\omega T_1 - 2\mathrm{j}}{-2\omega T_1 - 2\mathrm{j}} \mathrm{e}^{\mathrm{j}\omega t} \tag{4.104}$$

$$y(t) = Z_1 A_1 \left[\frac{-\omega T_1 + \mathrm{j}}{2(\omega^2 T_1^2 + 1)} \mathrm{e}^{-\mathrm{j}\omega t} + \frac{-\omega T_1 - \mathrm{j}}{2(\omega^2 T_1^2 + 1)} \mathrm{e}^{\mathrm{j}\omega t} \right] \tag{4.105}$$

$$y(t) = Z_1 A_1 \left[\frac{-\omega T_1}{(\omega^2 T_1^2 + 1)} \frac{\mathrm{e}^{-\mathrm{j}\omega t} + \mathrm{e}^{\mathrm{j}\omega t}}{2} + \frac{1}{(\omega^2 T_1^2 + 1)} \frac{\mathrm{e}^{\mathrm{j}\omega t} - \mathrm{e}^{-\mathrm{j}\omega t}}{2} \right] \tag{4.106}$$

Applying the Euler identities (3.15) yields

$$y(t) = Z_1 A_1 \left[-\frac{\omega T_1}{\omega^2 T_1^2 + 1} \cos \omega t + \frac{1}{\omega^2 T_1^2 + 1} \sin \omega t \right] \tag{4.107}$$

Finally, using the trigonometric identity

$$\sin(\omega t + \varphi) = \sin \varphi \cos \omega t + \cos \varphi \sin \omega t$$

gives

$$y(t) = Z_1 A_1 \left[\frac{\sqrt{\omega^2 T_1^2 + 1}}{\omega^2 T_1^2 + 1} \sin(\omega t + \varphi) \right] \tag{4.108}$$

where $\varphi = -\arctan \omega T_1$.

If we set in (4.97) $s = \mathrm{j}\omega$, then

$$G(\mathrm{j}\omega) = \frac{Z_1}{T_1 \mathrm{j}\omega + 1} \tag{4.109}$$

$$|G(\mathrm{j}\omega)| = \frac{Z_1}{\sqrt{\omega^2 T_1^2 + 1}} \tag{4.110}$$

which is the same as the amplitude in (4.108) divided by A_1. Thus $y(t)$ can also be written as

$$y(t) = A_1 |G(\mathrm{j}\omega)| \sin(\omega t + \varphi) \tag{4.111}$$

It follows from (4.108) that the output amplitude is a function of the input amplitude A_1, input frequency ω, and the system properties. Thus,

$$A_1 |G(\mathrm{j}\omega)| = A_1 f(\omega, Z_1, T_1). \tag{4.112}$$

For the given system with the constants Z_1 and T_1, it is clear that increasing input frequency results in decreasing output amplitude. The phase lag is given as

$$\varphi = -\arctan T_1 \omega \tag{4.113}$$

and the influence of the input frequency ω to φ is opposite to amplitude.

www The simulation of $u(t)$ and $y(t)$ from the equations (4.98) and (4.108) for $Z_1 = 0.4$, $T_1 = 5.2\,\mathrm{min}$, $A_1 = 5°C$, and $\omega = 0.2\,\mathrm{rad/min}$ is shown in Fig. 4.17.

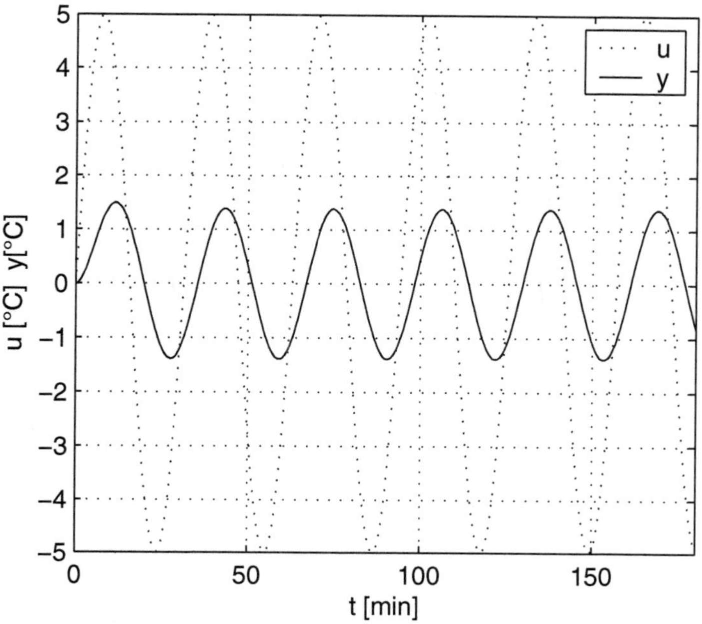

Fig. 4.17. Ultimate response of the heat exchanger to sinusoidal input

4.3.2 Definition of Frequency Responses

A time periodic function $f(t)$ with the period T_f satisfying the Dirichlet conditions can be expanded into the Fourier expansion

$$f(t) = \frac{2}{T_f}\left[\frac{a_0}{2} + \sum_{n=1}^{\infty}(a_n \cos n\omega_f t + b_n \sin n\omega_f t)\right] \tag{4.114}$$

where $\omega_f = 2\pi/T_f$ is the basic circular frequency. The coefficients a_0, a_n, b_n, $(n = 1, 2, \ldots)$ are given as

$$a_0 = \int_{-T_f/2}^{T_f/2} f(\tau)\mathrm{d}\tau$$

$$a_n = \int_{-T_f/2}^{T_f/2} f(\tau)\cos\frac{2\pi n\tau}{T_f}\mathrm{d}\tau$$

$$b_n = \int_{-T_f/2}^{T_f/2} f(\tau)\sin\frac{2\pi n\tau}{T_f}\mathrm{d}\tau$$

Using the Euler identity, the Fourier expansion can be rewritten as

$$f(t) = \sum_{n=-\infty}^{\infty} c_n \mathrm{e}^{\mathrm{j}n\omega_f t} \tag{4.115}$$

where

$$c_n = \frac{1}{T_f} \int_{-T_f/2}^{T_f/2} f(\tau) \mathrm{e}^{-\mathrm{j}n\omega_f \tau} \mathrm{d}\tau$$

Any nonperiodic time function can be assumed as periodic with the period approaching infinity. If we define $\omega = n\omega_f$ then

$$f(t) = \frac{1}{2\pi} \sum_{n=-\infty}^{\infty} \left\{ \int_{-T_f/2}^{T_f/2} f(\tau) \mathrm{e}^{-\mathrm{j}\omega\tau} \mathrm{d}\tau \right\} \mathrm{e}^{\mathrm{j}\omega t} [(n+1)\omega_f - n\omega_f] \tag{4.116}$$

If $T_f \to \infty$ and $[(n+1)\omega_f - n\omega_f] = \Delta\omega_f \to 0$ then

$$f(t) = \frac{1}{2\pi} \int_{-\infty}^{\infty} F(\mathrm{j}\omega) \mathrm{e}^{\mathrm{j}\omega t} \mathrm{d}\omega \tag{4.117}$$

or

$$F(\mathrm{j}\omega) = \int_{-\infty}^{\infty} f(t) \mathrm{e}^{-\mathrm{j}\omega t} \mathrm{d}t \tag{4.118}$$

The equations (4.117), (4.118) are *the Fourier integral equations* that describe the influence between the original time function and its frequency transform $F(\mathrm{j}\omega)$. Compared to the Fourier expansion they describe an expansion to the continuous spectrum with the infinitesimally small difference $\mathrm{d}\omega$ of two neighbouring harmonic signals.

The integral (4.118) expresses a transformation (or operator) that assigns to the function $f(t)$ the function $F(\mathrm{j}\omega)$. This transformation is called *the Fourier transform* of $f(t)$. If we know the transformed $F(\mathrm{j}\omega)$, the original function $f(t)$ can be found from (4.117) by the inverse Fourier transform. The important condition of the Fourier transform is the existence of the integrals (4.117), (4.118).

Complex transfer function, or *frequency transfer function* $G(\mathrm{j}\omega)$ is the Fourier transform corresponding to the transfer function $G(s)$. For $G(\mathrm{j}\omega)$ holds

$$G(\mathrm{j}\omega) = \frac{Y(\mathrm{j}\omega)}{U(\mathrm{j}\omega)} \tag{4.119}$$

If the frequency transfer function exists, it can be easily obtained from the system transfer function by formal exchange of the argument s,

$$G(\mathrm{j}\omega) = G(s)_{s=\mathrm{j}\omega} \tag{4.120}$$

or

$$G(\mathrm{j}\omega) = \frac{b_m(\mathrm{j}\omega)^m + b_{m-1}(\mathrm{j}\omega)^{m-1} + \cdots + b_0}{a_n(\mathrm{j}\omega)^n + b_{n-1}(\mathrm{j}\omega)^{n-1} + \cdots + a_0} \tag{4.121}$$

The frequency transfer function of a singlevariable system can be obtained from

$$G(\mathrm{j}\omega) \equiv \int_0^\infty g(t)\mathrm{e}^{-\mathrm{j}\omega t}\mathrm{d}t. \tag{4.122}$$

Analogously, for multivariable systems follows

$$\boldsymbol{G}(\mathrm{j}\omega) \equiv \int_0^\infty \boldsymbol{g}(t)\mathrm{e}^{-\mathrm{j}\omega t}\mathrm{d}t. \tag{4.123}$$

If the transfer function matrix $\boldsymbol{G}(s)$ is stable, then *frequency transfer function matrix* exists as the Fourier transform of the impulse response matrix and can be calculated from (see equations (3.46), (4.8))

$$\boldsymbol{G}(\mathrm{j}\omega) = \boldsymbol{C}(\mathrm{j}\omega \boldsymbol{I} - \boldsymbol{A})^{-1}\boldsymbol{B} + \boldsymbol{D} \tag{4.124}$$

Frequency transfer function is a complex variable corresponding to a real variable ω that characterises the forced oscillations of the output $y(t)$ for the harmonic input $u(t)$ with frequency ω. Harmonic functions can be mathematically described as

$$\bar{u} = A_1\mathrm{e}^{\mathrm{j}(\omega t+\varphi_1)} \tag{4.125}$$
$$\bar{y} = A_2\mathrm{e}^{\mathrm{j}(\omega t+\varphi_2)}. \tag{4.126}$$

The ratio of these functions is a complex variable $G(\mathrm{j}\omega)$ defined by (4.122). Thus it can be written as

$$\frac{\bar{y}}{\bar{u}} = \frac{A_2}{A_1}\mathrm{e}^{\mathrm{j}(\varphi_2-\varphi_1)} = G(\mathrm{j}\omega). \tag{4.127}$$

The *magnitude* of $G(\mathrm{j}\omega)$

$$|G(\mathrm{j}\omega)| = A(\omega) \tag{4.128}$$

is given as the ratio A_2/A_1 of output and input variable magnitudes. The *phase angle*

$$\varphi(\omega) = \varphi_2 - \varphi_1 \tag{4.129}$$

is determined as the phase shift between the output and input variable and is given in units of radians as a function of ω. $G(\mathrm{j}\omega)$ can be written in the polar form

$$G(\mathrm{j}\omega) = A(\omega)\mathrm{e}^{\mathrm{j}\varphi(\omega)} \tag{4.130}$$

The graph of $G(\mathrm{j}\omega)$

$$G(\mathrm{j}\omega) = |G(\mathrm{j}\omega)|\mathrm{e}^{\mathrm{j}\arg G(\mathrm{j}\omega)} = \Re[G(\mathrm{j}\omega)] + \mathrm{j}\Im[G(\mathrm{j}\omega)] \tag{4.131}$$

in the complex plane is called the *Nyquist diagram*. The magnitude and phase angle can be expressed as follows:

$$|G(j\omega)| = \sqrt{\{\Re[G(j\omega)]\}^2 + \{\Im[G(j\omega)]\}^2} \tag{4.132}$$

$$|G(j\omega)| = \sqrt{G(j\omega)G(-j\omega)} \tag{4.133}$$

$$\tan\varphi = \frac{\Im[G(j\omega)]}{\Re[G(j\omega)]} \tag{4.134}$$

$$\varphi = \arctan\frac{\Im[G(j\omega)]}{\Re[G(j\omega)]} \tag{4.135}$$

Essentially, the Nyquist diagram is a polar plot of $G(j\omega)$ in which frequency ω appears as an implicit parameter.

The function $A = A(\omega)$ is called *magnitude frequency response* and the function $\varphi = \varphi(\omega)$ *phase angle frequency response.* Their plots are usually given with logarithmic axes for frequency and magnitude and are referred to as Bode plots.

Let us investigate the logarithm of $A(\omega)\exp[j\varphi(\omega)]$

$$\ln G(j\omega) = \ln A(\omega) + j\varphi(\omega) \tag{4.136}$$

The function

$$\ln A(\omega) = f_1(\log\omega) \tag{4.137}$$

defines the *magnitude logarithmic amplitude frequency response* and is shown in the graphs as

$$L(\omega) = 20\log A(\omega) = 20\;0.434\ln A(\omega). \tag{4.138}$$

L is given in decibels (dB) which is the unit that comes from the acoustic theory and merely rescales the amplitude ratio portion of a Bode diagram.

Logarithmic phase angle frequency response is defined as

$$\varphi(\omega) = f_2(\log\omega) \tag{4.139}$$

www **Example 4.7:** *Nyquist and Bode diagrams for the heat exchanger as the first order system*

The process transfer function of the heat exchanger was given in (4.97). $G(j\omega)$ is given as

$$\begin{aligned} G(j\omega) &= \frac{Z_1}{T_1 j\omega + 1} = \frac{Z_1(T_1 j\omega - 1)}{(T_1 j\omega)^2 + 1} \\ &= \frac{Z_1}{(T_1\omega)^2 + 1} - j\frac{Z_1 T_1\omega}{(T_1\omega)^2 + 1} \\ &= \frac{Z_1}{\sqrt{(T_1\omega)^2 + 1}} e^{-j\arctan T_1\omega} \end{aligned}$$

The magnitude and phase angle are of the form

$$A(\omega) = \frac{Z_1}{\sqrt{(T_1\omega)^2 + 1}}$$

$$\varphi(\omega) = -\arctan T_1\omega$$

Nyquist and Bode diagrams of the heat exchanger for $Z_1 = 0.4$, $T_1 = 5.2\,\mathrm{s}$ are shown in Figs. 4.18, 4.19, respectively.

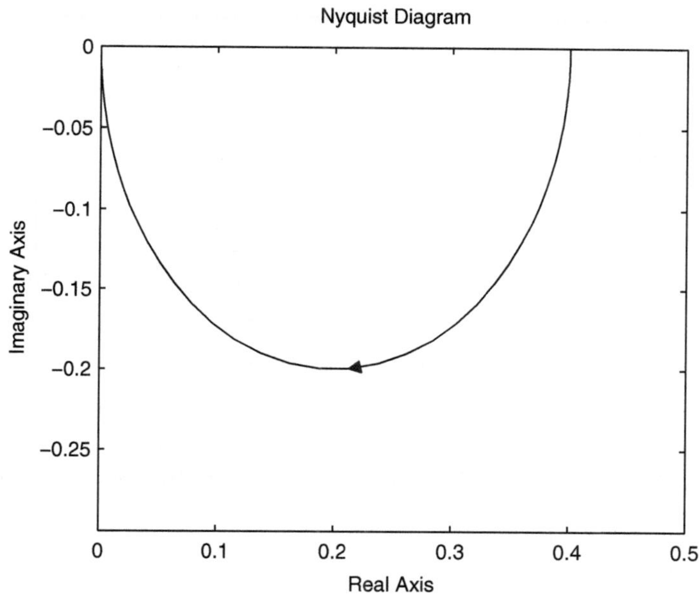

Fig. 4.18. The Nyquist diagram for the heat exchanger

4.3.3 Frequency Characteristics of a First Order System

In general, the dependency $\Im[G(j\omega)]$ on $\Re[G(j\omega)]$ for a first order system described by (4.97) can easily be found from the equations

$$u = \Re[G(j\omega)] = \frac{Z_1}{(T_1\omega)^2 + 1} \tag{4.140}$$

$$v = \Im[G(j\omega)] = -\frac{Z_1 T_1 \omega}{(T_1\omega)^2 + 1} \tag{4.141}$$

Equating the terms $T_1\omega$ in both equations yields

$$(v - 0)^2 - \left(u - \frac{Z_1}{2}\right) = \left(\frac{Z_1}{2}\right)^2 \tag{4.142}$$

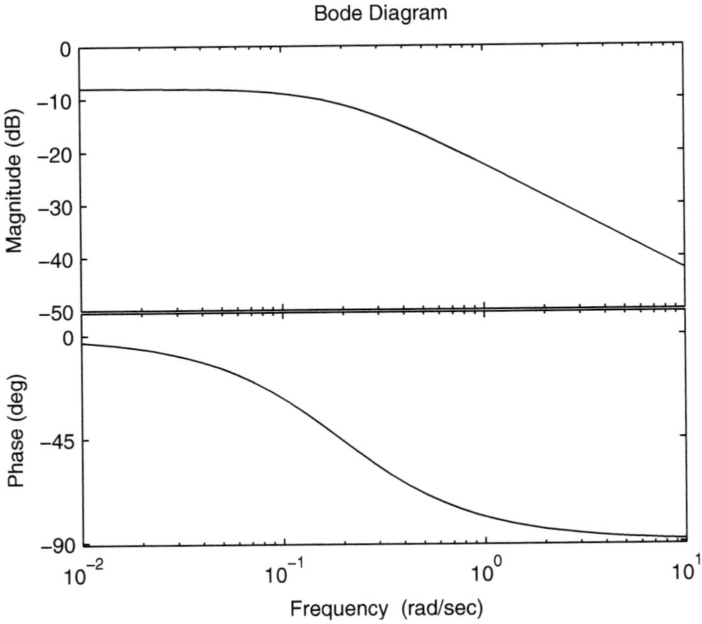

Fig. 4.19. The Bode diagram for the heat exchanger

which is the equation of a circle.

Let us denote $\omega_1 = 1/T_1$. The transfer function (4.97) can be written as

$$G(s) = \frac{\omega_1 Z_1}{s + \omega_1}. \tag{4.143}$$

The magnitude is given as

$$A(\omega) = \frac{Z_1 \omega_1}{\sqrt{\omega_1^2 + \omega^2}} \tag{4.144}$$

and its logarithm as

$$L = 20 \log Z_1 + 20 \log \omega_1 - 20 \log \sqrt{\omega_1^2 + \omega^2}. \tag{4.145}$$

This curve can easily be sketched by finding its asymptotes. If ω approaches zero, then

$$L \to 20 \log Z_1 \tag{4.146}$$

and if it approaches infinity, then

$$\sqrt{\omega_1^2 + \omega^2} \to \sqrt{\omega^2} \tag{4.147}$$

$$L \to 20 \log Z_1 + 20 \log \omega_1 - 20 \log \omega. \tag{4.148}$$

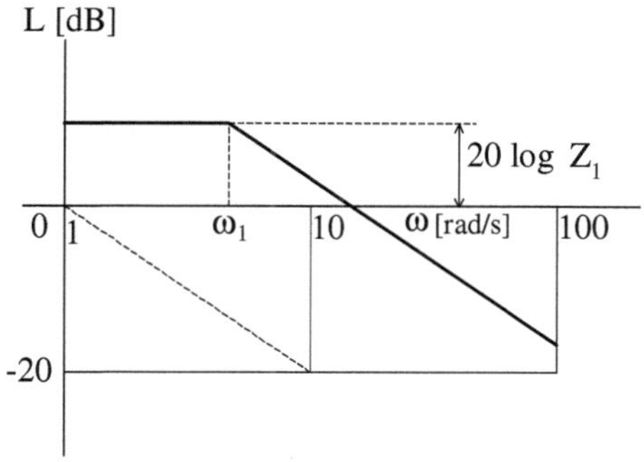

Fig. 4.20. Asymptotes of the magnitude plot for a first order system

This is the equation of an asymptote that for $\omega = \omega_1$ is equal to $20 \log Z_1$. The slope of this asymptote is -20 dB/decade (Fig 4.20).

The asymptotes (4.146) and (4.148) introduce an error δ for $\omega < \omega_1$:

$$\delta = 20 \log \omega_1 - 20 \log \sqrt{\omega_1^2 + \omega^2} \tag{4.149}$$

and for $\omega > \omega_1$:

$$\delta = 20 \log \omega_1 - 20 \log \sqrt{\omega_1^2 + \omega^2} - [20 \log \omega_1 - 20 \log \omega] \tag{4.150}$$

The tabulation of errors for various ω is given in Table 4.2.

Table 4.2. The errors of the magnitude plot resulting from the use of asymptotes

ω	$\frac{1}{5}\omega_1$	$\frac{1}{4}\omega_1$	$\frac{1}{2}\omega_1$	ω_1	$2\omega_1$	$4\omega_1$	$5\omega_1$
δ(dB)	0.17	-0.3	-1	-3	-1	-0.3	-0.17

A phase angle plot can also be sketched using asymptotes and tangent in its inflex point (Fig 4.21).

We can easily verify the following characteristics of the phase angle plot:

If $\omega = 0$, then $\varphi = 0$,
If $\omega = \infty$, then $\varphi = -\pi/2$,
If $\omega = 1/T_1$, then $\varphi = -\pi/4$,

and it can be shown that the curve has an inflex point at $\omega = \omega_1 = 1/T_1$. This frequency is called the *corner frequency*. The slope of the tangent can be calculated if we substitute for $\omega = 10^z$ ($\log \omega = z$) into $\varphi = \arctan(-T_1\omega)$:

$$\dot{\varphi} = \frac{1}{1 + x^2}, \quad x = -T_1\omega$$

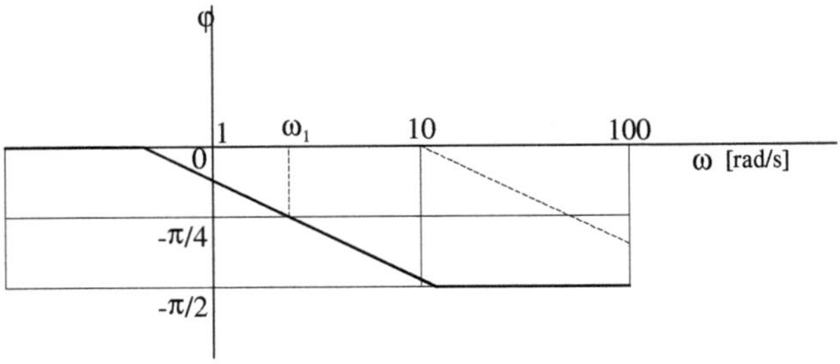

Fig. 4.21. Asymptotes of phase angle plot for a first order system

$$\frac{d\varphi}{dz} = \frac{-2.3}{1+(-T_1 10^z)^2} T_1 10^z$$
$$\frac{d\varphi}{d\log\omega} = \frac{-2.3}{1+(-T_1\omega)^2} T_1\omega$$

for $\omega = 1/T_1$

$$\frac{d\varphi}{d\log\omega} = -1.15 \text{ rad/decade}$$

-1.15 rad corresponds to $-66°$. The tangent crosses the asymptotes $\varphi = 0$ and $\varphi = -\pi/2$ with error of $11°40'$.

4.3.4 Frequency Characteristics of a Second Order System

Consider an underdamped system of the second order with the transfer function

$$G(s) = \frac{1}{T_k^2 s^2 + 2\zeta T_k s + 1}, \quad \zeta < 1. \tag{4.151}$$

Its frequency transfer function is given as

$$G(\mathrm{j}\omega) = \frac{\frac{1}{T_k^2}}{\sqrt{\left(\frac{1}{T_k^2} - \omega^2\right)^2 + \left(\frac{2\zeta}{T_k}\right)^2 \omega^2}} \exp\left(\mathrm{j}\arctan\frac{-\frac{2\zeta}{T_k}\omega}{\frac{1}{T_k^2} - \omega^2}\right) \tag{4.152}$$

$$A(\omega) = \frac{\frac{1}{T_k^2}}{\sqrt{\left(\frac{1}{T_k^2} - \omega^2\right)^2 + \left(\frac{2\zeta}{T_k}\right)^2 \omega^2}} \tag{4.153}$$

$$\varphi(\omega) = \arctan\frac{-\frac{2\zeta}{T_k}\omega}{\frac{1}{T_k^2} - \omega^2}. \tag{4.154}$$

The magnitude plot has a maximum for $\omega = \omega_k$ where $T_k = 1/\omega_k$ (resonant frequency). If $\omega = \infty$, $A = 0$. The expression

$$M = \frac{A(\omega_k)}{A(0)} = \frac{A_{\max}}{A(0)} \tag{4.155}$$

is called the *resonance coefficient.*

If the system gain is Z_1, then

$$L(\omega) = 20 \log |G(\mathrm{j}\omega)| = 20 \log \left| \frac{Z_1}{\frac{(\mathrm{j}\omega)^2}{\omega_k^2} + 2\frac{\zeta}{\omega_k}\mathrm{j}\omega + 1} \right| \tag{4.156}$$

$$L(\omega) = 20 \log \left| \frac{Z_1}{T_k^2(\mathrm{j}\omega)^2 + 2\zeta T_k \mathrm{j}\omega + 1} \right| \tag{4.157}$$

$$L(\omega) = 20 \log Z_1 - 20 \log \sqrt{(1 - T_k^2\omega^2)^2 + (2\zeta T_k \omega)^2} \tag{4.158}$$

It follows from (4.158) that the curve $L(\omega)$ for $Z_1 \neq 1$ is given by summation of $20 \log Z_1$ to normalised L for $Z_1 = 1$. Let us therefore investigate only the case $Z_1 = 1$. From (4.158) follows

$$L(\omega) = -20 \log \sqrt{(1 - T_k^2\omega^2)^2 + (2\zeta T_k \omega)^2}. \tag{4.159}$$

In the range of low frequencies ($\omega \ll 1/T_k$) holds approximately

$$L(\omega) \approx -20 \log \sqrt{1} = 0. \tag{4.160}$$

For high frequencies ($\omega \gg 1/T_k$) and $T_k^2\omega^2 \gg 1$ and $(2\zeta T_k\omega)^2 \ll (T_k^2\omega^2)^2$ holds

$$L(\omega) \approx -20 \log (T_k\omega)^2 = -2\ 20 \log T_k\omega = -40 \log T_k\omega. \tag{4.161}$$

Thus, the magnitude frequency response can be approximated by the curve shown in Fig 4.22. Exact curves deviate with an error δ from this approximation. For $0.38 \leq \zeta \leq 0.7$ the values of δ are maximally ± 3 dB.

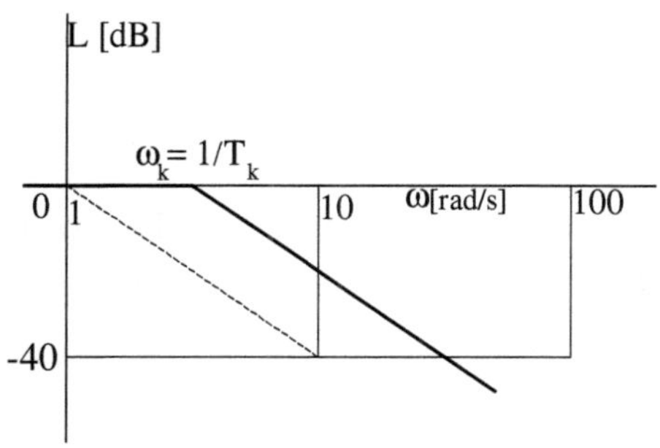

Fig. 4.22. Asymptotes of magnitude plot for a second order system

The phase angle plot is described by the equation

$$\varphi(\omega) = -\arctan \frac{2\zeta T_k}{1 - T_k^2\omega^2}. \tag{4.162}$$

At the corner frequency $\omega_k = 1/T_k$ this gives $\varphi(\omega) = -90°$.

www

Bode diagrams of the second order systems with $Z_1 = 1$ and $T_k = 1\,\text{min}$ are shown in Fig. 4.23.

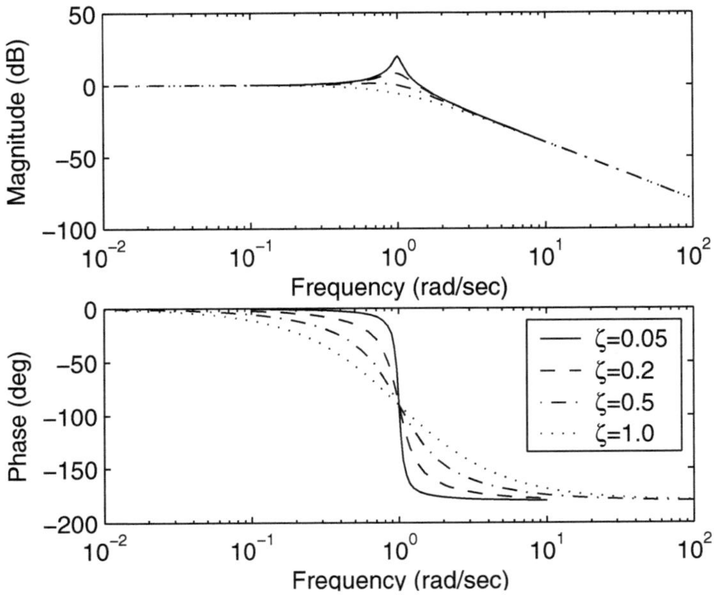

Fig. 4.23. Bode diagrams of an underdamped second order system ($Z_1 = 1$, $T_k = 1$)

4.3.5 Frequency Characteristics of an Integrator

The transfer function of an integrator is

$$G(s) = \frac{Z_1}{s} \tag{4.163}$$

where $Z_1 = 1/T_I$ and T_I is the time constant of the integrator.

We note, that integrator is an astatic system. Substitution for $s = \mathrm{j}\omega$ yields

$$G(\mathrm{j}\omega) = \frac{Z_1}{\mathrm{j}\omega} = -\mathrm{j}\frac{Z_1}{\omega} = \frac{Z_1}{\omega}\mathrm{e}^{-\mathrm{j}\frac{\pi}{2}}. \tag{4.164}$$

From this follows

$$A(\omega) = \frac{Z_1}{\omega}, \tag{4.165}$$

$$\varphi(\omega) = -\frac{\pi}{2}. \tag{4.166}$$

The corresponding Nyquist diagram is shown in Fig. 4.24. The curve coincides with the negative imaginary axis. The magnitude is for increasing ω decreasing. The phase angle is independent of frequency. Thus, output variable is always delayed to input variable for 90°.

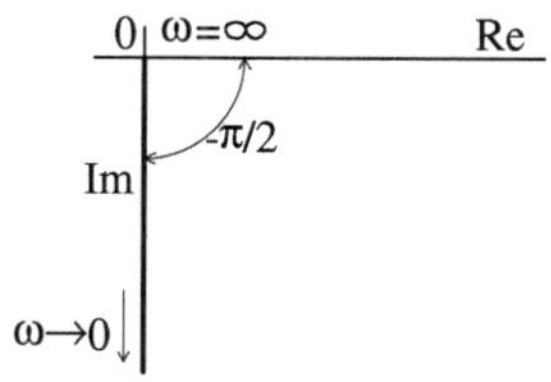

Fig. 4.24. The Nyquist diagram of an integrator

Magnitude curve is given by the expression

$$L(\omega) = 20 \log A(\omega) = 20 \log \frac{Z_1}{\omega} \tag{4.167}$$

$$L(\omega) = 20 \log Z_1 - 20 \log \omega. \tag{4.168}$$

The phase angle is constant and given by (4.166).

If $\omega_2 = 10\omega_1$, then

$$20 \log \omega_2 = 20 \log 10\omega_1 = 20 + 20 \log \omega_1, \tag{4.169}$$

thus the slope of magnitude plot is -20dB/decade.

Fig. 4.25 shows Bode diagram of the integrator with $T_I = 5\,\mathrm{s}$. The values of $L(\omega)$ are given by the summation of two terms as given by (4.168). www

4.3.6 Frequency Characteristics of Systems in a Series

Consider a system with the transfer function

$$G(s) = G_1(s)G_2(s)\ldots G_n(s). \tag{4.170}$$

Its frequency response is given as

$$G(\mathrm{j}\omega) = \prod_{i=1}^{n} A_i(\omega)\mathrm{e}^{\mathrm{j}\varphi_i(\omega)} \tag{4.171}$$

$$G(\mathrm{j}\omega) = \exp\left(\mathrm{j}\sum_{i=1}^{n}\varphi_i(\omega)\right)\prod_{i=1}^{n} A_i(\omega), \tag{4.172}$$

and

$$20 \log A(\omega) = \sum_{i=1}^{n} 20 \log A_i(\omega), \tag{4.173}$$

$$\varphi(\omega) = \sum_{i=1}^{n} \varphi_i(\omega). \tag{4.174}$$

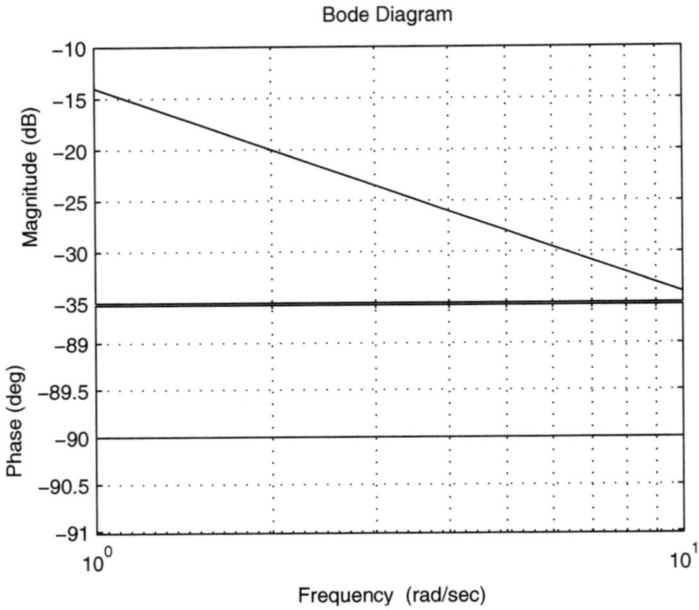

Fig. 4.25. Bode diagram of an integrator

It is clear from the last equations that magnitude and phase angle plots are obtained as the sum of individual functions of systems in series.

www

Example 4.8: *Nyquist and Bode diagrams for a third order system*
Consider a system with the transfer function

$$G(s) = \frac{Z_3}{s(T_1 s + 1)(T_2 s + 1)}.$$

The function $G(\mathrm{j}\omega)$ is then given as

$$G(\mathrm{j}\omega) = \frac{Z_3}{\mathrm{j}\omega(T_1\mathrm{j}\omega + 1)(T_2\mathrm{j}\omega + 1)}.$$

Consequently, for magnitude follows

$$\begin{aligned} L(\omega) &= 20\log\frac{Z_3}{\omega\sqrt{(T_1\omega)^2+1}\sqrt{(T_2\omega)^2+1}} \\ &= 20\log Z_3 - 20\log\omega - 20\log\sqrt{(T_1\omega)^2+1} - 20\log\sqrt{(T_2\omega)^2+1} \end{aligned}$$

and for phase angle:

$$\varphi(\omega) = -\frac{\pi}{2} - \arctan(T_1\omega) - \arctan(T_2\omega)$$

Nyquist and Bode diagrams for the system with $Z_3 = 0.5$, $T_1 = 2\,\mathrm{s}$, and $T_2 = 3\,\mathrm{s}$ are given in Figs. 4.26 and 4.27.

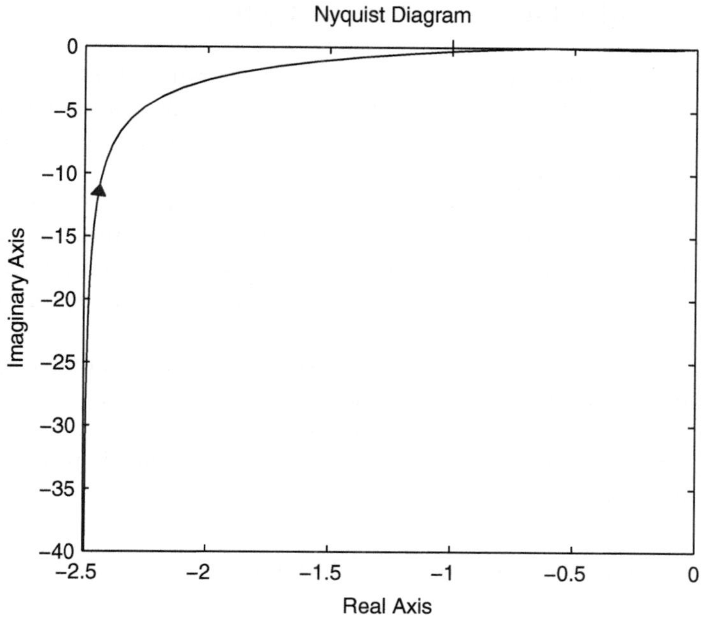

Fig. 4.26. The Nyquist diagram for the third order system

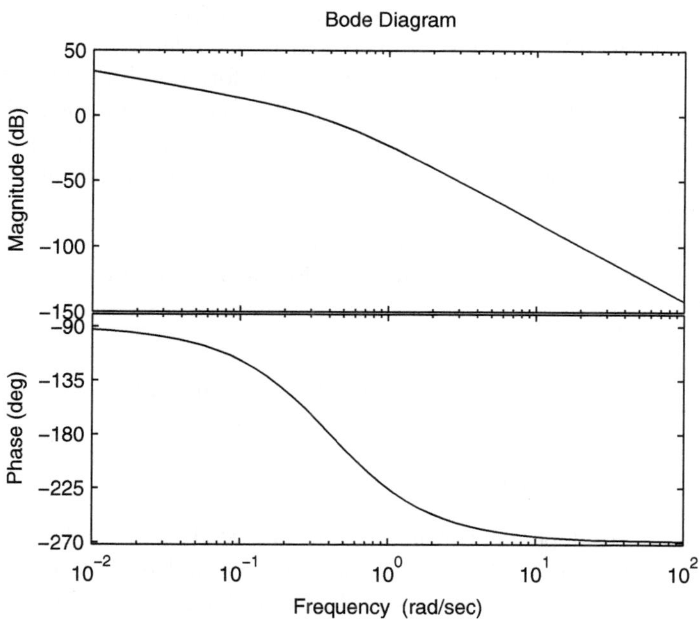

Fig. 4.27. Bode diagram for the third order system

4.4 Statistical Characteristics of Dynamic Systems

Dynamic systems are quite often subject to input variables that are not functions exactly specified by time as opposed to step, impulse, harmonic or other *standard* functions. A *concrete (deterministic)* time function has at any time a completely determined value.

Input variables may take different random values through time. In these cases, the only characteristics that can be determined is probability of its influence at certain time. This does not imply from the fact that the input influence cannot be foreseen, but from the fact that a large number of variables and their changes influence the process simultaneously.

The variables that at any time are assigned to a real number by some statement from a space of possible values are called *random*.

Before investigating the behaviour of dynamic systems with random inputs let us now recall some facts about random variables, stochastic processes, and their probability characteristics.

4.4.1 Fundamentals of Probability Theory

Let us investigate an *event* that is characterised by some conditions of existence and it is known that this event may or may not be realised within these conditions. This *random* event is characterised by its *probability*. Let us assume that we make N experiments and that in m cases, the event A has been realised. The fraction m/N is called *the relative occurrence*. It is the experimental characteristics of the event. Performing different number of experiments, it may be observed, that different values are obtained. However, with the increasing number of experiments, the ratio approaches some constant value. This value is called *probability* of the random event A and is denoted by $P(A)$.

There may exist events with probability equal to one (sure events) and to zero (impossible events). For all other, the following inequality holds

$$0 < P(A) < 1 \tag{4.175}$$

Events A and B are called *disjoint* if they are mutually exclusive within the same conditions. Their probability is given as

$$P(A \cup B) = P(A) + P(B) \tag{4.176}$$

An event A is independent from an event B if $P(A)$ is not influenced when B has or has not occurred. When this does not hold and A is *dependent* on B then $P(A)$ changes if B occurred or not. Such a probability is called *conditional probability* and is denoted by $P(A|B)$.

When two events A, B are independent, then for the probability holds

$$P(A|B) = P(A) \tag{4.177}$$

Let us consider two events A and B where $P(B) > 0$. Then for the conditional probability $P(A|B)$ of the event A when B has occurred holds

$$P(A|B) = \frac{P(A \cap B)}{P(B)} \tag{4.178}$$

For independent events we may also write

$$P(A \cap B) = P(A)P(B) \tag{4.179}$$

4.4.2 Random Variables

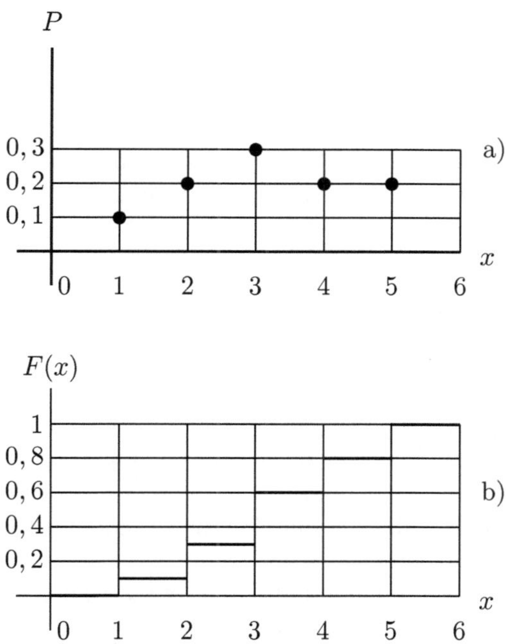

Fig. 4.28. Graphical representation of the law of distribution of a random variable and of the associated distribution function

Let us consider discrete random variables. Any random variable can be assigned to any real value from a given set of possible outcomes. A *discrete random variable* ξ is assigned a real value from a finite set of values $x_1, x_2, \ldots, x_n$. A discrete random variable is determined by the set of finite values and their corresponding probabilities P_i $(i = 1, 2, \ldots, n)$ of their occurrences. The table

$$\xi = \begin{cases} x_1, x_2, \ldots, x_n \\ P_1, P_2, \ldots, P_n \end{cases} \tag{4.180}$$

represents the *law of distribution* of a random variable. An example of the graphical representation for some random variable is shown in Fig. 4.28a. Here, the values of probabilities are assigned to outcomes of some random variable with values x_i. The random variable can attain any value of x_i $(i = 1, 2, \ldots, n)$. It is easily confirmed that

$$\sum_{i=1}^{n} P_i = 1 \tag{4.181}$$

Aside from the law of distribution, we may use another variable that characterises the probability of a random variable. It is denoted by $F(x)$ and defined as

$$F(x) = \sum_{x_i \le x} P_i \tag{4.182}$$

and called *cumulative distribution function*, or simply distribution function of a variable ξ. This function completely determines the distribution of all real values of x. The symbol $x_i \le x$ takes into account all values of x_i less or equal to x. $F(x)$ is a probability of event $\xi \le x$ written as

$$F(x) = P(\xi \le x) \tag{4.183}$$

Further, $F(x)$ satisfies the inequality

$$0 \le F(x) \le 1 \tag{4.184}$$

When the set of all possible outcomes of a random variable ξ is reordered such that $x_1 \le x_2 \le \cdots \le x_n$, the from the probability definition follows that $F(x) = 0$ for any $x < x_1$. Similarly, $F(x) = 1$ for any $x > x_n$. Graphical representation of the distribution function for a random variable in Fig. 4.28a is shown in Fig. 4.28b.

Although the distribution function characterises a completely random variable, for practical reasons there are also defined other characteristics given by some non-random values. Among the possible, its expected value, variance, and standard deviation play an important role.

The *expected value* of a discrete random variable is given as

$$\mu = E\{\xi\} = \sum_{i=1}^{n} x_i P_i \tag{4.185}$$

In the case of uniform distribution law the expected value (4.185) can be written as

$$\mu = \frac{1}{n} \sum_{i=1}^{n} x_i \tag{4.186}$$

For a play-cube tossing yields

$$\mu = \sum_{i=1}^{6} x_i P_i = (1\frac{1}{6} + 2\frac{1}{6} + 3\frac{1}{6} + 4\frac{1}{6} + 5\frac{1}{6} + 6\frac{1}{6}) = 3.5$$

The properties of the expected value are the following:

Constant Z

$$E\{Z\} = Z \tag{4.187}$$

Multiplication by a constant Z

$$E\{Z\xi\} = ZE\{\xi\} \tag{4.188}$$

Summation

$$E\{\xi + \eta\} = E\{\xi\} + E\{\eta\} \tag{4.189}$$

Multiplication of independent random variables

$$E\{\xi\eta\} = E\{\xi\} E\{\eta\} \tag{4.190}$$

If ξ is a random variable and μ is its expected value then the variable $(\xi - \mu)$ that denotes the deviation of a random variable from its expected value is also a random variable.

Variance of a random variable ξ is the expected value of the squared deviation $(\xi - \mu)$

$$\sigma^2 = D[\xi] = E\left\{(\xi - \mu)^2\right\} = \sum_{i=1}^{n} (x_i - \mu)^2 P_i \tag{4.191}$$

Whereas the expected value is a parameter in the neighbourhood of which all values of a random variable are located, variance characterises the distance of the values from μ. If the variance is small, then the values far from the expected value are less probable.

Variance can easily be computed from the properties of expected value:

$$\sigma^2 = E\left\{\xi^2 - 2\xi E\{\xi\} + (E\{\xi\})^2\right\} = E\left\{\xi^2\right\} - (E\{\xi\})^2, \tag{4.192}$$

i.e. variance is given as the difference between the expected value of the squared random variable and squared expected value of random variable. Because the following holds always

$$E\left\{\xi^2\right\} \geq (E\{\xi\})^2, \tag{4.193}$$

variance is always positive, i.e.

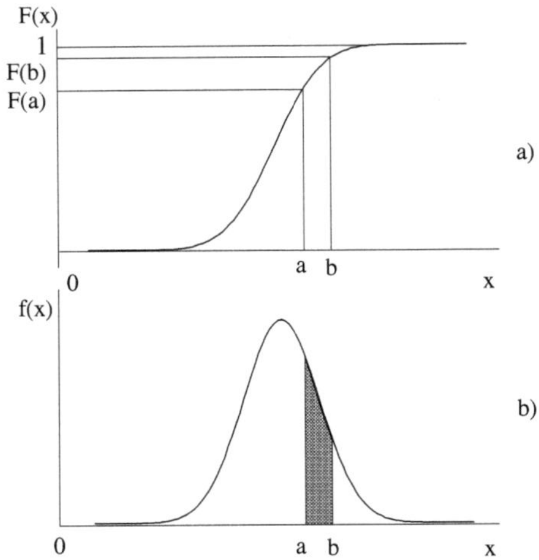

Fig. 4.29. Distribution function and corresponding probability density function of a continuous random variable

$$\sigma^2 \geq 0 \tag{4.194}$$

The square root of the variance is called *standard deviation* of a random variable

$$\sigma = \sqrt{D[\xi]} = \sqrt{E\{\xi^2\} - (E\{\xi\})^2} \tag{4.195}$$

A *continuous random variable* can be assigned to any real value within some interval if its distribution function $F(x)$ is continuous on this interval. The *distribution function of a continuous random variable* ξ

$$F(x) = P(\xi < x) \tag{4.196}$$

is the probability the random variable is less than x. A typical plot of such a distribution function is shown in Fig. 4.29a. The following hold for $F(x)$:

$$\lim_{x \to \infty} F(x) = 1 \tag{4.197}$$

$$\lim_{x \to -\infty} F(x) = 0 \tag{4.198}$$

The probability that a random variable attains some specified value is infinitesimally small. On the contrary, the probability of random variable lying in a some interval (a, b) is finite and can be calculated as

$$P(a \leq \xi < b) = F(b) - F(a) \tag{4.199}$$

The probability that a continuous random variable is between x and $x+\mathrm{d}x$ is given as

$$P(x \le \xi < x + \mathrm{d}x) = \frac{\mathrm{d}F(x)}{\mathrm{d}x}\mathrm{d}x \tag{4.200}$$

where the variable

$$f(x) = \frac{\mathrm{d}F(x)}{\mathrm{d}x} \tag{4.201}$$

is called *probability density.* Figure 4.29b shows an example of $f(x)$. Thus, the distribution function $F(x)$ may be written as

$$F(x) = \int_{-\infty}^{x} f(x)\mathrm{d}x \tag{4.202}$$

Because $F(x)$ is non-decreasing, the probability density function must be positive

$$f(x) \ge 0 \tag{4.203}$$

The probability that a random variable is within an interval (a, b) calculated from its density function is given as the surface under curve $f(x)$ within the given interval. Thus, we can write

$$P(a \le \xi < b) = \int_{a}^{b} f(x)\mathrm{d}x \tag{4.204}$$

Correspondingly, when the interval comprises of all real values, yields

$$\int_{-\infty}^{\infty} f(x)\mathrm{d}x = 1 \tag{4.205}$$

Expected value of a continuous random variable is determined as

$$\mu = E\{\xi\} = \int_{-\infty}^{\infty} x f(x)\mathrm{d}x \tag{4.206}$$

A random variable can be characterised by the equation

$$E\{\xi^m\} = \int_{-\infty}^{\infty} x^m f(x)\mathrm{d}x \tag{4.207}$$

which determines *m-th moment* of a random variable ξ. The first moment is the expected value. The second moment is given as

$$E\{\xi^2\} = \int_{-\infty}^{\infty} x^2 f(x)\mathrm{d}x \tag{4.208}$$

Central m-th moment is of the form

$$E\{(\xi - \mu)^m\} = \int_{-\infty}^{\infty} (x - \mu)^m f(x)\mathrm{d}x \tag{4.209}$$

Variance of a continuous random variable ξ can be expressed as follows

$$\sigma^2 = E\left\{(\xi - \mu)^2\right\} = \int_{-\infty}^{\infty} (x - \mu)^2 f(x)\mathrm{d}x \tag{4.210}$$

$$\sigma^2 = E\left\{\xi^2\right\} - (E\left\{\xi\right\})^2 \tag{4.211}$$

The *standard deviation* is its square root

$$\sigma = \sqrt{E\left\{\xi^2\right\} - (E\left\{\xi\right\})^2} \tag{4.212}$$

Normal distribution for a continuous random variable is given by the following density function

$$f(x) = \frac{1}{\sigma\sqrt{2\pi}} \mathrm{e}^{-\frac{(x-\mu)^2}{2\sigma^2}} \tag{4.213}$$

Let us now consider two independent continuous random variables ξ_1, ξ_2 defined in the same probability space. Their *joint density* function is given by the product

$$f(x_1, x_2) = f_1(x_1) f_2(x_2) \tag{4.214}$$

where $f_1(x_1), f_2(x_2)$ are density functions of the variables ξ_1, ξ_2.

Similarly as for one random variable, we can introduce the moments (if they exist) also for two random variables, for example by

$$E\left\{\xi_1^r, \xi_2^s\right\} = \int_{-\infty}^{\infty}\int_{-\infty}^{\infty} x_1^r x_2^s f(x_1, x_2)\mathrm{d}x_1\mathrm{d}x_2 \tag{4.215}$$

Correspondingly, the central moments are defined as

$$E\left\{(\xi_1 - \mu_1)^r(\xi_2 - \mu_2)^s\right\} = \int_{-\infty}^{\infty}\int_{-\infty}^{\infty} (x_1 - \mu_1)^r (x_2 - \mu_2)^s f(x_1, x_2)\mathrm{d}x_1\mathrm{d}x_2 \tag{4.216}$$

where $\mu_1 = E\left\{\xi_1\right\}, \mu_2 = E\left\{\xi_2\right\}$.

Another important property characterising two random variables is their *covariance* defined as

$$\begin{aligned} \mathrm{Cov}\,(\xi_1, \xi_2) &= E\left\{(\xi_1 - \mu_1)(\xi_2 - \mu_2)\right\} \\ &= \int_{-\infty}^{\infty}\int_{-\infty}^{\infty} (x_1 - \mu_1)(x_2 - \mu_2) f(x_1, x_2)\mathrm{d}x_1\mathrm{d}x_2 \end{aligned} \tag{4.217}$$

If ξ_1, ξ_2 have finite variances, then the number

$$r(\xi_1, \xi_2) = \frac{\mathrm{Cov}\,(\xi_1, \xi_2)}{\sigma_1\sigma_2} \tag{4.218}$$

is called the *correlation coefficient* ($\sigma_1 = \sqrt{D[\xi_1]}, \sigma_2 = \sqrt{D[\xi_2]}$).

Random variables ξ_1, ξ_2 are uncorrelated if

$$\mathrm{Cov}\,(\xi_1, \xi_2) = 0 \tag{4.219}$$

Integrable random variables ξ_1, ξ_2 with integrable term $\xi_1\xi_2$ are uncorrelated if and only if

$$E\{\xi_1, \xi_2\} = E\{\xi_1\} E\{\xi_2\} \tag{4.220}$$

This follows from the fact that $\mathrm{Cov}\,(\xi_1, \xi_2) = E\{\xi_1, \xi_2\} - E\{\xi_1\} E\{\xi_2\}$. Thus the multiplicative property of probabilities extends to expectations.

If ξ_1, ξ_2 are independent integrable random variables then they are uncorrelated.

A vector of random variables $\boldsymbol{\xi} = (\xi_1, \dots, \xi_n)^T$ is usually computationally characterised only by its expected value $E\{\boldsymbol{\xi}\}$ and the covariance matrix $\mathrm{Cov}\,(\boldsymbol{\xi})$.

The expected value $E\{\boldsymbol{\xi}\}$ of a vector $\boldsymbol{\xi}$ is given as the vector of expected values of the elements ξ_i.

The *covariance matrix* $\mathrm{Cov}\,(\boldsymbol{\xi})$ of a vector $\boldsymbol{\xi}$ with the expected value $E\{\boldsymbol{\xi}\}$ is the expected value of the matrix $(\boldsymbol{\xi} - E\{\boldsymbol{\xi}\})(\boldsymbol{\xi} - E\{\boldsymbol{\xi}\})^T$, hence

$$\mathrm{Cov}\,(\boldsymbol{\xi}) = E\left\{(\boldsymbol{\xi} - E[\boldsymbol{\xi}])(\boldsymbol{\xi} - E\{\boldsymbol{\xi}\})^T\right\} \tag{4.221}$$

A covariance matrix is a symmetric positive (semi-)definite matrix that contains in i-th row and j-th column covariances of the random variables ξ_i, ξ_j:

$$\mathrm{Cov}\,(\xi_i, \xi_j) = E\{(\xi_i - E\{\xi_i\})(\xi_j - E\{\xi_j\})\} \tag{4.222}$$

Elements of a covariance matrix determine a degree of correlation between random variables where

$$\mathrm{Cov}\,(\xi_i, \xi_j) = \mathrm{Cov}\,(\xi_j, \xi_i) \tag{4.223}$$

The main diagonal of a covariance matrix contains variances of random variables ξ_i :

$$\mathrm{Cov}\,(\xi_i, \xi_i) = E\{(\xi_i - E\{\xi_i\})(\xi_i - E\{\xi_i\})\} = \sigma_i^2 \tag{4.224}$$

4.4.3 Stochastic Processes

When dealing with dynamic systems, some phenomenon can be observed as a function of time. When some physical variable (temperature in a CSTR) is measured under the same conditions in the same time t_m several times, the results may resemble trajectories shown in Fig. 4.30. The trajectories 1,2,3 are all different. It is impossible to determine the trajectory 2 from the trajectory 1 and from 2 we are unable to predict the trajectory 3. This is the reason that it is not interesting to investigate time functions independently but rather their

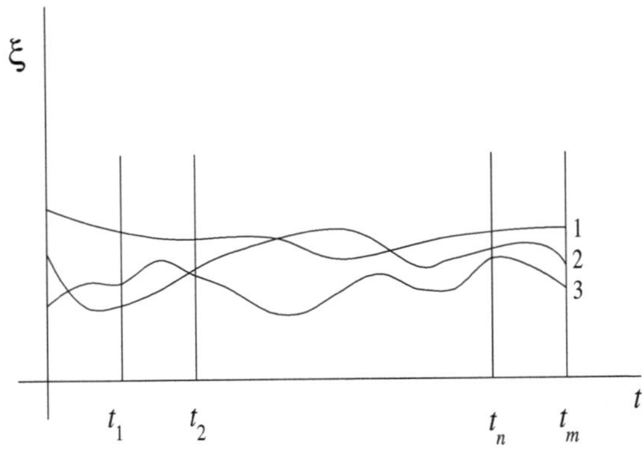

Fig. 4.30. Realisations of a stochastic process

large sets. If their number approaches infinity, we speak about a *stochastic (random) process.*

A *stochastic process* is given as a set of time-dependent random variables $\xi(t)$. Thus, the concept of a random variable ξ is broadened to a random function $\xi(t)$. It might be said that a stochastic process is such a function of time whose values are at any time instant random variables. A random variable in a stochastic process yields random values not only as an outcome of an experiment but also as a function of time. A random variable corresponding to some experimental conditions and changing in time that belongs to the set of random variables $\xi(t)$ is called the *realisation of a stochastic process.*

A stochastic process in some fixed time instants $t_1, t_2, \ldots, t_n$ depends only on the outcome of the experiment and changes to a corresponding random variable with a given density function. From this follows that a stochastic process can be determined by a set of density functions that corresponds to random variables $\xi(t_1), \xi(t_2), \ldots, \xi(t_n)$ in the time instants $t_1, t_2, \ldots, t_n$. The density function is a function of time and is denoted by $f(x,t)$. For any time t_i $(i = 1, 2, \ldots, n)$ exists the corresponding density function $f(x_i, t_i)$.

Consider a time t_1 and the corresponding random variable $\xi(t_1)$. The probability that $\xi(t_1)$ will be between x_1 and $x_1 + \mathrm{d}x_1$ is given as

$$P(x_1 \le \xi(t_1) < x_1 + \mathrm{d}x_1) = f_1(x_1, t_1)\mathrm{d}x_1 \tag{4.225}$$

where $f_1(x_1, t_1)$ is the density function in time t_1 (*one-dimensional density function*).

Now consider two time instants t_1 and t_2. The probability that a random variable $\xi(t_1)$ will be in time t_1 between x_1 and $x_1+\mathrm{d}x_1$ and in time t_2 between x_2 and $x_2 + \mathrm{d}x_2$ can be calculated as

$$P(x_1 \le \xi(t_1) < x_1 + \mathrm{d}x_1; x_2 \le \xi(t_2) < x_2 + \mathrm{d}x_2) = f_2(x_1, t_1; x_2, t_2)\mathrm{d}x_1\mathrm{d}x_2 \tag{4.226}$$

where $f_2(x_1, t_1; x_2, t_2)$ is *two-dimensional density function* and determines the relationship between the values of a stochastic process $\xi(t)$ in the time instants t_1 and t_2.

Sometimes, also *n-dimensional density function* $f_2(x_1, t_1; x_2, t_2; \ldots; x_n, t_n)$ is introduced and is analogously defined as a probability that a process $\xi(t)$ passes through n points with deviation not greater than $\mathrm{d}x_1, \mathrm{d}x_2, \ldots, \mathrm{d}x_n$.

A stochastic process is statistically completely determined by the density functions $f_1, \ldots, f_n$ and the relationships among them.

The simplest stochastic process is a totally independent stochastic process (*white noise*). For this process, any random variables at any time instants are mutually independent. For this process holds

$$f_2(x_1, t_1; x_2, t_2) = f(x_1, t_1) f(x_2, t_2) \tag{4.227}$$

as well as

$$f_n(x_1, t_1; x_2, t_2; \ldots; x_n, t_n) = f(x_1, t_1) f(x_2, t_2) \ldots f(x_n, t_n) \tag{4.228}$$

Based on the one-dimensional density function, the expected value of a stochastic process is given by

$$\mu(t) = E\{\xi(t)\} = \int_{-\infty}^{\infty} x f_1(x, t) \mathrm{d}x \tag{4.229}$$

In (4.229) the index of variables of f_1 is not given as it can be arbitrary.

Variance of a stochastic process can be written as

$$D[\xi(t)] = \int_{-\infty}^{\infty} [x - \mu(t)]^2 f_1(x, t) \mathrm{d}x \tag{4.230}$$

$$D[\xi(t)] = E\{\xi^2(t)\} - (E\{\xi(t)\})^2 \tag{4.231}$$

Expected value of a stochastic process $\mu(t)$ is a function of time and it is the mean value of all realisations of a stochastic process. The variance $D[\xi(t)]$ gives information about dispersion of realisations with respect to the mean value $\mu(t)$.

Based on the information given by the two-dimensional density function, it is possible to find an influence between the values of a stochastic process at times t_1 and t_2. This is given by the *auto-correlation function* of the form

$$R_\xi(t_1, t_2) = E\{\xi(t_1)\xi(t_2)\} = \int_{-\infty}^{\infty} \int_{-\infty}^{\infty} x_1 x_2 f_2(x_1, t_1; x_2, t_2) \mathrm{d}x_1 \mathrm{d}x_2 \tag{4.232}$$

The *auto-covariance function* is given as

$$\begin{aligned} \mathrm{Cov}_\xi(t_1, t_2) &= E\{(\xi(t_1) - \mu(t_1))(\xi(t_2) - \mu(t_2))\} \\ &= \int_{-\infty}^{\infty} \int_{-\infty}^{\infty} [x_1 - \mu(t_1)][x_2 - \mu(t_2)] f_2(x_1, t_1; x_2, t_2) \mathrm{d}x_1 \mathrm{d}x_2 \end{aligned} \tag{4.233}$$

For the auto-correlation function follows

$$R_\xi(t_1, t_2) = \mathrm{Cov}_\xi\,(t_1, t_2) - \mu(t_1)\mu(t_2) \tag{4.234}$$

Similarly, for two stochastic processes $\xi(t)$ and $\eta(t)$, we can define the *correlation function*

$$R_{\xi\eta}(t_1, t_2) = E\left\{\xi(t_1)\eta(t_2)\right\} \tag{4.235}$$

and the *covariance function*

$$\mathrm{Cov}_{\xi\eta}\,(t_1, t_2) = E\left\{(\xi(t_1) - \mu(t_1))(\eta(t_2) - \mu_\eta(t_2))\right\} \tag{4.236}$$

If a stochastic process with normal distribution is to be characterised, it usually suffices to specify its mean value and the correlation function. However, this does not hold in the majority of cases.

When replacing the arguments t_1, t_2 in Equations (4.232), (4.234) by t and τ then

$$R_\xi(t, \tau) = E\left\{\xi(t)\xi(\tau)\right\} \tag{4.237}$$

and

$$\mathrm{Cov}_\xi\,(t, \tau) = E\left\{(\xi(t) - \mu(t))(\xi(\tau) - \mu(\tau))\right\} \tag{4.238}$$

If $t = \tau$ then

$$\mathrm{Cov}_\xi\,(t, t) = E\left\{(\xi(t) - \mu(t))^2\right\} \tag{4.239}$$

where $\mathrm{Cov}_\xi\,(t, t)$ is equal to the variance of the random variable ξ. The abbreviated form $\mathrm{Cov}_\xi\,(t) = \mathrm{Cov}_\xi\,(t, t)$ is also often used.

Consider now mutually dependent stochastic processes $\xi_1(t), \xi_2(t), \dots \xi_n(t)$ that are elements of *stochastic process vector* $\boldsymbol{\xi}(t)$. In this case, the mean values and auto-covariance function are often sufficient characteristics of the process.

The *vector mean value* of the vector $\boldsymbol{\xi}(t)$ is given as

$$\boldsymbol{\mu}(t) = E\left\{\boldsymbol{\xi}(t)\right\} \tag{4.240}$$

The expression

$$\mathrm{Cov}_{\boldsymbol{\xi}}\,(t_1, t_2) = E\left\{(\boldsymbol{\xi}(t_1) - \boldsymbol{\mu}(t_1))(\boldsymbol{\xi}(t_2) - \boldsymbol{\mu}(t_2))^T\right\} \tag{4.241}$$

or

$$\mathrm{Cov}_{\boldsymbol{\xi}}\,(t, \tau) = E\left\{(\boldsymbol{\xi}(t) - \boldsymbol{\mu}(t))(\boldsymbol{\xi}(\tau) - \boldsymbol{\mu}(\tau))^T\right\} \tag{4.242}$$

is the corresponding *auto-covariance matrix* of the stochastic process vector $\boldsymbol{\xi}(t)$.

The auto-covariance matrix is symmetric, thus

$$\mathrm{Cov}_{\boldsymbol{\xi}}(\tau, t) = \mathrm{Cov}_{\boldsymbol{\xi}}^T(t, \tau) \tag{4.243}$$

If a stochastic process is normally distributed then the knowledge about its mean value and covariance is sufficient for obtaining any other process characteristics.

For the investigation of stochastic processes, the following expression is often used

$$\bar{\mu} = \lim_{T \to \infty} \frac{1}{2T} \int_{-T}^{T} \xi(t) \mathrm{d}t \tag{4.244}$$

$\bar{\mu}$ is not time dependent and follows from observations of the stochastic process in a sufficiently large time interval and $\xi(t)$ is any realisation of the stochastic process. In general, the following expression is used

$$\bar{\mu^m} = \lim_{T \to \infty} \frac{1}{2T} \int_{-T}^{T} [\xi(t)]^m \mathrm{d}t \tag{4.245}$$

For $m = 2$ this expression gives $\bar{\mu^2}$.

Stochastic processes are divided into *stationary* and *non-stationary*. In the case of a stationary stochastic process, all probability densities $f_1, f_2, \dots f_n$ do not depend on the start of observations and onedimensional probability density is not a function of time t. Hence, the mean value (4.229) and the variance (4.230) are not time dependent as well.

Many stationary processes are *ergodic*, i.e. the following holds with probability equal to one

$$\mu = \int_{-\infty}^{\infty} x f_1(x) \mathrm{d}x = \bar{\mu} = \lim_{T \to \infty} \frac{1}{2T} \int_{-T}^{T} \xi(t) \mathrm{d}t \tag{4.246}$$

$$\mu^2 = \bar{\mu^2}, \qquad \mu^m = \bar{\mu^m} \tag{4.247}$$

The usual assumption in practice is that stochastic processes are stationary and ergodic.

The properties (4.246) and (4.247) show that for the investigation of statistical properties of a stationary and ergodic process, it is only necessary to observe its one realisation in a sufficiently large time interval.

Stationary stochastic processes have a two-dimensional density function f_2 independent of the time instants t_1, t_2, but dependent on $\tau = t_2 - t_1$ that separates the two random variables $\xi(t_1)$, $\xi(t_2)$. As a result, the autocorrelation function (4.232) can be written as

$$R_\xi(\tau) = E\left\{\xi(t_1)\xi(t_2)\right\} = \int_{-\infty}^{\infty} \int_{-\infty}^{\infty} x_1 x_2 f_2(x_1, x_2, \tau) \mathrm{d}x_1 \mathrm{d}x_2 \tag{4.248}$$

For a stationary and ergodic process hold the equations (4.246), (4.247) and the expression $E\left\{\xi(t)\xi(t+\tau)\right\}$ can be written as

$$E\{\xi(t)\xi(t+\tau)\} = \overline{\xi(t)\xi(t+\tau)} = \lim_{T\to\infty} \frac{1}{2T}\int_{-T}^{T} \xi(t)\xi(t+\tau)\mathrm{d}t \tag{4.249}$$

Hence, the auto-correlation function of a stationary ergodic process is in the form

$$R_\xi(\tau) = \lim_{T\to\infty} \frac{1}{2T}\int_{-T}^{T} \xi(t)\xi(t+\tau)\mathrm{d}t \tag{4.250}$$

Auto-correlation function of a process determines the influence of a random variable between the times $t+\tau$ and t. If a stationary ergodic stochastic process is concerned, its auto-correlation function can be determined from any of its realisations.

The auto-correlation function $R_\xi(\tau)$ is symmetric

$$R_\xi(-\tau) = R_\xi(\tau) \tag{4.251}$$

For $\tau = 0$ the auto-correlation function is determined by the expected value of the square of the random variable

$$R_\xi(0) = E\left\{\xi^2(t)\right\} = \overline{\xi(t)\xi(t)} \tag{4.252}$$

For $\tau \to \infty$ the auto-correlation function is given as the square of the expected value. This can easily be proved.

$$R_\xi(\tau) = \overline{\xi(t)\xi(t+\tau)} = \int_{-\infty}^{\infty}\int_{-\infty}^{\infty} x_1 x_2 f_2(x_1, x_2, \tau)\mathrm{d}x_1\mathrm{d}x_2 \tag{4.253}$$

For $\tau \to \infty$, $\xi(t)$ and $\xi(t+\tau)$ are mutually independent. Using (4.227) that can be applied to a stochastic process yields

$$R_\xi(\infty) = \int_{-\infty}^{\infty} x_1 f(x_1)\mathrm{d}x_1 \int_{-\infty}^{\infty} x_2 f(x_2)\mathrm{d}x_2 = \mu^2 = (\bar{\mu})^2 \tag{4.254}$$

The value of the auto-correlation function for $\tau = 0$ is in its maximum and holds

$$R_\xi(0) \geq R_\xi(\tau) \tag{4.255}$$

The *cross-correlation function* of two *mutually ergodic* stochastic processes $\xi(t)$, $\eta(t)$ can be given as

$$E\{\xi(t)\eta(t+\tau)\} = \overline{\xi(t)\eta(t+\tau)} \tag{4.256}$$

or

$$R_{\xi\eta}(\tau) = \int_{-\infty}^{\infty}\int_{-\infty}^{\infty} x_1 y_2 f_2(x_1, y_2, \tau)\mathrm{d}x_1\mathrm{d}y_2 = \lim_{T\to\infty} \frac{1}{2T}\int_{-T}^{T} \xi(t)\eta(t+\tau)\mathrm{d}t \tag{4.257}$$

Consider now a stationary ergodic stochastic process with corresponding auto-correlation function $R_\xi(\tau)$. This auto-correlation function provides information about the stochastic process in the time domain. The same information can be obtained in the frequency domain by taking the Fourier transform of the auto-correlation function. The Fourier transform $S_\xi(\omega)$ of $R_\xi(\tau)$ is given as

$$S_\xi(\omega) = \int_{-\infty}^{\infty} R_\xi(\tau)\mathrm{e}^{-\mathrm{j}\omega\tau}\mathrm{d}\tau \tag{4.258}$$

Correspondingly, the auto-correlation function $R_\xi(\tau)$ can be obtained if $S_\xi(\omega)$ is known using the inverse Fourier transform

$$R_\xi(\tau) = \frac{1}{2\pi}\int_{-\infty}^{\infty} S_\xi(\omega)\mathrm{e}^{\mathrm{j}\omega\tau}\mathrm{d}\omega \tag{4.259}$$

$R_\xi(\tau)$ and $S_\xi(\omega)$ are non-random characteristics of stochastic processes. $S_\xi(\omega)$ is called *power spectral density* of a stochastic process. This function has large importance for investigation of transformations of stochastic signals entering linear dynamical systems.

The power spectral density is an even function of ω:

$$S_\xi(-\omega) = S_\xi(\omega) \tag{4.260}$$

For its determination, the following relations can be used.

$$S_\xi(\omega) = 2\int_{0}^{\infty} R_\xi(\tau)\cos\omega\tau\mathrm{d}\tau \tag{4.261}$$

$$R_\xi(\tau) = \frac{1}{\pi}\int_{0}^{\infty} S_\xi(\omega)\cos\omega\tau\mathrm{d}\omega \tag{4.262}$$

The *cross-power spectral density* $S_{\xi\eta}(\omega)$ of two mutually ergodic stochastic processes $\xi(t), \eta(t)$ with zero means is the Fourier transform of the associated cross-correlation function $R_{\xi\eta}(\tau)$:

$$S_{\xi\eta}(\omega) = \int_{-\infty}^{\infty} R_{\xi\eta}(\tau)\mathrm{e}^{-\mathrm{j}\omega\tau}\mathrm{d}\tau \tag{4.263}$$

The inverse relation for the cross-correlation function $R_{\xi\eta}(\tau)$ if $S_{\xi\eta}(\omega)$ is known, is given as

$$R_{\xi\eta}(\tau) = \frac{1}{2\pi}\int_{-\infty}^{\infty} S_{\xi\eta}(\omega)\mathrm{e}^{\mathrm{j}\omega\tau}\mathrm{d}\omega \tag{4.264}$$

If we substitute in (4.249), (4.259) for $\tau = 0$ then the following relations can be obtained

$$E\left\{\xi(t)^2\right\} = R_\xi(0) = \lim_{T\to\infty}\frac{1}{2T}\int_{-T}^{T} \xi^2(t)\mathrm{d}t \tag{4.265}$$

$$E\left\{\xi(t)^2\right\} = R_\xi(0) = \frac{1}{2\pi}\int_{-\infty}^{\infty} S_\xi(\omega)\mathrm{d}\omega = \frac{1}{\pi}\int_{0}^{\infty} S_\xi(\omega)\mathrm{d}\omega \tag{4.266}$$

The equation (4.265) describes energetical characteristics of a process. The right hand side of the equation can be interpreted as the average power of the process. The equation (4.266) determines the power as well but expressed in terms of power spectral density. The average power is given by the area under the spectral density curve and $S_\xi(\omega)$ characterises power distribution of the signal according to the frequency. For $S_\xi(\omega)$ holds

$$S_\xi(\omega) \geq 0 \tag{4.267}$$

4.4.4 White Noise

Consider a stationary stochastic process with a constant power spectral density for all frequencies

$$S_\xi(\omega) = V \tag{4.268}$$

This process has a "white" spectrum and it is called *white noise.* Its power spectral density is shown in Fig. 4.31a. From (4.266) follows that the average power of white noise is indefinitely large, as

$$E\left\{\xi(t)^2\right\} = \frac{1}{\pi}V\int_0^\infty \mathrm{d}\omega \tag{4.269}$$

Therefore such a process does not exit in real conditions.

The auto-correlation function of the white noise can be determined from equation (4.262)

$$R_\xi(\tau) = \frac{1}{\pi}\int_0^\infty V\cos\omega\tau\mathrm{d}\omega = V\delta(\tau) \tag{4.270}$$

where

$$\delta(\tau) = \frac{1}{\pi}\int_0^\infty \cos\omega\tau\mathrm{d}\omega \tag{4.271}$$

because the Fourier transform of the delta function $F_\delta(\mathrm{j}\omega)$ is equal to one and the inverse Fourier transform is of the form

$$\begin{aligned}
\delta(\tau) &= \frac{1}{2\pi}\int_{-\infty}^\infty F_\delta(\mathrm{j}\omega)\mathrm{e}^{\mathrm{j}\omega\tau}\mathrm{d}\omega \\
&= \frac{1}{2\pi}\int_{-\infty}^\infty \mathrm{e}^{\mathrm{j}\omega\tau}\mathrm{d}\omega \\
&= \frac{1}{2\pi}\int_{-\infty}^\infty \cos\omega\tau\mathrm{d}\omega + \mathrm{j}\frac{1}{2\pi}\int_{-\infty}^\infty \sin\omega\tau\mathrm{d}\omega \\
&= \frac{1}{\pi}\int_0^\infty \cos\omega\tau\mathrm{d}\omega
\end{aligned} \tag{4.272}$$

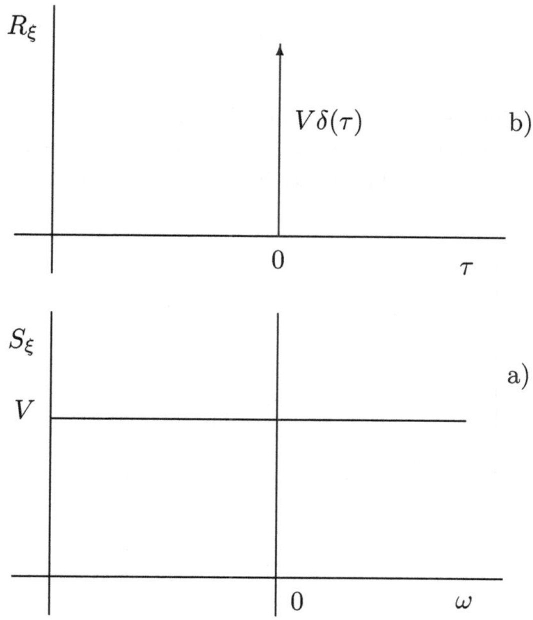

Fig. 4.31. Power spectral density and auto-correlation function of white noise

The auto-correlation function of white noise (Fig. 4.31b) is determined by the delta function and is equal to zero for any non-zero values of τ. White noise is an example of a stochastic process where $\xi(t)$ and $\xi(t+\tau)$ are independent.

A physically realisable white noise can be introduced if its power spectral density is constrained

$$\begin{aligned} S_\xi(\omega) &= V \text{ for } |\omega| < \omega_1 \\ S_\xi(\omega) &= 0 \text{ for } |\omega| > \omega_1 \end{aligned} \tag{4.273}$$

The associated auto-correlation function can be given as

$$R_\xi(\tau) = \frac{V}{\pi} \int_0^{\omega_1} \cos \omega\tau \mathrm{d}\omega = \frac{V}{\pi\tau} \sin \omega_1 \tau \tag{4.274}$$

The following relation also holds

$$\bar{\mu}^2 = D = \frac{V}{2\pi} \int_{-\omega_1}^{\omega_1} \mathrm{d}\omega = \frac{V\omega_1}{\pi} \tag{4.275}$$

Sometimes, the relation (4.268) is approximated by a continuous function. Often, the following relation can be used

$$S_\xi(\omega) = \frac{2aD}{\omega^2 + a^2} \tag{4.276}$$

The associated auto-correlation function is of the form

$$R_\xi(\tau) = \frac{1}{2\pi}\int_{-\infty}^{\infty} \frac{2aD}{\omega^2 + a^2} e^{j\omega\tau} d\omega = De^{-a|\tau|} \tag{4.277}$$

The figure 4.32 depicts power spectral density and auto-correlation function of this process. The equations (4.276), (4.277) describe many stochastic processes well. For example, if $a \gg 1$, the approximation is usually "very" good.

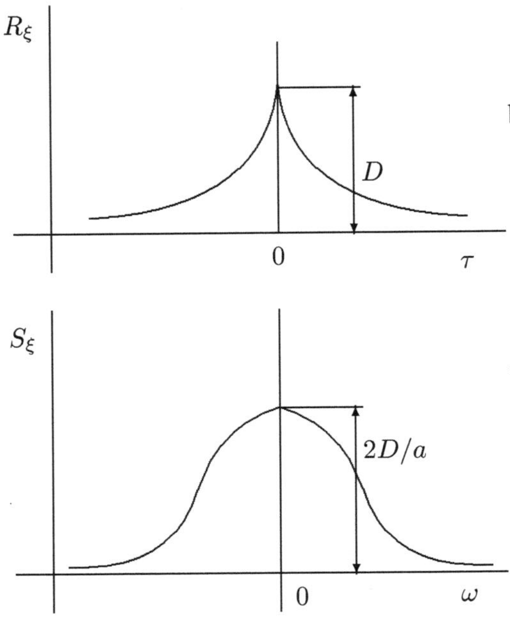

Fig. 4.32. Power spectral density and auto-correlation function of the process given by (4.276) and (4.277)

4.4.5 Response of a Linear System to Stochastic Input

Consider a continuous linear system with constant coefficients

$$\frac{d\boldsymbol{x}(t)}{dt} = \boldsymbol{A}\boldsymbol{x}(t) + \boldsymbol{B}\boldsymbol{\xi}(t) \tag{4.278}$$

$$\boldsymbol{x}(0) = \boldsymbol{\xi}_0 \tag{4.279}$$

where $\boldsymbol{x}(t) = [x_1(t), x_2(t), \ldots, x_n(t)]^T$ is the vector of state variables, $\boldsymbol{\xi}(t) = [\xi_1(t), \xi_2(t), \ldots, \xi_m(t)]^T$ is a stochastic process vector entering the system. $\boldsymbol{A}, \boldsymbol{B}$ are $n \times n, n \times m$ constant matrices, respectively. The initial condition $\boldsymbol{\xi}_0$ is a vector of random variables.

Suppose that the expectation $E\{\boldsymbol{\xi}_0\}$ and the covariance matrix $\mathrm{Cov}(\boldsymbol{\xi}_0)$ are known and given as

$$E\{\boldsymbol{\xi}_0\} = \boldsymbol{x}_0 \tag{4.280}$$

$$E\left\{(\boldsymbol{\xi}_0 - \boldsymbol{x}_0)(\boldsymbol{\xi}_0 - \boldsymbol{x}_0)^T\right\} = \mathrm{Cov}\,(\boldsymbol{\xi}_0) = \mathrm{Cov}_{\mathbf{0}} \tag{4.281}$$

Further, suppose that $\boldsymbol{\xi}(t)$ is independent on the initial condition vector $\boldsymbol{\xi}_0$ and that its mean value $\boldsymbol{\mu}(t)$ and its auto-covariance function $\mathrm{Cov}_{\boldsymbol{\xi}}\,(t,\tau)$ are known and holds

$$E\{\boldsymbol{\xi}(t)\} = \boldsymbol{\mu}(t), \qquad \text{for } t \geq 0 \tag{4.282}$$

$$E\left\{(\boldsymbol{\xi}(t) - \boldsymbol{\mu}(t))(\boldsymbol{\xi}(\tau) - \boldsymbol{\mu}(\tau))^T\right\} = \mathrm{Cov}_{\boldsymbol{\xi}}\,(t,\tau), \quad \text{for } t \geq 0, \tau \geq 0 \tag{4.283}$$

$$E\left\{(\boldsymbol{\xi}(t) - \boldsymbol{\mu}(t))(\boldsymbol{\xi}_0 - \boldsymbol{\mu}_0)^T\right\} \equiv 0, \quad \text{for } t \geq 0 \tag{4.284}$$

As $\boldsymbol{\xi}_0$ is a vector of random variables and $\boldsymbol{\xi}(t)$ is a vector of stochastic processes then $\boldsymbol{x}(t)$ is a vector of stochastic processes as well. We would like to determine its mean value $E\{\boldsymbol{x}(t)\}$, covariance matrix $\mathrm{Cov}_{\boldsymbol{x}}\,(t) = \mathrm{Cov}_{\boldsymbol{x}}\,(t,t)$, and auto-covariance matrix $\mathrm{Cov}_{\boldsymbol{x}}\,(t,\tau)$ for given $\boldsymbol{\xi}_0$ and $\boldsymbol{\xi}(t)$.

Any stochastic state trajectory can be determined for given initial conditions and stochastic inputs as

$$\boldsymbol{x}(t) = \boldsymbol{\Phi}(t)\boldsymbol{\xi}_0 + \int_0^t \boldsymbol{\Phi}(t-\alpha)\boldsymbol{B}\boldsymbol{\xi}(\alpha)\mathrm{d}\alpha \tag{4.285}$$

where $\boldsymbol{\Phi}(t) = \mathrm{e}^{\boldsymbol{A}t}$ is the system transition matrix.

Denoting

$$E\{\boldsymbol{x}(t)\} = \bar{\boldsymbol{x}}(t) \tag{4.286}$$

then the following holds

$$\bar{\boldsymbol{x}}(t) = \boldsymbol{\Phi}(t)\boldsymbol{x}_0 + \int_0^t \boldsymbol{\Phi}(t-\alpha)\boldsymbol{B}\boldsymbol{\mu}(\alpha)\mathrm{d}\alpha \tag{4.287}$$

This corresponds with the solution of the differential equation

$$\frac{\mathrm{d}\bar{\boldsymbol{x}}(t)}{\mathrm{d}t} = \boldsymbol{A}\bar{\boldsymbol{x}}(t) + \boldsymbol{B}\boldsymbol{\mu}(t) \tag{4.288}$$

with initial condition

$$\bar{\boldsymbol{x}}(0) = \boldsymbol{x}_0 \tag{4.289}$$

To find the covariance matrix and auto-correlation function, consider at first the deviation $\boldsymbol{x}(t) - \bar{\boldsymbol{x}}(t)$:

$$\boldsymbol{x}(t) - \bar{\boldsymbol{x}}(t) = \boldsymbol{\Phi}(t)[\boldsymbol{\xi}_0 - \bar{\boldsymbol{x}}_0] + \int_0^t \boldsymbol{\Phi}(t-\alpha)\boldsymbol{B}[\boldsymbol{\xi}(\alpha) - \boldsymbol{\mu}(\alpha)]\mathrm{d}\alpha \tag{4.290}$$

It is obvious that $\boldsymbol{x}(t) - \bar{\boldsymbol{x}}(t)$ is the solution of the following differential equation

$$\frac{d\boldsymbol{x}(t)}{dt} - \frac{d\bar{\boldsymbol{x}}(t)}{dt} = \boldsymbol{A}[\boldsymbol{x}(t) - \bar{\boldsymbol{x}}(t)] + \boldsymbol{B}[\boldsymbol{\xi}(t) - \boldsymbol{\mu}(t)] \tag{4.291}$$

with initial condition

$$\boldsymbol{x}(0) - \bar{\boldsymbol{x}}(0) = \boldsymbol{\xi}_0 - \boldsymbol{x}_0 \tag{4.292}$$

From the equation (4.290) for $\mathrm{Cov}_{\boldsymbol{x}}(t)$ follows

$$\begin{aligned}\mathrm{Cov}_{\boldsymbol{x}}(t) &= E\left\{(\boldsymbol{x}(t) - \bar{\boldsymbol{x}}(t))(\boldsymbol{x}(t) - \bar{\boldsymbol{x}}(t))^T\right\} \\ &= E\left\{\left\{\boldsymbol{\Phi}(t)[\boldsymbol{\xi}_0 - \boldsymbol{x}_0] + \int_0^t \boldsymbol{\Phi}(t-\alpha)\boldsymbol{B}[\boldsymbol{\xi}(\alpha) - \boldsymbol{\mu}(\alpha)]\mathrm{d}\alpha\right\} \times \right. \\ &\quad \left. \times \left\{\boldsymbol{\Phi}(t)[\boldsymbol{\xi}_0 - \boldsymbol{x}_0] + \int_0^t \boldsymbol{\Phi}(t-\beta)\boldsymbol{B}[\boldsymbol{\xi}(\beta) - \boldsymbol{\mu}(\beta)]\mathrm{d}\beta\right\}^T\right\}\end{aligned} \tag{4.293}$$

and after some manipulations,

$$\begin{aligned}\mathrm{Cov}_{\boldsymbol{x}}(t) &= \boldsymbol{\Phi}(t)E\left\{(\boldsymbol{\xi}_0 - \boldsymbol{x}_0)(\boldsymbol{\xi}_0 - \boldsymbol{x}_0)^T\right\}\boldsymbol{\Phi}^T(t) \\ &+ \int_0^t \boldsymbol{\Phi}(t)E\left\{(\boldsymbol{\xi}_0 - \boldsymbol{x}_0)(\boldsymbol{\xi}(\beta) - \boldsymbol{\mu}(\beta))^T\right\}\boldsymbol{B}^T\boldsymbol{\Phi}^T(t-\beta)\mathrm{d}\beta \\ &+ \int_0^t \boldsymbol{\Phi}(t-\alpha)\boldsymbol{B}E\left\{(\boldsymbol{\xi}(\alpha) - \boldsymbol{\mu}(\alpha))(\boldsymbol{\xi}_0 - \boldsymbol{x}_0)^T\right\}\boldsymbol{\Phi}^T(t)\mathrm{d}\alpha \\ &+ \int_0^t\int_0^t \boldsymbol{\Phi}(t-\alpha)\boldsymbol{B}E\left\{(\boldsymbol{\xi}(\alpha) - \boldsymbol{\mu}(\alpha))(\boldsymbol{\xi}(\beta) - \boldsymbol{\mu}(\beta))^T\right\}\boldsymbol{B}^T\boldsymbol{\Phi}^T(t-\beta)\mathrm{d}\beta\mathrm{d}\alpha\end{aligned} \tag{4.294}$$

Finally, using equations (4.281), (4.283), (4.284) yields

$$\begin{aligned}\mathrm{Cov}_{\boldsymbol{x}}(t) &= \boldsymbol{\Phi}(t)\mathrm{Cov}_{\boldsymbol{0}}\boldsymbol{\Phi}^T(t) \\ &\quad + \int_0^t\int_0^t \boldsymbol{\Phi}(t-\alpha)\boldsymbol{B}\mathrm{Cov}_{\boldsymbol{\xi}}(\alpha,\beta)\,\boldsymbol{B}^T\boldsymbol{\Phi}^T(t-\beta)\mathrm{d}\beta\mathrm{d}\alpha\end{aligned} \tag{4.295}$$

Analogously, for $\mathrm{Cov}_{\boldsymbol{x}}(t,\tau)$ holds

$$\begin{aligned}\mathrm{Cov}_{\boldsymbol{x}}(t,\tau) &= \boldsymbol{\Phi}(t)\mathrm{Cov}_{\boldsymbol{0}}\boldsymbol{\Phi}^T(t) \\ &\quad + \int_0^t\int_0^\tau \boldsymbol{\Phi}(t-\alpha)\boldsymbol{B}\mathrm{Cov}_{\boldsymbol{\xi}}(\alpha,\beta)\,\boldsymbol{B}^T\boldsymbol{\Phi}^T(\tau-\beta)\mathrm{d}\beta\mathrm{d}\alpha\end{aligned} \tag{4.296}$$

Consider now a particular case when the system input is a white noise vector, characterised by

$$\begin{aligned}&E\left\{(\boldsymbol{\xi}(t) - \boldsymbol{\mu}(t))(\boldsymbol{\xi}(\tau) - \boldsymbol{\mu}(\tau))^T\right\} = \boldsymbol{V}(t)\delta(t-\tau) \\ &\qquad\qquad \text{for } t \geq 0, \tau \geq 0, \boldsymbol{V}(t) = \boldsymbol{V}^T(t) \geq 0\end{aligned} \tag{4.297}$$

The state covariance matrix $\mathrm{Cov}_{\boldsymbol{x}}(t)$ can be determined if the auto-covariance matrix $\mathrm{Cov}_{\boldsymbol{x}}(t,\tau)$ of the vector white noise $\boldsymbol{\xi}(t)$

$$\mathrm{Cov}_{\boldsymbol{\xi}}(\alpha,\beta) = \boldsymbol{V}(\alpha)\delta(\alpha-\beta) \tag{4.298}$$

is used in the equation (4.295) that yields

$$\begin{aligned}\mathrm{Cov}_{\boldsymbol{x}}(t) &= \boldsymbol{\Phi}(t)\mathrm{Cov}_{\mathbf{0}}\boldsymbol{\Phi}^T(t) \\ &\quad + \int_0^t \int_0^t \boldsymbol{\Phi}(t-\alpha)\boldsymbol{B}\boldsymbol{V}(\alpha)\delta(\alpha-\beta)\boldsymbol{B}^T\boldsymbol{\Phi}^T(t-\beta)\mathrm{d}\beta\mathrm{d}\alpha \quad (4.299) \\ &= \boldsymbol{\Phi}(t)\mathrm{Cov}_{\mathbf{0}}\boldsymbol{\Phi}^T(t) \\ &\quad + \int_0^t \boldsymbol{\Phi}(t-\alpha)\boldsymbol{B}\boldsymbol{V}(\alpha)\boldsymbol{B}^T\boldsymbol{\Phi}^T(t-\alpha)\mathrm{d}\alpha \quad (4.300)\end{aligned}$$

The covariance matrix $\mathrm{Cov}_{\boldsymbol{x}}(t)$ of the state vector $\boldsymbol{x}(t)$ is the solution of the matrix differential equation

$$\frac{\mathrm{dCov}_{\boldsymbol{x}}(t)}{\mathrm{d}t} = \boldsymbol{A}\mathrm{Cov}_{\boldsymbol{x}}(t) + \mathrm{Cov}_{\boldsymbol{x}}(t)\boldsymbol{A}^T + \boldsymbol{B}\boldsymbol{V}(t)\boldsymbol{B}^T \quad (4.301)$$

with initial condition

$$\mathrm{Cov}_{\boldsymbol{x}}(0) = \mathrm{Cov}_{\mathbf{0}} \quad (4.302)$$

The auto-covariance matrix $\mathrm{Cov}_{\boldsymbol{x}}(t,\tau)$ of the state vector $\boldsymbol{x}(t)$ is given by applying (4.298) to (4.296). After some manipulations follows

$$\begin{aligned}\mathrm{Cov}_{\boldsymbol{x}}(t,\tau) &= \boldsymbol{\Phi}(t-\tau)\mathrm{Cov}_{\boldsymbol{x}}(t) \quad \text{for } t>\tau \\ \mathrm{Cov}_{\boldsymbol{x}}(t,\tau) &= \mathrm{Cov}_{\boldsymbol{x}}(t)\boldsymbol{\Phi}(\tau-t) \quad \text{for } \tau>t\end{aligned} \quad (4.303)$$

If a linear continuous system with constant coefficients is asymptotically stable and it is observed from time $-\infty$ and if the system input is a stationary white noise vector, then $\boldsymbol{x}(t)$ is a stationary stochastic process.

The mean value

$$E\{\boldsymbol{x}(t)\} = \bar{\boldsymbol{x}} \quad (4.304)$$

is the solution of the equation

$$\mathbf{0} = \boldsymbol{A}\bar{\boldsymbol{x}} + \boldsymbol{B}\boldsymbol{\mu} \quad (4.305)$$

where $\boldsymbol{\mu}$ is a vector of constant mean values of stationary white noises at the system input.

The covariance matrix

$$E\left\{(\boldsymbol{x}(t)-\bar{\boldsymbol{x}})(\boldsymbol{x}(t)-\bar{\boldsymbol{x}})^T\right\} = \mathrm{Cov}_{\boldsymbol{x}} \quad (4.306)$$

is a constant matrix and is given as the solution of

$$\mathbf{0} = \boldsymbol{A}\mathrm{Cov}_{\boldsymbol{x}} + \mathrm{Cov}_{\boldsymbol{x}}\boldsymbol{A}^T + \boldsymbol{B}\boldsymbol{V}\boldsymbol{B}^T \quad (4.307)$$

where $\boldsymbol{V}$ is a symmetric positive-definite constant matrix defined as

$$E\left\{(\boldsymbol{\xi}(t)-\boldsymbol{\mu})(\boldsymbol{\xi}(t)-\boldsymbol{\mu})^T\right\} = \boldsymbol{V}\delta(t-\tau) \quad (4.308)$$

The auto-covariance matrix

$$E\left\{(\boldsymbol{x}(t_1)-\bar{\boldsymbol{x}})(\boldsymbol{x}(t_2)-\bar{\boldsymbol{x}})^T\right\} = \mathrm{Cov}_{\boldsymbol{x}}(t_1,t_2) \equiv \mathrm{Cov}_{\boldsymbol{x}}(t_1-t_2,0) \tag{4.309}$$

is in the case of stationary processes dependent only on $\tau = t_1 - t_2$ and can be determined as

$$\begin{aligned} \mathrm{Cov}_{\boldsymbol{x}}(\tau,0) &= \mathrm{e}^{\boldsymbol{A}\tau}\mathrm{Cov}_{\boldsymbol{x}} && \text{for } \tau > 0 \\ \mathrm{Cov}_{\boldsymbol{x}}(\tau,0) &= \mathrm{Cov}_{\boldsymbol{x}}\mathrm{e}^{-\boldsymbol{A}^T\tau} && \text{for } \tau < 0 \end{aligned} \tag{4.310}$$

Example 4.9: *Analysis of a first order system*

Consider the mixing process example from page 67 given by the state equation

$$\frac{\mathrm{d}x(t)}{\mathrm{d}t} = ax(t) + b\xi(t)$$

where $x(t)$ is the output concentration, $\xi(t)$ is a stochastic input concentration, $a = -1/T_1$, $b = 1/T_1$, and $T_1 = V/q$ is the time constant defined as the ratio of the constant tank volume V and constant volumetric flow q.

Suppose that

$$x(0) = \xi_0$$

where ξ_0 is a random variable.

Further assume that the following probability characteristics are known

$$\begin{aligned} E\{\xi_0\} &= x_0 \\ E\left\{(\xi_0 - x_0)^2\right\} &= \mathrm{Cov}_0 \\ E\{\xi(t)\} &= \mu \qquad \text{for } t \geq 0 \\ E\{(\xi(t)-\mu)(\xi(\tau)-\mu)\} &= V\delta(t-\tau) \quad \text{for } t,\tau \geq 0 \\ E\{(\xi(t)-\mu)(\xi_0 - x_0)\} &\equiv 0 \quad \text{for } t \geq 0 \end{aligned}$$

The task is to determine the mean value $E\{x(t)\}$, variance $\mathrm{Cov}_x(t)$, and auto-covariance function in the stationary case $\mathrm{Cov}_x(\tau,0)$.

The mean value $E\{x(t)\}$ is given as

$$\bar{x} = \mathrm{e}^{at}x_0 - \frac{b}{a}\left(1-\mathrm{e}^{at}\right)\mu$$

As $a < 0$, the output concentration for $t \to \infty$ is an asymptotically stationary stochastic process with the mean value

$$\bar{x}_\infty = -\frac{b}{a}\mu$$

The output concentration variance is determined from (4.300) as

$$\mathrm{Cov}_x(t) = \mathrm{e}^{2at}\mathrm{Cov}_0 - \frac{b^2}{2a}\left(1 - \mathrm{e}^{2at}\right)V$$

Again, for $t \to \infty$ the variance is given as

$$\lim_{t\to\infty} \mathrm{Cov}_x(t) = -\frac{b^2 V}{2a}$$

The auto-covariance function in the stationary case can be written as

$$\mathrm{Cov}_x(\tau, 0) = -\mathrm{e}^{a|\tau|}\frac{b^2 V}{2a}$$

4.4.6 Frequency Domain Analysis of a Linear System with Stochastic Input

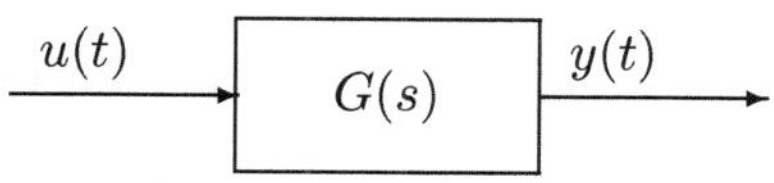

Fig. 4.33. Block-scheme of a system with transfer function $G(s)$

Consider a continuous linear system with constant coefficients (Fig. 4.33). The system response to a stochastic input signal is a stochastic process determined by its auto-correlation function and power spectral density. The probability characteristics of the stochastic output signal can be found if the process input and system characteristics are known.

Let $u(t)$ be any realisation of a stationary stochastic process in the system input and $y(t)$ be the associated system response

$$y(t) = \int_{-\infty}^{\infty} g(\tau_1)u(t-\tau_1)\mathrm{d}\tau_1 \tag{4.311}$$

where $g(t)$ is the impulse response. The mean value of $y(t)$ can be determined in the same way as

$$E\{y(t)\} = \int_{-\infty}^{\infty} g(\tau_1)E\{u(t-\tau_1)\}\,\mathrm{d}\tau_1 \tag{4.312}$$

Analogously to (4.311) which determines the system output in time t, in another time $t+\tau$ holds

$$y(t+\tau) = \int_{-\infty}^{\infty} g(\tau_2)E\{u(t+\tau-\tau_2)\}\,\mathrm{d}\tau_2 \tag{4.313}$$

The auto-correlation function of the output signal is thus given as

$$R_{yy}(\tau) = E\{y(t)y(y+\tau)\}$$
$$= E\left\{\left\{\int_{-\infty}^{\infty} g(\tau_1)u(t-\tau_1)\mathrm{d}\tau_1\right\}\left\{\int_{-\infty}^{\infty} g(\tau_2)u(t+\tau-\tau_2)\mathrm{d}\tau_2\right\}\right\} \tag{4.314}$$

or

$$R_{yy}(\tau) = \int_{-\infty}^{\infty}\int_{-\infty}^{\infty} g(\tau_1)g(\tau_2)E\{u(t-\tau_1)u(t+\tau-\tau_2)\}\,\mathrm{d}\tau_1\mathrm{d}\tau_2 \tag{4.315}$$

As the following holds

$$E\{u(t-\tau_1)u(t+\tau-\tau_2)\} = E\{u(t-\tau_1)u\{(t-\tau_1)+(\tau+\tau_1-\tau_2)\}\} \tag{4.316}$$

then it follows, that

$$R_{yy}(\tau) = \int_{-\infty}^{\infty}\int_{-\infty}^{\infty} g(\tau_1)g(\tau_2)R_{uu}(\tau+\tau_1-\tau_2)\mathrm{d}\tau_1\mathrm{d}\tau_2 \tag{4.317}$$

where $R_{uu}(\tau+\tau_1-\tau_2)$ is the input auto-correlation function with the argument $(\tau+\tau_1-\tau_2)$.

The mean value of the squared output signal is given as

$$\overline{y^2(t)} = R_{yy}(0) = \int_{-\infty}^{\infty}\int_{-\infty}^{\infty} g(\tau_1)g(\tau_2)R_{uu}(\tau_1-\tau_2)\mathrm{d}\tau_1\mathrm{d}\tau_2 \tag{4.318}$$

The output power spectral density is given as the Fourier transform of the associated auto-correlation function as

$$S_{yy}(\omega) = \int_{-\infty}^{\infty} R_{yy}(\tau)\mathrm{e}^{-\mathrm{j}\omega\tau}\mathrm{d}\tau$$
$$= \int_{-\infty}^{\infty}\int_{-\infty}^{\infty}\int_{-\infty}^{\infty} g(\tau_1)g(\tau_2)R_{uu}[\tau+(\tau_1-\tau_2)]\mathrm{e}^{-\mathrm{j}\omega\tau}\mathrm{d}\tau_1\mathrm{d}\tau_2\mathrm{d}\tau \tag{4.319}$$

Multiplying the subintegral term by $(\mathrm{e}^{\mathrm{j}\omega\tau_1}\mathrm{e}^{-\mathrm{j}\omega\tau_2})(\mathrm{e}^{-\mathrm{j}\omega\tau_1}\mathrm{e}^{\mathrm{j}\omega\tau_2}) = 1$ yields

$$S_{yy}(\omega) = \int_{-\infty}^{\infty} g(\tau_1)\mathrm{e}^{\mathrm{j}\omega\tau_1}\mathrm{d}\tau_1 \int_{-\infty}^{\infty} g(\tau_2)\mathrm{e}^{-\mathrm{j}\omega\tau_2}\mathrm{d}\tau_2$$
$$\times \int_{-\infty}^{\infty} R_{uu}[\tau+(\tau_1-\tau_2)]\mathrm{e}^{-\mathrm{j}\omega(\tau+\tau_1-\tau_2)}\mathrm{d}\tau \tag{4.320}$$

Now we introduce a new variable $\tau' = \tau+\tau_1-\tau_2$, yielding

$$S_{yy}(\omega) = \left[\int_{-\infty}^{\infty} g(\tau_1)\mathrm{e}^{\mathrm{j}\omega\tau_1}\mathrm{d}\tau_1\right]\left[\int_{-\infty}^{\infty} g(\tau_2)\mathrm{e}^{-\mathrm{j}\omega\tau_2}\mathrm{d}\tau_2\right]$$
$$\times \left[\int_{-\infty}^{\infty} R_{uu}(\tau')\mathrm{e}^{-\mathrm{j}\omega\tau'}\mathrm{d}\tau'\right] \tag{4.321}$$

The last integral is the input power spectral density

$$S_{uu}(\omega) = \int_{-\infty}^{\infty} R_{uu}(\tau')\mathrm{e}^{-\mathrm{j}\omega\tau'}\mathrm{d}\tau' \tag{4.322}$$

The second integral is the Fourier transform of the impulse function $g(t)$, i.e. it is the frequency transfer function of the system.

$$G(\mathrm{j}\omega) = \int_{-\infty}^{\infty} g(\tau_2)\mathrm{e}^{-\mathrm{j}\omega\tau_2}\mathrm{d}\tau_2 \tag{4.323}$$

Finally, the following holds for the first integral

$$G(-\mathrm{j}\omega) = \int_{-\infty}^{\infty} g(\tau_1)\mathrm{e}^{\mathrm{j}\omega\tau_1}\mathrm{d}\tau_1 \tag{4.324}$$

Hence, from (4.320)–(4.324) follows

$$S_{yy}(\omega) = |G(\mathrm{j}\omega)|^2 S_{uu}(\omega) \tag{4.325}$$

with

$$|G(\mathrm{j}\omega)|^2 = G(-\mathrm{j}\omega)G(\mathrm{j}\omega).$$

If the power spectral density $S_{yy}(\omega)$ is known then the mean value of the squared output variable is in the form

$$\overline{y^2(t)} = \frac{1}{\pi}\int_0^{\infty} S_{yy}(\omega)\mathrm{d}\omega = \frac{1}{\pi}\int_0^{\infty} |G(\mathrm{j}\omega)|^2 S_{uu}(\omega)\mathrm{d}\omega \tag{4.326}$$

4.5 References

Time responses of linear systems to unit step and unit impulse are dealt with in the majority of control books, for example

J. Mikleš and V. Hutla. *Theory of Automatic Control.* Alfa, Bratislava, 1986. (in Slovak).

L. B. Koppel. *Introduction to Control Theory with Application to Process Control.* Prentice Hall, Englewood Cliffs, New Jersey, 1968.

G. Stephanopoulos. *Chemical Process Control, An Introduction to Theory and Practice.* Prentice Hall, Englewood Cliffs, New Jersey, 1984.

D. E. Seborg, T. F. Edgar, and D. A. Mellichamp. *Process Dynamics and Control.* Wiley, New York, 1989.

K. Reinisch. *Kybernetische Grundlagen und Beschreibung kontinuericher Systeme.* VEB Verlag Technik, Berlin, 1974.

J. Mikleš. *Theory of Automatic Control of Processes in Chemical Technology, Part I.* ES SVŠT, Bratislava, 1978. (in Slovak).

S. Kubík, Z. Kotek, V. Strejc, and J. Štecha. *Theory of Automatic Control I.* SNTL/ALFA, Praha, 1982. (in Czech).

V. A. Besekerskij and E. P. Popov. *Theory of Automatic Control Systems.* Nauka, Moskva, 1975. (in Russian).

A. A. Feldbaum and A. G. Butkovskij. *Methods in Theory of Automatic Control.* Nauka, Moskva, 1971. (in Russian).

Y. Z. Cypkin. *Foundations of Theory of Automatic Systems.* Nauka, Moskva, 1977. (in Russian).

H. Unbehauen. *Regelungstechnik I.* Vieweg, Braunschweig/Wiesbaden, 1986.

J. Mikleš, P. Dostál, and A. Mészáros. *Control of Processes in Chemical Technology.* STU, Bratislava, 1994. (in Slovak).

Numerical methods for the solution of a system response – simulations can be found in

J. Olehla, M. Olehla, et al. *BASIC for Microcomputers.* NADAS, Praha, 1998. (in Czech).

M. Jamshidi, M. Tarokh, and B. Shafai. *Computer-aided Analysis and Design of Linear Control Systems.* Prentice Hall, Englewood Cliffs, New Jersey, 1992.

J. Penny and G. Lindfield. *Numerical Methods Using MATLAB.* ELLIS HORWOOD, London, 1995.

W. H. Press, S. A. Teukolsky, W. T. Vetterling, and B. P. Flannery. *Numerical Recipes in C. The Art of Scientific Computing.* Cambridge University Press, 1992.

A. Grace, A. J. Laub, J. N. Little, and C. Thompson. *Control System Toolbox for Use with MATLAB.* The Mathworks, Inc., Natich, MA, 1990.

Simulations of the liquid storage system used coefficients from

J. Danko, M. Ondrovičová, and A. Mészáros. Experimental model of a hydraulic system. *Automatizace*, (12):478 – 481, 1995. (in Slovak).

Frequency responses are quite common in many control books:

K. Reinisch. *Kybernetische Grundlagen und Beschreibung kontinuericher Systeme.* VEB Verlag Technik, Berlin, 1974.

J. Mikleš. *Theory of Automatic Control of Processes in Chemical Technology, Part I.* ES SVŠT, Bratislava, 1978. (in Slovak).

S. Kubík, Z. Kotek, V. Strejc, and J. Štecha. *Theory of Automatic Control I.* SNTL/ALFA, Praha, 1982. (in Czech).

V. A. Besekerskij and E. P. Popov. *Theory of Automatic Control Systems.* Nauka, Moskva, 1975. (in Russian).

Y. Z. Cypkin. *Foundations of Theory of Automatic Systems.* Nauka, Moskva, 1977. (in Russian).

H. Unbehauen. *Regelungstechnik I.* Vieweg, Braunschweig/Wiesbaden, 1986.

Statistical characteristics of dynamical systems can be found in:

V. A. Ivanov, V. S. Medvedev, B. K. Čemodanov, and A. S. Juščenko. *Mathematical Foundations of Theory of Automatic Control. Vol II.* Vyššaja škola, Moskva, 1977. (in Russian).

P. S. Maybeck. *Stochastic Models, Estimation, and Control*, volume 1. Academic Press, New York, 1979.

J. A. Seinfeld and L. Lapidus. *Mathematical Methods in Chemical Engineering, Vol. 3, Process Modeling, Estimation, and Identification.* Prentice Hall, Inc., New Jersey, 1980.

H. Unbehauen. *Regelungstechnik III.* Vieweg, Braunschweig/Wiesbaden, 1988.

4.6 Exercises

Exercise 4.1:
Consider a system with the transfer function given as

$$G(s) = \frac{0.6671s + 3.0610}{s^2 + 4.0406s + 5.4345}$$

Find:
1. response of the system to the unit impulse,
2. response of the system to the unit step.

Exercise 4.2:
Consider the system from the previous exercise. Plot:
1. the Nyquist diagram,
2. the Bode diagram.

Exercise 4.3:
Consider a system with the transfer function given as

$$R(s) = \frac{-0.0887s + 1.774}{1.25s}$$

Find:
1. response of the system to the unit step,
2. the Nyquist diagram,
3. the Bode diagram.

Exercise 4.4:
Consider a system with the transfer function given as

$$L(s) = G(s)R(s)$$

where $G(s)$ is the first exercise and $R(s)$ is in the third exercise. Find:

1. the Nyquist diagram,
2. the Bode diagram.

Exercise 4.5:
Consider a system with the transfer function given as

$$T(s) = \frac{G(s)R(s)}{1 + G(s)R(s)}$$

where $G(s)$ is in the first exercise and $R(s)$ is in the third exercise. Find:
1. response of the system to the unit step,
2. the Bode diagram.

Exercise 4.6:
Consider a system with the transfer function given as

$$S(s) = \frac{1}{1 + G(s)R(s)}$$

where $G(s)$ is in the first exercise and $R(s)$ is in the third exercise. Find its Bode diagram.

5

Discrete-Time Process Models

The aim of this chapter is to explain the reason why discrete-time process models are used. It has a close relation to computer control of processes. The chapter demonstrates how discrete-time processes and discrete transfer functions are obtained. Basic properties of discrete-time systems are examined.

5.1 Computer Controlled and Sampled Data Systems

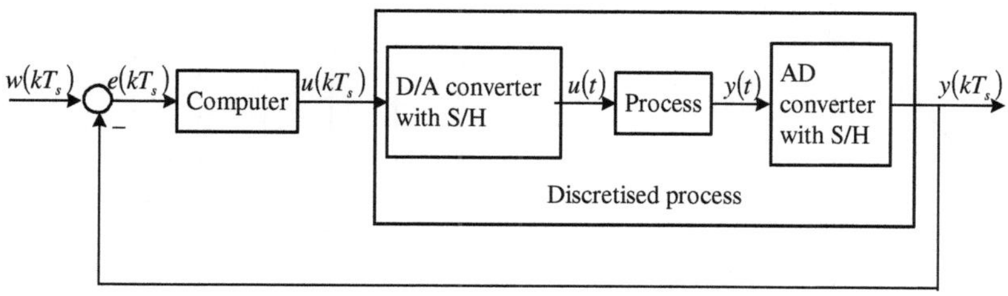

Fig. 5.1. A digital control system controlling a continuous-time process

Computer controlled process control indicates that the control law is calculated by computer. A feedback scheme of such a control is shown in Fig. 5.1. This is a simplified scheme of direct digital control. The scheme in Fig. 5.1 contains four basic blocks: computer, digital-to-analog (D/A) converter, controlled process, and analog-to-digital (A/D) converter. Both converters contain sample-and-hold (S/H) device (or simply sampler) that holds each sample constant until a new information arrives. The control error $e(kT_s)$ is given as the difference between the setpoint signal $w(kT_s)$ and controlled process output $y(kT_s)$ in digital form in sampling times specified by the *sampling period* T_s. The computer interprets the signal $e(kT_s)$ as a sequence of numbers and given the control law, it generates a new sequence of control signals $u(kT_s)$.

The discretised process represents a system with input being the sequence $[u(kT_s)]$ and output being the sequence $[y(kT_s)]$. This system is characterised by a discrete-time model that defines relations between sequences $[u(kT_s)]$ and $[y(kT_s)]$. A/D converter implements transformation of a continuous-time signal to a sequence of numbers (Fig. 5.2). D/A converter with a sampler implements transformation of a digital signal to a continuous-time signal that is constant within one sampling period (Fig. 5.3). A/D and D/A converters operate synchronously.

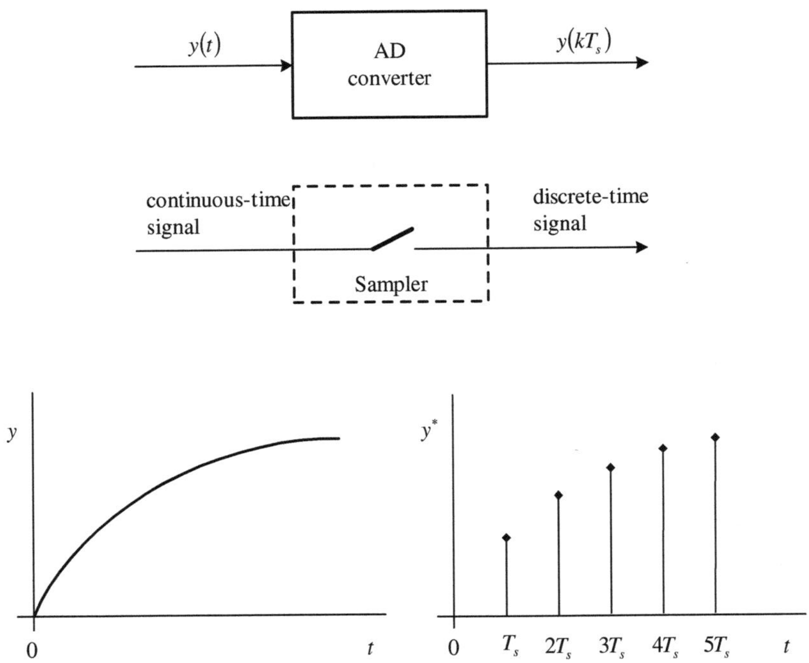

Fig. 5.2. Transformation of a continuous-time to a discrete-time signal

A possible realisation of a sampler is zero-order hold with the transfer function of the form

$$G(s) = \frac{1 - e^{-T_s s}}{s} \tag{5.1}$$

When processes are controlled digitally, it is important to choose a suitable sampling period T_s that will capture the process dynamics correctly. There are two contradictory requirements on its choice. When computational load is considered, the sampling period as large as possible should be chosen. On the other hand, large sampling times imply loss of information.

In any case, the sampling period cannot be larger than some critical value at which important information about the continuous-time signal can be lost.

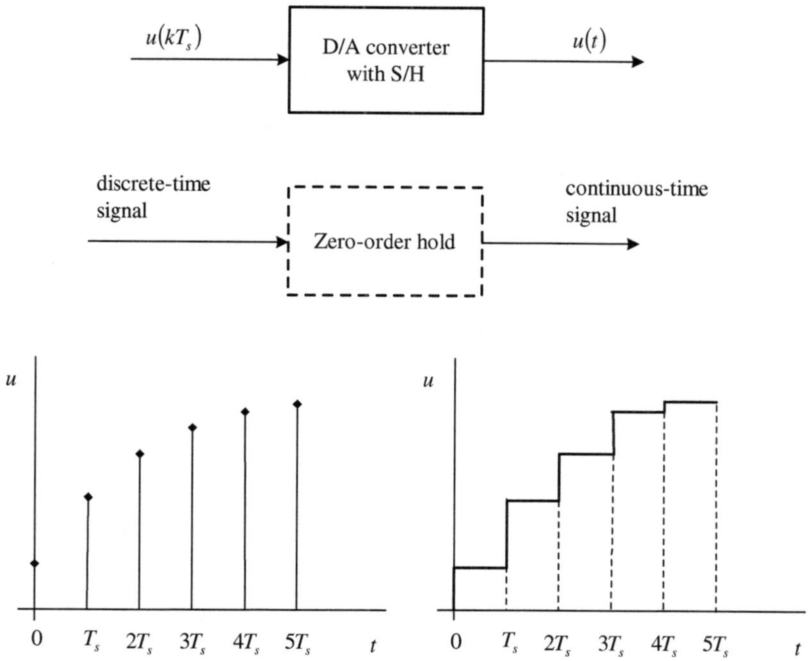

Fig. 5.3. Continuous-time signal step reconstruction from its discrete-time counterpart

If a continuous-time sine signal is considered, loss of information occurs when the following inequality does not hold

$$T_s < \frac{T_{\sin}}{2} \tag{5.2}$$

where $T_{\sin}$ is the oscillation period of the sine signal. This can be seen in Fig. 5.4.

If (5.2) is not valid, the phenomenon occurring with the sine signal can also occur with other sampled continuous-time signals.

If this phenomenon is studied using magnitude frequency spectra then it will be shown by aliasing of individual spectral densities.

Let us now investigate properties of an ideal sampler shown in Fig. 5.5. Its output variable y^* can be represented as a sequence of modulated δ functions. A periodical sequence of δ functions can be represented as

$$\delta^*(t) = \sum_{k=-\infty}^{\infty} \delta(t - kT_s) \tag{5.3}$$

Let us define the sampling frequency

$$\omega_s = \frac{2\pi}{T_s} \tag{5.4}$$

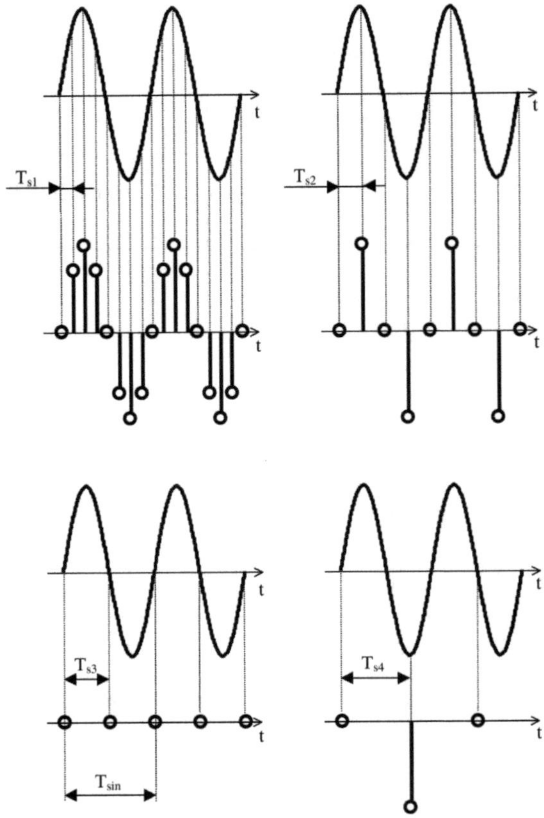

Fig. 5.4. A possible loss of information caused by sampling

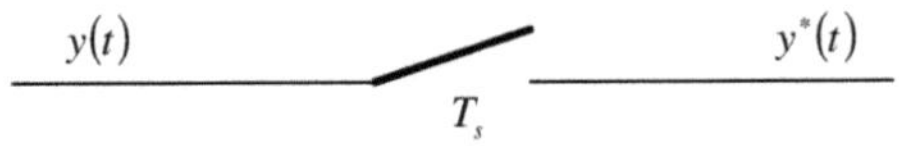

Fig. 5.5. An ideal sampler

Then this sequence can be expressed using the exponential form of the Fourier series

$$\delta^*(t) = \frac{\omega_s}{2\pi} \sum_{n=-\infty}^{\infty} \mathrm{e}^{-\mathrm{j}n\omega_s t} \tag{5.5}$$

The output variable of the ideal sampler can then be written as

$$y^*(t) = y(t)\delta^*(t) \quad y(t) \equiv 0, t < 0 \tag{5.6}$$

Substituting $\delta^*(t)$ from (5.5) into equation (5.6) yields

$$y^*(t) = \frac{\omega_s}{2\pi} y(t) \sum_{n=-\infty}^{\infty} \mathrm{e}^{-\mathrm{j}n\omega_s t} \tag{5.7}$$

or

$$y^*(t) = \frac{1}{T_s} y(t) \sum_{n=-\infty}^{\infty} \mathrm{e}^{-\mathrm{j}n\omega_s t} \tag{5.8}$$

The Fourier transform of this function if $y(0) = 0$ is given as

$$\int_0^\infty y^*(t)\mathrm{e}^{-\mathrm{j}\omega t}\mathrm{d}t = \frac{1}{T_s}\int_0^\infty y(t)\mathrm{e}^{-\mathrm{j}\omega t}\left(\sum_{n=-\infty}^{\infty} \mathrm{e}^{-\mathrm{j}n\omega_s t}\right)\mathrm{d}t \tag{5.9}$$

Next, the following holds

$$Y^*(\mathrm{j}\omega) = \frac{1}{T_s}\sum_{n=-\infty}^{\infty}\int_0^\infty \mathrm{e}^{-\mathrm{j}(\omega+n\omega_s)t} y(t)\,\mathrm{d}t \tag{5.10}$$

$$Y^*(\mathrm{j}\omega) = \frac{1}{T_s}\sum_{n=-\infty}^{\infty} Y[\mathrm{j}(\omega + n\omega_s)] \tag{5.11}$$

as for the Fourier integral holds

$$\int_0^\infty y(t)\mathrm{e}^{-\mathrm{j}(\omega+n\omega_s)t}\mathrm{d}t = Y[\mathrm{j}(\omega + n\omega_s)] \tag{5.12}$$

Substituting s for $\mathrm{j}\omega$ in equation (5.11) gives

$$Y^*(s) = \frac{1}{T_s}\sum_{n=-\infty}^{\infty} Y(s + \mathrm{j}n\omega_s) \tag{5.13}$$

It is clear from (5.11) that the transform $Y(\mathrm{j}\omega)$ is only $n\omega_s$ shifted form of the function $y(t)$. If the spectral density function of the variable $y(t)$ is $|Y(\mathrm{j}\omega)|$ (see Fig. 5.7) then the spectral density of the sampled signal $y^*(t)$ is given as

$$|Y^*(\mathrm{j}\omega)| = \left|\frac{1}{T_s}\sum_{n=-\infty}^{\infty} Y[\mathrm{j}(\omega + n\omega_s)]\right| \tag{5.14}$$

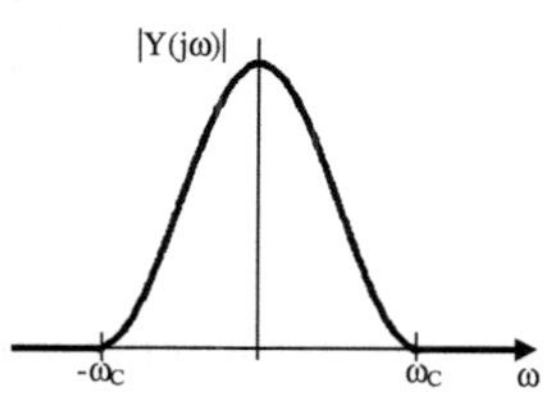

Fig. 5.6. Spectral density of a signal $y(t)$

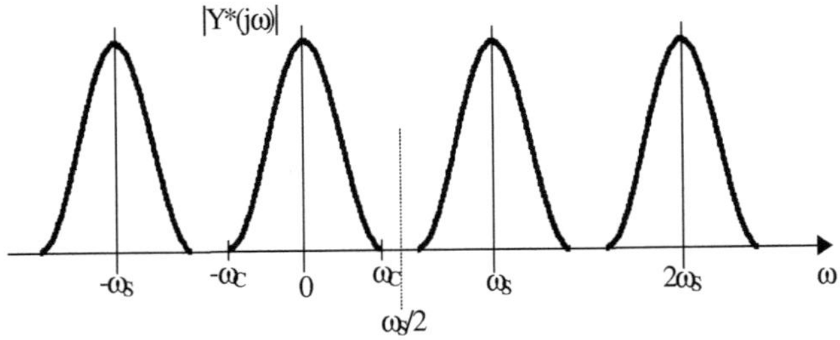

Fig. 5.7. Spectral density of a sampled signal $y^*(t)$

Two cases can occur. In the first one the frequency ω_c is smaller than or equal to half of the sampling period, hence

$$\omega_c \leq \frac{\omega_s}{2} \tag{5.15}$$

In this case the spectral density of $|Y^*(j\omega)|$ is composed of spectra of $|Y(j\omega)|$ shifted to the right of $n\omega_s$ and that are non-overlapping.

The second case occurs if the frequency ω_c is larger than half of the sampling period

$$\omega_c > \frac{\omega_s}{2}$$

Here the spectral density of $|Y^*(j\omega)|$ consists of spectra $|Y(j\omega)|$ shifted to the right of $n\omega_s$ and overlapping. Hence, the spectral density of the signal $|Y^*(j\omega)|$ is distorted.

The previous analysis has shown that it is imperative for non-overlapping spectral densities that the sampling period obeys the relation

$$\omega_s \geq 2\omega_c \tag{5.16}$$

Overlapping of the spectra can be removed when a suitable anti-aliasing filter is used for the original continuous-time signal before sampling.

The sampling period choice is rather a problem of experience than some exact procedure. In general is has a strong influence on dynamic properties of controlled system, as well as on whole closed-loop system.

Consider a dynamical system of the first order of the form

$$T\frac{\mathrm{d}y(t)}{\mathrm{d}t} + y(t) = u(t) \tag{5.17}$$

where $y(t)$ is the output variable, $u(t)$ is the input variable, and T is the process time constant. The sampling period can be chosen based on the relation

$$\frac{T}{5} < T_s < \frac{T}{2}$$

Consider a dynamical system of the second order of the form

$$T_k^2 \frac{\mathrm{d}^2 y(t)}{\mathrm{d}t^2} + 2\zeta T_k \frac{\mathrm{d}y(t)}{\mathrm{d}t} + y(t) = u(t) \tag{5.18}$$

where T_k is its time constant andζ the damping coefficient. The sampling period is usually chosen within the interval

$$0.25 \leq \frac{T_s}{T_k} < 1.5 \quad 0.7 \leq \zeta \leq 1 \tag{5.19}$$

If $\zeta > 1$ then an approximate model of the form (5.17) can be used. If a closed-loop system is considered, the overall damping coefficient should be larger than 0.7.

5.2 $\mathcal{Z}$ – Transform

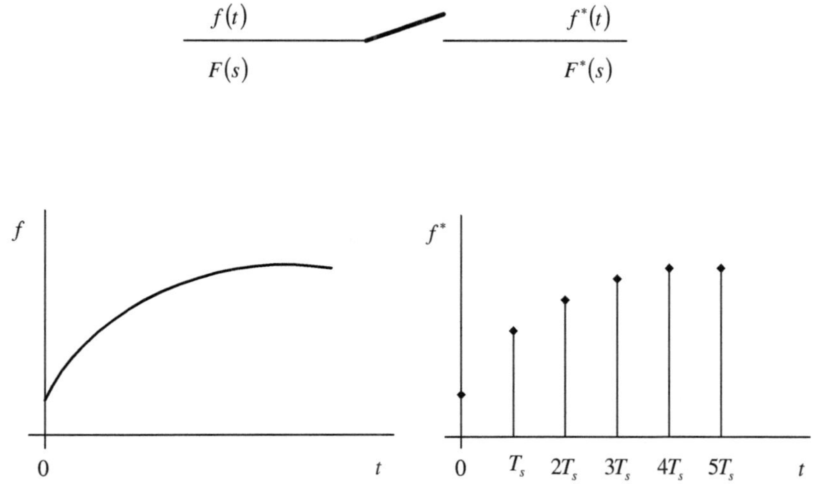

Fig. 5.8. An ideal sampler and an impulse modulated signal

Let us again consider an ideal sampler shown in Fig 5.8. This sampler implements transformation of a continuous-time signal $f(t)$ to an impulse modulated signal. Individual impulses appear on the sampler output in the sampling times kT_s, $k = 0, 1, 2, \ldots$ and are equal to functions $f(kT_s)$, $k = 0, 1, 2, \ldots$. This impulse modulated signal containing a sequence of impulses is denoted by $f^*(t)$. The signal $f^*(t)$ can be expressed as

$$f^*(t) = \sum_{k=0}^{\infty} f(kT_s)\delta(t - kT_s) \tag{5.20}$$

The Laplace transform of this function is

$$\mathcal{L}\{f^*(t)\} = F^*(s) = \sum_{k=0}^{\infty} f(kT_s)\mathrm{e}^{-kT_s s} \tag{5.21}$$

Let us introduce a new variable in equation (5.21)

$$z = \mathrm{e}^{T_s s} \tag{5.22}$$

Then we can write

$$\mathcal{L}\{f^*(t)\} = \sum_{k=0}^{\infty} f(kT_s)z^{-k} \tag{5.23}$$

$\mathcal{Z}$-transform of the signal $f(t)$ or $f^*(t)$ is defined by the expression

$$F(z) = \mathcal{Z}\{f(t)\} = \sum_{k=0}^{\infty} f(kT_s)z^{-k} \tag{5.24}$$

or

$$\mathcal{Z}\{f(t)\} = F^*(s)\big|_{z=\mathrm{e}^{T_s s}} \tag{5.25}$$

$\mathcal{Z}$-transform is mathematically equivalent to the Laplace transform and differs only in the argument. $\mathcal{Z}$-transform exists only if some z exists such that the series in (5.24) converges.

Consider the unit step function $1(t)$ defined as

$$1(t) = \begin{cases} 1, & t \geq 0 \\ 0, & t < 0 \end{cases} \tag{5.26}$$

Sampled unit step function is given as

$$f(kT_s) = 1, \quad k \geq 0 \tag{5.27}$$

and thus

$$\mathcal{Z}\{1(t)\} = \sum_{k=0}^{\infty} z^{-k} \tag{5.28}$$

If $|z| > 1$ then the series in (5.28) is convergent and $\mathcal{Z}\{1(t)\}$ is

$$\mathcal{Z}\{1(t)\} = \frac{1}{1 - z^{-1}} \tag{5.29}$$

or

$$\mathcal{Z}\{1(t)\} = \frac{z}{z - 1} \tag{5.30}$$

Consider now an exponential function

$$f(t) = \mathrm{e}^{-at}1(t) \tag{5.31}$$

where $a = 1/T$ and T is a time constant. Then

$$\mathcal{L}\{f^*(t)\} = \sum_{k=0}^{\infty} \mathrm{e}^{-akT_s}\mathrm{e}^{-kT_s s} \tag{5.32}$$

Its $\mathcal{Z}$-transform is given as

$$\mathcal{Z}\left\{\mathrm{e}^{-at}1(t)\right\} = \sum_{k=0}^{\infty} \left(\mathrm{e}^{-aT_s}z^{-1}\right)^k \tag{5.33}$$

If $\left|z^{-1}\mathrm{e}^{-aT_s}\right| < 1$ then

$$\mathcal{Z}\left\{\mathrm{e}^{-at}1(t)\right\} = \frac{1}{1 - z^{-1}\mathrm{e}^{-aT_s}} = \frac{z}{z - \mathrm{e}^{-aT_s}} \tag{5.34}$$

$\mathcal{Z}$-transforms of other functions can be calculated in a similar way. Table 5.1 lists some functions and their Laplace and $\mathcal{Z}$ transformations.

$\mathcal{Z}$-transform is a linear operation.

$\mathcal{Z}$-transform of a delayed function is given as

$$\mathcal{Z}\{f(t - kT_s)\} = z^{-k}F(z) \tag{5.35}$$

where k is a positive integer and $f(t) = 0$ for $t < 0$.

For an initial function value holds

$$\lim_{k\to 0} f(kT_s) = \lim_{z\to\infty} \frac{z-1}{z}F(z) \tag{5.36}$$

For a final function value holds

$$\lim_{k\to\infty} f(kT_s) = \lim_{z\to 1}(1 - z^{-1})F(z) \tag{5.37}$$

Reciprocally, given the $\mathcal{Z}$-transform of a function, we can find the function values in sampling times. The inverse $\mathcal{Z}$-transform is symbolically written as

$$\mathcal{Z}^{-1}\{F(z)\} = [f(0), f(T_s), f(2T_s), \ldots] \tag{5.38}$$

or

$$\mathcal{Z}^{-1}\{F(z)\} = [f(kT_s)] = f^*(t) \tag{5.39}$$

However, using the inverse $\mathcal{Z}$-transform, only values of f in sampling times are found. It is not possible to find the sampling time. Also, a unique function $F(z)$ can be found for several continuous-time functions $f(t)$.

Discrete functions can practically be found from $\mathcal{Z}$-transforms using the polynomial division or the expansion as a sum of partial fractions.

$f(t)$	$F(s)$	$F(z)$
$\delta(t)$	1	1
$1(t)$	$\frac{1}{s}$	$\frac{z}{z-1}$
$1(t)-1(t-T_s)$	$\frac{1-e^{-T_s s}}{s}$	1
$t1(t)$	$\frac{1}{s^2}$	$\frac{T_s z}{(z-1)^2}$
$e^{-at}1(t)$	$\frac{1}{s+a}$	$\frac{z}{z-e^{-aT_s}}$
$te^{-at}1(t)$	$\frac{1}{(s+a)^2}$	$\frac{T_s z e^{-aT_s}}{(z-e^{-aT_s})^2}$
$e^{-at}\sin(\omega t)1(t)$	$\frac{\omega}{(s+a)^2+\omega^2}$	$\frac{ze^{-aT_s}\sin(\omega T_s)}{z^2-2ze^{-aT_s}\cos(\omega T_s)+e^{-2aT_s}}$
$e^{-at}\cos(\omega t)1(t)$	$\frac{s+a}{(s+a)^2+\omega^2}$	$\frac{z^2-ze^{-aT_s}\cos(\omega T_s)}{z^2-2ze^{-aT_s}\cos(\omega T_s)+e^{-2aT_s}}$
$\left[\frac{1}{ab}+\frac{1}{a(a-b)}e^{-at}+\frac{1}{b(b-a)}e^{-bt}\right]1(t)$	$\frac{1}{s(s+a)(s+b)}$	$\frac{z}{ab(z-1)}+\frac{z}{a(a-b)(z-e^{-aT_s})}+\frac{z}{b(b-a)(z-e^{-bT_s})}$
$\left[\frac{c}{ab}+\frac{c-a}{a(a-b)}e^{-at}+\frac{c-b}{b(b-a)}e^{-bt}\right]1(t)$	$\frac{s+c}{s(s+a)(s+b)}$	$\frac{cz}{ab(z-1)}+\frac{z(c-a)}{a(a-b)(z-e^{-aT_s})}+\frac{z(c-b)}{b(b-a)(z-e^{-bT_s})}$

Table 5.1. Laplace and $\mathcal{Z}$-transforms of some common functions

Polynomial Division

The $\mathcal{Z}$-transform definition gives

$$F(z) = \sum_{k=0}^{\infty} f(k) z^{-k} \tag{5.40}$$

From this follows that

$$F(z) = f(0) + f(1)z^{-1} + f(2)z^{-2} + \cdots \tag{5.41}$$

To simplify expressions, we have replaced $f(kT_s)$ by $f(k)$. If we can somehow obtain the function $F(z)$ as

$$F(z) = c_o + c_1 z^{-1} + c_2 z^{-2} + \cdots \tag{5.42}$$

then by equating (5.41) and (5.42) follows

$$c_0 = f(0),\; c_1 = f(1),\; c_2 = f(2), \cdots \tag{5.43}$$

If rational functions are considered, equation (5.42) can be found simply by polynomial division.

Partial Fraction Expansion

The inverse $\mathcal{Z}$-transform using partial fraction expansion is similar to the inverse Laplace transform.

A function $F(z)$ can be expressed as a ratio of two polynomials of the form

$$F(z) = \frac{M(z^{-1})}{N(z^{-1})} \tag{5.44}$$

where

$$M(z^{-1}) = m_0 + m_1 z^{-1} + \cdots + m_m z^{-m} \tag{5.45}$$

$$N(z^{-1}) = 1 + n_1 z^{-1} + \cdots + n_n z^{-n} \tag{5.46}$$

Let us assume that $m < n$ and that the polynomial $N(z)$ has n unequal real roots

$$N(z^{-1}) = (1 - z_1 z^{-1})(1 - z_2 z^{-1}) \cdots (1 - z_n z^{-1}) \tag{5.47}$$

Partial fraction expansion gives

$$F(z) = \frac{K_1}{1 - z_1 z^{-1}} + \frac{K_2}{1 - z_2 z^{-1}} + \cdots + \frac{K_n}{1 - z_n z^{-1}} \tag{5.48}$$

The corresponding time domain function $f(kT_s)$ is

$$f(k) = \mathcal{Z}^{-1}\left\{\frac{K_1}{1 - z_1 z^{-1}}\right\} + \mathcal{Z}^{-1}\left\{\frac{K_2}{1 - z_2 z^{-1}}\right\} + \cdots + \mathcal{Z}^{-1}\left\{\frac{K_n}{1 - z_n z^{-1}}\right\} \tag{5.49}$$

or

$$f(kT_s) = K_1 z_1^k + K_2 z_2^k + \cdots + K_n z_n^k \tag{5.50}$$

If the denominator of $F(z)$ has complex conjugate roots, partial fraction expansion gives fractions with complex denominators. This can be handled similarly as in case of the inverse Laplace transform.

Example 5.1: *Inverse $\mathcal{Z}$-transform using partial fraction expansion*

Consider a function $F(z)$ of the form

$$F(z) = \frac{z}{z^2 - 1.4z + 0.4}$$

or

$$F(z) = \frac{z^{-1}}{1 - 1.4z^{-1} + 0.4z^{-2}}$$

The denominator polynomial has roots $z_1^{-1} = 1$ and $z_2^{-1} = 2.5$. $F(z)$ can be written using partial fractions as

$$F(z) = \frac{z^{-1}}{(1 - z^{-1})(1 - 0.4z^{-1})} = \frac{K_1}{1 - z^{-1}} + \frac{K_2}{1 - 0.4z^{-1}}$$

Multiplying $F(z)$ by $1 - z^{-1}$ and substituting for $z^{-1} = 1$ yields

$$K_1 = \frac{5}{3}$$

Multiplying $F(z)$ by $1 - 0.4z^{-1}$ and substituting for $z^{-1} = 2.5$ yields

$$K_2 = -\frac{5}{3}$$

Table 5.1 gives a corresponding time domain function (step function) for expression $(5/3)(1 - z^{-1})$

$$(5/3)\,1(kT_s)$$

The term $(-5/3)(1 - 0.4z^{-1})$ is in time domain

$$-\frac{5}{3}\mathrm{e}^{-akT_s}$$

where $\mathrm{e}^{-aT_s} = 0.4 = 2/5$ and thus

$$aT_s = -\ln 0.4 = 0.916$$

Therefore, the time domain function corresponding to $F(z)$ is given as

$$f(kT_s) = \frac{5}{3}\left(1 - \mathrm{e}^{-0.916k}\right) = \frac{5}{3}\left(1 - \left(\frac{2}{5}\right)^k\right), \qquad k = 0, 1, 2, \ldots$$

Analytical form of the corresponding continuous-time exponential function can be found only if the sampling period T_s is known.

5.3 Discrete-Time Transfer Functions

Let us calculate a transient response of a combined discrete-time and continuous-time system shown in Fig. 5.9. The signal

$$u^*(\tau) = \sum_{k=0}^{\infty} u(kT_s)\delta(\tau - kT_s) \tag{5.51}$$

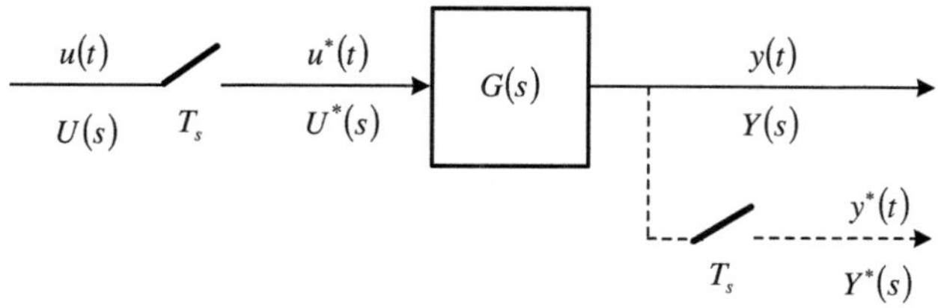

Fig. 5.9. Sample-and-hold device in series with a continuous-time system

is on the input of the continuous-time system with transfer function $G(s)$.

The system response $y(t)$ is given by the convolution integral

$$y(t) = \int_0^t g(t-\tau) \sum_{k=0}^{\infty} u(kT_s)\delta(\tau - kT_s)\mathrm{d}\tau \tag{5.52}$$

where τ is time and $g(t) = \mathcal{L}^{-1}\{G(s)\}$. Equation (5.52) for $0 \le \tau \le t$ can be written as

$$y(t) = \sum_{k=0}^{\infty} g(t - kT_s)u(kT_s) \tag{5.53}$$

Let us further assume only information about $y(t)$ in sampling instants. Therefore, the output of the process enters a fictitious sampler that is ideally synchronised with the input sampler $u^*(t)$. The output sampler gives the signal $y^*(t)$ and its values are the same as the values of $y(t)$ in the sampling instants. For $t = jT_s$ holds

$$y(jT_s) = \sum_{k=0}^{\infty} g(jT_s - kT_s)u(kT_s) \tag{5.54}$$

Applying the $\mathcal{Z}$-transform yields

$$Y(z) = \sum_{j=0}^{\infty} \sum_{k=0}^{\infty} g(jT_s - kT_s)u(kT_s)z^{-j} \tag{5.55}$$

If $j - k = i$, then

$$Y(z) = \sum_{i=-k}^{\infty} \sum_{k=0}^{\infty} g(iT_s)u(kT_s)z^{-(i+k)} \tag{5.56}$$

$g(t)$ is the response of the continuous-time system to the unit impulse with zero initial conditions, hence $g(iT_s) = 0$, $i < 0$. The first summation term on the right-hand side of (5.56) is thus simplified (k falls out) and summations can be written separately:

$$Y(z) = \sum_{i=0}^{\infty} g(iT_s)z^{-i} \sum_{k=0}^{\infty} u(kT_s)z^{-k} \quad (5.57)$$

The $\mathcal{Z}$-transform definition given by (5.24) gives

$$Y(z) = G(z)U(z) \quad (5.58)$$

$$G(z) = \mathcal{Z}\{g(t)\} = \sum_{i=0}^{\infty} g(iT_s)z^{-i} \quad (5.59)$$

The function $G(z)$ is referred to as *discrete-time transfer function* of the system with continuous-time transfer function $G(s)$.

The term

$$U(z) = \mathcal{Z}\{u(t)\} = \sum_{k=0}^{\infty} u(kT_s)z^{-k} \quad (5.60)$$

is $\mathcal{Z}$-transform of the input signal $u(t)$.

In the rest of the book will $G(z)$ be simply referred to as transfer function.

Equation (5.58) seems to be the same as its continuous-time counterpart using the Laplace transform. However, $Y(z)$ only indicates information about $y(t)$ in sampling times. $G(z)$ does not relate input and output signals at times between sampling times. A possible way to have information between sampling times is to employ a generalised $\mathcal{Z}$-transform.

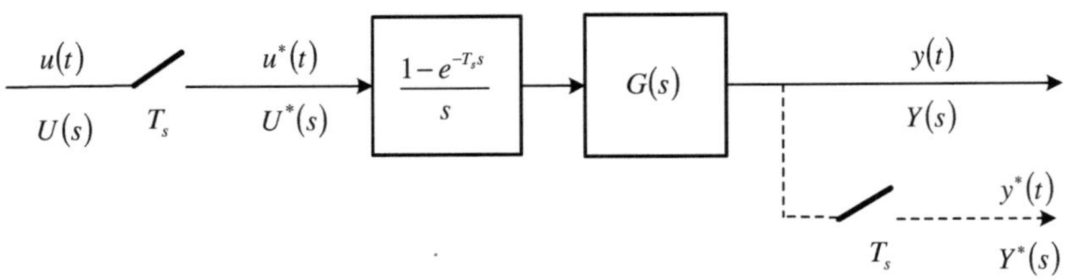

Fig. 5.10. Sampler and zero-order hold in series with a continuous-time system $G(s)$

When the sample-and-hold device in Fig 5.9 is assumed to be zero-order hold (Fig. 5.10) then the relation between $G(s)$ and $G(z)$ is

$$G(z) = (1 - z^{-1})\mathcal{Z}\left\{\mathcal{L}^{-1}\left\{\left[\frac{G(s)}{s}\right]\right\}\right\} \quad (5.61)$$

Example 5.2: *$\mathcal{Z}$-transform of a continuous-time first order system with zero-order hold*

Let us find discrete-time transfer function of a continuous-time system given by

$$G(s) = G_1(s)G_2(s)$$

where

$$G_1(s) = \frac{1 - \mathrm{e}^{-sT_s}}{s}, \quad G_2(s) = \frac{Z_1}{T_1 s + 1}$$

Applying equation (5.61) gives

$$G(z) = (1 - z^{-1})\mathcal{Z}\left\{\mathcal{L}^{-1}\left\{\frac{Z_1}{s(T_1 s + 1)}\right\}\right\}$$

In this case $\mathcal{Z}$-transform of the term $Z_1/(s(T_1 s + 1))$ is needed. Partial fraction expansion gives

$$\frac{Z_1}{s(T_1 s + 1)} = Z_1\left(\frac{1}{s} - \frac{1}{s + \frac{1}{T_1}}\right)$$

The right-hand side term corresponds to continuous-time function $Z_1(1 - \mathrm{e}^{-t/T_1})$. Table 5.1 gives

$$\mathcal{Z}\left\{Z_1\left(1 - \mathrm{e}^{-\frac{t}{T_1}}\right)\right\} = Z_1\left(\frac{z}{z-1} - \frac{z}{z - \mathrm{e}^{-(T_s/T_1)}}\right)$$

Finally

$$G(z^{-1}) = \frac{b_1 z^{-1}}{1 + a_1 z^{-1}}$$

where $b_1 = Z_1(1 - \mathrm{e}^{-(T_s/T_1)})$, $a_1 = \mathrm{e}^{-(T_s/T_1)}$.

Example 5.3: *$\mathcal{Z}$-transform of a continuous-time time-delayed first order system with zero-order hold*

Let us find discrete-time transfer function of a continuous-time system given by

$$G(s) = \frac{Z_1 \mathrm{e}^{-T_d s}}{T_1 s + 1}$$

where T_d is a time delay. Let us suppose that

$$T_d = dT_s$$

is an integer multiple of the sampling period. The discrete-time transfer function is of the form

$$G(z^{-1}) = \frac{z^{-d}(b_1 z^{-1})}{1 + a_1 z^{-1}}$$

where b_1, a_1 are the same as in the previous example.

Example 5.4: *$\mathcal{Z}$-transform of a continuous-time second order system with zero-order hold*

Let us find discrete-time transfer function of a continuous-time system given by

$$G(s) = \frac{Z_2}{(T_1 s + 1)(T_2 s + 1)}, \qquad T_1 \neq T_2$$

The discrete-time transfer function is of the form

$$G(z) = (1 - z^{-1}) \mathcal{Z} \left\{ \mathcal{L}^{-1} \left\{ \frac{Z_2}{s(T_1 s + 1)(T_2 s + 1)} \right\} \right\}$$

Partial fraction expansion and Table 5.1 give

$$G(z) = \frac{Z_2}{T_1 T_2} \frac{(z-1)}{z} \left[\frac{T_1 T_2 z}{z - 1} + \frac{z T_1^2 T_2}{(T_2 - T_1)(z - \mathrm{e}^{-(T_s/T_1)})} + \frac{z T_1 T_2^2}{(T_1 - T_2)(z - \mathrm{e}^{-(T_s/T_2)})} \right]$$

After some simplifications we find the final form

$$G(z^{-1}) = \frac{b_1 z^{-1} + b_2 z^{-2}}{1 + a_1 z^{-1} + a_2 z^{-2}}$$

where

$$b_1 = Z_2 \left[-\left(\mathrm{e}^{-\frac{T_s}{T_1}} + \mathrm{e}^{-\frac{T_s}{T_2}} \right) - \frac{T_1(1 + \mathrm{e}^{-\frac{T_s}{T_2}})}{T_2 - T_1} + \frac{T_2(1 + \mathrm{e}^{-\frac{T_s}{T_1}})}{T_2 - T_1} \right]$$

$$b_2 = Z_2 \left[\mathrm{e}^{-\frac{T_s}{T_1}} \mathrm{e}^{-\frac{T_s}{T_2}} + \frac{T_1 \mathrm{e}^{-\frac{T_s}{T_2}}}{T_2 - T_1} - \frac{T_2 \mathrm{e}^{-\frac{T_s}{T_1}}}{T_2 - T_1} \right]$$

$$a_1 = -\left(\mathrm{e}^{-\frac{T_s}{T_1}} + \mathrm{e}^{-\frac{T_s}{T_2}} \right)$$

$$a_2 = \mathrm{e}^{-\frac{T_s}{T_1}} \mathrm{e}^{-\frac{T_s}{T_2}}$$

5.4 Input-Output Discrete-Time Models – Difference Equations

A general discrete-time linear model can be written in time domain as

$$y(k) = -\sum_{i=1}^{n} a_i y(k-i) + \sum_{i=1}^{m} b_i u(k-d-i) \tag{5.62}$$

Let us now define a shift operator q^{-1} as

$$q^{-1}y(k) = y(k-1) \tag{5.63}$$

Equation (5.62) can then be rewritten as

$$y(k) = -\sum_{i=1}^{n} a_i q^{-i} y(k) + \sum_{i=1}^{m} b_i q^{-(d+i)} u(k) \tag{5.64}$$

or

$$A(q^{-1})y(k) = q^{-d}B(q^{-1})u(k) \tag{5.65}$$

where

$$A(q^{-1}) = 1 + a_1 q^{-1} + a_2 q^{-2} + \cdots + a_n q^{-n} \tag{5.66}$$
$$B(q^{-1}) = b_1 q^{-1} + b_2 q^{-2} + \cdots + b_m q^{-m} \tag{5.67}$$

Equation (5.65) can also be written as

$$\frac{y(k)}{u(k)} = \frac{q^{-d}B(q^{-1})}{A(q^{-1})} \tag{5.68}$$

Hence, we can define a function

$$G(q^{-1}) = \frac{q^{-d}B(q^{-1})}{A(q^{-1})} \tag{5.69}$$

that formally coincides with the discrete-time transfer function $G(z)$ if we replace q^{-1} for z^{-1}

$$G(z^{-1}) = \frac{z^{-d}B(z^{-1})}{A(z^{-1})} \tag{5.70}$$

5.4.1 Direct Digital Control

Direct digital control can be represented by the block scheme shown in Fig. 5.11.

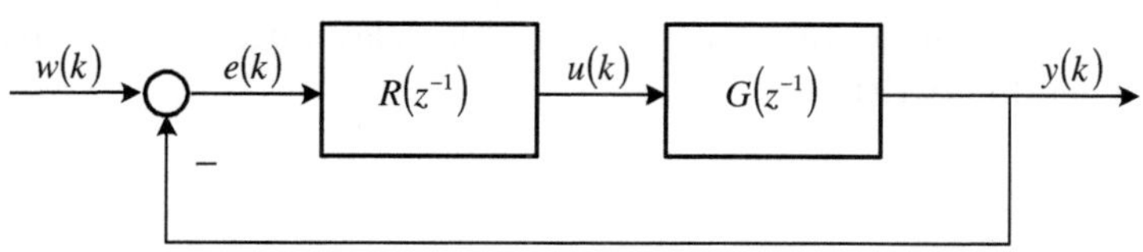

Fig. 5.11. Direct digital control

Difference equation used for determination of the manipulated variable (discrete-time control law) is given as

$$u(k) = -\sum_{i=1}^{r} p_i u(k-i) + \sum_{i=0}^{r} q_i e(k-i) \tag{5.71}$$

where p_i and q_i are constants. Computer that implements the control law remembers the past values of control errors and of manipulated variables. It then calculates the control law from (5.71) in sampling times. Constants $p_i (i = 1, 2, \ldots, r)$ and $q_i (i = 0, 1, \ldots, r)$ are calculated in control design procedure.

If the shift operator q^{-1} is used then (5.71) can be given as

$$u(k) = -\sum_{i=1}^{r} p_i q^{-i} u(k) + \sum_{i=0}^{r} q_i q^{-i} e(k) \tag{5.72}$$

or

$$P(q^{-1})u(k) = Q(q^{-1})e(k) \tag{5.73}$$

where

$$P(q^{-1}) = 1 + p_1 q^{-1} + p_2 q^{-2} + \cdots + p_r q^{-r} \tag{5.74}$$
$$Q(q^{-1}) = q_0 + q_1 q^{-1} + q_2 q^{-2} + \cdots + q_r q^{-r} \tag{5.75}$$

If we write equation (5.73) as

$$\frac{u(k)}{e(k)} = \frac{Q(q^{-1})}{P(q^{-1})} \tag{5.76}$$

we can see that we get a function

$$R(q^{-1}) = \frac{Q(q^{-1})}{P(q^{-1})} \tag{5.77}$$

This is formally equivalent to the discrete-time transfer function if q^{-1} is replaced by z^{-1}

$$R(z^{-1}) = \frac{Q(z^{-1})}{P(z^{-1})} \tag{5.78}$$

Equations (5.70), (5.78) define *transfer functions.*

Discrete-time transfer function without time delay with input u and output y can in general be written as

$$G(z) = \frac{b_0 z^m + b_1 z^{m-1} + \cdots + b_m}{z^n + a_1 z^{n-1} + \cdots + a_n} \tag{5.79}$$

The corresponding difference equation is

$$\begin{aligned} y(k+n) + a_1 y(k+n-1) + \cdots + a_n y(k) \\ = b_0 u(k+m) + b_1 u(k+m-1) + \cdots + b_m u(k) \end{aligned} \tag{5.80}$$

From this follows that a system with a transfer function is only realisable if the degree of the numerator polynomial is smaller than or equal to the degree of the denominator polynomial

$$n \geq m \tag{5.81}$$

System poles are roots of the denominator of $G(z)$, i. e. roots of the characteristic polynomial.

System zeros are poles of the inverse system.

5.5 State-Space Discrete-Time Models

Consider a continuous-time system described by state-space equations

$$\frac{\mathrm{d}\boldsymbol{x}(t)}{\mathrm{d}t} = \boldsymbol{A}\boldsymbol{x}(t) + \boldsymbol{B}\boldsymbol{u}(t) \tag{5.82}$$

$$\boldsymbol{y}(t) = \boldsymbol{C}(t) + \boldsymbol{D}\boldsymbol{u}(t) \tag{5.83}$$

System order of this system is n, the system has m inputs and r outputs. $\boldsymbol{A}, \boldsymbol{B}, \boldsymbol{C}, \boldsymbol{D}$ are constant matrices of appropriate dimensions.

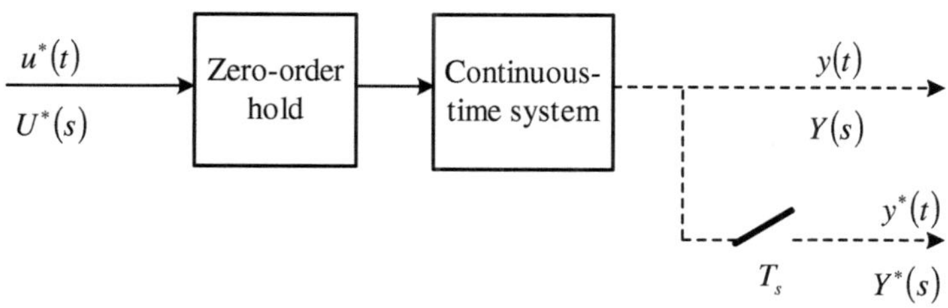

Fig. 5.12. Zero-order hold in series with a continuous-time system

If the continuous-time system given by (5.82), (5.83) is in series with the zero-order hold (Fig. 5.12) then

$$\boldsymbol{x}(t_{k+1}) = \mathrm{e}^{\boldsymbol{A}(t_{k+1}-t_k)}\boldsymbol{x}(t_k) + \left[\int_{t_k}^{t_{k+1}} \mathrm{e}^{\boldsymbol{A}(t_{k+1}-\tau)}\mathrm{d}\tau\right]\boldsymbol{B}\boldsymbol{u}(t_k) \tag{5.84}$$

The zero-order hold ensures that the input signal $\boldsymbol{u}(t)$, $t_k \leq t < t_{k+1}$, $k = \ldots, -1, 0, 1, \ldots$, is piece-wise constant. If the sampling time is constant

$$T_s = t_{k+1} - t_k \tag{5.85}$$

then the sampled data system equations are given as

$$\boldsymbol{x}(t_{k+1}) = \boldsymbol{\Phi}(t_{k+1}, t_k)\boldsymbol{x}(t_k) + \boldsymbol{\Gamma}(t_{k+1}, t_k)\boldsymbol{u}(t_k) \tag{5.86}$$

$$\boldsymbol{y}(t_k) = \boldsymbol{C}\boldsymbol{x}(t_k) + \boldsymbol{D}\boldsymbol{u}(t_k) \tag{5.87}$$

where

$$\boldsymbol{\Phi}(t_{k+1}, t_k) = \mathrm{e}^{\boldsymbol{A}T_s} \tag{5.88}$$

$$\boldsymbol{\Gamma}(t_{k+1}, t_k) = \left(\int_0^{T_s} \mathrm{e}^{\boldsymbol{A}\tau}\mathrm{d}\tau\right)\boldsymbol{B} \tag{5.89}$$

Let us denote $t_k = kT_s$. Then equations (5.86), (5.87) can be simplified as

$$\boldsymbol{x}(k+1) = \boldsymbol{\Phi}\boldsymbol{x}(k) + \boldsymbol{\Gamma}\boldsymbol{u}(k) \tag{5.90}$$

$$\boldsymbol{y}(k) = \boldsymbol{C}\boldsymbol{x}(k) + \boldsymbol{D}\boldsymbol{u}(k) \tag{5.91}$$

Matrices $\boldsymbol{\Phi}$, $\boldsymbol{\Gamma}$ are constant. This follows from (5.88), (5.89). Matrix $\boldsymbol{D}$ is in majority of discrete-time systems equal to zero.

The input-output model of a system can be easily derived from state-space description using the following procedure:

Using the relation

$$q\boldsymbol{x}(k) = \boldsymbol{x}(k+1) \tag{5.92}$$

we can write

$$(q\boldsymbol{I} - \boldsymbol{\Phi})\boldsymbol{x}(k) = \boldsymbol{\Gamma}\boldsymbol{u}(k) \tag{5.93}$$

$$\boldsymbol{y}(k) = \boldsymbol{C}(q\boldsymbol{I} - \boldsymbol{\Phi})^{-1}\boldsymbol{\Gamma}\boldsymbol{u}(k) + \boldsymbol{D}\boldsymbol{u}(k) \tag{5.94}$$

If (5.94) is rewritten as

$$\boldsymbol{y}(k) = \boldsymbol{G}(q)\boldsymbol{u}(k) \tag{5.95}$$

then

$$\boldsymbol{G}(q) = \boldsymbol{C}(q\boldsymbol{I} - \boldsymbol{\Phi})^{-1}\boldsymbol{\Gamma} + \boldsymbol{D} \tag{5.96}$$

or

$$\boldsymbol{G}(q^{-1}) = \boldsymbol{C}(\boldsymbol{I} - q^{-1}\boldsymbol{\Phi})^{-1}q^{-1}\boldsymbol{\Gamma} + \boldsymbol{D} \tag{5.97}$$

Consider now a singlevariable case where

$$G(q) = \frac{B(q)}{A(q)} \tag{5.98}$$

If the *system order* is n and polynomials $B(q)$, $A(q)$ are coprime then the A polynomial has degree n. It follows from (5.98) that the polynomial A is the characteristic polynomial of the matrix $\boldsymbol{\Phi}$ and the input-output model can be written as

$$y(k) = -\sum_{i=1}^{n} a_i y(k-i) + \sum_{i=0}^{n} b_i u(k-i) \tag{5.99}$$

The coefficient b_0 is equal to zero in direct digital control, i. e. there is no direct feed-through. The output signal is measured before the input signal $u(k)$ is calculated. $y(k)$ cannot be influenced by $u(k)$.

Of course, the same holds as with continuous-time systems: one input-output system is equivalent to infinitely many state-space models.

Consider now a non-singular matrix $\boldsymbol{T}$ and define a new state vector as

$$\boldsymbol{z}(k) = \boldsymbol{T}\boldsymbol{x}(k) \tag{5.100}$$

Further we can write

$$\boldsymbol{z}(k+1) = \boldsymbol{T}\boldsymbol{\Phi}\boldsymbol{T}^{-1}\boldsymbol{z}(k) + \boldsymbol{T}\boldsymbol{\Gamma}\boldsymbol{u}(k) \tag{5.101}$$

$$\boldsymbol{y}(k) = \boldsymbol{C}\boldsymbol{T}^{-1}\boldsymbol{z}(k) + \boldsymbol{D}\boldsymbol{u}(k) \tag{5.102}$$

The characteristic equation of the original and transformed system is of the form

$$\det(\lambda\boldsymbol{I} - \boldsymbol{\Phi}) = 0 \tag{5.103}$$

This follows from the fact that

$$\det(\lambda\boldsymbol{I} - \boldsymbol{T}\boldsymbol{\Phi}\boldsymbol{T}^{-1}) = \det\boldsymbol{T}\det(\lambda\boldsymbol{I} - \boldsymbol{\Phi})\det\boldsymbol{T}^{-1} \tag{5.104}$$

$$= \det(\lambda\boldsymbol{I} - \boldsymbol{\Phi}) \tag{5.105}$$

If $\boldsymbol{\Phi}$ has distinct eigenvalues $\lambda_i, i = 1, 2, \ldots, n$ then there exists a matrix $\boldsymbol{T}$ such that

$$\boldsymbol{T}\boldsymbol{\Phi}\boldsymbol{T}^{-1} = \begin{pmatrix} \lambda_1 & 0 & \cdots & 0 \\ 0 & \lambda_2 & \cdots & 0 \\ \vdots & & & \vdots \\ 0 & 0 & \cdots & \lambda_n \end{pmatrix} \tag{5.106}$$

Hence, the transformed system is diagonal.

If the form of the characteristic equation of the matrix $\boldsymbol{\Phi}$ of a singlevariable system is

$$\lambda^n + a_{n-1}\lambda^{n-1} + \cdots + a_0 = 0 \tag{5.107}$$

then the system can be transformed into

$$\boldsymbol{z}(k+1) = \begin{pmatrix} 0 & 1 & \cdots & 0 \\ \vdots & & & \vdots \\ 0 & 0 & \cdots & 1 \\ -a_0 & -a_1 & \cdots & -a_{n-1} \end{pmatrix} \boldsymbol{z}(k) + \begin{pmatrix} 0 \\ \vdots \\ 0 \\ 1 \end{pmatrix} u(k) \tag{5.108}$$

$$y(k) = \begin{pmatrix} b_0 & b_1 & \ldots & b_{n-1} \end{pmatrix} \boldsymbol{z}(k) \tag{5.109}$$

that is called the *controllable canonical form.*

If the characteristic equation of the matrix $\boldsymbol{\Phi}$ of a singlevariable system is of the form (5.107) then the system can be transformed into

$$\boldsymbol{z}(k+1) = \begin{pmatrix} -a_{n-1} & 1 \cdots & 0 \\ \vdots & & \vdots \\ -a_1 & 0 \cdots & 1 \\ -a_0 & 0 \cdots & 0 \end{pmatrix} \boldsymbol{z}(k) + \begin{pmatrix} b_{n-1} \\ \vdots \\ b_1 \\ b_0 \end{pmatrix} u(k) \tag{5.110}$$

$$y(k) = \begin{pmatrix} 1 \; 0 \cdots & 0 \end{pmatrix} \boldsymbol{z}(k) \tag{5.111}$$

This is called the *observable canonical form.*

To solve the state-space equations

$$\boldsymbol{x}(k+1) = \boldsymbol{\Phi x}(k) + \boldsymbol{\Gamma u}(k) \tag{5.112}$$

$$\boldsymbol{y}(k) = \boldsymbol{C x}(k) \tag{5.113}$$

we can use $\mathcal{Z}$-transform. Transformed equation (5.112) is of the form

$$z[\boldsymbol{X}(z) - \boldsymbol{x}(0)] = \boldsymbol{\Phi X}(z) + \boldsymbol{\Gamma U}(z) \tag{5.114}$$

Further, we can write

$$\boldsymbol{X}(z) = (z\boldsymbol{I} - \boldsymbol{\Phi})^{-1} z\boldsymbol{x}(0) + (z\boldsymbol{I} - \boldsymbol{\Phi})^{-1} \boldsymbol{\Gamma U}(z) \tag{5.115}$$

$$\boldsymbol{Y}(z) = \boldsymbol{C}(z\boldsymbol{I} - \boldsymbol{\Phi})^{-1} z\boldsymbol{x}(0) + \boldsymbol{C}(z\boldsymbol{I} - \boldsymbol{\Phi})^{-1} \boldsymbol{\Gamma U}(z) \tag{5.116}$$

where

$$\boldsymbol{G}(z) = \boldsymbol{C}(z\boldsymbol{I} - \boldsymbol{\Phi})^{-1} \boldsymbol{\Gamma} \tag{5.117}$$

is the *discrete-time transfer function matrix.*

The solution $\boldsymbol{x}(k)$, $\boldsymbol{y}(k)$ can be found using the inverse $\mathcal{Z}$-transform with equations (5.115), (5.116). However, there is a much simpler way. Let us write $\boldsymbol{x}(1), \boldsymbol{x}(2)$ as

$$\boldsymbol{x}(1) = \boldsymbol{\Phi x}(0) + \boldsymbol{\Gamma u}(0) \tag{5.118}$$

$$\boldsymbol{x}(2) = \boldsymbol{\Phi x}(1) + \boldsymbol{\Gamma u}(1) \tag{5.119}$$

$$= \boldsymbol{\Phi}^2 \boldsymbol{x}(0) + \boldsymbol{\Phi \Gamma u}(0) + \boldsymbol{B u}(1) \tag{5.120}$$

Further continuation gives

$$\boldsymbol{x}(k) = \boldsymbol{\Phi}^k \boldsymbol{x}(0) + \sum_{i=0}^{k-1} \boldsymbol{\Phi}^{k-i-1} \boldsymbol{\Gamma u}(i), \qquad k \geq 1 \tag{5.121}$$

5.6 Properties of Discrete-Time Systems

5.6.1 Stability

A discrete-time linear system of the form

$$\boldsymbol{x}(k+1) = \boldsymbol{\Phi}\boldsymbol{x}(k) \tag{5.122}$$

is asymptotically stable in large in origin if and only if for a given symmetric positive definite matrix $\boldsymbol{Q}$ there exists a symmetric positive definite matrix $\boldsymbol{P}$ that satisfies solution of the Laypunov equation

$$\boldsymbol{\Phi}^T\boldsymbol{P}\boldsymbol{\Phi} - \boldsymbol{P} = -\boldsymbol{Q} \tag{5.123}$$

The proof of sufficiency is very simple. Assume that a function $V(\boldsymbol{x})$ exists

$$V(\boldsymbol{x}) = \boldsymbol{x}^T\boldsymbol{P}\boldsymbol{x} \tag{5.124}$$

such that

$$\begin{aligned} V(\boldsymbol{x}) &> 0; && \boldsymbol{x} \neq \boldsymbol{0} \\ V(\boldsymbol{0}) &= 0 && \\ V(\boldsymbol{x}) &\to \infty && \text{for} \quad \|\boldsymbol{x}\| \to \infty \end{aligned} \tag{5.125}$$

$$\Delta V(\boldsymbol{x}) = \boldsymbol{x}^T(k+1)\boldsymbol{P}\boldsymbol{x}(k+1) - \boldsymbol{x}^T(k)\boldsymbol{P}\boldsymbol{x}(k) \tag{5.126}$$

$$\Delta V(\boldsymbol{x}) = \boldsymbol{x}^T(k)(\boldsymbol{\Phi}^T\boldsymbol{P}\boldsymbol{\Phi} - \boldsymbol{P})\boldsymbol{x}(k) \tag{5.127}$$

$$\Delta V(\boldsymbol{x}) = -\boldsymbol{x}^T(k)\boldsymbol{Q}\boldsymbol{x}(k) \tag{5.128}$$

Thus,

$$\Delta V(\boldsymbol{x}) < 0 \qquad \boldsymbol{x} \neq \boldsymbol{0} \tag{5.129}$$

as the matrix $\boldsymbol{Q}$ is positive definite. This concludes the proof.

An alternative way to check stability can be the method that determines eigenvalues of matrix $\boldsymbol{\Phi}$.

If input-output models are studied, it can be stated that a singlevariable discrete-time system is stable if and only if all roots of the system denominator

$$1 + a_1 z^{-1} + \cdots + a_n z^{-n} = 0 \tag{5.130}$$

are within unit circle with radius in origin. This follows from the equation

$$z_i = \mathrm{e}^{s_i T_s} \tag{5.131}$$

where $z_i, i = 1, 2, \ldots, n$ are roots of the characteristic equation. The condition of negative s_i is identical with the condition

$$|z_i| < 1 \tag{5.132}$$

5.6.2 Controllability

A discrete-time linear system with constant coefficients

$$\boldsymbol{x}(k+1) = \boldsymbol{\Phi}\boldsymbol{x}(k) + \boldsymbol{\Gamma}\boldsymbol{u}(k) \tag{5.133}$$

$$\boldsymbol{y}(k) = \boldsymbol{C}\boldsymbol{x}(k) \tag{5.134}$$

is completely controllable if and only if the *controllability matrix*

$$\boldsymbol{Q}_{Rd} = (\boldsymbol{\Gamma}\ \boldsymbol{\Phi}\boldsymbol{\Gamma}\ \boldsymbol{\Phi}^2\boldsymbol{\Gamma}\ \cdots\ \boldsymbol{\Phi}^{n-1}\boldsymbol{\Gamma}) \tag{5.135}$$

is of full rank n.

5.6.3 Observability

A discrete-time linear system with constant coefficients (5.133), (5.134) is completely observable if and only if the *observability matrix*

$$\boldsymbol{Q}_{Pd} = \begin{pmatrix} \boldsymbol{C} \\ \boldsymbol{C}\boldsymbol{\Phi} \\ \boldsymbol{C}\boldsymbol{\Phi}^2 \\ \vdots \\ \boldsymbol{C}\boldsymbol{\Phi}^{n-1} \end{pmatrix} \tag{5.136}$$

is of full rank n.

5.6.4 Discrete-Time Feedback Systems – Control Performance

The stability issue explained above is important in feedback control of discrete-time systems. Stability is the principal requirement in feedback control design and is included within the general problem of control performance. If stability is guaranteed, the next requirement is to remove the steady-state control error.

Let us explain briefly the problem. Consider the feedback discrete-time system shown in Fig. 5.13.

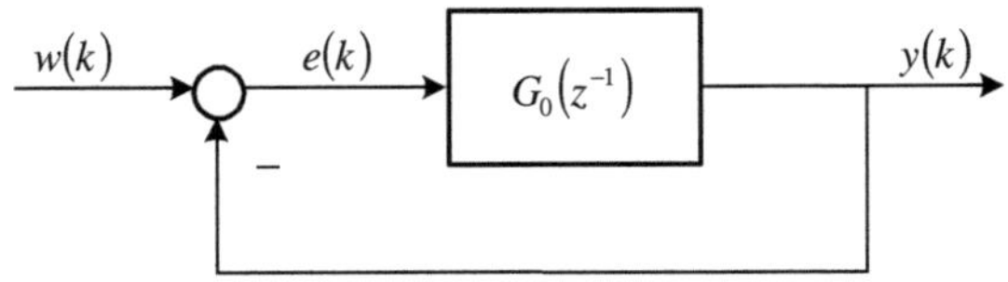

Fig. 5.13. A feedback discrete-time system

The feedforward path includes a discrete-time controller $R(z^{-1})$ and the controlled process including D/A converter, the process, and A/D converter,

i. e. the discretised process $G(z^{-1})$. The open-loop transfer function is given as

$$G_o(z^{-1}) = G(z^{-1})R(z^{-1}) \tag{5.137}$$

and can be written as

$$G_o(z^{-1}) = \frac{B_o(z^{-1})}{A_o(z^{-1})} \tag{5.138}$$

where

$$B_o(z^{-1}) = b_{o1}z^{-1} + b_{o2}z^{-2} + \cdots + b_{om_o}z^{-m_o} \tag{5.139}$$

The closed-loop transfer function with the input w and output y is

$$G_c(z^{-1}) = \frac{G_o(z^{-1})}{1 + G_o(z^{-1})} \tag{5.140}$$

or

$$G_c(z^{-1}) = \frac{B_o(z^{-1})}{A_o(z^{-1}) + B_o(z^{-1})} \tag{5.141}$$

The output in steady-state for a unit step on input can be derived by substituting for $z = 1$ into the closed-loop transfer function. To remove the steady-state control error, the following has to hold in the steady-state

$$G_c(1) = 1 \tag{5.142}$$

From the equation

$$G_c(1) = \frac{\sum_{i=1}^{mo} b_i}{A_o(1) + \sum_{i=1}^{m_o} b_i} \tag{5.143}$$

follows the condition for the zero steady-state control error

$$\sum_{i=1}^{m_o} b_{oi} \neq 0, \qquad A_o(1) = 0 \tag{5.144}$$

From the requirement $A_o(1) = 0$ yields that $A_o(z^{-1})$ has to be of the form

$$A_o(z^{-1}) = (1 - z^{-1})\bar{A}_o(z^{-1}) \tag{5.145}$$

where

$$\bar{A}_o(z^{-1}) = 1 + a_1 z^{-1} + \cdots + a_{n_o-1}z^{-n_o-1} \tag{5.146}$$

From this follows that if the open-loop transfer function contains a digital integrator, the feedback control will guarantee the zero steady-state control error.

5.7 Examples of Discrete-Time Process Models

This section explains a general procedure how discrete-time process models can be obtained on some concrete examples.

5.7.1 Discrete-Time Tank Model

Consider a tank filled with a liquid shown in Fig. 2.1 on page 16. Linearised state-space model of the tank is given as

$$\frac{\mathrm{d}x}{\mathrm{d}t} = ax + bu \tag{5.147}$$

$$y = x \tag{5.148}$$

where

$$x = h - h^s$$

$$u = q_0 - q_0^s$$

$$a = -\frac{k_{11}}{2F\sqrt{h^s}}, \qquad b = \frac{1}{F}$$

t is time, h – liquid level in the tank, q_0 – inlet volumetric flow rate, F – cross-sectional area of the tank, h^s – steady-state level, q_0^s – steady-state flow rate, and k_{11} is a constant.

Let us find a model of the tank that makes it possible to find transient response of the tank in times t_k for the input variable given as

$$u(k) = u(t_k); \quad t_k < t \leq t_{k+1}; \quad k = 0, 1, 2, \dots \tag{5.149}$$

The solution can be found as follows. Comparing equations (5.147) and (5.148) with the general state-space model gives

$$\boldsymbol{A} = a, \quad \boldsymbol{B} = b, \quad \boldsymbol{C} = 1, \quad \boldsymbol{D} = 0 \tag{5.150}$$

The discrete-time state matrix $\boldsymbol{\Phi}(t)$ for $\boldsymbol{A} = a$ is

$$\boldsymbol{\Phi}(t) = \mathcal{L}^{-1}\left\{(s-a)^{-1}\right\} = \mathcal{L}^{-1}\left\{\frac{1}{s-a}\right\} = \mathrm{e}^{at} \tag{5.151}$$

Hence,

$$\boldsymbol{\Phi}(T_s) = \mathrm{e}^{aT_s} \tag{5.152}$$

From (5.89) follows

$$\boldsymbol{\Gamma} = \left(\int_0^{T_s} \mathrm{e}^{a\tau}\mathrm{d}\tau\right) b = \frac{b}{a}\left(\mathrm{e}^{aT_s} - 1\right) \tag{5.153}$$

Finally, equation (5.90) gives

$$x(k+1) = \mathrm{e}^{aT_s}x(k) + \frac{b}{a}(\mathrm{e}^{aT_s} - 1)u(k) \tag{5.154}$$

The process output $y = x$ can be found from this equation in sampling times for $u(k)$.

5.7.2 Discrete-Time Model of Two Tanks in Series

Consider two tanks in series shown in Fig 3.10 on page 86. The behaviour of the second tank is influenced by that of the first tank.

Let us define new variables

$$
\begin{aligned}
x_1 &= h_1 - h_1^s, \qquad x_2 = h_2 - h_2^s \\
y &= x_2, \qquad u = q_0 - q_0^s \\
T_1 &= \frac{2F_1\sqrt{h_1^s}}{k_{11}}, \qquad T_2 = \frac{2F_2\sqrt{h_2^s}}{k_{22}}, \qquad Z = \frac{2\sqrt{h_2^s}}{k_{22}}
\end{aligned}
$$

The symbol definitions are evident from Fig. 3.10 and from Section 5.7.1. The superscript $(.)^s$ denotes the variable in steady-state.

The linearised state-space model of the tanks is characterised by matrices

$$
\boldsymbol{A} = \begin{pmatrix} -\frac{1}{T_1} & 0 \\ \frac{Z}{T_2} & -\frac{1}{T_2} \end{pmatrix}, \quad \boldsymbol{B} = \begin{pmatrix} \frac{1}{T_1} \\ 0 \end{pmatrix}, \quad \boldsymbol{C} = (0\ 1) \tag{5.155}
$$

The corresponding input-output model of the process is given by the differential equation

$$
T_1T_2\frac{\mathrm{d}^2y}{\mathrm{d}t^2} + (T_1+T_2)\frac{\mathrm{d}y}{\mathrm{d}t} + y = Zu \tag{5.156}
$$

The state transition matrix is given as

$$
\boldsymbol{\Phi}(t) = \mathcal{L}^{-1}\left\{\left[\begin{pmatrix} s & 0 \\ 0 & s \end{pmatrix} - \begin{pmatrix} 0 & 1 \\ -a_0 & -a_1 \end{pmatrix}\right]^{-1}\right\} \tag{5.157}
$$

$$
= \mathcal{L}^{-1}\left\{\begin{pmatrix} s & -1 \\ a_0 & s+a \end{pmatrix}^{-1}\right\} \tag{5.158}
$$

$$
= \mathcal{L}^{-1}\left\{\begin{array}{cc} \frac{s+a_1}{s^2+a_1s+a_0} & \frac{1}{s^2+a_1s+a_0} \\ \frac{-a_0}{s^2+a_1s+a_0} & \frac{s}{s^2+a_1s+a_0} \end{array}\right\} \tag{5.159}
$$

$$
= \begin{pmatrix} \frac{T_1}{T_1-T_2}\mathrm{e}^{-\frac{t}{T_1}} + \frac{T_2}{T_2-T_1}\mathrm{e}^{-\frac{t}{T_2}} & \frac{T_1T_2}{T_1-T_2}\mathrm{e}^{-\frac{t}{T_1}} + \frac{T_1T_2}{T_2-T_1}\mathrm{e}^{-\frac{t}{T_2}} \\ \frac{1}{T_2-T_1}\mathrm{e}^{-\frac{t}{T_1}} + \frac{1}{T_1-T_2}\mathrm{e}^{-\frac{t}{T_2}} & \frac{T_2}{T_2-T_1}\mathrm{e}^{-\frac{t}{T_1}} + \frac{T_1}{T_1-T_2}\mathrm{e}^{-\frac{t}{T_2}} \end{pmatrix} \tag{5.160}
$$

$$
\boldsymbol{\Phi}(T_s) = \begin{pmatrix} \mathrm{e}^{-\frac{T_s}{T_1}} & 0 \\ \frac{ZT_1}{T_2-T_1}\left(\mathrm{e}^{-\frac{T_s}{T_2}} - \mathrm{e}^{-\frac{T_s}{T_1}}\right) & \mathrm{e}^{-\frac{T_s}{T_2}} \end{pmatrix} \tag{5.161}
$$

From (5.89) follows

$$\boldsymbol{\Gamma} = \frac{1}{T_1}\int_0^{T_s} \begin{pmatrix} \mathrm{e}^{-\frac{\tau}{T_1}} \\ \frac{ZT_1}{T_2-T_1}\left(\mathrm{e}^{-\frac{\tau}{T_2}} - \mathrm{e}^{-\frac{\tau}{T_1}}\right) \end{pmatrix} \mathrm{d}\tau \tag{5.162}$$

$$= \begin{pmatrix} 1-\mathrm{e}^{-\frac{T_s}{T_1}} \\ \frac{ZT_2}{T_2-T_1}\left(1-\mathrm{e}^{-\frac{T_s}{T_2}}\right) + \frac{ZT_1}{T_1-T_2}\left(1-\mathrm{e}^{-\frac{T_s}{T_1}}\right) \end{pmatrix} \tag{5.163}$$

Therefore, the discrete-time state-space representation can be given as (see also (5.90))

$$\begin{pmatrix} x_1(k+1) \\ x_2(k+1) \end{pmatrix} = \begin{pmatrix} \mathrm{e}^{-\frac{T_s}{T_1}} & 0 \\ \frac{ZT_1}{T_2-T_1}\left(\mathrm{e}^{-\frac{T_s}{T_2}} - \mathrm{e}^{-\frac{T_s}{T_1}}\right) & \mathrm{e}^{-\frac{T_s}{T_2}} \end{pmatrix} \begin{pmatrix} x_1(k) \\ x_2(k) \end{pmatrix}$$
$$+ \begin{pmatrix} 1-\mathrm{e}^{-\frac{T_s}{T_1}} \\ \frac{ZT_2}{T_2-T_1}\left(1-\mathrm{e}^{-\frac{T_s}{T_2}}\right) + \frac{ZT_1}{T_1-T_2}\left(1-\mathrm{e}^{-\frac{T_s}{T_1}}\right) \end{pmatrix} u(k) \tag{5.164}$$

This equation specifies $y = x_2$ in the sampling times if the input variable $u(k)$ is a piece-wise constant.

5.7.3 Steady-State Discrete-Time Model of Heat Exchangers in Series

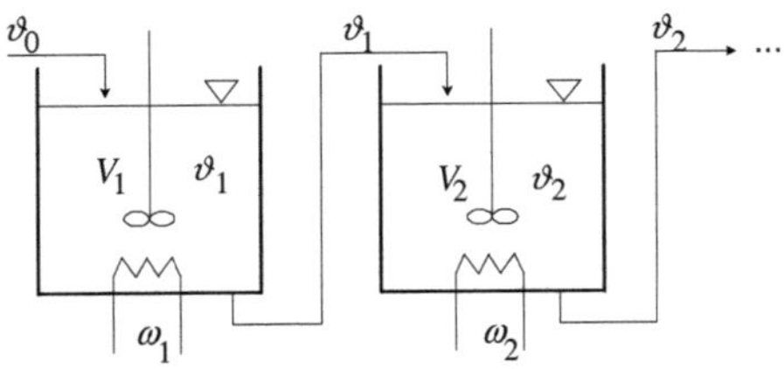

Fig. 5.14. Series of heat exchangers

Consider a series of n heat exchangers where liquid is heated (Fig. 5.14). We assume that heat flows from heat sources into liquid are independent from liquid temperature. Further an ideal liquid mixing and zero heat losses are assumed. We neglect accumulation ability of exchangers walls. Hold-ups of exchangers, as well as flow rates, liquid specific heat capacity are constant.

Under these assumptions for the first heat exchanger holds

$$V_1\rho c_p \frac{\mathrm{d}\vartheta_1}{\mathrm{d}t} = q\rho c_p\vartheta_0 - q\rho c_p\vartheta_1 + \omega_1 \tag{5.165}$$

where t is time,
ϑ_1 – outlet temperature of the heat exchanger,
ϑ_0 – inlet temperature,

ω_1 – heat input,
V_1 – volume of liquid in the exchanger,
ρ – liquid density,
c_p – liquid specific heat capacity.

The condition for steady-state is $\mathrm{d}\vartheta_1/\mathrm{d}t = 0$. This yields for the first heat exchanger

$$q\rho c_p(\vartheta_0 - \vartheta_1) + \omega_1 = 0 \tag{5.166}$$

Let us define state variables

$$x_m = \vartheta_m, \qquad m = 1, 2, \ldots, n \tag{5.167}$$

and input variables

$$u_m = -\frac{1}{q\rho c_p}\omega_m, \qquad m = 1, \ldots, n \tag{5.168}$$

The first exchanger is in the steady-state given as

$$x_1 = x_0 + u_1 \tag{5.169}$$

The $m + 1$th exchanger is in the steady-state given as

$$x_{m+1} = x_m + u_{m+1} \tag{5.170}$$

All heat exchangers can thus in steady-state be written as

$$x_1 = x_0 + u_1 \tag{5.171}$$

$$x_2 = x_0 + u_1 + u_2 \tag{5.172}$$

$$\vdots$$

$$x_m = x_0 + \sum_{i=1}^{m} u_i \tag{5.173}$$

$$\vdots$$

$$x_n = x_0 + \sum_{i=1}^{n} u_i \tag{5.174}$$

Variable m in (5.171)–(5.174) denotes one of the heat exchangers. This points out possibilities to generalise theory of discrete-time systems.

5.8 References

The problem of computer control and discrete-time systems has been frequently dealt with. We give here only some works – the ones that fundamentally contributed to further advance in the field.

Y. Z. Cypkin. *Foundations of Theory of Automatic Systems.* Nauka, Moskva, 1977. (in Russian).

Y. Z. Cypkin. *Grundlagen der Theorie automatischer Systeme.* VEB Verlag Technik, Berlin, 1981.

E. I. Jury. *Digital Control Design.* GIFML, Moskva, 1963. (in Russian).

B. Hanuš. *Introduction to linear discrete control,* volume I. ES VSŠT, Liberec, 1972. (in Czech).

M. Šalamon. *Control Theory.* ES SVŠT, Bratislava, 1979. (in Slovak).

V. Strejc and kol. *Control Design with Digital Computers.* NČSAV, 1965. (in Czech).

H. Unbehauen. *Regelungstechnik II.* Vieweg, Braunschweig/Wiesbaden, 1987.

K. Ogata. *Discrete-time control systems.* Prentice Hall, Englewood Cliffs, New Jersey, 1987.

R. Isermann. *Digitale Regelsysteme.* Springer Verlag, Berlin, 1977.

J. Mikleš and V. Hutla. *Theory of Automatic Control.* Alfa, Bratislava, 1986. (in Slovak).

J. Lunze. *Regelungstechnik 2.* Springer Verlag, Berlin, 1997.

5.9 Exercises

Exercise 5.1:
Derive discrete-time state-space model of interacting tanks in series process shown in Fig. 2.2 on page 17. Consider input variable q_0 and output variable h_2. Zero-order hold is assumed.

Exercise 5.2:
Derive discrete-time transfer functions of the process shown in Fig. 2.2. Assume zero-order hold.

Exercise 5.3:
Derive state-space model of the system

$$y(k+2) + 3.1y(k+1) + 2.2y(k) = u(k)$$

Exercise 5.4:
Find a general solution $\boldsymbol{y}(k)$ for an input $\boldsymbol{u}(k)$ and a system

$$\begin{aligned} -y_1(k+2) - 9y_2(k+1) + y_1(k) &= u_1(k) \\ y_2(k+1) + 3y_2(k) &= u_2(k) \end{aligned}$$

Exercise 5.5:
Find the state transition matrix $\boldsymbol{\Phi}(T_s)$ for the system in Exercise 5.3.

Exercise 5.6:
Find the discrete-time transfer function matrix for the system in Exercise 5.4.

6

Process Identification

Processes in chemical technology can be modelled in various ways. The most natural representation is based on the state-space description resulting from material and energy balances. The models mostly used for control purposes are in form of linear differential or difference equations. We often assume that their parameters are known and constant. However, in experimental conditions it is often necessary to measure or estimate parameters from process input and output signals with carefully chosen input signal types. In this case we will speak about parameter estimation or *system identification.*

The first part of the chapter deals with classification of the identification methods, the mostly used model structures, and guidelines for identification. The second part is devoted to identification from step responses that is mainly used to gain some basic information about the process. The third part will discuss least-squares estimation and its recursive version to identify parameters of continuous and discrete time systems.

6.1 Introduction

While mathematical models based in mass and energy balances construct model by the means of analysis, identification strives to describe the process according the information gained from relation between process inputs and outputs. Also, analysis methods result mainly in state-space model descriptions, whereas identification produces usually input-output models.

Identification Procedure

A general procedure for process estimation includes the following steps:

Determination of the model structure This often makes use of empirical experience about the process, or information from some basic experiments. It is necessary to choose the model type (linear vs. nonlinear), as well as its complexity. There are various criteria that can be selected: quality, flexibility, or the model price. The choice of the structure still remains more art than a systematic procedure.

Parameter estimation There are lots of procedures for parameter estimation. It depends on the type and characteristics of the process input, as well as the desired model structure.

Model verification There are several important aspects at this stage. A suitable model should agree with the experimental data, it should describe the process accurately, and it should meet the purpose it was obtained for. Further, it can be verified whether the parameters obtained are within physical limits. It is also possible to reduce the model and compare it with the original model to see if a simpler model suffices.

Classification of Identification Methods

There are several possibilities to classify identification methods:

A Passive or active experiment identification. This is usually determined by the given process technology and demands of the experiment. This indicates whether it is permitted to generate special signals on inputs or it is just possible to collect typical process inputs and outputs.

B From the point of view of the mathematics, it is possible to distinguish the following methods:
- deterministic,
- stochastic.

Deterministic methods assume exact knowledge about the process inputs and outputs and do not consider random sources and influences.

Stochastic methods include for example the least squares method and its modifications, maximum likelihood method, the Bayesian approach, etc. Any of these methods assumes some properties of random disturbances and some knowledge about them (the least demanding from this point of view are the least squares methods).

Of course, the choice of the method is not arbitrary. It depends on signal to noise ration, disturbance properties, etc.

C From the processing point of view we can divide the methods into:
- single-shot (batch) methods,
- recursive methods.

Batch methods can further be divided into the manual and computer processed. Batch methods are not suitable for computer processing because it is difficult to make algorithm from them, as they depend on some arbitrary performance evaluation (e. g. inflexion point at the step response). Computer batch methods process some data not suitable for manual calculation (least-squares, numerical integration, calculation of correlation functions) and are sometimes called off-line methods.

Recursive methods process the experimental data gradually and estimated parameters are improved from each experiment. They can be used in on-line control of processes and lay foundations for adaptive control.

6.1.1 Models of Linear Dynamic Systems

This section describes basic kinds of linear models employed in identification. We will distinguish between deterministic, stochastic and continuous, discrete models.

Deterministic Models

Mostly used are continuous models described by a linear differential equation with constant coefficients of the form

$$a_n \frac{\mathrm{d}^n y(t)}{\mathrm{d}t^n} + a_{n-1} \frac{\mathrm{d}^{n-1} y(t)}{\mathrm{d}t^{n-1}} + \cdots + y(t) = b_m \frac{\mathrm{d}^m u(t)}{\mathrm{d}t^m} + \cdots + b_0 u(t) \tag{6.1}$$

or

$$A(p)y(t) = B(p)u(t) \tag{6.2}$$

$$A(p) = 1 + a_1 p + \cdots + a_{n-1} p^{n-1} + a_n p^n \tag{6.3}$$

$$B(p) = b_0 + b_1 p + \cdots + b_{m-1} p^{m-1} + b_m p^m \tag{6.4}$$

where $p = \mathrm{d}/\mathrm{d}t$ is the differentiation operator. The identification aim is to estimate the coefficients of the polynomials A, B for given degrees n, m. Very often the time delay T_d is also estimated, hence

$$A(p)y(t) = B(p)u(t - T_d) \tag{6.5}$$

Alternatively, the following equation will be considered in the case of linear discrete-time deterministic models

$$y(k) = -a_1 y(k-1) - \cdots - a_n y(k-n) + b_1 u(k-1) + \cdots + b_m u(k-m) \tag{6.6}$$

or

$$A(q^{-1})y(k) = B(q^{-1})u(k) \tag{6.7}$$

$$A(q^{-1}) = 1 + a_1 q^{-1} + \cdots + a_{n-1} q^{-n+1} + a_n q^{-n} \tag{6.8}$$

$$B(q^{-1}) = b_1 q^{-1} + \cdots + b_{m-1} q^{-m+1} + b_m q^{-m} \tag{6.9}$$

where q is the shift operator ($q^{-1}y(k) = y(k-1)$). The identification aim is again to estimate the coefficients of the polynomials A, B.

Stochastic Models

When stochastic models are considered, we assume a random factor $\xi(t)$ in the equation (6.1), or (6.6). Mostly, white noise ξ will be assumed, but also more complex cases can be discussed. Stochastic models given below are described in discrete-time.

ARMA Process

Consider a stationary process $v(k)$ that can be represented as white noise filtered by a linear process

$$v(k) = \frac{C(q^{-1})}{A(q^{-1})}\xi(k) \tag{6.10}$$

where

$$C(q^{-1}) = 1 + c_1 q^{-1} + \cdots + c_{n_c} q^{-n_c}$$
$$A(q^{-1}) = 1 + a_1 q^{-1} + \cdots + a_{n_a} q^{-n_a}$$

Hence,

$$\begin{aligned} v(k) = &-a_1 v(k-1) - \cdots - a_{n_a} v(k-n_a) \\ &+ \xi(k) + c_1\xi(k-1) + \cdots + c_{n_c}\xi(k-n_c) \end{aligned} \tag{6.11}$$

$$A(q^{-1})v(k) = C(q^{-1})\xi(k) \tag{6.12}$$

Stochastic process described by this equation is called ARMA. It can be split into two parts: AR (autoregressive), when $n_c = 0$

$$v(k) + a_1 v(k-1) + \cdots + a_{n_a} v(k-n_a) = \xi(k), \quad Av = \xi \tag{6.13}$$

and MA (moving average), when $n_a = 0$

$$v(k) = \xi(k) + c_1\xi(k-1) + \cdots + c_{n_c}\xi(k-n_c), \quad v = C\xi \tag{6.14}$$

ARX Model

It is assumed that the white noise enters directly the process equation

$$Ay = Bu + \xi \tag{6.15}$$

$$\begin{aligned} y(k) = &-a_1 y(k-1) - \cdots - a_{n_a} y(k-n_a) \\ &+ b_1 u(k-1) + \cdots + b_{n_b} u(k-n_b) + \xi(k) \end{aligned} \tag{6.16}$$

This model is denoted ARX, because AR describes the part $A(q)y$ and X is an extra input. The special case when $n_a = 0$ is called FIR (finite impulse response) model.

ARX model is due to its simplicity and linearity with respect to identified parameters most often used in identification.

ARMAX Model

In this model the random source enters the system as MA model, hence

$$Ay = Bu + C\xi \tag{6.17}$$

$$\begin{aligned} y(k) = &-a_1 y(k-1) - \cdots - a_{n_a} y(k-n_a) \\ &+ b_1 u(k-1) + \cdots + b_{n_b} u(k-n_b) \\ &+ \xi(k) + c_1\xi(k-1) + \cdots + c_{n_c}\xi(k-n_c) \end{aligned} \tag{6.18}$$

Frequently a modification of ARMAX model is used, the so called ARIMAX (ARMAX Integrated) or CARIMA (Controlled ARIMA) model. In this case a nonstationary disturbance is assumed when the white noise is filtered by an integrator

$$Ay = Bu + \frac{C}{\Delta}e \tag{6.19}$$

where $\Delta = 1 - q^{-1}$ corresponds to the discrete integrator.

Output Error Model

The white noise is added to the output variable (measurement error)

$$y = \frac{B}{A}u + \xi \tag{6.20}$$

$$\begin{aligned} w(k) &= -a_1 w(k-1) - \cdots - a_{n_a} w(k-n_a) \\ &\quad + b_1 u(k-1) + \cdots + b_{n_b} u(k-n_b) \end{aligned} \tag{6.21}$$

$$y(k) = w(k) + \xi(k) \tag{6.22}$$

6.2 Identification from Step Responses

The broad variety of methods based on step response identification belong to deterministic methods as the input signal is deterministic and random sources are not considered in the process description. The results of the identification are coefficients of the equation (6.5). Methods in this category aim at first estimates of the process and provide information about an approximate process gain, dominant time constant, and time delay.

The input signal used is a step change of one of the process inputs when all other inputs are held constant. It is necessary that the controlled process is in a steady state before the step change. The measured process response is a real step response that needs to be further normalised for unit step change and for zero initial conditions.

As the identified process can in general be nonlinear, it might be advantageous to record several step responses with different step changes and signs of the input signals. To obtain the normalised step response, the following formula can then be used

$$\hat{y}_i = \frac{\sum_{k=1}^{N} \Delta u_k y_{ik}}{\sum_{k=1}^{N} (\Delta u_k)^2} \tag{6.23}$$

where i is ith point of the step response,
k - kth measurement, $k = 1, \dots, N$,
Δu_k - step change of the input at the kth measurement,
y_{ik} - output value at the kth measurement in ith interval,
$\hat{y}_i$ - final value of the step response in time i.

6.2.1 First Order System

Consider a first order approximation of the identified process

$$G(s) = \frac{Z}{Ts + 1} e^{-T_d s} \tag{6.24}$$

where Z is the process gain, T time constant, and T_d time delay that need to be determined. The step response corresponding to this transfer function can be obtained via the inverse Laplace transform of the output as

$$y(t) = \begin{cases} 0 & t < T_d \\ Z\left(1 - e^{-\dfrac{t - T_d}{T}}\right) & t \geq T_d \end{cases} \tag{6.25}$$

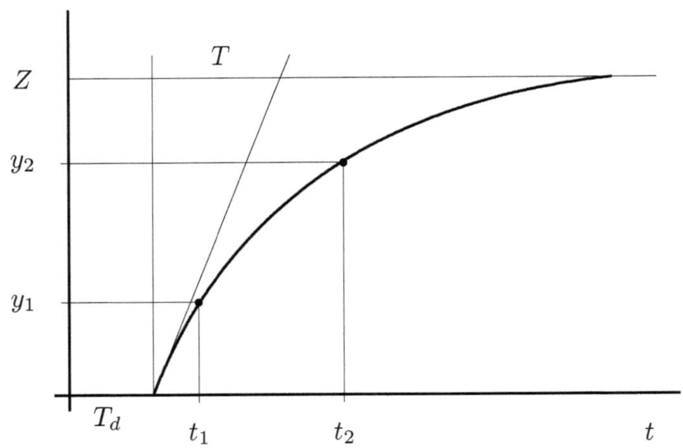

Fig. 6.1. Step response of the first order system

We assume the normalised step response. The process static gain is given as the new steady-state output $Z = y(\infty)$ because the normalised input is the unit step change (Fig. 6.1). If we assume that two points t_1, y_1 and t_2, y_2 from the step response are known then it follows from (6.25)

$$y_1 = Z\left(1 - e^{-\dfrac{t_1 - T_d}{T}}\right) \tag{6.26}$$

$$y_2 = Z\left(1 - e^{-\dfrac{t_2 - T_d}{T}}\right) \tag{6.27}$$

After some simple manipulations we get

$$T = \frac{t_2 - t_1}{\ln \frac{Z-y_1}{Z-y_2}} \tag{6.28}$$

$$T_d = \frac{t_2 x - t_1}{x - 1}, \quad x = \frac{\ln \frac{Z-y_1}{Z}}{\ln \frac{Z-y_2}{Z}} \tag{6.29}$$

The value of the time constant T can also be obtained from the step response as it is shown in Fig. 6.1, or as the time when the output reaches 63% from its new steady state.

Example 6.1: *Approximation with the first order system* www

Consider step response of dimensionless deviation output concentration x_1 in a CSTR to step change of $\Delta q_c = 10$ shown in Fig. 6.2. To obtain the normalised step response, the original one was divided by the step change value and shown in Fig. 6.2 (right) by a solid line. Two points were chosen: [0.85; 0.0082] and [2.18; 0.0224]. The process gain was obtained as $Z = 0.027$ from $y(\infty)$. From the points t_1, y_1 and t_2, y_2 were calculated the time constant $T = 0.94$ and the time delay $T_d = 0.51$. The resulting transfer function is then of the form

$$G(s) = \frac{0.027}{0.94s + 1} e^{-0.51s}$$

The approximated step response is shown in the same figure as the variable x_1^n by a dashed line. Both curves coincide at the measured points. However, there are significant discrepancies elsewhere. We can see that this procedure serves only for a crude estimate of process parameters.

6.2.2 Underdamped Second Order System

If the process step response exhibits underdamped behaviour (Fig. 6.3) it is possible to approximate it by a second order system with a transfer function of the form

$$G(s) = \frac{Z}{T^2 s^2 + 2T\zeta s + 1} = \frac{Z\omega_0^2}{s^2 + 2\omega_0 \zeta s + \omega_0^2} \tag{6.30}$$

The identification task is to determine the process gain Z, natural frequency ω_0, time constant $T = 1/\omega_0$, and damping ζ. Given are points $[t_1, y_1]$, $[t_2, y_2]$ and the steady state output $y(\infty)$.

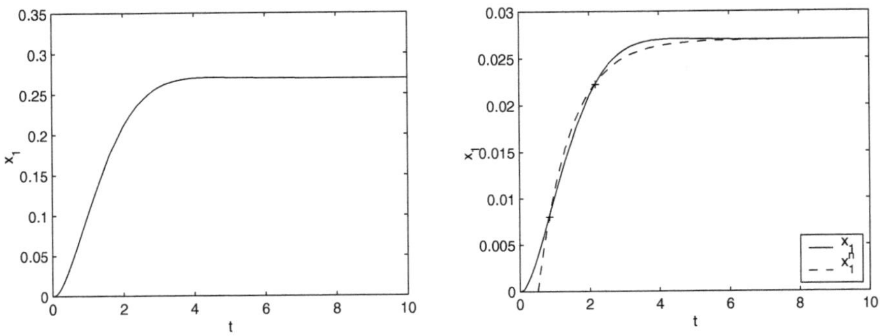

Fig. 6.2. Measured step response of a chemical reactor using the input change $\Delta u = 10$ (left), approximation of step response by the first order step response (right)

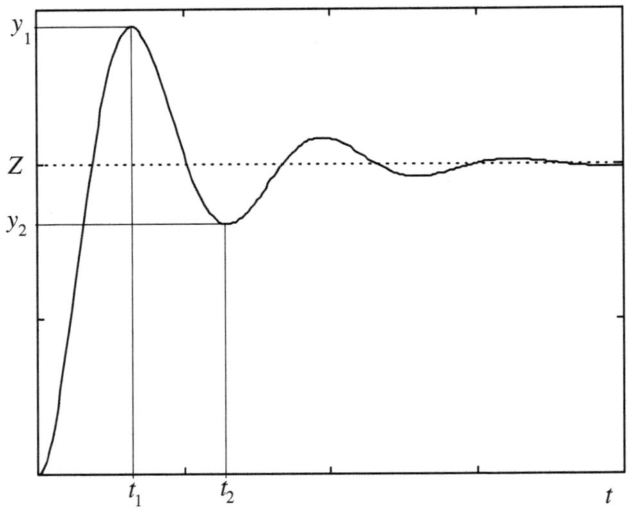

Fig. 6.3. Step response of the second order underdamped system

For the derivation of relations between the known and estimated variables we will use the fact that the derivative of the step response with respect to times is in the points t_n (local extrema) zero.

The process static gain is as in the previous case given as the new steady-state value of the process output $Z = y(\infty)$.

The step response is of the form (see the Laplace transform table)

$$y(t) = Z\left[1 - \frac{1}{P}\mathrm{e}^{-\zeta\omega_0 t}\sin(\omega_0 Pt + \varphi)\right], \quad \varphi = \arccos\zeta, \quad P = \sqrt{1-\zeta^2} \tag{6.31}$$

The derivative of $y(t)$ with respect to time is given as

$$\dot{y}(t) = Z\frac{\omega_0}{P}\mathrm{e}^{-\zeta\omega_0 t}\left[\zeta\sin(\omega_0 Pt + \varphi) - P\cos(\omega_0 Pt + \varphi)\right] \tag{6.32}$$

$$= Z\frac{\omega_0}{P}\mathrm{e}^{-\zeta\omega_0 t}\sin(\omega_0 Pt) \tag{6.33}$$

where the formula for $\sin(a - \varphi)$ was used. The following holds at the local extrema

$$\dot{y}(t_n) = 0 \Longleftrightarrow \sin(\omega_0 Pt_n) = 0 \Rightarrow t_n = \frac{n\pi}{\omega_0 P} \tag{6.34}$$

Substituting for t_n into the process output equation (6.31) yields

$$y(t_n) = Z\left[1 - \frac{1}{P}\mathrm{e}^{-\frac{1}{P}n\pi\zeta}\sin(n\pi + \varphi)\right] \tag{6.35}$$

$$= Z\left[1 - (-1)^n\left(\mathrm{e}^{-\frac{1}{P}\pi\zeta}\right)^n\right] \tag{6.36}$$

$$= Z(1 - (-1)^n M^n)\,, \quad M = \mathrm{e}^{-\frac{1}{P}\pi\zeta} \tag{6.37}$$

The identification procedure is then as follows:

1. $Z = y(\infty)$,
2. $y_1 = Z(1 + M), y_2 = Z(1 - M^2) \Rightarrow M = \dfrac{y_1 - y_2}{y_1}$,
3. $M = \mathrm{e}^{-\frac{1}{P}\pi\zeta} \Rightarrow \zeta = \left|\dfrac{\ln M}{\sqrt{\pi^2 + \ln^2 M}}\right|$,
4. $t_1 = \dfrac{\pi}{\omega_0 P}, t_2 = \dfrac{2\pi}{\omega_0 P} \Rightarrow \omega_0 = \dfrac{\pi}{(t_2 - t_1)\sqrt{1 - \zeta^2}}, \quad T = 1/\omega_0.$

Example 6.2: *Underdamped system* www

Consider a measured step response shown in the left part of Fig. 6.4 that has been measured from the steady-state characterised by the input variable at the value $u_0 = 0.2$ changed to the value $u_\infty = -0.3$. Such a step response can be obtained for example from a U-tube manometer by a step change of the measured pressure.

The measured step response is first shifted to the origin by a value of $y_0 = -2.3608$ and then normalised – divided by the step change of the input $\Delta u = 0.5$. The obtained normalised step response y is shown by a solid line on the right side of Fig. 6.4.

The values of the first maximum and minimum are found as $[15.00; 0.38]$ and $[30.50; 0.32]$, respectively. The above described identification procedure yielded the following values of the estimated parameters: $Z = 0.33$, $\zeta = 0.51$, and $T = 4.22$. The approximated transfer function is then of the form

$$G(s) = \frac{0.33}{17.8084s^2 + 4.3044s + 1}$$

The step response y^n of the approximated transfer function is shown by a dashed line on the right side of Fig. 6.4.

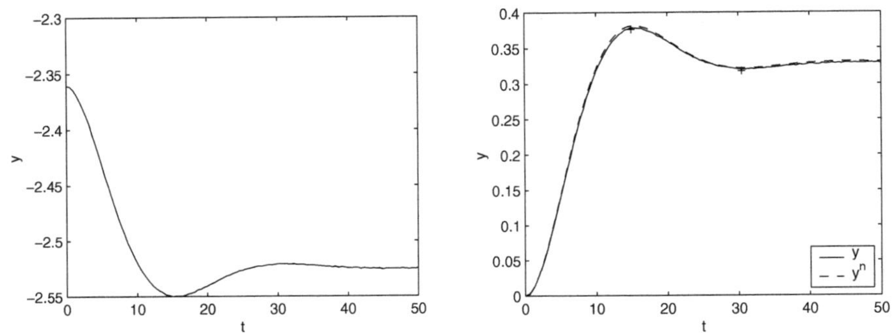

Fig. 6.4. Measured step response of an underdamped system (left). Normalised and approximated step response (right)

6.2.3 System of a Higher Order

Strejc Method

Consider an approximation of a process by the nth order system

$$G(s) = \frac{Z}{(Ts+1)^n}\mathrm{e}^{-T_d s} \tag{6.38}$$

where Z is the process gain, T time constant, T_d time delay, and n the system order that all are to be estimated. Two process characteristics are determined from the step response: times T_n, T_u (see page 257) from the tangent at the inflexion point.

Consider now properties of the normalised step response if $T_d = 0$ (Fig. 6.5)

$$G(s) = \frac{1}{(Ts+1)^n} \tag{6.39}$$

The following facts will be used for the derivation:

1. The tangent of the step response in the inflexion point $[t_i, y_i]$ is given by the equation of a line $p : y = a + bt$,
2. The line p passes through points $[T_u, 0]$, $[T_u + T_n, 1]$, $[t_i, y_i]$. The following holds: $b = 1/T_n$ and $a = -T_u b$,
3. The slope of the line p is given as $b = \dot{y}(t_i)$,
4. $\ddot{y}(t_i) = 0$ (inflexion point).

The step response can by obtained by the means of the inverse Laplace transform and is given as

$$y(t) = 1 - \mathrm{e}^{-\frac{t}{T}} \sum_{k=0}^{n-1} \frac{1}{k!}\left(\frac{t}{T}\right)^k \tag{6.40}$$

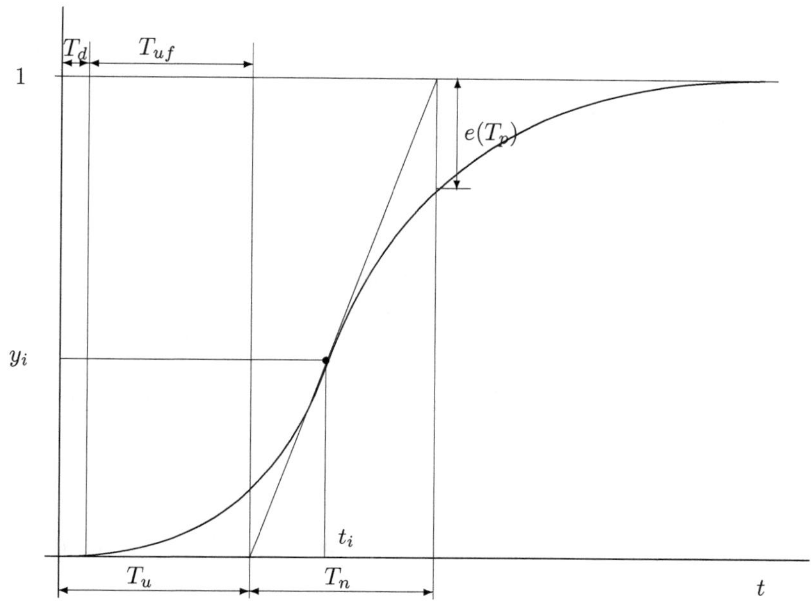

Fig. 6.5. Step response of a higher order system

We calculate the first and the second derivative of the output y with respect to time

$$\dot{y}(t) = \frac{1}{T^n(n-1)!} t^{n-1} \mathrm{e}^{-\frac{t}{T}} \tag{6.41}$$

$$\ddot{y}(t) = \frac{1}{T^n} \left[\frac{t^{n-2}}{(n-2)!} - \frac{1}{T} \frac{t^{n-1}}{(n-1)!} \right] \mathrm{e}^{-\frac{t}{T}} \tag{6.42}$$

In the inflexion point holds $\ddot{y}(t_i) = 0$, hence

$$t_i = T(n-1) \tag{6.43}$$

Evaluating $\dot{y}$ at time t_i gives

$$\dot{y}(t_i) = \frac{(n-1)^{n-1}}{T(n-1)!} \mathrm{e}^{-(n-1)} \tag{6.44}$$

It can be seen from the figure that $\dot{y}(t_i) = 1/T_n$, thus

$$\frac{T}{T_n} = \frac{(n-1)^{n-1}}{(n-1)!} \mathrm{e}^{-(n-1)} = g(n) \tag{6.45}$$

We can see that this function depends only on n.

Further, it can be shown that the relation between T_u and T_n is again only a function of n

$$\frac{T_u}{T_n} = \mathrm{e}^{-(n-1)} \left[\frac{(n-1)^n}{(n-1)!} + \sum_{k=0}^{n-1} \frac{1}{k!}(n-1)^k \right] - 1 = f(n) \tag{6.46}$$

The following table can then be constructed:

n	1	2	3	4	5	6
$f(n)$	0.000	0.104	0.218	0.319	0.410	0.493
$g(n)$	1.000	0.368	0.271	0.224	0.195	0.161

The identification procedure is then as follows:

1. The values of $Z = y(\infty)$, T_{us}, T_n are read from the step response,
2. The quotient $f_s = T_{us}/T_n$ is calculated,
3. The degree n_0 is chosen from the table in such a way that the following holds

   $$f(n_0) \leq f_s < f(n_0 + 1),$$

4. Time delay T_d can be determined as the difference between the real and theoretical time T_u

   $$T_d = [f_s - f(n_0)]T_n$$

   because $T_u = T_n f(n)$,
5. The process time constant T can be read from the row of $g(n)$ for the corresponding n_0. T is obtained from the definition of g

   $$T = T_n g(n_0).$$

Broida method

The assumption that all time constants of the process are the same may not always hold. Broida considered the transfer function of the form

$$G(s) = \frac{Z}{\prod_{k=1}^{n}(\frac{T}{k}s + 1)} \mathrm{e}^{-T_d s} \tag{6.47}$$

In the same way as in the Strejc method, the table for $f(n), g(n)$ can be derived.

n	1	2	3	4	5	6
$f(n)$	0.000	0.096	0.192	0.268	0.331	0.385
$g(n)$	1.000	0.500	0.440	0.420	0.410	0.400

The identification procedure is the same as with the Strejc method.

WWW **Example 6.3:** *Higher order system approximation*
Consider again identification of the step response of the chemical reactor from Example 6.1. Inflexion point has been determined, tangent drawn

that crosses it, and times $T_u = 0.2451$ a $T_n = 1.9979$ were read (Fig. 6.6 left).

Both Strejc and Broida methods were applied. Resulting transfer functions are given as

$$G_{\text{Strejc}}(s) = \frac{0.027}{0.54s^2 + 1.47s + 1}\mathrm{e}^{-0.037s}$$

$$G_{\text{Broida}}(s) = \frac{0.027}{0.5s^2 + 1.5s + 1}\mathrm{e}^{-0.053s}$$

The step responses together with the original one are shown in Fig. 6.6 (right). We can notice that both methods approximate the process dynamics better than the first order transfer function. However, neither Strejc nor Broida methods can cover small overshoot as both are suitable only for aperiodic and monotone step responses.

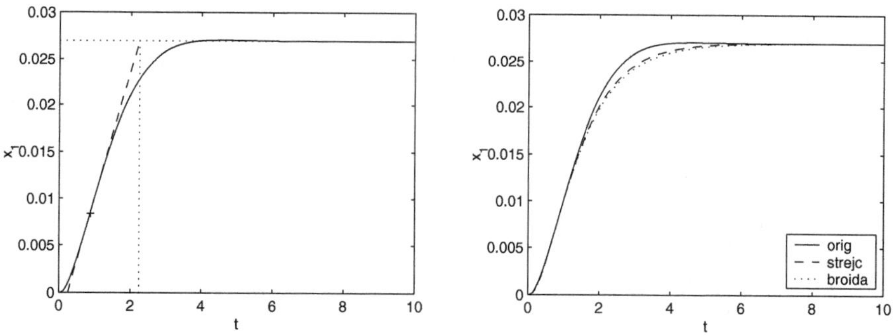

Fig. 6.6. Normalised step response of the chemical reactor with lines drawn to estimate T_u, T_n (left), approximated step response (right)

6.3 Least Squares Methods

The most often used methods nowadays are based on minimisation of squares of errors between measured and estimated process output – the least squares method (RL). Main advantages include the possibility of recursivity of calculations, possibility to track time variant parameters, and various modifications to make the algorithm robust. Originally the method was derived for estimation of parameters of discrete-time systems. We will also show its modification for continuous-time systems.

Consider the ARX model (see page 224).

$$y(k) = -a_1 y(k-1) - \cdots - a_{n_a} y(k-n_a) + b_1 u(k-1) + \cdots + b_{n_b} u(k-n_b) + \xi(k) \tag{6.48}$$

We will introduce the notation

$$\begin{aligned}\boldsymbol{\theta}^T &= (a_1, \dots, a_{n_a}, b_1, \dots, b_{n_b}) \\ \boldsymbol{z}^T(k) &= (-y(k-1), \dots, -y(k-n_a), u(k-1), \dots, u(k-n_b))\end{aligned}$$

Hence

$$y(k) = \boldsymbol{\theta}^T \boldsymbol{z}(k) + \xi(k) \tag{6.49}$$

The identification aim is to determine the parameter vector $\boldsymbol{\theta}$ based on information of measured process output $y(k)$ and the data vector $\boldsymbol{z}(k)$.

Assume that the equation (6.49) holds at times $1, \dots, k$

$$\begin{aligned}y(1) &= -\textstyle\sum_{i=1}^{n_a} a_i y(1-i) + \sum_{i=1}^{n_b} b_i u(1-i) + \xi(1) = \boldsymbol{z}^T(1)\boldsymbol{\theta} + \xi(1) \\ y(2) &= -\textstyle\sum_{i=1}^{n_a} a_i y(2-i) + \sum_{i=1}^{n_b} b_i u(2-i) + \xi(2) = \boldsymbol{z}^T(2)\boldsymbol{\theta} + \xi(2) \\ &\ \vdots \\ y(k) &= -\textstyle\sum_{i=1}^{n_a} a_i y(k-i) + \sum_{i=1}^{n_b} b_i u(k-i) + \xi(k) = \boldsymbol{z}^T(k)\boldsymbol{\theta} + \xi(k)\end{aligned} \tag{6.50}$$

where $k > n_a + n_b$ and denote

$$\boldsymbol{Y} = \begin{pmatrix} y(1) \\ y(2) \\ \vdots \\ y(k) \end{pmatrix}, \quad \boldsymbol{\xi} = \begin{pmatrix} \xi(1) \\ \xi(2) \\ \vdots \\ \xi(k) \end{pmatrix} \tag{6.51}$$

$$\boldsymbol{Z} = \begin{pmatrix} \boldsymbol{z}^T(1) \\ \boldsymbol{z}^T(2) \\ \vdots \\ \boldsymbol{z}^T(k) \end{pmatrix} \tag{6.52}$$

We will look for such an estimate $\hat{\boldsymbol{\theta}}$ that minimises sum of squares of errors between measured and modelled outputs, i. e.

$$I(\boldsymbol{\theta}) = \sum_{i=1}^{k} (y(i) - \boldsymbol{z}^T(i)\boldsymbol{\theta})^2 \tag{6.53}$$

This equation can be rewritten in vector notation

$$I(\boldsymbol{\theta}) = (\boldsymbol{Y} - \boldsymbol{Z}\boldsymbol{\theta})^T (\boldsymbol{Y} - \boldsymbol{Z}\boldsymbol{\theta}) \tag{6.54}$$

Minimum is found if gradient of this function with respect to $\boldsymbol{\theta}$ is equal to zero

$$(\boldsymbol{Z}^T \boldsymbol{Z})\hat{\boldsymbol{\theta}} = \boldsymbol{Z}^T \boldsymbol{Y} \tag{6.55}$$

If the matrix $\boldsymbol{Z}^T\boldsymbol{Z}$ is invertible, then

$$\hat{\boldsymbol{\theta}} = (\boldsymbol{Z}^T\boldsymbol{Z})^{-1}\boldsymbol{Z}^T\boldsymbol{Y} \tag{6.56}$$

The matrix $\boldsymbol{P} = (\boldsymbol{Z}^T\boldsymbol{Z})^{-1}$ is called the covariance matrix if the stochastic part has unit variance.

In general, the estimate of $\hat{\boldsymbol{\theta}}$ is unbiased (its expectation is equal to $\boldsymbol{\theta}$) if $\xi(k)$ is white noise.

6.3.1 Recursive Least Squares Method

In recursive least squares (RLS) estimated parameters are improved with each new data. This means that the estimate $\hat{\boldsymbol{\theta}}(k)$ can be obtained by some simple manipulations from the estimate $\hat{\boldsymbol{\theta}}(k-1)$ based on the data available up to time $k-1$.

Characteristic features of recursive methods are as follows:

- Their requirements for computer memory are very modest as not all measured data up to current time are needed.
- They form important part of adaptive systems where actual controllers are based on current process estimates.
- They are easily modifiable for real-time data treatment and for time variant parameters.

To understand better derivation of a recursive identification method, let us consider the following example.

Example 6.4: *Recursive mean value calculation*

Consider a model of the form

$$y(k) = a + \xi(k)$$

where $\xi(k)$ is a disturbance with standard deviation of one. It is easy to show that the best estimate of a based on information up to time k in the sense of least squares is given as the mean of all measurements

$$\hat{a}(k) = \frac{1}{k}\sum_{i=1}^{k} y(i)$$

This equation can be rewritten as

$$\begin{aligned}\hat{a}(k) &= \frac{1}{k}\left[\sum_{i=1}^{k-1} y(i) + y(k)\right] = \frac{1}{k}[(k-1)\hat{a}(k-1) + y(k)] \\ &= \hat{a}(k-1) + \frac{1}{k}[y(k) - \hat{a}(k-1)]\end{aligned}$$

The result says that the estimate of the parameter a in time k is equal to the estimate at time $k-1$ plus some correction term. The correction term depends linearly on the error between $\hat{a}(k-1)$ and its prediction at

time k, i. e. $y(k)$. The proportionality factor is $1/k$ which means that the magnitude of changes will be decreasing with increasing time as the value of $\hat{a}(k-1)$ will approach the true value of a.
Similarly, it can be shown that the covariance matrix is given as $P(k) = 1/k$. This relation can recursively be rewritten as

$$P(k) = \frac{P(k-1)}{1+P(k-1)}$$

The following matrix inversion lemma will be used to derive recursive least squares method.

Theorem 6.1. *Assume that* $\boldsymbol{M} = \boldsymbol{A} + \boldsymbol{B}\boldsymbol{C}^{-1}\boldsymbol{D}$. *If the matrices* $\boldsymbol{A}, \boldsymbol{C}$ *are nonsingular, then*

$$\boldsymbol{M}^{-1} = \boldsymbol{A}^{-1} - \boldsymbol{A}^{-1}\boldsymbol{B}(\boldsymbol{D}\boldsymbol{A}^{-1}\boldsymbol{B}+\boldsymbol{C})^{-1}\boldsymbol{D}\boldsymbol{A}^{-1} \tag{6.57}$$

Proof. Consider inversion of a matrix $\boldsymbol{X}$

$$\boldsymbol{X}^{-1} = \begin{pmatrix} \boldsymbol{A} & \boldsymbol{B} \\ -\boldsymbol{D} & \boldsymbol{C} \end{pmatrix}^{-1} = \begin{pmatrix} \boldsymbol{P}_1 & \boldsymbol{P}_2 \\ \boldsymbol{P}_3 & \boldsymbol{P}_4 \end{pmatrix} = \boldsymbol{P} \tag{6.58}$$

Because $\boldsymbol{X}\boldsymbol{P} = \boldsymbol{I}$, multiplication of all submatrices gives

$$\boldsymbol{A}\boldsymbol{P}_1 + \boldsymbol{B}\boldsymbol{P}_3 = \boldsymbol{I}, \qquad \boldsymbol{A}\boldsymbol{P}_2 + \boldsymbol{B}\boldsymbol{P}_4 = \boldsymbol{0} \tag{6.59}$$

$$-\boldsymbol{D}\boldsymbol{P}_1 + \boldsymbol{C}\boldsymbol{P}_3 = \boldsymbol{0}, \qquad -\boldsymbol{D}\boldsymbol{P}_2 + \boldsymbol{C}\boldsymbol{P}_4 = \boldsymbol{I} \tag{6.60}$$

We assume that $\boldsymbol{A}, \boldsymbol{C}$ are nonsingular. Hence

$$\begin{array}{ll} \boldsymbol{P}_3 = \boldsymbol{C}^{-1}\boldsymbol{D}\boldsymbol{P}_1, & \boldsymbol{P}_2 = -\boldsymbol{A}^{-1}\boldsymbol{B}\boldsymbol{P}_4 \\ \boldsymbol{P}_1 = [\boldsymbol{A}+\boldsymbol{B}\boldsymbol{C}^{-1}\boldsymbol{D}]^{-1}, & \boldsymbol{P}_4 = [\boldsymbol{C}+\boldsymbol{D}\boldsymbol{A}^{-1}\boldsymbol{B}]^{-1} \end{array} \tag{6.61}$$

As it also holds that $\boldsymbol{P}\boldsymbol{X} = \boldsymbol{I}$ we can write

$$\boldsymbol{P}_1\boldsymbol{A} - \boldsymbol{P}_2\boldsymbol{D} = \boldsymbol{I} \Rightarrow \boldsymbol{P}_1 = \boldsymbol{A}^{-1} + \boldsymbol{P}_2\boldsymbol{D}\boldsymbol{A}^{-1} \tag{6.62}$$

Comparison of the last three equations yields the desired result. □

If we consider $\boldsymbol{C} = 1, \boldsymbol{B} = \boldsymbol{b}, \boldsymbol{D} = \boldsymbol{b}^T$ then the matrix inversion lemma yields

$$\boldsymbol{M}^{-1} = (\boldsymbol{A} + \boldsymbol{b}\boldsymbol{b}^T)^{-1} = \boldsymbol{A}^{-1} - \boldsymbol{A}^{-1}\boldsymbol{b}(\boldsymbol{b}^T\boldsymbol{A}^{-1}\boldsymbol{b}+1)^{-1}\boldsymbol{b}^T\boldsymbol{A}^{-1} \tag{6.63}$$

where the term that has to be inverted is only a scalar.

To derive the RLS method assume that the parameter estimate at time k denoted by $\hat{\boldsymbol{\theta}}(k)$ given by (6.56) and the covariance matrix $\boldsymbol{P}(k) = (\boldsymbol{Z}^T(k)\boldsymbol{Z}(k))^{-1}$ are known. The aim is to derive recursive relations for $\hat{\boldsymbol{\theta}}(k+1)$ and $\boldsymbol{P}(k+1)$. When the measurement at time $k+1$ is known then

$$\boldsymbol{Y}(k+1) = \begin{pmatrix} \boldsymbol{Y}(k) \\ y(k+1) \end{pmatrix}$$

$$\boldsymbol{Z}(k+1) = \begin{pmatrix} \boldsymbol{Z}(k) \\ \boldsymbol{z}^T(k+1) \end{pmatrix}, \quad \boldsymbol{Z}^T(k+1) = \left(\boldsymbol{Z}^T(k)\ \boldsymbol{z}(k+1)\right)$$

The covariance matrix $\boldsymbol{P}(k+1)$ is given as

$$\boldsymbol{P}(k+1) = (\boldsymbol{Z}^T(k+1)\boldsymbol{Z}(k+1))^{-1} \tag{6.64}$$

$$= \left[\left(\boldsymbol{Z}^T(k)\ \boldsymbol{z}(k+1)\right)\begin{pmatrix} \boldsymbol{Z}(k) \\ \boldsymbol{z}^T(k+1) \end{pmatrix}\right]^{-1} \tag{6.65}$$

$$= \left[\boldsymbol{Z}^T(k)\boldsymbol{Z}(k) + \boldsymbol{z}(k+1)\boldsymbol{z}^T(k+1)\right]^{-1} \tag{6.66}$$

$$= \left[\boldsymbol{P}(k)^{-1} + \boldsymbol{z}(k+1)\boldsymbol{z}^T(k+1)\right]^{-1} \tag{6.67}$$

Using the matrix inversion lemma yields

$$\boldsymbol{P}(k+1) = \boldsymbol{P}(k) - \boldsymbol{P}(k)\boldsymbol{z}(k+1)[\boldsymbol{z}^T(k+1)\boldsymbol{P}(k)\boldsymbol{z}(k+1)+1]^{-1}\boldsymbol{z}^T(k+1)\boldsymbol{P}(k) \tag{6.68}$$

Denote $\gamma(k+1) = [\boldsymbol{z}^T(k+1)\boldsymbol{P}(k)\boldsymbol{z}(k+1)+1]^{-1}$. This can be manipulated as

$$\gamma(k+1) = 1 - \gamma(k+1)\boldsymbol{z}^T(k+1)\boldsymbol{P}(k)\boldsymbol{z}(k+1) \tag{6.69}$$

This relation will later be used.

Therefore, the covariance matrix $\boldsymbol{P}$ update is given as

$$\boldsymbol{P}(k+1) = \boldsymbol{P}(k) - \gamma(k+1)\boldsymbol{P}(k)\boldsymbol{z}(k+1)\boldsymbol{z}^T(k+1)\boldsymbol{P}(k) \tag{6.70}$$

Derivation of a new parameter estimate $\hat{\boldsymbol{\theta}}(k+1)$ is similar and makes use of (6.69), (6.70):

$$\begin{aligned}
\hat{\boldsymbol{\theta}}_{k+1} &= \boldsymbol{P}(k+1)\boldsymbol{Z}^T(k+1)\boldsymbol{Y}(k+1) \\
&= \boldsymbol{P}(k+1)\left(\boldsymbol{Z}^T(k)\ \boldsymbol{z}(k+1)\right)\begin{pmatrix} \boldsymbol{Y}(k) \\ y(k+1) \end{pmatrix} \\
&= \boldsymbol{P}(k+1)[\boldsymbol{Z}^T(k)\boldsymbol{Y}(k) + \boldsymbol{z}(k+1)y(k+1)] \\
&= [\boldsymbol{P}(k) - \gamma(k+1)\boldsymbol{P}(k)\boldsymbol{z}(k+1)\boldsymbol{z}^T(k+1)\boldsymbol{P}(k)] \times \\
&\quad [\boldsymbol{Z}^T(k)\boldsymbol{Y}(k) + \boldsymbol{z}(k+1)y(k+1)] \\
&= \hat{\boldsymbol{\theta}}(k) - \gamma(k+1)\boldsymbol{P}(k)\boldsymbol{z}(k+1)\boldsymbol{z}^T(k+1)\hat{\boldsymbol{\theta}}(k) \\
&\quad +\boldsymbol{P}(k)\boldsymbol{z}(k+1)y(k+1) \\
&\quad -\gamma(k+1)\boldsymbol{P}(k)\boldsymbol{z}(k+1)\boldsymbol{z}^T(k+1)\boldsymbol{P}(k)\boldsymbol{z}(k+1)y(k+1) \\
&= \hat{\boldsymbol{\theta}}(k) - \gamma(k+1)\boldsymbol{P}(k)\boldsymbol{z}(k+1)\boldsymbol{z}^T(k+1)\hat{\boldsymbol{\theta}}(k) \\
&\quad +\boldsymbol{P}(k)\boldsymbol{z}(k+1)[1 - \gamma(k+1)\boldsymbol{z}^T(k+1)\boldsymbol{P}(k)\boldsymbol{z}(k+1)]y(k+1) \\
&= \hat{\boldsymbol{\theta}}(k) - \gamma(k+1)\boldsymbol{P}(k)\boldsymbol{z}(k+1)\boldsymbol{z}^T(k+1)\hat{\boldsymbol{\theta}}(k) \\
&\quad +\gamma(k+1)\boldsymbol{P}(k)\boldsymbol{z}(k+1)y(k+1) \\
&= \hat{\boldsymbol{\theta}}(k) + \gamma(k+1)\boldsymbol{P}(k)\boldsymbol{z}(k+1)[y(k+1) - \boldsymbol{z}^T(k+1)\hat{\boldsymbol{\theta}}(k)]
\end{aligned} \tag{6.71}$$

To conclude, equations for recursive least squares are given as

$$\begin{aligned} \epsilon(k+1) &= y(k+1) - \boldsymbol{z}^T(k+1)\hat{\boldsymbol{\theta}}(k) \\ \gamma(k+1) &= [1 + \boldsymbol{z}^T(k+1)\boldsymbol{P}(k)\boldsymbol{z}(k+1)]^{-1} \\ \boldsymbol{L}(k+1) &= \gamma(k+1)\boldsymbol{P}(k)\boldsymbol{z}(k+1) \\ \boldsymbol{P}(k+1) &= \boldsymbol{P}(k) - \gamma(k+1)\boldsymbol{P}(k)\boldsymbol{z}(k+1)\boldsymbol{z}^T(k+1)\boldsymbol{P}(k) \\ \hat{\boldsymbol{\theta}}(k+1) &= \hat{\boldsymbol{\theta}}(k) + \boldsymbol{L}(k+1)\epsilon(k+1) \end{aligned} \tag{6.72}$$

Every recursive algorithm needs some initial conditions. In our case these are $\hat{\boldsymbol{\theta}}(0)$ and $\boldsymbol{P}(0)$. The value of $\boldsymbol{P}(0)$ can be thought as uncertainty of the estimate $\hat{\boldsymbol{\theta}}(0)$. Both initial conditions have some influence on convergence properties of RLS. As it will be shown below, the cost function minimised by RLS is not given by (6.53), but as

$$I_{k+1}(\boldsymbol{\theta}) = [\boldsymbol{\theta} - \hat{\boldsymbol{\theta}}(0)]^T\boldsymbol{P}^{-1}(0)[\boldsymbol{\theta} - \hat{\boldsymbol{\theta}}(0)] + \sum_{i=1}^{k+1}[y(i) - \boldsymbol{z}^T(i)\boldsymbol{\theta}]^2 \tag{6.73}$$

We can notice that the original cost function includes a term characterising effects of initial conditions. To minimise these effects we usually choose $\hat{\boldsymbol{\theta}}(0) = \boldsymbol{0}$ and $\boldsymbol{P}(0) = c\boldsymbol{I}$ where c is some large constant, for example $10^5 - 10^{10}$.

For completeness we will give the proof of above mentioned statements

Theorem 6.2. *Minimisation of the cost function (6.73) leads to (6.72).*

Proof. The cost function (6.73) can be rewritten to the vector notation

$$\begin{aligned} I_{k+1}(\boldsymbol{\theta}) = {} & [\boldsymbol{\theta} - \hat{\boldsymbol{\theta}}(0)]^T\boldsymbol{P}^{-1}(0)[\boldsymbol{\theta} - \hat{\boldsymbol{\theta}}(0)] \\ & + [\boldsymbol{Y}(k+1) - \boldsymbol{Z}(k+1)\boldsymbol{\theta}]^T[\boldsymbol{Y}(k+1) - \boldsymbol{Z}(k+1)\boldsymbol{\theta}] \end{aligned} \tag{6.74}$$

Setting its partial derivative with respect to $\boldsymbol{\theta}$ equal to zero gives

$$[\boldsymbol{Z}^T(k+1)\boldsymbol{Z}(k+1) + \boldsymbol{P}^{-1}(0)]\boldsymbol{\theta} = \boldsymbol{P}^{-1}(0)\hat{\boldsymbol{\theta}}(0) + \boldsymbol{Z}^T(k+1)\boldsymbol{Y}(k+1) \tag{6.75}$$

Data up to time $k+1$ gives $\hat{\boldsymbol{\theta}}(k+1)$, thus

$$\hat{\boldsymbol{\theta}}(k+1) = [\boldsymbol{Z}^T(k+1)\boldsymbol{Z}(k+1) + \boldsymbol{P}^{-1}(0)]^{-1}[\boldsymbol{P}^{-1}(0)\hat{\boldsymbol{\theta}}(0) + \boldsymbol{Z}^T(k+1)\boldsymbol{Y}(k+1)] \tag{6.76}$$

Recursion of (6.67) gives the relation

$$\boldsymbol{P}^{-1}(k+1) = \boldsymbol{P}^{-1}(0) + \boldsymbol{Z}^T(k+1)\boldsymbol{Z}(k+1) \tag{6.77}$$

and for the estimate of the parameters $\hat{\boldsymbol{\theta}}(k+1)$ holds

$$\hat{\boldsymbol{\theta}}(k+1) = \boldsymbol{P}(k+1)[\boldsymbol{P}^{-1}(0)\hat{\boldsymbol{\theta}}(0) + \boldsymbol{Z}^T(k+1)\boldsymbol{Y}(k+1)] \tag{6.78}$$

$$= \boldsymbol{P}(k+1)[\boldsymbol{P}^{-1}(0)\hat{\boldsymbol{\theta}}(0) + \boldsymbol{Z}^T(k)\boldsymbol{Y}(k) + \boldsymbol{z}(k+1)y(k+1)] \tag{6.79}$$

Equation (6.76) shifted to time k is given as

$$\hat{\boldsymbol{\theta}}(k) = \boldsymbol{P}(k)[\boldsymbol{P}^{-1}(0)\hat{\boldsymbol{\theta}}(0) + \boldsymbol{Z}^T(k)\boldsymbol{Y}(k)] \tag{6.80}$$

and equation (6.79) can be rewritten as

$$\hat{\boldsymbol{\theta}}(k+1) = \boldsymbol{P}(k+1)[\boldsymbol{P}^{-1}(k)\hat{\boldsymbol{\theta}}(k) + \boldsymbol{z}(k+1)y(k+1)] \tag{6.81}$$

Substituting for $\boldsymbol{P}^{-1}(k)$ from (6.67) yields

$$\begin{aligned} \hat{\boldsymbol{\theta}}(k+1) &= \hat{\boldsymbol{\theta}}(k) - \boldsymbol{P}(k+1)[\boldsymbol{z}(k+1)\boldsymbol{z}^T(k+1)\hat{\boldsymbol{\theta}}(k) + \boldsymbol{z}(k+1)y(k+1)] \quad (6.82)\\ &= \hat{\boldsymbol{\theta}}(k) + \boldsymbol{P}(k+1)\boldsymbol{z}(k+1)\epsilon(k+1) \quad (6.83) \end{aligned}$$

We can see that this relation would be equivalent to the last of equations (6.72) if

$$\boldsymbol{P}(k+1)\boldsymbol{z}(k+1) = \gamma(k+1)\boldsymbol{P}(k)\boldsymbol{z}(k+1) \tag{6.84}$$

Substituting for $\boldsymbol{P}(k+1)$ from (6.70) gives

$$\begin{aligned} &[\boldsymbol{P}(k) - \gamma(k+1)\boldsymbol{P}(k)\boldsymbol{z}(k+1)\boldsymbol{z}^T(k+1)\boldsymbol{P}(k)]\boldsymbol{z}(k+1)\\ &= \gamma(k+1)\boldsymbol{P}(k)\boldsymbol{z}(k+1)\\ &\qquad \gamma(k+1)\boldsymbol{P}(k)\boldsymbol{z}(k+1)[1 + \boldsymbol{z}^T(k+1)\boldsymbol{P}(k)\boldsymbol{z}(k+1)] = \boldsymbol{P}(k)\boldsymbol{z}(k+1) \end{aligned}$$

The term in square brackets is nothing else than $\gamma^{-1}(k+1)$ which finishes the proof. □

Convergence of Parameters

Let us assume that the identified system model structure is the same as the true system. The question then arises whether the data set is informative enough and in consequence whether the identification procedure yields a unique value of the identified vector $\hat{\boldsymbol{\theta}}$.

Whether a model described by a difference equation is identifiable correctly by the RLS method depends on nonsingularity of its covariance matrix and on the input signal $u(k)$. The covariance matrix can be given as

$$\boldsymbol{P}^{-1}(k) = \sum_{i=1}^{k} \boldsymbol{z}(i)\boldsymbol{z}^T(i) \tag{6.85}$$

The necessary condition for invertibility of this matrix is that $k \geq n$ where n is dimension of the data vector $\boldsymbol{z}(k)$. The sufficient condition is connected to the idea of *persistence of excitation* (PE) of the process input. To understand this concept let us consider the FIR class of models (see page 224) of the form

$$y(k) = \sum_{i=1}^{n} b_i u(k-i) + \xi(k) \tag{6.86}$$

The parameter and data vectors are of the form

$$\boldsymbol{\theta}^T = [b_1, \ldots, b_n] \tag{6.87}$$

$$\boldsymbol{z}(k)^T = [u(k-1), \ldots, u(k-n)] \tag{6.88}$$

Comparing with (6.85) gives the condition for persistence of excitation of the input $u(k)$: the process input $u(k)$ is *persistently exciting of order* n if

$$m_1\boldsymbol{I} > \sum_{i=k}^{k+l} [u(i-1), \ldots, u(i-n)]^T [u(i-1), \ldots, u(i-n)] > m_2\boldsymbol{I} \tag{6.89}$$

where $m_1, m_2 > 0$ and l is a positive integer (necessary condition implies $l \geq n$).

We can then state

Theorem 6.3. *RLS for the FIR system* (6.86) *converges to* $\boldsymbol{\theta}$ *if*

1. *the process input* $u(k)$ *is persistently exciting of order at least* n.

If ARX model is considered of the form

$$y(k) = -\sum_{i=1}^{n} a_i y(k-i) + \sum_{i=1}^{n} b_i u(k-i) + \xi(k) \tag{6.90}$$

then the convergence conditions are the following:

Theorem 6.4. *RLS for the ARX system (6.90) converges to* $\boldsymbol{\theta}$ *if*

1. *polynomials* A, B *are coprime,*
2. *the system is stable,*
3. *the process input* $u(k)$ *is persistently exciting of order at least* $2n$.

In general, parameter estimation using RLS as described above can be used only for stable systems. This was stated in the previous theorem. For unstable systems it is necessary to stabilise the controlled process at first by some controller. However, in that case parameter convergence cannot be guaranteed as $u(k)$ is generated as a linear combination of the data vector $\boldsymbol{z}(k)$

$$u(k) = -\boldsymbol{K}\boldsymbol{z}(k) \tag{6.91}$$

If the control law is given by the previous relation, convergence of parameters cannot be assured. However, the problem can be solved by addition of an external signal to the closed-loop system that will be persistently exciting.

Theorem 6.5. *RLS for the system (6.90) converges to* $\boldsymbol{\theta}$ *if*

1. *polynomials* A, B *are coprime,*
2. *the process input* $u(k)$ *is generated from the feedback control law of the form*

$$u(k) = -\boldsymbol{K}\boldsymbol{z}(k) + v(k) \tag{6.92}$$

3. *external signal* $v(k)$ *is persistently exciting of order at least* $4n$.

6.3.2 Modifications of Recursive Least Squares

Equation (6.67) for the covariance matrix update gives

$$\boldsymbol{P}^{-1}(k+1) = \boldsymbol{P}^{-1}(k) + \boldsymbol{z}(k+1)\boldsymbol{z}^T(k+1) \tag{6.93}$$

This relation can be generalised to the form

$$\boldsymbol{P}^{-1}(k+1) = \lambda_1(k)\boldsymbol{P}^{-1}(k) + \lambda_2(k)\boldsymbol{z}(k+1)\boldsymbol{z}^T(k+1) \tag{6.94}$$

where $0 < \lambda_1(k) \leq 1, 0 \leq \lambda_2(k) < 2$. We note that λ_1, λ_2 have opposite effects. λ_1 increases the covariance matrix and λ_2 decreases it. This modification also leads to change in $\gamma(k+1)$, as well as in $\boldsymbol{L}(k+1)$.

Recursive formula for the covariance matrix is then given as

$$\gamma(k+1) = \frac{1}{\lambda_1(k)/\lambda_2(k) + \boldsymbol{z}^T(k+1)\boldsymbol{P}(k)\boldsymbol{z}(k+1)} \tag{6.95}$$

$$\boldsymbol{P}(k+1) = \frac{1}{\lambda_1(k)}\left[\boldsymbol{P}(k) - \gamma(k+1)\boldsymbol{P}(k)\boldsymbol{z}(k+1)\boldsymbol{z}^T(k+1)\boldsymbol{P}(k)\right] \tag{6.96}$$

Various strategies for choices of λ_i give different modifications of the identification algorithm.

Decreasing gain The basic setting is $\lambda_1 = \lambda_2 = 1$. The identification gain decreases and the covariance matrix increases. This choice is suitable for identification of constant parameters.

Constant exponential forgetting if $\lambda_1 < 1$, $\lambda_2 = 1$. Typical values of λ_1 are between 0.95 and 0.99. The cost function can then be written as

$$I(k) = \sum_{i=1}^{k} \lambda_1^{k-i}\epsilon_i^2 \tag{6.97}$$

We can see that the effect of λ_1 is in forgetting the older data as the most important are the most recent data. This identification algorithm is suitable in case when the identified parameters change slowly.

Increasing exponential forgetting In this case is $\lambda_2 = 1$ and exponential forgetting λ_1 is given as

$$\lambda_1(k) = \lambda_0\lambda_1(k-1) + 1 - \lambda_0 \tag{6.98}$$

with typical initial conditions

$$\lambda_1(0) = \lambda_0 \in \langle 0.95, 0.99\rangle \tag{6.99}$$

This form of exponential forgetting asymptotically converges to one and only initial data are forgotten.

This modification is used for systems with constant parameters. Initial data are considered uncertain and are forgotten.

Varying exponential forgetting In this case is $\lambda_2 = 1$ and exponential forgetting λ_1 is given as

$$\lambda_1(k) = 1 - k\gamma(k)\epsilon^2(k) \tag{6.100}$$

where the constant k is a small positive number (for example 0.001).
In this case the algorithm works as follows: if the identified process changes, the prediction error ϵ^2 will increase and results in a decrease of λ_1. Older data will be forgotten more quickly. If ϵ^2 decreases, the process is well identified, λ_1 approaches the its upper limit and the rate of forgetting will be slower.

Constant trace In this case are λ_i chosen in such a way that the trace of the covariance matrix will be constant

$$\mathrm{tr}\boldsymbol{P}(k+1) = \mathrm{tr}\boldsymbol{P}(k) = ng \tag{6.101}$$

where n is the number of identified parameters and $g = 0.1 - 4$ is initial gain.
This modification is suitable for estimation of time-varying parameters.

Constant gain In this case is $\lambda_1 = 1, \lambda_2 = 0$ and the covariance matrix is given as

$$\boldsymbol{P}(k+1) = \boldsymbol{P}(k) = \boldsymbol{P}(0) \tag{6.102}$$

This algorithm can be used for identification of a small number of parameters (≤ 3) if the signal-to-noise ratio is small. Convergence of parameters is usually smaller but the algorithm can be easily implemented.

Also used are combinations of the above mentioned methods, e. g. constant trace with exponential forgetting. These implementations are suitable for estimation of time-varying parameters if initial estimates are poor.

There is one drawback of recursive methods with exponential forgetting. If there are no new data ($\boldsymbol{z}(k+1) = \boldsymbol{z}(k)$) it can happen that the covariance matrix increases and is no longer positive definite. The algorithm will break down (so called bursting effect). In general it is recommended to check the trace of the covariance matrix.

A method has been developed to improve stability that forgets only in that direction from which some new information comes. The formulas describing this method are as follows:

$$r(k) = \boldsymbol{z}(k+1)^T\boldsymbol{P}(k)\boldsymbol{z}(k+1) \tag{6.103}$$

$$\boldsymbol{L}(k+1) = \frac{\boldsymbol{P}(k)\boldsymbol{z}(k+1)}{1+r(k)} \tag{6.104}$$

$$\beta(k) = \begin{cases} \lambda_1(k) - \frac{1-\lambda_1(k)}{r(k)} & \text{if } r(k) > 0 \\ 1 & \text{if } r(k) = 0 \end{cases} \tag{6.105}$$

$$\boldsymbol{P}(k+1) = \boldsymbol{P}(k) - \frac{\boldsymbol{P}(k)\boldsymbol{z}(k+1)\boldsymbol{z}(k+1)^T\boldsymbol{P}(k)}{\beta(k)^{-1} + r(k)} \tag{6.106}$$

When the factorisation of the covariance matrix $\boldsymbol{P} = \boldsymbol{L}\boldsymbol{D}\boldsymbol{L}^T$ is employed (where $\boldsymbol{L}$ is a lower triangular and $\boldsymbol{D}$ diagonal matrix), this method is known under the name LDDIF and is used throughout the book in adaptive control of processes.

Another approach to estimate better time-varying parameters consists in a direct modification of the covariance matrix $\boldsymbol{P}$. It is clear that the main diagonal contains information about dispersion of the parameters. If the parameters are time varying, it is possible to increase dispersions and thus speed-up adaptation to new parameter values

$$\boldsymbol{P}(k+1) = \boldsymbol{P}(k+1) + \delta\boldsymbol{I} \tag{6.107}$$

where $\delta < 0.01$.

Another possibility is to selectively turn off RLS method if the parameter estimates are correct. Similarly to the case of the variable exponential forgetting, the prediction error is checked. If it is small then the parameters will be held constant. We define a parameter α

$$\alpha(k) = \begin{cases} 1 & \text{if } \gamma(k)\epsilon^2(k) > \varepsilon^2 > 0 \\ 0 & \text{otherwise} \end{cases} \tag{6.108}$$

where ε is a small positive number. Then equations (6.72) are of the form

$$\begin{aligned} \boldsymbol{P}(k+1) &= \boldsymbol{P}(k) - \alpha(k+1)\gamma(k+1)\boldsymbol{P}(k)\boldsymbol{z}(k+1)\boldsymbol{z}^T(k+1)\boldsymbol{P}(k) \\ \hat{\boldsymbol{\theta}}(k+1) &= \hat{\boldsymbol{\theta}}(k) + \alpha(k+1)\boldsymbol{L}(k+1)\epsilon(k+1) \end{aligned} \tag{6.109}$$

This will guarantee that the covariance matrix and parameter estimates will not change if the process output and the model output agree.

Example 6.5: *Second order system identification* WWW

For the second order system with the transfer function of the form

$$G(z) = \frac{b_1 z^{-1} + b_2 z^{-2}}{1 + a_1 z^{-1} + a_2 z^{-2}}$$

is the data vector given as

$$\boldsymbol{z}^T(k) = [-y(k-1), -y(k-2), u(k-1), u(k-2)]$$

in each sampling time. The parameter vector corresponding to the data vector is then given as

$$\boldsymbol{\theta} = [a_1, a_2, b_1, b_2]$$

Consider for example $a_1 = 0.5$, $a_2 = 0.1$, $b_1 = 0.4$, $b_2 = 0.2$. The Simulink schema for identification is shown in Fig. 6.7 and trajectories of estimated parameters are in Fig. 6.8. We can notice that parameters converge to

nominal values. The block denoted as `rls` is the S-function for recursive least squares and its code is shown in Program 6.1.
Alternatively, numerically robust implementation of the scheme in Fig. 6.7 using blocks from Simulink library IDTOOL with recursive least squares algorithm LDDIF is shown in Fig. 6.9.

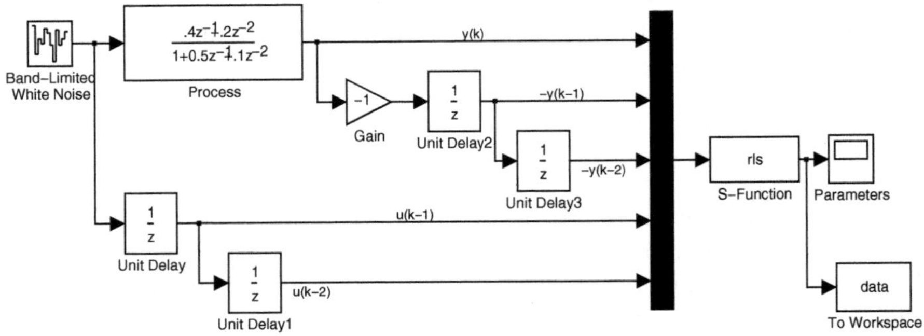

Fig. 6.7. Simulink schema for parameter estimation of the second order discrete system

Program 6.1 (S-function for RLS (`rls.m`))

```
function [sys,x0,str,ts] = rls(t,x,u,flag,tsamp,n)
% tsamp - sampling time
% n: length of the input vector
%     (n = num. of parameters + 1)
% u = [y; z]

  switch flag,
  case 0,
    [sys,x0,str,ts]=mdlInitializeSizes(tsamp,n-1);
  case 2,
    sys=mdlUpdate(t,x,u,n-1);
  case 3,
    sys=mdlOutputs(t,x,u,n-1);
  case 9,
    sys=mdlTerminate(t,x,u);
  otherwise
    error(['Unhandled flag = ',num2str(flag)]);
end

function [sys,x0,str,ts]=mdlInitializeSizes(tsamp,n)
  sizes = simsizes;
```

```
  sizes.NumContStates  = 0;
  sizes.NumDiscStates  = n*(n+1);
  sizes.NumOutputs     = n;
  sizes.NumInputs      = n+1;
  sizes.DirFeedthrough = 0;
  sizes.NumSampleTimes = 1;

  sys = simsizes(sizes);

  theta=(1:n)'*0.01;
  p=eye(n,n)*1e6;
  x0  = [theta;p(:)];
  str = [];
  ts  = [tsamp 0];

function sys=mdlUpdate(t,x,u,n)
  p =zeros(n,n);
  theta=zeros(n,1);

  theta(:)=x(1:n);
  p(:)=x(n+1:n+n*n);

  y=u(1);
  z=u(2:n+1);
  % begin
  e=y-z'*theta;
  gamma=1/(1+z'*p*z);
  l=gamma*p*z;
  p=p-gamma*p*z*z'*p;
  theta=theta+l*e;
  % end
  sys = [theta(:);p(:)];

function sys=mdlOutputs(t,x,u,n)
  sys=x(1:n);

function sys=mdlTerminate(t,x,u)
  sys = [];
```

6.3.3 Identification of a Continuous-time Transfer Function

Consider a differential equation of a linear continuous-time system

$$A(p)y(t) = B(p)u(t) \tag{6.110}$$

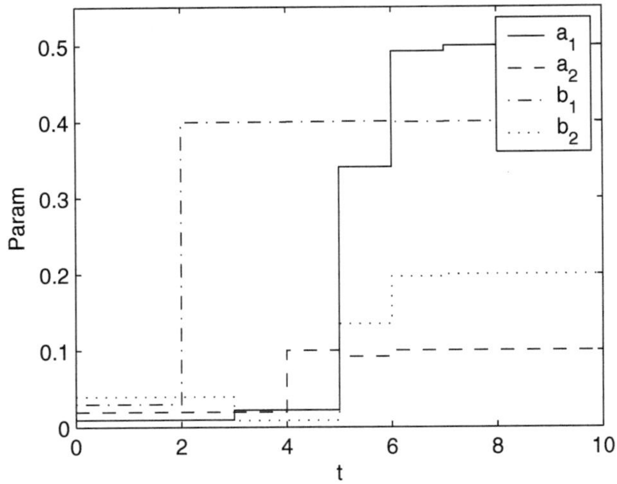

Fig. 6.8. Estimated parameter trajectories for the second order discrete system

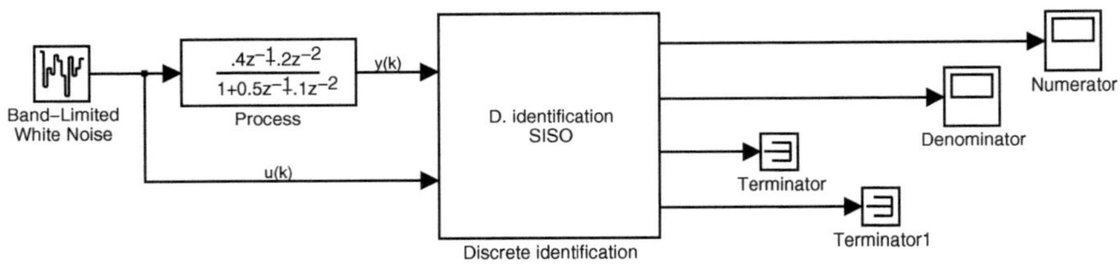

Fig. 6.9. Alternate Simulink schema for parameter estimation of the second order discrete-time system using blocks from library IDTOOL

where $p = \mathrm{d}/\mathrm{d}t$ is the derivative operator and the polynomials are given as

$$A(p) = a_0 + a_1 p + \cdots + a_{n-1} p^{n-1} + p^n \tag{6.111}$$

$$B(p) = b_0 + b_1 p + \cdots + b_m p^m \tag{6.112}$$

We assume that the process is strictly proper, i. e. the degree of the polynomial $B(p)$ is lower than the degree of the polynomial $A(p)$.

If derivatives of inputs and outputs were available, we could directly estimate coefficients of the polynomials a_i, b_i. However, the derivatives are usually not measurable. Therefore, equation (6.110) will be divided by a stable polynomial $C(p)$

$$\frac{A(p)}{C(p)} y(t) = \frac{B(p)}{C(p)} u(t) \tag{6.113}$$

$$A(p) y_f(t) = B(p) u_f(t) \tag{6.114}$$

where

$$y_f(t) = \frac{1}{C(p)} y(t), \qquad u_f(t) = \frac{1}{C(p)} u(t) \tag{6.115}$$

We can see that it is also possible to estimate the parameters from (6.114). All necessary derivatives of filtered variables are available from (6.115) under the conditions that the degree of the polynomial C is greater or equal to the degree of the polynomial A.

If the degrees are equal, the filter is of a minimum realisation. If the degree of C is greater, the filter is strictly proper.

The structure of the C polynomial is usually chosen as

$$C(p) = (1 + c_0 p)^{n_a} \tag{6.116}$$

where c_0 is the time constant of the filter and its value should be smaller than the smallest time constant of A polynomial.

If signals u_f, y_f and their derivatives will be measured at sampling times $t_k = kT_s$ then the identification problem can be posed as follows

$$y_f^{(n)}(k) = \boldsymbol{\theta}^T \boldsymbol{z}(k) + \xi(k) \tag{6.117}$$

$$\boldsymbol{\theta}^T = (a_0, \ldots, a_n, b_0, \ldots, b_m) \tag{6.118}$$

$$\boldsymbol{z}^T(k) = (-\dot{y}_f(k), \ldots, -y_f^{(n-1)}(k), u_f(k), \dot{u}_f(k), \ldots, u_f^{(m)}(k)) \tag{6.119}$$

The block scheme of this procedure is shown in Fig. 6.10.

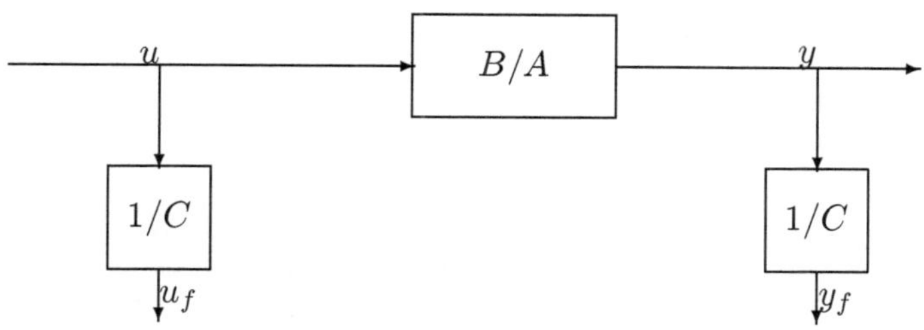

Fig. 6.10. Block scheme of identification of a continuous-time transfer function

Example 6.6: *Second order system* WWW

Consider identification of a continuous-time second order system of the form

$$\ddot{y} + a_1\dot{y} + a_0 y = b_0 u + b_1 \dot{u}$$

We introduce a stable second order filter $C(p)$

$$C(p) = (1 + c_0 p)^2$$

and filtered variables

$$C(p) y_f = y, \qquad C(p) u_f = u$$

If state variables are $x_1 = y_f$, $x_2 = \dot{y}_f$ then the output filter equation is in the state-space form given as

$$\begin{pmatrix} \dot{x}_1 \\ \dot{x}_2 \end{pmatrix} = \begin{pmatrix} 0 & 1 \\ -\frac{1}{c_0^2} & -\frac{2}{c_0} \end{pmatrix} \begin{pmatrix} x_1 \\ x_2 \end{pmatrix} + \begin{pmatrix} 0 \\ \frac{1}{c_0^2} \end{pmatrix} y$$

$$\begin{pmatrix} z_1 \\ z_2 \\ z_3 \end{pmatrix} = \begin{pmatrix} -\frac{1}{c_0^2} & -\frac{2}{c_0} \\ -1 & 0 \\ 0 & -1 \end{pmatrix} \begin{pmatrix} x_1 \\ x_2 \end{pmatrix} + \begin{pmatrix} \frac{1}{c_0^2} \\ 0 \\ 0 \end{pmatrix} y$$

where the state-space outputs are $z_1 = \ddot{y}_f$, $z_2 = -y_f$, $z_3 = -\dot{y}_f$.
We can define states for the input filter similarly as $x_3 = u_f$, $x_4 = \dot{u}_f$ and the state-space description is then given as

$$\begin{pmatrix} \dot{x}_3 \\ \dot{x}_4 \end{pmatrix} = \begin{pmatrix} 0 & 1 \\ -\frac{1}{c_0^2} & -\frac{2}{c_0} \end{pmatrix} \begin{pmatrix} x_3 \\ x_4 \end{pmatrix} + \begin{pmatrix} 0 \\ \frac{1}{c_0^2} \end{pmatrix} u,$$

$$\begin{pmatrix} z_4 \\ z_5 \end{pmatrix} = \begin{pmatrix} 1 & 0 \\ 0 & 1 \end{pmatrix} \begin{pmatrix} x_3 \\ x_4 \end{pmatrix} + \begin{pmatrix} 0 \\ 0 \end{pmatrix} u$$

where the state-space outputs are $z_4 = u_f$, $z_5 = \dot{u}_f$.
The differential equation of the system is now of the form

$$\ddot{y}_f = -a_0 y_f - a_1 \dot{y}_f + b_0 u_f + b_1 \dot{u}_f$$

$$z_1 = a_0 z_2 + a_1 z_3 + b_0 z_4 + b_1 z_5$$

Hence

$$\boldsymbol{\theta}^T = (a_0, a_1, b_0, b_1)$$

$$\boldsymbol{z}^T = (z_2, z_3, z_4, z_5)$$

Consider for example $a_0 = 2$, $a_1 = 3$, $b_0 = 1$, $b_1 = -1$. We choose a second order filter with the time constant $c_0 = 0.25$ and the sampling period for identification $T_s = 1$. State filters for output $\boldsymbol{A}, \boldsymbol{B}, \boldsymbol{C}_y, \boldsymbol{D}_y$ and for input $\boldsymbol{A}, \boldsymbol{B}, \boldsymbol{C}_u, \boldsymbol{D}_u$ are given as

$$\boldsymbol{A} = \begin{pmatrix} 0 & 1 \\ -16 & -8 \end{pmatrix}, \quad \boldsymbol{B} = \begin{pmatrix} 0 \\ 16 \end{pmatrix}, \quad \boldsymbol{C}_y = \begin{pmatrix} -16 & -8 \\ -1 & 0 \\ 0 & -1 \end{pmatrix}, \quad \boldsymbol{D}_y = \begin{pmatrix} 16 \\ 0 \\ 0 \end{pmatrix}$$

$$\boldsymbol{C}_u = \begin{pmatrix} 1 & 0 \\ 0 & 1 \end{pmatrix}, \quad \boldsymbol{D}_u = \begin{pmatrix} 0 \\ 0 \end{pmatrix}$$

Simulink schema is shown in Fig. 6.11 and trajectories of estimated parameters are in Fig. 6.12. We can see that convergence has been achieved within a few sampling steps.
Alternatively, numerically robust implementation of the scheme shown in Fig. 6.11 using blocks from Simulink library IDTOOL with recursive least squares algorithm LDDIF is shown in Fig. 6.13.

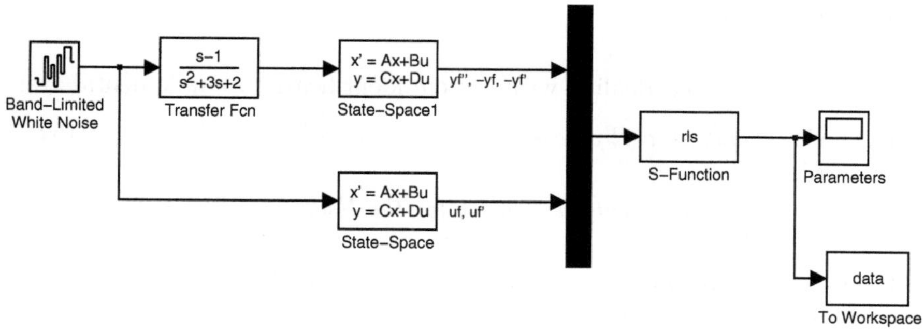

Fig. 6.11. Simulink schema for estimation of parameters of the second order continuous-time system

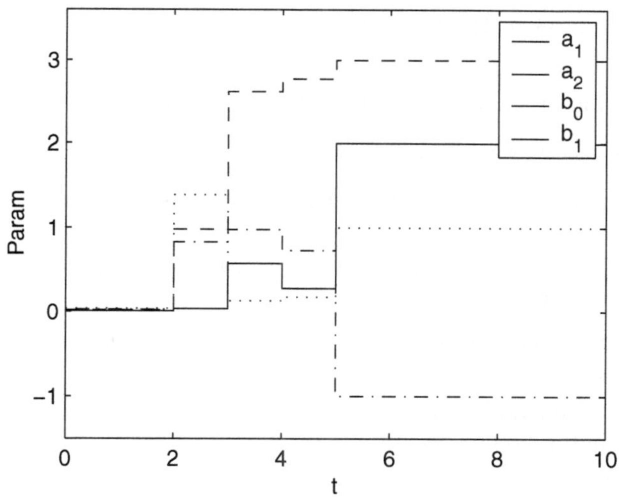

Fig. 6.12. Estimated parameter trajectories for the second order continuous-time system

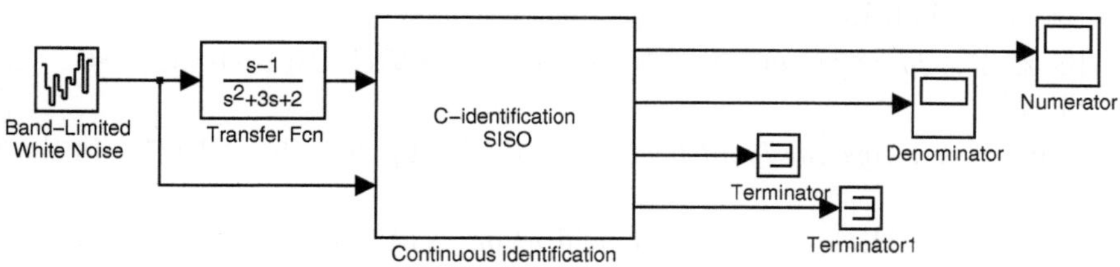

Fig. 6.13. Alternate Simulink schema for parameter estimation of the second order continuous-time system using blocks from library IDTOOL

6.4 References

There are many books dealing with system identification. Fundamental are:

P. Eykhoff. *Trends and Progress in System Identification.* Pergamon Press, Oxford, 1981.

L. Ljung. *System Identification: Theory for the User.* MIT Press, Cambridge, Mass., 1987.

L. Ljung and T. Söderström. *Theory and Practice of Recursive Identification.* MIT Press, Cambridge, Mass., 1983.

T. Söderström and P. Stoica. *System Identification.* Prentice Hall, Inc., 1989.

H. Unbehauen and G. P. Rao. *Identification of Continuous Systems.* North-Holland, Amsterdam, 1987.

The least squares method was derived by Gauss two centuries ago:

K. F. Gauss. *Theoria motus corporum celestium.* English translation: Theory of the Motion of the Heavenly Bodies, Dover, New York, 1963, 1809.

For further reading we also recommend:

V. Strejc. Approximation aperiodischer Übertragungscharakteristiken. *Regelungstechnik,* 7:124 – 128, 1959.

K. J. Åström. *Introduction to Stochastic Control Theory.* Academic Press, New York, 1970.

G. J. Bierman. *Factorization Methods for Discrete Sequential Estimation.* Academic Press, New York, 1977.

V. Bobál, J. Böhm, and J. Fessl. *Digital Self-tuning Controllers.* Springer Verlag, London, 2005.

T. R. Fortescue, L. S. Kershenbaum, and B. E. Ydsie. Implementation of self tuning regulators with variable forgetting factors. *Automatica,* 17: 831–835, 1981.

G. C. Goodwin and K. S. Sin. *Adaptive Filtering Prediction and Control.* Prentice Hall, Inc., Englewood Cliffs, New Jersey, 1984.

P. Hudzovič. *Identifcation and Modelling.* ES SVŠT, Bratislava, 1986. (in Slovak).

B. Rohál'-Il'kiv. *System Identification.* SjF SVŠT, Bratislava, 1987. (in Slovak).

M. Fikar and J. Mikleš. *System Identification.* STU Press, 1999. (in Slovak).

The method LDDIF used for identification in this book is in more detail described in:

R. Kulhavý and M. Kárný. Tracking of slowly varying parameters by directional forgetting. In *Proc. 9th IFAC World Congr., Budapest,* pages 79–83, 1984.

S. Bittanti, P. Bolzern, and M. Campi. Exponential convergence of a modified directional forgetting identification algorithm. *Systems & Control Letters*, 14:131–137, 1990.

Ľ. Čirka and M. Fikar. Identification tool for Simulink. Technical report KAMF9803, Department of Process Control, FCT STU, Bratislava, Slovakia, 1998.

Identification Toolbox for Simulink IDTOOL that contains blocks simplifying identification of continuous and discrete time singlevariable and multivariable systems is freely available from the authors web site `http://www.kirp.chtf.stuba.sk/~fikar/`.

6.5 Exercises

Exercise 6.1:
Explain why it is possible to approximate time constant T for the first order system using the tangent line from Fig. 6.1.

Exercise 6.2:
Approximate the periodic system step response in Fig. 6.3 using the Strejc and Broida methods.

Exercise 6.3:
Derive equations (6.40), (6.46) by the Strejc method.

Exercise 6.4:
Derive the Broida method of the step response approximation.

Exercise 6.5:
Derive relations (6.95) a (6.96) in recursive least squares method.

Exercise 6.6:
Design an identification scheme for processes with transfer functions of the form:

$$G_1(z^{-1}) = \frac{b_1 z^{-1}}{1 + a_1 z^{-1}} z^{-2} \tag{6.120}$$

$$G_2(s) = \frac{b_0}{(Ts + 1)^2} \tag{6.121}$$

7

The Control Problem and Design of Simple Controllers

Generally speaking control system design involves a series of steps that produce a control system with a satisfactory behaviour of a controlled process. If a fairly comprehensive list of steps is to be considered, a step by step design is as follows:

1. Study of the controlled process and definition of control objectives.
2. Process modelling.
3. Determination of process properties and behaviour.
4. Selection (pairing) of manipulated and controlled variables.
5. Determination of structure of the control system.
6. Determination of the regulator type.
7. Design of performance specifications based on control objectives.
8. Controller design.
9. Analysis of the closed-loop system.
10. Simulations of the closed-loop system, repeat from 2 if necessary.
11. Implementation of the controller on chosen hardware and software.
12. Testing and validation on-line.

The aim of this chapter is to explain steps 7 and 8, why control is used, and what are its specifications and demands. Next, structure and design of simple controllers are introduced, performance specifications are discussed. It is supposed that the controlled system is sufficiently precisely described by a linear model and the controller is linear as well.

7.1 Closed-Loop System

A general scheme of a closed-loop system is shown in Fig. 7.1a and consists of four parts: controller, actuator, controlled process, and measurement device. To simplify the scheme, the actuator and the measurement device are usually included into the controlled process. In that case, the scheme in Fig. 7.1b is obtained.

a

b

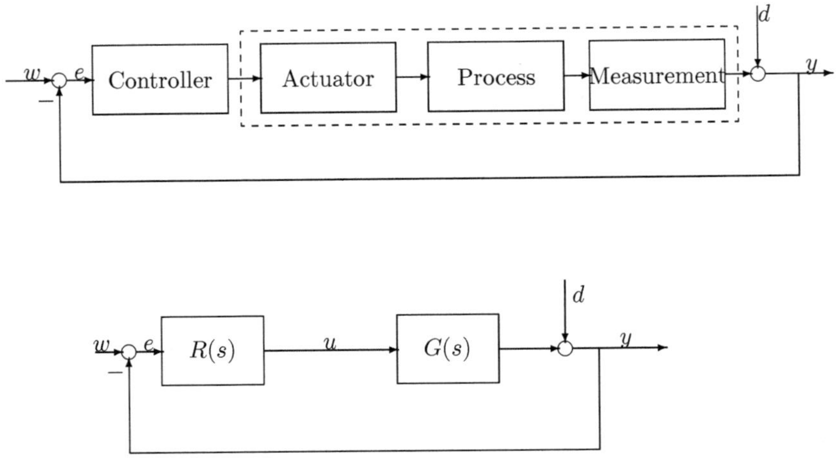

Fig. 7.1. Detailed and simplified closed-loop system scheme

The signals in this scheme are w – setpoint, e – control error, u – manipulated variable, d – disturbance, and y – controlled output. The controller is described by the transfer function $R(s)$ and the controlled process by the transfer function $G(s)$.

The study of dynamical properties of closed-loop systems involves responses to setpoint and disturbance variables. Applying block diagram algebra to fig. 7.1b yields for the output variable

$$Y(s) = D(s) + R(s)G(s)[W(s) - Y(s)] \tag{7.1}$$

Hence,

$$Y(s) = \frac{1}{1 + R(s)G(s)} D(s) + \frac{R(s)G(s)}{1 + R(s)G(s)} W(s) \tag{7.2}$$

From this expression follows for regulation and tracking:

1. For $D(s) = 0$ is the transfer function between w and y given as

$$G_{yw}(s) = \frac{R(s)G(s)}{1 + R(s)G(s)} \tag{7.3}$$

2. For $W(s) = 0$ is the transfer function between d and y given as

$$G_{yd}(s) = \frac{1}{1 + R(s)G(s)} \tag{7.4}$$

Both transfer functions contain common partial transfer function $S(s)$ of the form

$$S(s) = \frac{1}{1 + G_o} \tag{7.5}$$

that is called the closed-loop sensitivity and $G_o = RG$. If we set $W(s) = 0, D(s) = 0$, disconnect the loop at any place, and define new output and input signals y_o, y_i (Fig. 7.2) we get transfer function

$$\frac{Y_o}{Y_i} = -GR = -G_o \tag{7.6}$$

We can see that the transfer function G_o represents dynamics of the open-loop system.

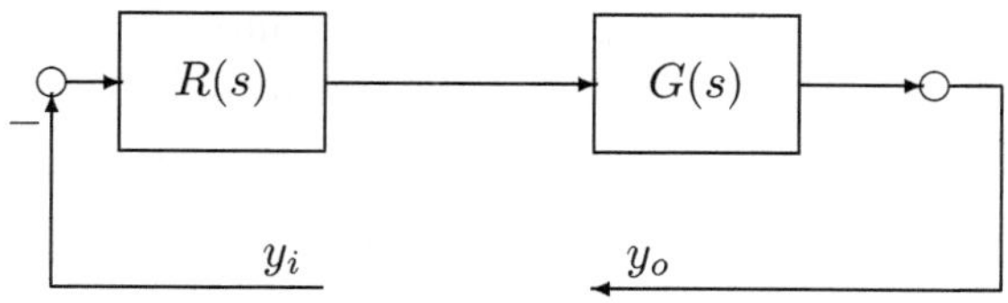

Fig. 7.2. Open-loop system

7.1.1 Feedback Control Problem Definition

The control problem design constitutes one of the most important tasks in the process control. If we put aside the problem of controller structure selection and technical realisation of the closed-loop system, the problem reduces to determination of the controller for the given controlled process.

The requirements on closed-loop system are as follows:

1. Stability.
2. Disturbance rejection.
3. Controlled variable should follow the setpoint as fast and exact as possible.
4. Robustness to parameter changes.

The items 2 and 3 can mathematically be described as

$$G_{yd}(s) = 0 \tag{7.7}$$
$$G_{yw}(s) = 1 \tag{7.8}$$

Example 7.1: *Non-realisability of an ideal controller*

Consider for simplicity a controlled process described by a first order differential equation and with the transfer function of the form

$$G(s) = \frac{Z}{Ts + 1}$$

and a controller as another dynamical system with the transfer function $R(s)$. An ideal controller according to (7.7), (7.8) is then given from

$$\frac{\frac{Z}{Ts+1}R(s)}{1+\frac{Z}{Ts+1}R(s)} = 1, \qquad \frac{1}{1+\frac{Z}{Ts+1}R(s)} = 0$$

We can see that both conditions cannot be fulfilled for any controller. Their validity is only for $R(s) \to \infty$. However, such a controller would be unsuitable because:

- an arbitrarily small control error would cause a very high value of the control signal. However, minimum and maximum values of the control signal always exist in practice. This would mean a bang-bang type of controller.
- the same gain would be applied to disturbances.
- the controller gain would be too high and instability of the closed-loop system could occur for higher order systems.

7.2 Steady-State Behaviour

The steady-state behaviour of the closed-loop system describes its properties after all transient effects have finished. To analyse the control accuracy, we analyse the value of the steady-state control error. For purposes of the analysis assume the open-loop transfer function of the form

$$G_o(s) = \frac{Z_o}{s^k}\frac{B_o(s)}{A_o(s)}\mathrm{e}^{-T_d s} \tag{7.9}$$

where

$$B_o(s) = b_0 + b_1 s + \cdots + b_{m_o} s^{m_o} \tag{7.10}$$

$$A_o(s) = 1 + a_1 s + \cdots + a_{n_o-k} s^{n_o-k}, \quad n_o \geq m_o \tag{7.11}$$

and k is a nonnegative integer defining the system type. Its increased value adds pure integrators to the system. Based on the stability analysis it can be shown that it is desirable to have the smallest gain Z_o and the smallest k in the open-loop system.

For the steady-state behaviour examination assume that the $A_o(s)$ is stable.

If we set

$$E(s) = W(s) - Y(s) \tag{7.12}$$

then from (7.2) follows

$$E(s) = \frac{1}{1+G_o(s)}(W(s) - D(s)) \tag{7.13}$$

This relation shows that impacts of disturbances and setpoints are (up to sign) the same and that it is not necessary to study them both.

Let us assume that the variable y_i acting as the open-loop system input (see Fig. 7.2) is given as $y_i(t) = y_0 t^i 1(t)$, $i = 0, 1, 2$ (step, ramp, ...) with the Laplace transform $Y_i(s) = y_0 i!/s^{i+1}$. The control error is then given as

$$E(s) = \frac{1}{1 + G_o(s)} Y_i(s) \tag{7.14}$$

Its steady-state value is in the time domain given by the Final value theorem

$$\lim_{t \to \infty} e(t) = \lim_{s \to 0} sE(s) = \lim_{s \to 0} i! \frac{s^{k-i}}{s^k + Z_o} y_0 \tag{7.15}$$

It is then clear that steady-state control error is a function of the input signal and the open-loop system properties: gain Z_o and the coefficient k. For the zero steady-state control error holds $k > i$, i. e. the higher order of the input variable, the more integrators the open-loop system should contain. If $k = i$, non-zero steady-state control error results and its value is inversely proportional to the gain Z_o. As the gain is given as the product of gains of the controlled system and the controller, the lowest value of the steady-state control error is attained with the highest controller gain. Hence, the control design problem is in opposition to requirements on stability. The least desired state is $k < i$ when $e(\infty) = \infty$.

7.3 Control Performance Indices

The third requirement on closed-loop system in Chapter 7.1.1 dictates that the controlled output follows the reference value as precisely as possible. Several performance indices can describe this issue more quantitatively.

7.3.1 Time Domain

For performance specifications in time domain it is suitable to choose a standard input variable behaviour and to evaluate the controlled output. Most often a step response is chosen. A typical closed-loop output trajectory is shown in Fig. 7.3.

Performance quantities characterising the transient response of the closed-loop system are then as follows:

- Maximum overshoot $e_{\max}$ defines (in percent) the normalised maximum control error after the process output reaches setpoint for the first time. It can mathematically be written as

$$e_{\max} = \frac{y_{\max} - y_\infty}{y_\infty} 100 \tag{7.16}$$

  If the transient response is aperiodic and without overshoot then $e_{\max} = 0$. The recommended value for majority of cases is smaller than 25%.

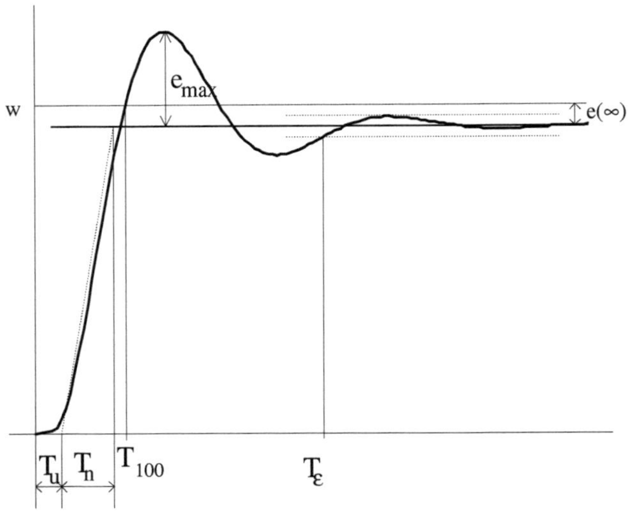

Fig. 7.3. A typical response of a controlled process to a step change of a setpoint variable

- Settling time T_ϵ is given as the time after which the absolute value of the control error remains in some prescribed region. Usually the region is characterised in percent of the maximum control error with the mean value being the process output new steady state. The recommended value is about 1% – 5%. The lower value of T_ϵ, the better control performance results.
- Damping ratio – ratio of the first and the second maximum value of the controlled output. The value of 0.3 or less can be suitable.
- Time T_u determines the point obtained as the intersection of the time axis and tangent of the response at the inflexion point.
- Time T_n determines the point obtained as the intersection of the settling value of the controlled output with the tangent of the response at the inflexion point. Both times T_u, T_n are used in identification from the process step response.
- Rise time is the time required for the response to rise from 0% to n% of its final value. For underdamped responses time T_{100} is considered (time when the output reaches its final value for the first time. For overdamped systems time T_{90} is commonly used as time between 10% and 90% of the new steady state output. However, a too small value can cause high overshoot and settling time.
- Dominant time constant T_{DOM} can be thought of as an alternative to the settling time and can be defined as the largest time constant of the process. If for example the first order system contains a term $\mathrm{e}^{-\gamma t}$ then the dominant time constant is $T_{\mathrm{DOM}} = 1/\gamma$. For higher order systems T_{DOM} can be approximated from a similar second order system with a damping

coefficient ζ and the natural undamped frequency ω_n. In this case it can be shown that

$$T_{\mathrm{DOM}} \geq \frac{1}{\zeta\omega_n} \tag{7.17}$$

- Steady-state control error $e(\infty)$ defines deviation between the setpoint w and the controlled output in the new steady state. It is non-zero for controllers without integral action.

Time indices define speed of control, other characterise control quality.

Similar characteristics can be defined for disturbance rejection (Fig. 7.4).

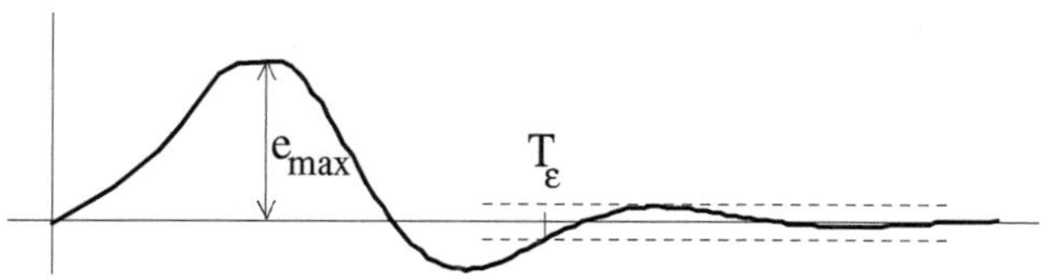

Fig. 7.4. A typical response of the controlled process to a step change of disturbance

7.3.2 Integral Criteria

A large number of control quality specifications can lead to a difficult control design. Usually, it is easier to define a single performance index. One possibility is for example to define a weighted mean value of several quantities, e. g.

$$I = a_1 T_u + a_2 e_{\max} + a_3 T_\epsilon \tag{7.18}$$

and then to minimise the index I. An issue then remains how to choose weighting coefficients $a_1, a_2, a_3, \ldots$.

Another possibility is to use integral criteria when an area (or some function of it) of control error is minimised. The advantage of using integral criteria is that the entire transient response is evaluated, not only some of its characteristics. From the mathematical point of view a general cost function can be defined as

$$I_k = \int_0^\infty f_k[e(t)]\mathrm{d}t \tag{7.19}$$

where the function f_k can contain various terms. Suitable candidates should lead to a higher value of I_k when the control quality deteriorates.

Most often, the following functions are used:

- $f_k = e(t)$: (IE = integral of error) is the simplest function. The advantage are simple calculations. However, it is only suitable for overdamped transient responses as otherwise positive and negative parts cancel themselves and the criterion can be zero even if the transient response is highly oscillatory.
- $f_k = |e(t)|$: (IAE = integral absolute value of error) removes the drawback of the previous approach and is also suitable for oscillatory responses. However, from the computational point of view, it can be difficult to implement as the absolute value is non-differentiable.
- $f_k = |e(t)|t$: (ITAE = integral time multiplied absolute value of error) the time term in the cost function penalises the settling time, as well as reduces the large initial error in the cost function. Otherwise it is the same as the previous one.
- $f_k = e^2(t)$: (ISE = integral squared value of error) combines advantages of a simple surface and of the absolute value. Large values of control error are penalised more than small values and lead to controllers with larger settling time than the IAE cost. Squared error is mathematically convenient for analytical purposes.
- $f_k = e^2(t)t$: improves the previous cost and decreases the settling time.
- $f_k = e^2(t) + \phi \dot{e}^2(t)$: by penalising a square of derivative suppresses oscillatory behaviour typical for ISE cost. Dampens large changes of the control error and thus also large values of the manipulated input and its change. The choice of the penalisation factor ϕ can be difficult.
- $f_k = e^2(t) + \phi u^2(t)$: can suitably penalise the manipulated variable and simply regulate the ratio between speed and robustness of the controller.

In each case it has been assumed that $e(t)$ and $u(t)$ are stable and converge to zero, so that the cost functions converge. If this is not correct, it is necessary to work with deviations of the variables from their steady state or to implement some special strategy.

7.3.3 Control Quality and Frequency Indices

Control quality can also be characterised by frequency domain indices. Some of them measure directly quality, others describe *relative stability*, i. e. quantify how far is the closed-loop system away from instability.

- Gain margin measures relative stability and is defined as the amount of gain that can be inserted in the loop before the closed-loop system reaches instability. Mathematically spoken it is the magnitude of the reciprocal of the open-loop transfer function evaluated at the frequency ω_π at which the phase angle of the frequency transfer function is $-180°$

$$GM = \frac{1}{|G_o(\omega_\pi)|} \tag{7.20}$$

where $\arg G_o(\omega_\pi) = -180° = -\pi$ and ω_π is called the phase crossover frequency.

- Phase margin is a measure of relative stability and it is defined as 180° plus the phase angle of the open-loop at the frequency ω_1 at unity open-loop gain

$$\phi_{PM} = 180° + \varphi[G_o(\omega_1)] \tag{7.21}$$

where $|G_o(\omega_1)| = 1$ and ω_1 is called the gain crossover frequency. Recommended values are between $30° - 65°$.
Gain and phase margins are graphically shown in Bode plots in Fig. 7.5.

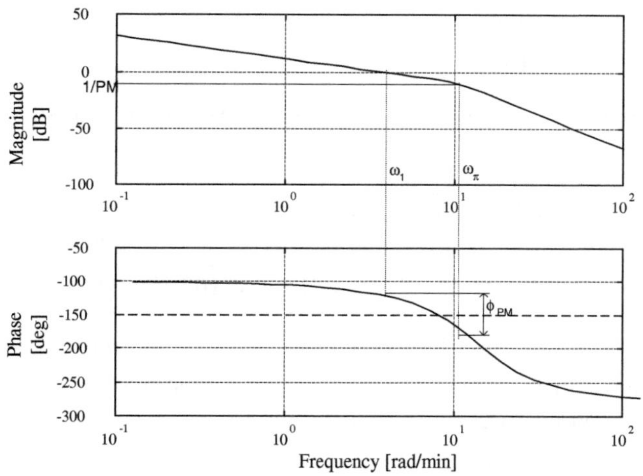

Fig. 7.5. Gain and phase margins for a typical continuous system. Intersection for ω_1 is at the value 0 dB

- Bandwidth ω_b. Loosely spoken it is defined as the frequency range over which the system responds satisfactorily. Often it is defined as the frequency range where the magnitude is approximately constant compared to the value at some specified frequency and differs by not more than -3 dB. However, there are also other definitions. A good approximation is the frequency ω_1 defined by (7.21).
- Cutoff rate is the frequency rate at which the magnitude decreases beyond the cutoff frequency ω_c. For example it can be chosen as 6 dB / decade.
- Resonance peak M_p is the maximum of the magnitude frequency response.
- Resonant frequency ω_p is the frequency at which the resonance peak occurs.

These indices are shown in Fig. 7.6.

Example 7.2: *Gain and phase margins*
Let us consider a system with the transfer function

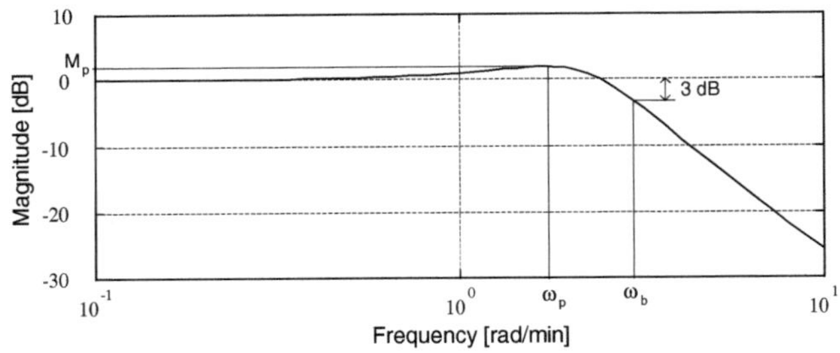

Fig. 7.6. Bandwidth, resonant peak, and resonance frequency in magnitude frequency response

$$G(s) = \frac{432}{s(s^2 + 13s + 115)}$$

Substituting $s = \mathrm{j}\omega$, the magnitude and phase of the transfer function is given as

$$|G(\mathrm{j}\omega)| = \left| \frac{432}{(\mathrm{j}\omega)^3 + 13(\mathrm{j}\omega)^2 + 115(\mathrm{j}\omega)} \right|$$
$$= \frac{432}{|-13\omega^2 + \mathrm{j}(115\omega - \omega^3)|} = \frac{432}{\sqrt{169\omega^4 + (115\omega - \omega^3)^2}}$$
$$\varphi = \varphi(1) - \varphi(-13\omega^2 + \mathrm{j}(115\omega - \omega^3)) = -\arctan\frac{115\omega - \omega^3}{13\omega}$$

To find the gain margin it is necessary to determine the frequency when the phase is equal to $\varphi = -\pi$. This gives

$$-\pi = -\arctan\frac{115\omega_\pi - \omega_\pi^3}{13\omega_\pi} \Rightarrow \frac{115\omega_\pi - \omega_\pi^3}{13\omega_\pi} = \tan\pi \Rightarrow \omega_\pi = \sqrt{115}$$

Gain margin is given by the reciprocal of the magnitude of the transfer function at this frequency (7.20), hence

$$GM = \frac{1}{|G(\mathrm{j}\omega_\pi)|} = \frac{1}{0.2897} = 3.45$$

For phase margin it is necessary to find the frequency when the magnitude is equal to one

$$\frac{432}{\sqrt{169\omega_1^4 + (115\omega_1 - \omega_1^3)^2}} = 1$$

Manipulating this equation yields the 6th order equation

$$\omega_1^6 - 61\omega_1^4 + 115^2\omega_1^2 - 432^2 = 0$$

Solving this equation gives a single real positive root $\omega_1 = 3.858$. From the phase at this frequency follows from (7.21) $\phi_{PM} = 63°$.
Fig. 7.5 illustrates graphical procedure to find the indices for this system.

Example 7.3: *Determination of the resonant peak*

Consider a system with the transfer function

$$G(s) = \frac{5}{s^2 + 2s + 5}$$

We will find values of the resonant peak M_p, the corresponding resonance frequency ω_p, and bandwidth ω_b. As these are calculated from the magnitude frequency resonance, this will be derived at first

$$|G(\mathrm{j}\omega)| = \frac{5}{|-\omega^2 + 2\mathrm{j}\omega + 5|} = \frac{5}{\sqrt{\omega^4 - 6\omega^2 + 25}}$$

The maximum can be found by setting the derivative equal to zero

$$\omega_p(\omega_p^2 - 3) = 0$$

Only the second root has meaning and thus $\omega_p = \sqrt{3}$. The resonant peak is then

$$M_p = |G(\mathrm{j}\omega_p)| = \frac{5}{\sqrt{\omega_p^4 - 6\omega_p^2 + 25}} = 1.25$$

Bandwidth can be determined if the magnitude of $G(s)$ decreases by 3 dB as compared to the value at the beginning ($0.707G(0)$). As $G(0) = 1$, it holds

$$|G(\mathrm{j}\omega_b)| = \frac{\sqrt{2}}{2} \Rightarrow \omega_b = 2.97.$$

Fig. 7.6 illustrates graphical procedure to find the indices.

7.3.4 Poles

Very often specification of the control quality can be described by the locations of poles. Of course, stability dictates the left half plane for the pole locations. A suitable pole region is shown in Fig. 7.7.

In general, the closer the pole is near the origin, the larger settling time results. A real pole is coupled with aperiodic (overdamped) response. If the closed-loop system has more poles, some of them dominate the process dynamics – the so called *dominant poles*. These poles are at least five times nearer to the origin as the other. This means that for example when designing a pole placement control design it is necessary to characterise only 2-3 poles and the rest needed for physical realisation of the controller can be chosen within the far left half plane.

To clarify the relations and influences among several specifications let us consider the following example.

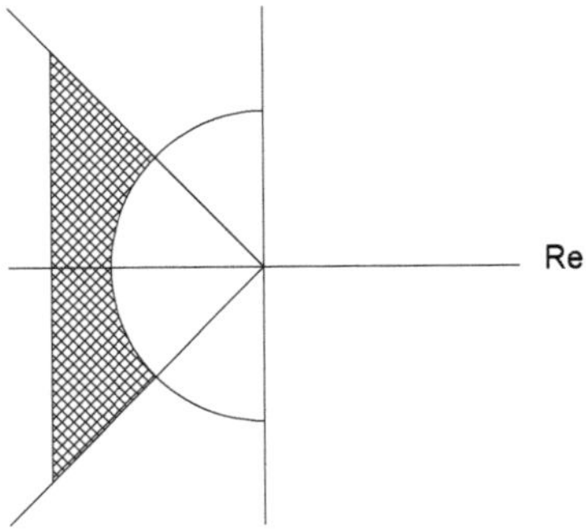

Fig. 7.7. Suitable region for pole locations of a closed-loop system

Example 7.4: *A second order system - quality indices*

Let us consider a closed-loop system with the transfer function of the form

$$G_{yw} = \frac{GR}{1+GR} = \frac{\omega_0^2}{s^2 + 2\zeta\omega_0 s + \omega_0^2} \tag{7.22}$$

where ζ is the damping coefficient and ω_0 is the natural undamped frequency. This transfer function can be obtained with a first order controlled system with gain Z and time constant T controlled by a controller with the transfer function $R = 1/T_I s$. In that case holds $\omega_0^2 = Z/TT_I$ a $2\zeta\omega_0 = 1/T$.

Assume that the damping coefficient is inside $\zeta \in (0,1)$. If the setpoint is the unit step change then the controlled output is given as

$$y(t) = 1 - \frac{1}{\sqrt{1-\zeta^2}} e^{-\zeta\omega_0 t} \sin\left(\omega_0 t\sqrt{1-\zeta^2} + \varphi\right), \quad \varphi = \arctan\frac{\sqrt{1-\zeta^2}}{\zeta} \tag{7.23}$$

Similarly, the transfer function between the output and a disturbance acting on the plant input is given as

$$G_{yd} = \frac{G}{1+GR} = \frac{Z}{T}\frac{s}{s^2 + 2\zeta\omega_0 s + \omega_0^2} \tag{7.24}$$

and the time response for the unit step change of the disturbance is

$$e(t) = \frac{Z}{\omega_0 T\sqrt{1-\zeta^2}} e^{-\zeta\omega_0 t} \sin\left(\omega_0 t\sqrt{1-\zeta^2}\right) \tag{7.25}$$

The closed-loop poles are complex conjugated and of the form

$$p_{1,2} = -\zeta\omega_0 \pm j\omega_0\sqrt{1-\zeta^2} \tag{7.26}$$

When the frequency ω_0 is assumed constant then poles are located in the complex plane on half circles with radius ω_0 centered in origin. If ζ is constant, then half lines from origin occur with slope being a function of ζ (Fig. 7.8).

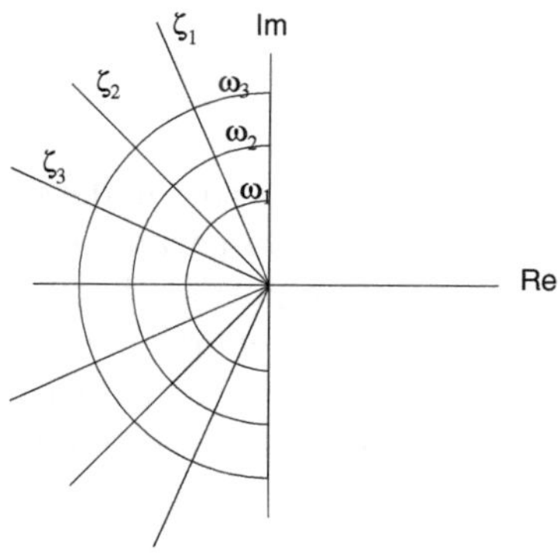

Fig. 7.8. Curves with constant values of ζ, ω_0 ($\zeta_1 < \zeta_2 < \zeta_3$, $\omega_1 < \omega_2 < \omega_3$)

The process is underdamped with a period

$$T_p = \frac{2\pi}{\omega_0\sqrt{1-\zeta^2}} \tag{7.27}$$

and damping coefficient

$$\zeta_d = \frac{1}{\omega_0} e^{\varphi/\tan\varphi} \tag{7.28}$$

Settling time T_ϵ for a band with accuracy $\pm p$ from the steady state is a complex function of parameters, but can be approximated as

$$T_\epsilon \approx \frac{\log\left(p\sqrt{1-\zeta^2}\right)}{\zeta\omega_0} \tag{7.29}$$

Maximum overshoot is given by the relation

$$e_{\max} = e^{-\pi\zeta\sqrt{1-\zeta^2}} = \sqrt{\zeta_d} \tag{7.30}$$

and occurs at the time

$$t(e_{\max}) = \frac{\pi}{\omega_0\sqrt{1-\zeta^2}} \tag{7.31}$$

If a step change of a disturbance at the plant input is considered, then for its maximum value holds

$$e_{\max} = \frac{Z}{T\omega_0\sqrt{1-\zeta^2}} \mathrm{e}^{\varphi \tan\varphi} \sin\varphi \tag{7.32}$$

$$t(e_{\max}) = \frac{1}{\omega_0\sqrt{1-\zeta^2}}\varphi \tag{7.33}$$

and the integral performance index IE can be derived as

$$I_{IE} = \frac{Z}{\omega_0^2 T} \tag{7.34}$$

These expressions can be used for better understanding of performance indices and their relations to controller parameters and vice versa. Integral cost function value is inversely proportional to ω_0^2. Maximum overshoot and all time indices are inversely proportional to ω_0. Transient response of the closed-loop system to setpoints or disturbances is improved by increasing ω_0. Further, maximum overshoot or damping coefficient increase with decreasing ζ.

Results presented in the example do not hold only for closed-loop systems with the given transfer function but can also be generalised for other systems as almost any process can be decomposed into a set of the first and second order systems. Moreover, from the dynamic point of view, in the majority of processes some subsystem dominates over the rest and the overall system is approximable by a second order system.

7.4 PID Controller

Among various controllers used in industry, the PID (proportional-integral-derivative) controller is the most often employed. There are estimates that more than 90% of controllers are of PID type. Its advantage is simplicity, robustness as well as the fact that it can be realised in various analogue (electrical, pneumatic), as well as digital versions.

The principle of the PID controller is processing of its input signal – control error in three branches that will be examined in more detail below.

7.4.1 Description of Components

Proportional Controller

Consider the most simple form of a feedback controller described be the expression

$$u(t) = \begin{cases} u_{\max} \text{ if } e(t) > 0 \\ u_{\min} \text{ if } e(t) < 0 \end{cases} \tag{7.35}$$

In this case the manipulated input u can have only two values depending on the control error e sign. This controller is called *on-off* controller. Its advantage is simplicity and easy implementation. As the control signal is not defined for $e = 0$ (see eq. (7.35)), it is usually modified by introducing some kind of hysteresis in the neighbourhood of zero.

Its drawbacks are oscillations of the controlled variable in the steady-state as even a small control error causes the maximum value of the manipulated input. A suitable improvement for this situation can be introduction of proportional behaviour for small control errors

$$u(t) = Z_R e(t) \tag{7.36}$$

The controller that realises this equation is called *proportional* (P controller) and Z_R is its gain.

From the practical point of view can any controller realise proportional action only in some limited range of control errors as the manipulated variable can always be only in the range between $u_{\min}$ and $u_{\max}$. Proportional behaviour of the controller can then be characterised either by its gain Z_R or by some range when the controller is linear – *proportionality band* P_p. Their relation is

$$u_{\max} - u_{\min} = Z_R P_p \tag{7.37}$$

If we consider the maximum range of the manipulated variable normalised, e. g. $u_{\max} - u_{\min} = 100\%$ then

$$Z_R = \frac{100}{P_p} \tag{7.38}$$

If the control error is too large in magnitude then the P controller implements on-off behaviour with the manipulated variable constrained to its maximum and minimum values.

From steady-state analysis of the P controller (see page 256) follows that if the controlled system does not contain at least one pure integrator, the closed-loop system will exhibit non-zero permanent control error. The simplest scheme to guarantee the zero steady-state control error is to introduce a suitable constant term u_b to the manipulated variable

$$u(t) = Z_R e(t) + u_b(t) \tag{7.39}$$

such that for $e(t) = 0$ holds $u = u_b$.

Example 7.5: *Proportional control of a heat exchanger* WWW

Consider a feedback control of a heat exchanger with the P controller described by the transfer function $R(s) = Z_R$. The heat exchanger can be given by the differential equation

$$T_1 y' + y = Z_1 u + \frac{Z_2}{Z_1} d$$

where u is the manipulated variable and d is a disturbance. For the simulation purposes assume that $Z_1 = 1, T_1 = 1$.
Consider at first the disturbance rejection case when the disturbance is a step change. ($W(s) = 0$, $D(s) = A/s$).
The closed-loop transfer function with respect to the disturbance is in this case given as

$$G_{yd} = \frac{GG_d}{1+GR} = \frac{\frac{Z_1}{T_1s+1}\frac{Z_2}{Z_1}}{1+Z_R\frac{Z_1}{T_1s+1}}$$

The output variable is then given as

$$Y(s) = \frac{\frac{Z_1}{T_1s+1}\frac{Z_2}{Z_1}}{1+Z_R\frac{Z_1}{T_1s+1}}\frac{A}{s} = \frac{AZ_2}{1+Z_RZ_1}\frac{1}{1+\frac{T_1}{1+Z_RZ_1}s}\frac{A}{s}$$

$$\frac{y(t)}{AZ_2} = \frac{1}{1+Z_RZ_1}\left(1-\mathrm{e}^{-\frac{t}{T_1/(1+Z_RZ_1)}}\right)$$

The heat exchanger output is stable for any value of Z_R and its new steady-state settles at the value

$$y(\infty) = \frac{AZ_2}{1+Z_RZ_1}$$

A permanent control error exists

$$e(\infty) = w(\infty) - y(\infty) = -\frac{AZ_2}{1+Z_RZ_1}$$

This confirms that by increasing the controller gain decreases the control error (Fig. 7.9).
The slope of the response in time $t = 0$ will be derived by evaluating the derivative

$$\frac{1}{AZ_2}\frac{\mathrm{d}y}{\mathrm{d}t} = \frac{1}{1+Z_RZ_1}\frac{1+Z_RZ_1}{T_1}\mathrm{e}^{-\frac{t}{T_1/(1+Z_RZ_1)}}$$

$$\left.\frac{1}{AZ_2}\frac{\mathrm{d}y}{\mathrm{d}t}\right|_{t=0} = \frac{1}{T_1}$$

It is clear that all step responses have the same slope in time $t = 0$ (Fig. 7.9).
Consider now tracking case when the setpoint is a step change ($W(s) = A/s$, $D(s) = 0$).
The closed-loop transfer function with respect to the setpoint is in this case given as

$$G_{yw} = \frac{GR}{1+GR} = \frac{Z_RZ_1}{T_1s+1+Z_RZ_1}$$

The output variable is then given as

$$Y(s) = \frac{Z_R Z_1}{T_1 s + 1 + Z_R Z_1} \frac{A}{s} = \frac{A Z_R Z_1}{1 + Z_R Z_1} \frac{1}{\frac{T_1}{1+Z_R Z_1} s + 1} \frac{1}{s}$$

$$\frac{y(t)}{A} = \frac{Z_R Z_1}{1 + Z_R Z_1} \left(1 - \mathrm{e}^{-\frac{t}{T_1/(1+Z_R Z_1)}}\right)$$

The heat exchanger output is stable for any value of Z_R and its new steady-state settles at the value

$$y(\infty) = \frac{A Z_R Z_1}{1 + Z_R Z_1}$$

A permanent control error exists

$$e(\infty) = w(\infty) - y(\infty) = A\left(1 - \frac{Z_R Z_1}{1 + Z_R Z_1}\right)$$

This again confirms that by increasing the controller gain decreases the control error.

The slope of the response in time $t = 0$ is given as

$$\frac{1}{A}\frac{\mathrm{d}y}{\mathrm{d}t} = \frac{Z_R Z_1}{1 + Z_R Z_1} \frac{1 + Z_R Z_1}{T_1} \mathrm{e}^{-\frac{t}{T_1/(1+Z_R Z_1)}}$$

$$\left.\frac{1}{A}\frac{\mathrm{d}y}{\mathrm{d}t}\right|_{t=0} = \frac{Z_R Z_1}{T_1}$$

Compared to the disturbance rejection case, here the speed of the response increases with the increasing controller gain (Fig. 7.10).

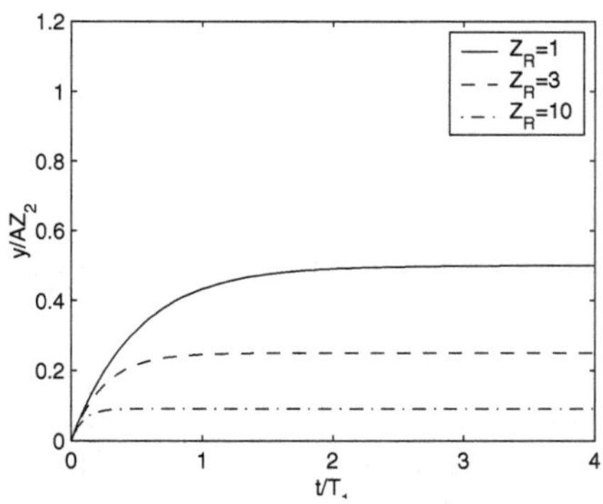

Fig. 7.9. Response of a heat exchanger to a step change of the d with a P controller

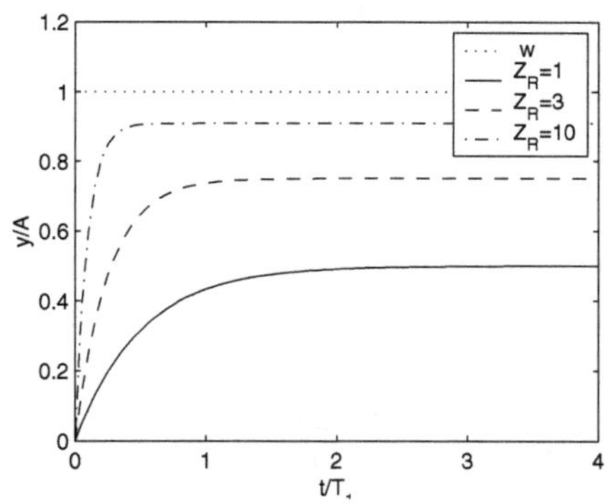

Fig. 7.10. Response of a heat exchanger to a step change of the w with a P controller

In both cases it can be noticed that the controller changes the closed-loop time constant from the open-loop value T_1 to $T_1/(1 + Z_R Z_1)$. If negative feedback is assumed, both gains Z_R a Z_1 have to be positive and the resulting time constant is smaller than the one of the controlled system – the closed-loop acts faster to a setpoint or disturbance changes as the controlled system. When pole locations are considered for the original ($s = -1/T_1$) and closed-loop system, the closed-loop pole is pushed more to the left half part of the complex plane. This also means that an unstable system can be stabilised by the P controller with a sufficiently high gain that causes the closed-loop pole moved to the left of the imaginary axis.

www

Example 7.6: *Proportional control of a higher order system*
Consider again proportional control of a system with the transfer function

$$G(s) = \frac{1}{(s+1)^4}$$

Fig. 7.11 shows simulations with various gains of the controller.

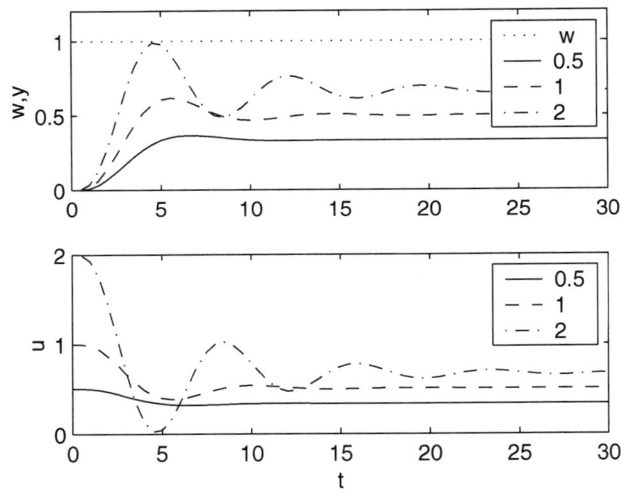

Fig. 7.11. Proportional control to the setpoint value $w = 1$ with various gains Z_R of the controller

Simulation results again illustrate typical features of proportional control. Permanent control error decreases with increasing controller gain. However, stability indices deteriorate and for a sufficiently high Z_R can the closed-loop system become instable.

Integral Control

The problem of the steady-state control error is not satisfactorily solved even with the modified proportional controller with the shift u_b. If the setpoint changes, it is necessary to find a new value of u_b. A more appropriate solution is to introduce a pure integrator – the manipulated variable changes while the control error is not zero. This can mathematically be defined as

$$\frac{\mathrm{d}u(t)}{\mathrm{d}t} = \frac{1}{T_I} e(t) \tag{7.40}$$

where the constant T_I is called integral time constant and determines the speed of the change of the manipulated variable in the case of the unit control error. The smaller value T_I the larger changes of u are generated.

Historically, the integral action (I controller) was discovered by improving equation (7.39) where the shift u_b was calculated using feedback with the first order filter with time constant T_I

$$U_b(s) = \frac{1}{T_I s + 1} U(s) \tag{7.41}$$

Combining (7.39), (7.41) and eliminating the intermediate variables yields

$$Z_R E(s) = T_I s U_b(s) \Leftrightarrow T_I \frac{\mathrm{d}u_b(t)}{\mathrm{d}t} = Z_R e(t) \tag{7.42}$$

$$U(s) = \frac{T_I s + 1}{T_I s} Z_R E(s) \Leftrightarrow u(t) = Z_R \left(e(t) + \frac{1}{T_I} \int_0^t e(\tau) \mathrm{d}\tau \right) \tag{7.43}$$

This represents integral action for $u_b(t)$ and the PI controller for $u(t)$.

Example 7.7: *Heat exchanger control with the PI controller* www

Consider feedback control of the heat exchanger with the PI controller defined with the transfer function $R(s) = Z_R[1 + 1/(T_I s)]$.

Transient response of the exchanger to a step change of the disturbance with magnitude A (setpoint is considered to be zero) in the Laplace transform is given as

$$Y(s) = \frac{\frac{Z_2}{Z_1}\frac{Z_1}{T_1 s+1}}{1 + Z_R \frac{T_I s+1}{T_I s}\frac{Z_1}{T_1 s+1}} \frac{A}{s} = \frac{A Z_2 T_I}{T_I T_1 s^2 + (T_I + Z_R T_I Z_1)s + Z_R Z_1}$$

This function is stable and converges to zero. Some simulations for $Z_R = 10$ are shown in Fig. 7.12.

The exchanger response to a step change of the setpoint with magnitude A (disturbance is considered to be zero) in the Laplace transform is

$$Y(s) = \frac{Z_1 Z_R (T_I s + 1)}{T_I T_1 s^2 + (T_I + Z_R T_I Z_1)s + Z_R Z_1} \frac{A}{s}$$

The steady-state control error is given as

$$e(\infty) = w(\infty) - y(\infty) = A - \lim_{s \to 0} sY(s) = A - A = 0$$

We can see that the controller removes the steady-state control error. Compared to the disturbance transfer function the numerator contains a stable zero $s = -1/T_I$ that amplifies with decreasing value T_I and thus approaches instability. Practical consequence will be increasing overshoot. Some simulation results are shown for $Z_R = 10$ in Fig. 7.13.

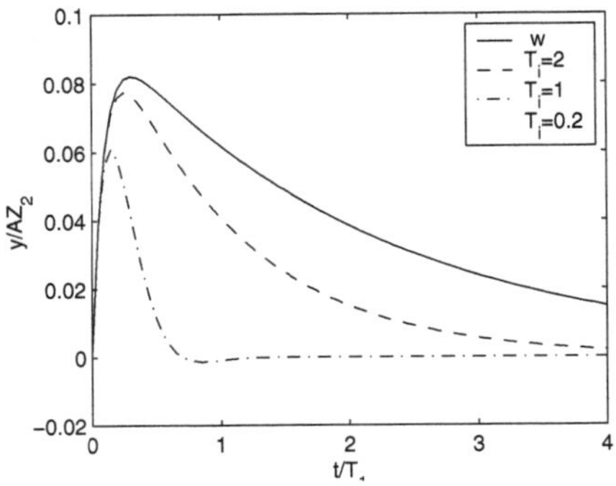

Fig. 7.12. Heat exchanger – disturbance rejection with the PI controller

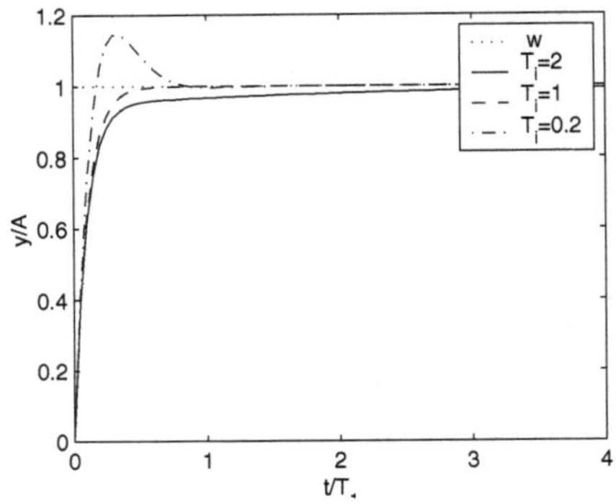

Fig. 7.13. Heat exchanger – setpoint tracking with the PI controller

Example 7.8: *Higher order system with the PI controller*

Consider again control of a higher order system with the transfer function of the form

$$G(s) = \frac{1}{(s+1)^4}$$

controlled by the PI controller with proportional gain $Z_R = 1$. Fig. 7.14 illustrated effects of changing the integral time constant.

Simulation results demonstrate typical features of PI control. The steady-state control error is zero, only with a large T_I the output variable converges only slowly. On the other side, with the decreasing T_I (increasing integral action) the controller becomes more aggressive and magnitudes of oscillations increase as well. The closed-loop system can eventually become instable for a sufficiently small T_I.

Derivative Control

Derivative action theoretically improves stability of the closed-loop system. P, I controllers do not influence the loop instantaneously but only after a certain time. Derivative controller defined by equation

$$u(t) = T_D \frac{\mathrm{d}e(t)}{\mathrm{d}t} \tag{7.44}$$

where T_D is derivative constant forecasts the future value of the control error. This can be shown by using the Taylor expansion of the term $e(t + T_D)$

$$e(t + T_D) \approx e(t) + T_D \frac{\mathrm{d}e(t)}{\mathrm{d}t} \tag{7.45}$$

Graphical representation of this equation is shown in Fig. 7.15.

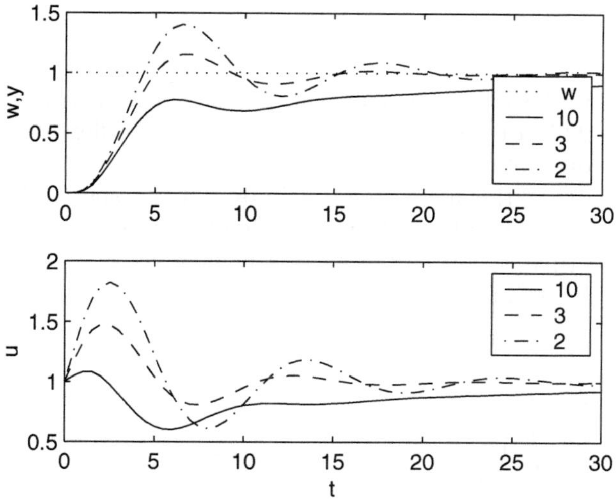

Fig. 7.14. PI control with setpoint $w = 1$ and several values of T_I

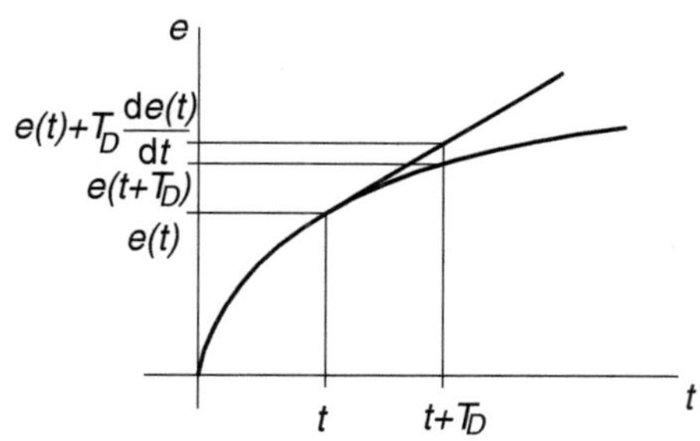

Fig. 7.15. Graphical representation of derivative controller effects

An ideal derivative controller is rather sensitive to noise in the controlled variable and the derivative action can lead to frequent and large changes of the manipulated variable. Besides, the ideal derivative action is not realisable as it is not proper.

This has lead to introduction of a filtered D controller with the transfer function

$$R(s) = \frac{T_D s}{1 + \frac{T_D}{N} s} \tag{7.46}$$

This is nothing else than the ideal D controller in series with a first order system with the time constant T_D/N. Typical values of N are between 5 and 20.

Example 7.9: *Higher order system PID control* www

Consider again control of a higher order system with the transfer function of the form

$$G(s) = \frac{1}{(s+1)^4}$$

controlled by the PID controller and assume constant parameters $Z_R = 3.76$, $T_I = 2.85$. Fig. 7.16 shows simulations with different values of the derivative constant.

We can see that the increasing T_D improves the speed of control and overshoot. Theoretically performance should always improve. However, as the derivative action has predictive behaviour, the presence of noise inhibits arbitrary increase of T_D.

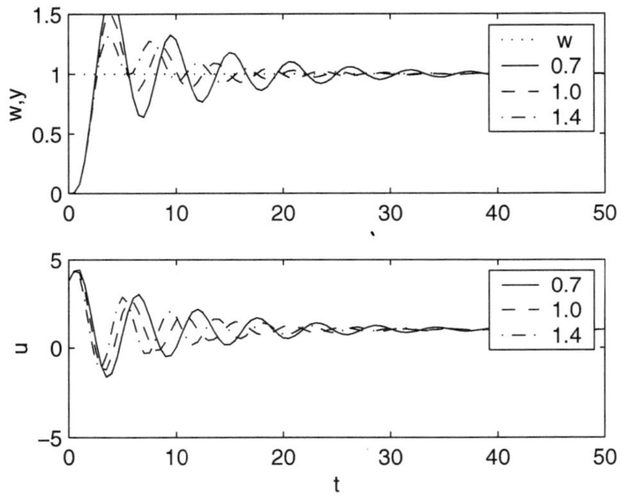

Fig. 7.16. PD control with setpoint $w = 1$ and several values of T_D

Loosely speaking PID controller parameters influence performance and stability in the following way:

- increase of P action increases speed and reduces stability,
- increase of I action decreases speed and improves stability,
- increase of D action increases speed and improves stability.

7.4.2 PID Controller Structures

The PID controller contains three components: proportional, integral, and derivative. Its realisations can be different depending on implementation (electrical, pneumatic, electronic, etc.)

Most often the following PID structures are used:

- without interaction

$$R(s) = Z_R\left(1 + \frac{1}{T_I s} + T_D s\right) \tag{7.47}$$

- with interaction (in series)

$$R(s) = Z_R\left(1 + \frac{1}{T_I s}\right)(1 + T_D s) \tag{7.48}$$

- parallel

$$R(s) = Z_R + \frac{1}{T_I s} + T_D s \tag{7.49}$$

It is important to remember that the controller constants are closely tied to its realisation and that alternative representation results in their different values. Design methods presented in Section 7.4.6 usually assume the structure without interaction (7.47).

7.4.3 Setpoint Weighting

The standard PID controller processes control error. A more flexible structure can be defined with a controller of the form

$$u(t) = Z_R\left(e_p(t) + \frac{1}{T_I}\int_0^t e(\tau)\mathrm{d}\tau + T_D\frac{\mathrm{d}e_d(t)}{\mathrm{d}t}\right) \tag{7.50}$$

where modified P and D variables are defined as

$$e_p(t) = bw(t) - y(t) \tag{7.51}$$

$$e_d(t) = cw(t) - y(t) \tag{7.52}$$

and b, c are user defined parameters. Both influence response to an abrupt change of setpoint w where the traditional PID controller reacts by a step change of proportional and derivative part. The parameter b controls the maximum overshoot that is the smallest with $b = 0$. Field tested values are between $0.3 - 0.8$. (see Example 7.15 on page 291). Similarly, the parameter c can be used to suppress large changes of the manipulated variable caused by derivative action.

In general, the classical controller described by equations (7.47)–(7.49) represents a one degree-of-freedom controller that reacts in exactly the same way to changes in disturbance or setpoint. Often however, closed-loop specifications are different and responses to setpoints and disturbances have to be specified independently. In that case a more suitable structure can be that of the two degree-of-freedom controller (Fig. 7.17). Here actually two controllers are used. The feedback part R_{fb} deals with the control error and the feedforward part R_{ff} processes only the setpoint. Their transfer functions can be specified as

$$R_{fb} = Z_R \left(1 + \frac{1}{T_I s} + T_D s\right) \tag{7.53}$$

$$R_{ff} = Z_R \left(b + \frac{1}{T_I s} + c T_D s\right) \tag{7.54}$$

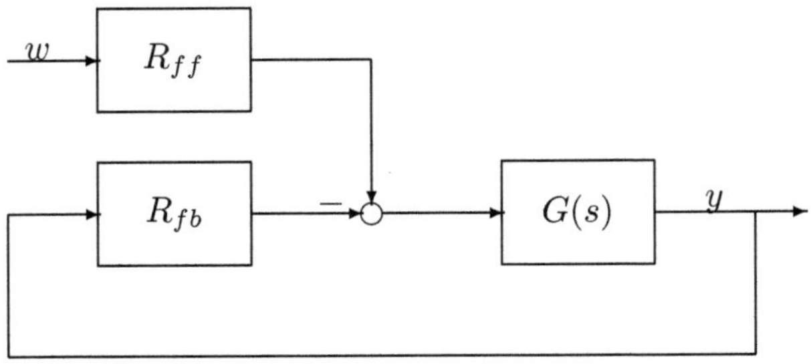

Fig. 7.17. Closed-loop system with a two degree-of-freedom controller

7.4.4 Simple Rules for Controller Selection

- Most often the best controller is the simplest, i. e. the proportional controller. It can be used if nonzero steady-state error is acceptable or if the controlled system contains pure integrator. In process control its typical use is in pressure control or level control.
- PI controller can be used if the simple proportional action is unacceptable and dynamics of the controlled system is simple. Its advantage is the zero steady-state error but at the price of larger overshoot and oscillatory behaviour. Moreover, further increase of the controller gain can lead to instability.
- PD controller – grace to derivative action does not exhibit large overshoot and settles fairly quickly. However, without integral action, the nonzero steady-state error results with the same value as that of the P controller. It is used mainly for processes with integrating behaviour or with large time constants.
- PID controller is suitable in remaining cases - it is robust, forecasts future process behaviour and in general its gain is smaller than that of the corresponding PI controller. Its drawback is an increased sensitivity to measurement noise.
- Problematic processes for PID controllers include systems with time delay, with time variant parameters or oscillatory processes.

7.4.5 Practical Aspects

Integrator Windup and Constraints Handling

Manipulated variables are always limited within the interval $(u_{\min}, u_{\max})$. For example valves can be somewhere between fully closed and fully open, the smallest motor speed is zero and the largest is given by its technical parameters, etc.

If a controller contains pure integrator(s) and the manipulated variable is on the constraint, the state corresponding to the integrator continues to integrate the control error. Hence, even if the control error sign finally changes, the manipulated variable does not change immediately, as the integral part dominates the controller. Classical symptoms of (*integrator windup*) are large overshoots caused by delayed activity of the controller. This phenomenon is illustrated in the next example.

Example 7.10: *Integrator windup* www

Consider again control of a higher order system with the transfer function of the form

$$G(s) = \frac{1}{(s+1)^4}$$

with the PI controller of the form

$$R(s) = 1 + \frac{1}{2s}$$

Assume step setpoint change $w = 1$ and constraints on the input $u \in \langle 0.9, 1.1 \rangle$. Simulation results are shown in Fig. 7.18a. Even if the controlled output y reaches setpoint at time $t = 7$, the manipulated variable remains at the upper constraint up to time $t = 20$. This implies a larger settling time. To understand better this phenomenon, state of the integral part is also shown in the figure. This variable increases even if the manipulated variable is on the upper constraint.

Improvement can be observed when feedback procedure described below is used. The simulation results are shown in Fig. 7.18b. Integrator value decreases rapidly and both control and the controlled output react correctly when constraints are active.

There are several ways how to eliminate or suppress the windup phenomenon. Often, it is caused by a sudden change of the setpoint variable. In that case it may be possible to change the setpoint smoothly and more slowly so that the control signal remains within limits.

Another possibility is to use velocity (incremental) algorithms when the controller generates only changes of the manipulated variable. Of course, if the manipulated variable hits the constraint, the controller output is zero.

The most commonly used method is the back calculation – if the manipulated variable is out of limits, the integral part is recalculated back so that

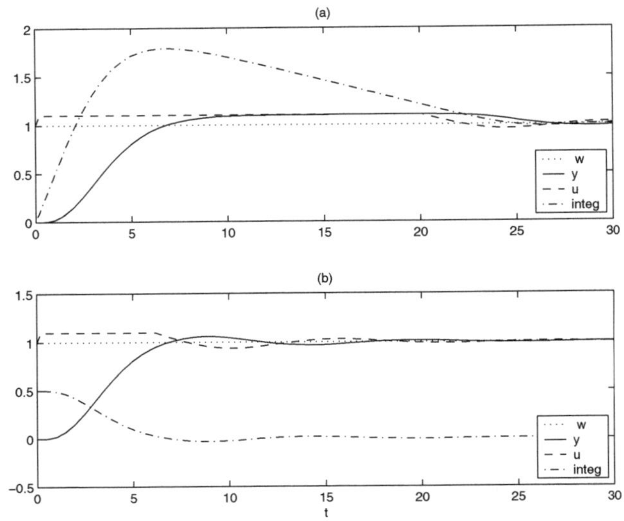

Fig. 7.18. Windup phenomenon. (a) Problem, (b) Solution

the control signal is on the constraint. In practice however, this is not realised in each calculation step but dynamically with a time constant T_t. The block scheme of this *anti-windup* setup is shown in Fig. 7.19.

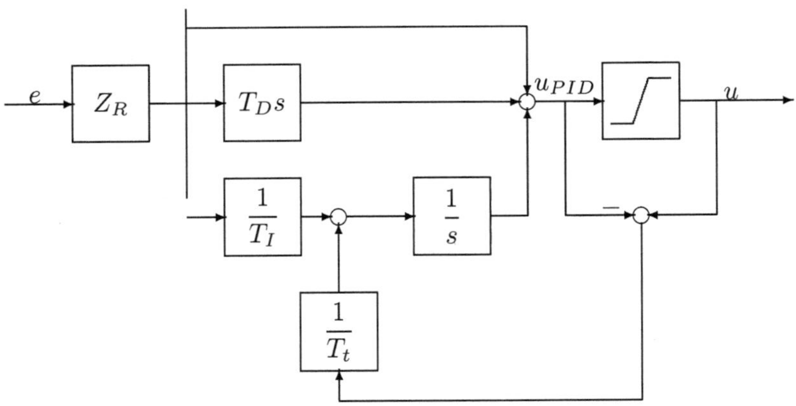

Fig. 7.19. Anti-windup controller

The modification consists in an additional feedback loop on the input of which is difference between calculated and really applied control signal. If the manipulated variable is within constraints, this error is zero and the original controller remains. If the error is nonzero, the integral signal is changed until it is again at the desired value.

The time constant T_t specifies speed of the integral rewind. At the first sight it seems that it should be the smallest possible. However, it has been found that in presence of noise the integrator could interrupt its activity. An empirical rule of thumb is the value about $T_t = \sqrt{T_I T_D}$.

Bumpless Transfer

Modern PID controllers can function in automatic and manual modes. The standard mode is automatic when the controller calculates the control signal. The manual mode serves to handle nonstandard situations when the control signal is specified by process operators. It is important to guarantee that change between the modes is without undesired transient effects, i. e. the new mode should start smoothly at the last calculated value of the old mode.

The bumpless transfer problem can be removed if incremental form of controllers is used. In other cases it is possible to use the anti-windup scheme (Fig. 7.19) where the controller follows the actually applied signal to the process.

Transient effects can also occur when controller parameters are changed. In that case it is necessary to recalculate the integral action.

Digital Implementation

The first PID controllers were constructed as thecontinuous-time pneumatic or (later) electric devices. With the recent advances of microprocessors controllers are mostly implemented digitally as discrete-time algorithms. In general, this brings many advantages as the microprocessor not only implements the control law but it can also trigger alarms, filter input variables, etc. However, there are some features and properties special to digital devices as such a controller does not process the analogue signal continuously but only at some time instants – sampling times.

The most important parameters of a digital device is its sampling period T_s. If it is chosen too large, some dynamic effects can escape. Practical rule of thumb is to choose it such that the interval $(5-10)T_s$ should cover the rise time of the plant. In the frequency domain this can correspond to 10 times the bandwidth. If there is no information about the controller process, it is recommended to choose the sampling time as small as possible. Theoretical issues on the choice of T_s are also discussed in Section 5.1.

The discretisation in itself covers approximation of all three parts of the PID controller when the original differential equation is transformed into a difference equation.

To derive a digital version of the PID controller assume the parallel form given by (7.47). Its discrete version can be obtained in various ways. For example by backward approximation of the derivative and integral parts as

$$\frac{\mathrm{d}e(t)}{\mathrm{d}t} \approx \frac{e(kT_s) - e((k-1)T_s)}{T_s}, \quad \int_0^t e(\tau)\mathrm{d}\tau \approx T_s \sum_{i=1}^{k} e(T_s(i-1)) \tag{7.55}$$

where the continuous time t has been replaced by its discrete time counterpart in sampling times $kT_s, k = 0, 1, \ldots$. This gives for the control law

$$u(kT_s) = Z_R \left\{ e(kT_s) + \frac{T_s}{T_I} \sum_{i=1}^{k} e(T_s(i-1)) + \frac{T_D}{T_s}[e(kT_s) - e((k-1)T_s)] \right\} \tag{7.56}$$

Another possibility is to use forward or trapezoidal approximation:

$$\int_0^t e(\tau)\mathrm{d}\tau \approx T_s \sum_{i=1}^{k} e(iT_s) \approx \frac{T_s}{2} \sum_{i=1}^{k} [e(iT_s) - e((i-1)T_s)] \tag{7.57}$$

If for example forward approximation of the part I is used then digital PID controller is given as

$$u(kT_s) = Z_R \left\{ e(kT_s) + \frac{T_s}{T_I} \sum_{i=1}^{k} e(iT_s)) + \frac{T_D}{T_s}[e(kT_s) - e((k-1)T_s)] \right\} \tag{7.58}$$

If the sampling time is sufficiently small, there are no large differences between approximations. Some precaution has to be taken in the case of forward or trapezoidal transformations when for small values of T_s can the approximations be unstable.

Equation (7.58) is not particularly suitable for practical implementation as it is necessary to know all values of past control errors. Therefore, recurrent versions are used that can be obtained by subtracting u in sampling times kT_s and $(k-1)T_s$:

$$u(kT_s) = \Delta u(kT_s) + u((k-1)T_s) \tag{7.59}$$

$$\begin{aligned} \Delta u(kT_s) = Z_R \bigg[& e(kT_s)\left(1 + \frac{T_s}{T_I} + \frac{T_D}{T_s}\right) - e((k-1)T_s)\left(1 + \frac{2T_D}{T_s}\right) \\ & + e((k-2)T_s)\frac{T_D}{T_s} \bigg] \end{aligned} \tag{7.60}$$

A general form of a digital PID controller can be given as

$$u(k) = q_0 e(k) + q_1 e(k-1) + q_2 e(k-2) + u(k-1) \tag{7.61}$$

The corresponding discrete time transfer function is then

$$R(z^{-1}) = \frac{q_0 + q_1 z^{-1} + q_2 z^{-2}}{1 - z^{-1}} \tag{7.62}$$

The presented technique can be used for an arbitrary transformation of a continuous-time PID controller to a digital form.

7.4.6 Controller Tuning

Analytical Methods

Standard Forms

The method of standard forms belongs to the class of analytical methods where transfer function of the controlled process is known. Standard forms refer to some known transfer function structures and coefficients where behaviour of the closed-loop system is known and optimal in some sense.

Consider the closed-loop system transfer function of the form

$$G_c(s) = \frac{a_0}{a_n s^n + a_{n-1}s^{n-1} + \cdots a_1 s + a_0} \tag{7.63}$$

that can be obtained if the open-loop transfer function contains one pure integrator and no zeros. Standard forms for this case are:

- Binomial form with a real stable pole $s = -\omega_0$ with mutiplicity of n

$$G_c(s) = \frac{\omega_0^n}{(s+\omega_0)^n} \tag{7.64}$$

  Different system orders n imply the following standard forms

$$\begin{gathered} s+\omega_0 \\ s^2 + 2\omega_0 s + \omega_0^2 \\ s^3 + 3\omega_0 s^2 + 3\omega_0^2 s + \omega_0^3 \\ s^4 + 4\omega_0 s^3 + 6\omega_0^2 s^2 + 4\omega_0^3 s + \omega_0^4 \end{gathered}$$

  This standard form provides relatively sluggish underdamped transient responses.
- Butterworth form where the poles are located in the left half of the complex plane on the semicircle with radius of ω_0. Normalised polynomials are then of the form

$$\begin{gathered} s+\omega_0 \\ s^2 + 1.4\omega_0 s + \omega_0^2 \\ s^3 + 2\omega_0 s^2 + 2\omega_0^2 s + \omega_0^3 \\ s^4 + 2.6\omega_0 s^3 + 3.4\omega_0^2 s^2 + 2.6\omega_0^3 s + \omega_0^4 \end{gathered}$$

- Minimual $t_{5\%}$ – the fastest transient reponse with maximum overshoot of 5%.

$$\begin{gathered} s+\omega_0 \\ s^2 + 1.4\omega_0 s + \omega_0^2 \\ s^3 + 1.55\omega_0 s^2 + 2.10\omega_0^2 s + \omega_0^3 \\ s^4 + 1.60\omega_0 s^3 + 3.15\omega_0^2 s^2 + 2.45\omega_0^3 s + \omega_0^4 \end{gathered}$$

- Minimum of ITAE cost function.

$$\begin{gathered} s+\omega_0 \\ s^2 + 1.4\omega_0 s + \omega_0^2 \\ s^3 + 1.75\omega_0 s^2 + 2.15\omega_0^2 s + \omega_0^3 \\ s^4 + 2.1\omega_0 s^3 + 3.4\omega_0^2 s^2 + 2.7\omega_0^3 s + \omega_0^4 \end{gathered}$$

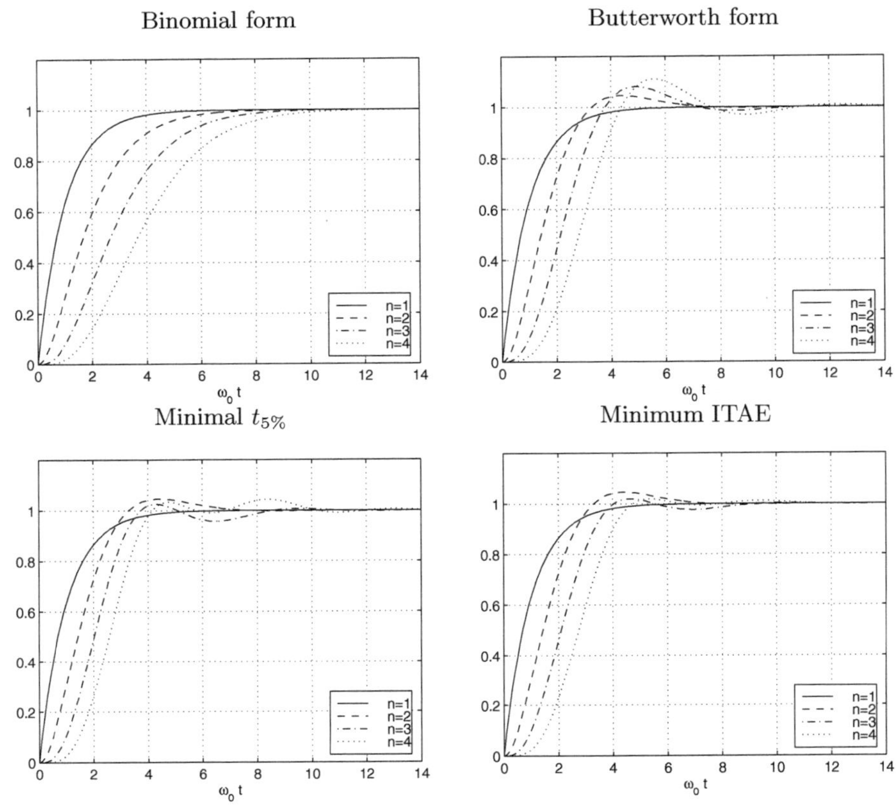

Fig. 7.20. Step responses of selected standard transfer functions

The above mentioned standard forms for various system orders and normalised with respect to ω_0 lead to step responses shown in Fig. 7.20.

Possible drawbacks of standard forms include a rather strict restriction on the open-loop system structure. The solution very often leads to a under or overdetermined system of equations and is practically useless for standard PID design. On the other side, denominators of standard forms are frequently used in pole placement control design.

The Naslin Method

Here the controller is designed based on the requirement on the maximum overshoot. The method considers relations between the coefficients of the closed-loop denominator (7.63) of the form

$$a_i^2 = \alpha a_{i-1} a_{i+1} \tag{7.65}$$

where the coefficient α is determined as a function of the desired overshoot from the table

α	1.7	1.8	1.9	2	2.2	2.4
$e_{\max}$	20	12	8	5	3	1

Example 7.11: *Naslin method*

The controlled system with the transfer function

$$G(s) = \frac{1}{12s^2 + 8s + 1}$$

is subject to unit step disturbance on the system input. Design PI controller using the Naslin method with maximum overshoot of 5%.
The transfer function between the disturbance and the process output in the closed-loop system can be derived using the block algebra

$$G_{yd}(s) = \frac{G(s)}{1 + G(s)R(s)}$$

Assume the PI controller with the structure

$$R(s) = Z_R \left(1 + \frac{1}{T_I s}\right)$$

The closed-loop denominator polynomial is then given as

$$12s^3 + 8s^2 + (1 + Z_R)s + \frac{Z_R}{T_I} = 0$$

If the maximum overshoot of 5% is specified then $\alpha = 2$ and two equations can be written

$$a_1^2 = 2a_0a_2 \Rightarrow (1 + Z_R)^2 = 16\frac{Z_R}{T_I}$$

$$a_2^2 = 2a_1a_3 \Rightarrow 64 = 24(1 + Z_R)$$

The resulting controller is then given as

$$R(s) = \frac{5}{3}\left(1 + \frac{4}{15s}\right)$$

The Strejc Method

To determine parameters of the PID controller of the form

$$R(s) = Z_R \left(1 + \frac{1}{T_I s} + T_D s\right) \tag{7.66}$$

using the Strejc method let us consider the controlled system of the form

$$F(s) = \frac{Z}{(Ts + 1)^n} \mathrm{e}^{-T_d s} \tag{7.67}$$

where Z is the process gain, T time cosntant, T_d time delay, and n the process order.

The controller parameters can then be read from Table 7.1. If only step response of the process is known, the transfer function can be estimated using the Strejc identification method described in Section 6.2.3.

We choose a suitable controller structure and read its parameters from the corresponding row of the table. If PID structure is chosen, the controlled process has to be at least of the third order. For P and PI controllers the system order has to be at least two. This follows from the controller gain Z_R calculations in Table 7.1.

Table 7.1. Controller tuning using Strejc method

Controller	Z_R	T_I	T_D
P	$\frac{1}{Z}\frac{1}{n-1}$		
PI	$\frac{1}{Z}\frac{n+2}{4(n-1)}$	$T\frac{n+2}{3}$	
PID	$\frac{1}{Z}\frac{7n+16}{16(n-2)}$	$T\frac{7n+16}{15}$	$T\frac{(n+1)(n+3)}{7n+16}$

www

Example 7.12: *The Strejc method*

Consider again the controlled process with parameters $Z = 1$, $T = 1$, $n = 4$. When PID controller structure is chosen, its parameters calculated from Table 7.1 are $Z_R = 1.38$, $T_I = 2.93$, $T_D = 0.80$. The simulation results are shown in Fig. 7.21.

Pole Placement

One drawback of standard forms methods is that the overall transfer function of the closed-loop system is prescribed. This narrows the applicability of the method considerably. In pole placement control design only the closed-loop denominator that determines stabilility is specified. The advantage of this approach is its usability for a broad range of systems. On the other side, as the closed-loop numerator cannot be specified, it can happen that its zeros can have impact on performance.

If the controlled system is of the first or second order and the controller of PID structure then the characteristic equation can be one of the following

$$s + \omega_0 = 0 \tag{7.68}$$

$$s^2 + 2\zeta\omega_0 s + \omega_0^2 = 0 \tag{7.69}$$

$$(s + \alpha\omega_0)(s^2 + 2\zeta\omega_0 s + \omega_0^2) = 0 \tag{7.70}$$

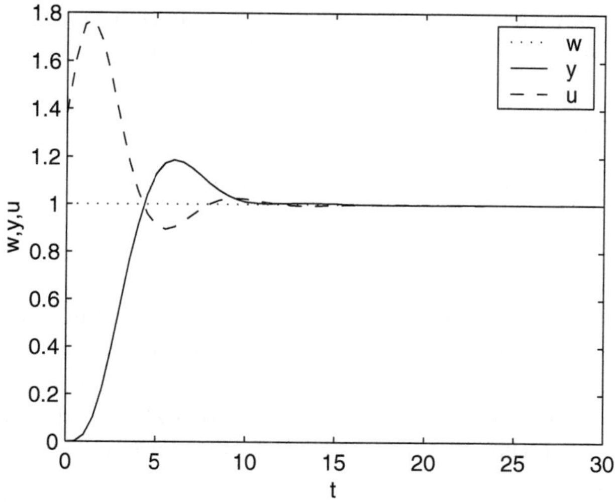

Fig. 7.21. The PID controller tuned according to Strejc

where ζ is relative damping, ω_0 natural undamped frequency, and α a coefficient. Specifying suitable values of parameters ζ, ω_0, α in (7.68)–(7.70) can lead to one of the standars forms. If higher order characteristic polynomial is to be considered, then it is usually taken with the structure of Butterworth polynomials or some other standard forms.

Consider the controlled system of the first order with transfer function

$$G(s) = \frac{Z}{Ts + 1} \tag{7.71}$$

controlled by the PID controller of the form

$$R(s) = Z_R\left(1 + \frac{1}{T_I s} + T_D s\right) = \frac{Z_R(T_I s + 1 + T_I T_D s^2)}{T_I s} \tag{7.72}$$

The closed-loop transfer function between the setpoint and output is given by equation (7.3) as

$$G_{yw}(s) = \frac{Z Z_R(T_I s + 1 + T_I T_D s^2)}{(T + Z Z_R T_D)s^2 + (1 + Z Z_R)s + Z Z_R/T_I} \tag{7.73}$$

The characteristic equation is of the second degree and thus it is possible to assign two poles. As the PID controller contains three tunable parameters, one of them is superfluous. Hence, we will in the further consider only the PI controller structure. In this case the characteristic equation can be written as

$$s^2 + \frac{1 + Z Z_R}{T} s + \frac{Z Z_R}{T T_I} = 0 \tag{7.74}$$

To place the poles, the standard characteristic equation (7.69) will be used. Comparing the coefficients at the corresponding powers of s yields

$$2\zeta\omega_0 = \frac{1+ZZ_R}{T}, \quad \omega_0^2 = \frac{ZZ_R}{TT_I} \tag{7.75}$$

Therefore, the controller parameters are given as

$$Z_R = \frac{2\zeta\omega_0 T - 1}{Z} \quad T_I = \frac{2\zeta\omega_0 T - 1}{T\omega_0^2} \tag{7.76}$$

The closed-loop system G_{yw} numerator contains a stable zero $s = -1/T_I$. If the integrating time constant is large, this zero is located near unstable region and can have impact on overshoot. To prevent this is is recommended either to lower the time constant or to use the two degree-of-freedom controller structure and weigh the setpoint influence by the parameter b (see (7.50)).

www

Example 7.13: *Pole placement design for a heat exchanger*

Let us consider minimisation of the ITAE performance index with requirement that the controlled output reaches the setpoint at $T_{100\%} = 0.5\,\text{min}$. We assume the heat exchanger time constant $T = 1\,\text{min}$ and the gain $Z = 1$.

Comparing the second order standard ITAE polynomial with equation (7.69) gives the value of $\zeta = 0.7$. The corresponding normalised step response specifies for time $T_{100\%}$ relation $\omega_0 T_{100\%} = 3.1$, hence $\omega_0 = 6.2$. The controller parameters are then

$$Z_R = 7.68, \quad T_I = 0.2$$

Simulation of the closed-loop system is shown in Fig. 7.22. It can be noticed that requirements on the closed-loop system were not exactly met as $T_{100\%}$ is smaller than specified. This is caused by the fact that the numerator is not constant but the first order polynomial. Smaller overshoot and thus larger $T_{100\%}$ can be achieved by modification of the controller with the parameter b.

Let us now consider the pole placement design for a second order system with the transfer function of the form

$$G(s) = \frac{b_1 s + b_0}{s^2 + a_1 s + a_0} \tag{7.77}$$

controlled by the PID controller

$$R(s) = Z_R\left(1 + \frac{1}{T_I s} + T_D s\right) = \frac{Z_R(T_I s + 1 + T_I T_D s^2)}{T_I s} \tag{7.78}$$

The closed-loop transfer function between the setpoint and output is given as

$$G_{yw}(s) = \frac{Z_R(T_I s + 1 + T_I T_D s^2)(b_1 s + b_0)}{T_I s(s^2 + a_1 s + a_0) Z_R(T_I s + 1 + T_I T_D s^2)(b_1 s + b_0)} \tag{7.79}$$

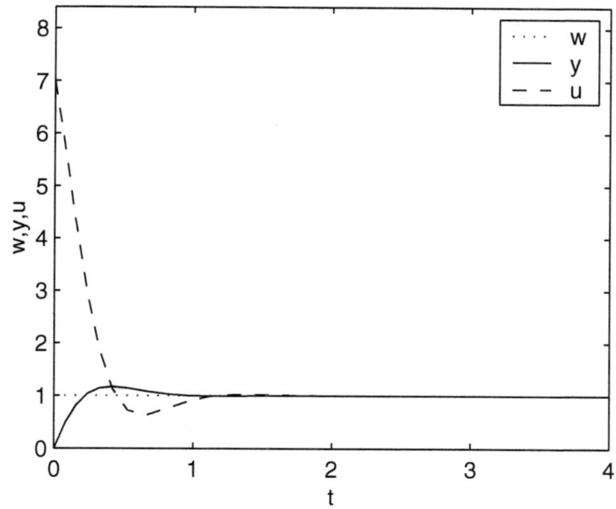

Fig. 7.22. Heat exchanger control using pole placement design

The characteristic equation is of the third order. Hence, it is possible to choose three poles. As the PID controller has three tunable parameters, it is suitable for this system. We can use the standard characteristic equation (7.70) with three parameters that specify an arbitrary type of the standard forms.

Comparing the coefficients at the corresponding powers of s yields the system of equations with three unknowns – controller parameters

$$a_1 + b_1 Z_R + b_0 Z_R T_D = \omega_0(\alpha + 2\zeta)(1 + b_1 Z_R T_D) = c_1(1 + b_1 Z_R T_D) \quad (7.80)$$

$$a_0 + b_0 Z_R + b_1 \frac{Z_R}{T_I} = \omega_0^2(1 + 2\alpha\zeta)(1 + b_1 Z_R T_D) = c_2(1 + b_1 Z_R T_D) \quad (7.81)$$

$$b_0 \frac{Z_R}{T_I} = \omega_0^3 \alpha(1 + b_1 Z_R T_D) = c_3(1 + b_1 Z_R T_D) \quad (7.82)$$

The equations are nonlinear as they contain multiplication and division of unknown variables. If a substitution with $\bar{T}_I = Z_R/T_I$, $\bar{T}_D = Z_R T_D$ will be introduced, a system of linear equations with unknown variables $Z_R, \bar{T}_I, \bar{T}_D$ will be obtained

$$\begin{pmatrix} b_1 & (b_0 - c_1 b_1) & 0 \\ b_0 & -c_2 b_1 & b_1 \\ 0 & -c_3 b_1 & b_0 \end{pmatrix} \begin{pmatrix} Z_R \\ \bar{T}_I \\ \bar{T}_D \end{pmatrix} = \begin{pmatrix} c_1 - a_1 \\ c_2 - a_0 \\ c_3 \end{pmatrix} \quad (7.83)$$

Solution of the equations and backsubstitution yield the controller parameters.

Experimental Methods

All analytical methods assume knowledge of the process transfer function. As this is rarely precisely known, the methods can be used only for the first estimate of the controller parameters. Next, these are applied to the process and manually retuned to yield the desired performance.

These drawbacks can partially be eliminated by using experimental methods that explore dynamic properites of the controlled process to find controller parameters.

Trial-and-Error Method

A typical tuning of PID controller parameters can be as follows:

- Integral and derivative actions are turned off (T_I is on maximum, T_D is zero) so that only proportional action with small controller gain results.
- The proportional gain is gradually increased until permanent oscillations result (undamped behaviour). A care has to be taken so that the manipulated variable is not on the constraints.
- The proportional gain is reduced to a half.
- Integral time constant is gradually reduced until again permanent oscillations occur. T_I is then set to $3T_{I,\mathrm{crit}}$.
- Derivative action is increased until again permanent oscillations occur. T_D is then set to a third of the critical value.

The method cannot always be used as permanent oscillations can compromise the safety of the technology. In any case these steps indicate influence of each controller part and their relation to controller tuning.

Ziegler-Nichols Methods

These are among the most used and widely spread methods for PID controller tuning that have been in use since 1950. The derivation is based on the first order controlled system with time delay and the controller parameters are optimised for underdamped transient response with the damping coefficient of about 25%. The procedure can either be applied using permanent oscillations or is based on the step response. The method of permanent oscillations is in fact only a modification of the trial-and-error method.

Method of Permanent Oscillations The controller is at first set up only with a proportional action and its gain is increased until permanent oscillations occur. The closed-loop system is only marginally stable at this point. Critical proportional gain Z_{Rk} amd critical time period of the oscillations T_k are obtained. These values are used for a given controller structure for the calculation of its parameters from Table 7.2. Several settings are given in the table – the usual Ziegler-Nichols settings as well as the settings for more robust controller with a smaller or zero overshoot.

Another possibility of getting permanent oscillations in the closed-loop system can be realised using a relay with hysteresis. In this case the process input is of rectangular shape. The advantage is that the output magnitude can be tuned. The relay output changes every time the control error changes

Table 7.2. Ziegler-Nichols controller tuning based on permanent oscillations. PID_1 sets smaller overshoot, PID_2 is underdamped

Regulátor	Z_R	T_I	T_D
P	$0.5Z_{Rk}$		
PI	$0.4Z_{Rk}$	$0.8T_k$	
PID	$0.6Z_{Rk}$	$0.5T_k$	$0.125T_k$
PID_1	$0.33Z_{Rk}$	$0.5T_k$	$T_k/3$
PID_2	$0.2Z_{Rk}$	$0.5T_k$	$T_k/3$

sign (Fig. 7.23). After a certain time the oscillations settle and it is possible to find out the value of T_k. If we consider the magnitude of the input $\pm u$ and the magnitude of the output $\pm y$ then it follows the theory of harmonic equilibrium that the critical controller gain is given as

$$Z_{Rk} = \frac{4u}{\pi y} \tag{7.84}$$

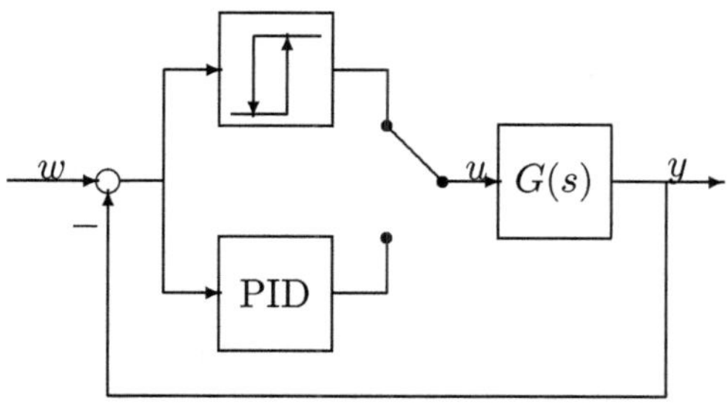

Fig. 7.23. Closed-Loop system with a relay

Step Response Method Often it is not possible to find the critical gain of the controller for safety reasons. In this case it can be easier to measure a step response of the process and to calculate the controller parameters from it using Table 7.3. Note that it is not necessary to measure the whole step response, only the values T_u and Z/T_n given from the slope of the curve in the inflexion point.

Table 7.3. The Ziegler-Nichols controller tuning based on the process step response

Regulátor	Z_R	T_I	T_D
P	$\frac{1}{Z}\frac{T_n}{T_u}$		
PI	$\frac{0.9}{Z}\frac{T_n}{T_u}$	$3.33T_u$	
PID	$\frac{1.2}{Z}\frac{T_n}{T_u}$	$2T_u$	$0.5T_u$

Example 7.14: *The Ziegler-Nichols method* www

Consider the fourth order system described by

$$G(s) = \frac{1}{(s+1)^4}$$

with the PID controller. The step response gives $T_u = 1.425$, $T_n = 4.463$ and thus $Z_R = 3.76, T_I = 2.85, T_D = 0.71$. Simulation results are shown in Fig. 7.24. The closed-loop response is fairly oscillatory. This is a common drawback of Ziegler-Nichols methods.

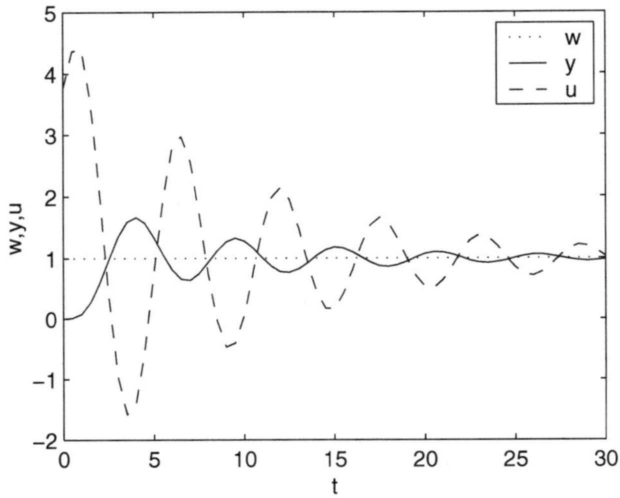

Fig. 7.24. PID control tuned with the Ziegler-Nichols method

Åström-Hägglund Methods

A possible disadvantage of Ziegler-Nichols methods is that they are based on only two process characteristics: either T_u, T_n for a step response or Z_{Rk}, T_k for permanent oscillations. Generally speaking, Ziegler-Nichols methods have the following in common:

- transient responses are fairly oscillatory,
- not suitable for tume-delay systems,
- difficult tuning.

Based on these observations, Åström a Hägglund have developed a new method that uses three process characteristics. The third one is optional – maximum of the sensitivity function M_s that specifies robustness of the controller. M_s is tunable and two choices exist: 1.4 and 2.0. The higher value of M_s characterises a faster controller.

The PID controller is assumed with setpoint weighting with the parameters: proportional gain Z_R, integral time constant T_I, derivative time constant T_D, and setpoint weighting b

$$u(t) = Z_R \left(bw(t) - y(t) + \frac{1}{T_I} \int_0^t e(s)\mathrm{d}s + T_D \frac{\mathrm{d}e(t)}{\mathrm{d}t} \right) \tag{7.85}$$

Similarly to the Ziegler-Nichols method, the authors have developed controller tuning based on frequency or step response.

Step Response To calculate the controller parameters, three process characteristics are used: process gain Z, time constant T, and time delay T_d found by step response identification. For the identification procedure of these parameters see Section 6.2.1. For the further derivation of the method, normalised parameters are introduced

$$a = Z\frac{T_d}{T} \tag{7.86}$$

$$\tau = \frac{T_d}{T_d + T} \tag{7.87}$$

If the controlled process is unstable with a pure integrator, it is recommended to measure its impulse response by applying a short pulse on the process input. From such a stable response fictitious values of Z', T', T_d' can be determined and normalised parameters from them

$$a = Z'(T' + T_d') \tag{7.88}$$

$$\tau = T' + T_d' \tag{7.89}$$

The controller parameters are normalised in the similar way to the process and are of the form aZ_R, T_I/T_d (or T_I/T for stable processes), and T_D/T_d.

Controller parameters have been obtained empirically from numerous simulations with different process models and interpolating the best results. The authors have shown that it is advantageous to relate the controller parameters to the normalised time delay τ in the functional form

$$f(\tau) = a_0 \exp(a_1\tau + a_2\tau^2) \tag{7.90}$$

PI and PID controller tuning for stable processes are given in Tables 7.4, 7.5, respectively. In the same manner, integrating processes are covered by Tables 7.6 (PI) and 7.7 (PID).

Example 7.15: *Åström-Hägglund method* WWW

Consider the fourth order system with the transfer function of the form

$$G(s) = \frac{1}{(s+1)^4}$$

controlled by PID controller. The process step response gives $T_d = 1.425$, $T = 4.463$. For the PID controller follow from Table. 7.5 parameter values

Table 7.4. PI controller tuning based on step response for stable processes. Parameters are given as functions of τ of the form (7.90)

	$M_s = 1.4$			$M_s = 2$		
	a_0	a_1	a_2	a_0	a_1	a_2
aZ	0.29	-2.7	3.7	0.78	-4.1	5.7
T_I/T_d	8.9	-6.6	3.0	8.9	-6.6	3.0
T_I/T	0.79	-1.4	2.4	0.79	-1.4	2.4
b	0.81	0.73	1.9	0.44	0.78	-0.45

Table 7.5. PID controller tuning based on step response for stable processes. Parameters are given as functions of τ of the form (7.90)

	$M_s = 1.4$			$M_s = 2$		
	a_0	a_1	a_2	a_0	a_1	a_2
aZ	3.8	-8.4	7.3	8.4	-9.6	9.8
T_I/T_d	5.2	-2.5	-1.4	3.2	-1.5	-0.93
T_I/T	0.46	2.8	-2.1	0.28	3.8	-1.6
T_D/T_d	0.89	-0.37	-4.1	0.86	-1.9	-0.44
T_D/T	0.077	5.0	-4.8	0.076	3.4	-1.1
b	0.40	0.18	2.8	0.22	0.65	0.051

Table 7.6. PI controller tuning based on step response for processes with pure integrator. Parameters are given as functions of τ of the form (7.90)

	$M_s = 1.4$			$M_s = 2$		
	a_0	a_1	a_2	a_0	a_1	a_2
aZ	0.41	-0.23	0.019	0.81	-1.1	0.76
T_I/T_d	5.7	1.7	-0.69	3.4	0.28	-0.089
b	0.33	2.5	-1.9	0.78	-1.9	1.2

Table 7.7. PID controller tuning based on step response for processes with pure integrator. Parameters are given as functions of τ of the form (7.90)

	$M_s = 1.4$			$M_s = 2$		
	a_0	a_1	a_2	a_0	a_1	a_2
aZ	5.6	-8.8	6.8	8.6	-7.1	5.4
T_I/T_d	1.1	6.7	-4.4	1.0	3.3	-2.3
T_D/T_d	1.7	-6.4	2.0	0.38	0.056	-0.60
b	0.12	6.9	-6.6	0.56	-2.2	1.2

$Z_R = 2.39$, $T_I = 3.58$, $T_D = 0.87$, and $b = 0.49$. Fig. 7.25 shows two simulations. The first uses the calculated controller, the second shows the influence of the setpoint weighting parameter $b = 1$. In contrast to the Ziegler-Nichols tuning control performance is better and oscillations are more heavily damped. When no setpoint weighting is used, larger overshoot results. However, compared to Example 7.14 on page 289, it is still smaller.

Frequency Response Like above, there are empirical procedures for PID controller tunung in the frequency domain. In this case it is necessary to find

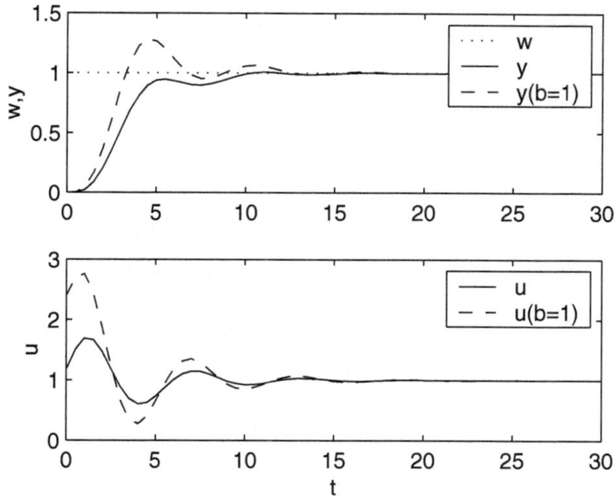

Fig. 7.25. PID control using the Åström-Hägglund method

values of T_k, Z_{Rk}, that are approximated as functions of the parameter $\kappa = 1/ZZ_R$.

Results for stable processes are given in Table 7.8 (PI controller) and in Table 7.9 (PID controller). When integrating processes are considered, it can be possible to find the parameters by setting $\kappa = 0$ for the PI controller. The case of the PID controller is more complex and the settings cannot be used.

Table 7.8. PI controller tuning based on frequency response for stable processes. Parameters are given as functions of κ of the form $a_0 \exp(a_1\kappa + a_2\kappa^2)$

	$M_s = 1.4$			$M_s = 2$		
	a_0	a_1	a_2	a_0	a_1	a_2
Z/Z_{Rk}	0.053	2.9	-2.6	0.13	1.9	-1.3
T_I/T_k	0.90	-4.4	2.7	0.90	-4.4	2.7
b	1.1	-0.0061	1.8	0.48	0.40	-0.17

Table 7.9. PID controller tuning based on frequency response for stable processes. Parameters are given as functions of κ of the form $a_0 \exp(a_1\kappa + a_2\kappa^2)$

	$M_s = 1.4$			$M_s = 2$		
	a_0	a_1	a_2	a_0	a_1	a_2
Z/Z_{Rk}	0.33	-0.31	-1.0	0.72	-1.6	1.2
T_I/T_k	0.76	-1.6	-0.36	0.59	-1.3	0.38
T_D/T_k	0.17	-0.46	-2.1	0.15	-1.4	0.56
b	0.58	-1.3	3.5	0.25	0.56	-0.12

7.5 References

This chapter is more or less a standard part of the majority books dealing with automatic control. At this place we mention particularly:

J. Mikleš and V. Hutla. *Theory of Automatic Control.* Alfa, Bratislava, 1986. (in Slovak).

H. Unbehauen. *Regelungstechnik I.* Vieweg, Braunschweig/Wiesbaden, 1986.

J.-P. Corriou. *Process Control. Theory and Applications.* Springer Verlag, 2004.

S. Kubík, Z. Kotek, V. Strejc, and J. Štecha. *Theory of Automatic Control I.* SNTL/ALFA, Praha, 1982. (in Czech).

PID controllers, their design, tuning, and implementation are covered by the works:

J. G. Ziegler and N. B. Nichols. Optimum settings for automatic controllers. *Trans. ACME*, 64(8):759–768, 1942.

K. J. Åström and T. J. Hägglund. *PID Controllers: Theory, Design, and Tuning.* Instrument Society of America, Research Triangle Park, 2 edition, 1995.

K. K. Tan, Q.-G. Wang, C. C. Hang, and T. J. Hägglund. *Advances in PID Control.* Springer Verlag, London, 1999.

V. Bobál, J. Böhm, and J. Fessl. *Digital Self-tuning Controllers.* Springer Verlag, London, 2005.

C.-C. Yu. *Autotuning of PID Controllers. Relay Feedback Approach.* Springer Verlag, London, 1999.

M. Huba. *Theory of Automatic Control 3. Constrained PID Control.* STU Press, Bratislava, 2006b. (in Slovak).

M. Huba. Constrained pole assignment control. In L. Menini, L. Zaccarian, and C. T. Abdallah, editors, *Current Trends in Nonlinear Systems and Control*, pages 395 – 398. Birkhäuser, Boston, 2006a.

7.6 Exercises

Exercise 7.1:
Consider a system with the transfer function

$$G(s) = \frac{1-2s}{s^2+s+1}$$

Find the following frequency indices:
1. gain and phase margins,
2. bandwidth,

3. resonant frequency and resonance peak.

Exercise 7.2:
Consider a system with the transfer function

$$G(s) = \frac{1}{(s+1)^3}$$

The task is:
1. to control the process and to find the critical P control gain, as well as the critical period,
2. to calculate both parameters analytically,
3. to design PI and PID controllers by the Ziegler-Nichols and Åström-Hägglund methods,
4. to compare the control performance.

Exercise 7.3:
Consider a controlled system with the transfer function

$$G(s) = \frac{1-2s}{s^2+1}$$

1. find the PID controller by the pole placement design method. The closed-loop poles should be placed using various standard forms with $\omega_0 = 1$,
2. compare the performance of the closed-loop system for all cases.

Exercise 7.4:
A continuous-time PID controller can be transformed to its discrete-time counterpart using the following substitutions:
1. backward approximation:
$$s = \frac{1-z^{-1}}{z^{-1}T_s}$$
2. forward approximation:
$$s = \frac{1-z^{-1}}{T_s}$$
3. the Tustin approximation:
$$s = \frac{2}{T_s}\frac{1-z^{-1}}{1+z^{-1}}$$

Derive the discrete-time controllers.

Exercise 7.5:
Consider a closed-loop system consisting of the controlled process

$$G(s) = \frac{1}{s^3+2s^2+3s+4}$$

and a controller with the structures P, I, PI. Find such intervals of controller parameters that the closed-loop system is stable.

8

Optimal Process Control

This chapter explains design of optimal feedback control based on state-space and input-output representations of linear systems. In the first part, dynamic programming and principle of minimum are presented. Then the pole placement feedback control design is derived. This serves as a basis for all advanced design methods. Based on the pole placement, stability of the closed-loop system is guaranteed. In the next parts it is shown how optimal pole locations can be found. This can be done either for minimisation of an integral cost or of H_2 norm. The Youla-Kučera parametrisation of all stabilising controllers is derived. Included are also state reconstruction and estimation.

The chapter deals mainly with continuous-time systems. However, discrete-time equivalents are also presented and differences in both domains are explained. In addition, dead-beat control is also presented.

In all cases, state-space control design is completed with the input-output design based on Diophantine equations.

8.1 Problem of Optimal Control and Principle of Minimum

There are many possible formulations of optimal control design in process control. One of them that leads to optimal feedback control of lumped continuous-time systems can be formulated as follows.

Consider a controlled process with mathematical model of the form

$$\frac{\mathrm{d}\boldsymbol{x}(t)}{\mathrm{d}t} = \boldsymbol{f}(\boldsymbol{x}(t), \boldsymbol{u}(t)), \quad \boldsymbol{x}(t_0) = \boldsymbol{x}_0 \tag{8.1}$$

where $\boldsymbol{x}$ is n-dimensional vector of state variables and $\boldsymbol{u}$ is m-dimensional vector of manipulated variables. $\boldsymbol{f}$ is n-dimensional vector function and it is assumed that it is continuous-time and continuously differentiable with respect to all its variables.

The optimal control design is to find such $\boldsymbol{u}(t)$ that the cost function

$$I(\boldsymbol{u}(t)) = G_{tf}(\boldsymbol{x}(\boldsymbol{t}_f)) + \int_{t_0}^{t_f} F(\boldsymbol{x}(t), \boldsymbol{u}(t))\mathrm{d}t \tag{8.2}$$

is minimal. It is assumed that there are no constraints on $\boldsymbol{x}$, $\boldsymbol{u}$. Functions G_{tf}, F are scalars and continuously differentiable.

Variations of this problem can cover additional requirements and constraints that can be included. This covers constraints on input and state variables, constraints on final state vector $\boldsymbol{x}(t_f)$. Some of the problems deal with fixed end time t_f, sometimes final time is also minimised.

Let us assume that the optimal control trajectory $\boldsymbol{u}^*(t)$ exists. If any other control trajectory $\boldsymbol{u}(t)$ is considered, it holds:

$$I\left[\boldsymbol{u}(t)\right] \geq I\left[\boldsymbol{u}^*(t)\right] \tag{8.3}$$

Equation (8.3) holds in general. However, it can happen that $\boldsymbol{u}^*(t)$ satisfying (8.3) does not exist. It is very difficult to prove the existence of optimal control. Therefore, applications of optimal control often use intuition and based on it assume that the optimal control trajectory exists. Its proof is based on the Hamilton-Jacobi equation.

Let us therefore assume that the optimal solution $\boldsymbol{u}^*(t)$ exists. We will derive conditions that this solutions has to satisfy. These are only necessary conditions and are based on the variational calculus. We assume a small change of $\boldsymbol{u}^*(t)$ leading to a new trajectory $\boldsymbol{u}(t)$ and study change of the cost function I.

Let $\boldsymbol{u}^*(t)$ be the optimal control and $\boldsymbol{x}^*(t)$ the corresponding optimal system response given by (8.1). Indefinitely small control variation $\delta\boldsymbol{u}(t)$ then produces state variation $\delta\boldsymbol{x}(t)$. The system response is then given as

$$\boldsymbol{x}(t) = \boldsymbol{x}^*(t) + \delta\boldsymbol{x}(t) \tag{8.4}$$

that is caused by the control

$$\boldsymbol{u}(t) = \boldsymbol{u}^*(t) + \delta\boldsymbol{u}(t) \tag{8.5}$$

Variation of the time derivative of state can be rewritten as

$$\delta\left(\frac{\mathrm{d}\boldsymbol{x}}{\mathrm{d}t}\right) = \frac{\mathrm{d}\boldsymbol{x}}{\mathrm{d}t} - \frac{\mathrm{d}\boldsymbol{x}^*}{\mathrm{d}t} \tag{8.6}$$

$$= \frac{\mathrm{d}\boldsymbol{x}^*}{\mathrm{d}t} + \frac{\mathrm{d}\left(\delta\boldsymbol{x}\right)}{\mathrm{d}t} - \frac{\mathrm{d}\boldsymbol{x}^*}{\mathrm{d}t} \tag{8.7}$$

$$= \frac{\mathrm{d}\left(\delta\boldsymbol{x}\right)}{\mathrm{d}t} \tag{8.8}$$

It follows from (8.8) that linear operators $\mathrm{d}/\mathrm{d}t$ and δ are commutative. The Taylor expansion in the neighbourhood of the optimal state gives

$$\boldsymbol{f}(\boldsymbol{x}, \boldsymbol{u}) = f(\boldsymbol{x}^*, \boldsymbol{u}^*) + \left(\frac{\partial \boldsymbol{f}}{\partial \boldsymbol{x}}\right)^* \delta\boldsymbol{x} + \left(\frac{\partial \boldsymbol{f}}{\partial \boldsymbol{u}}\right)^* \delta\boldsymbol{u} \tag{8.9}$$

where only linear terms were taken into account. Partial derivatives in (8.9) are evaluated for optimal trajectories $\boldsymbol{u}^*(t)$, $\boldsymbol{x}^*(t)$.

The terms

$$\left(\frac{\partial \boldsymbol{f}}{\partial \boldsymbol{x}}\right) = \begin{pmatrix} \frac{\partial f_1}{\partial x_1} & \frac{\partial f_1}{\partial x_2} & \cdots & \frac{\partial f_1}{\partial x_n} \\ \frac{\partial f_2}{\partial x_1} & \frac{\partial f_2}{\partial x_2} & \cdots & \frac{\partial f_2}{\partial x_n} \\ \vdots & \vdots & \cdots & \vdots \\ \frac{\partial f_n}{\partial x_1} & \frac{\partial f_n}{\partial x_2} & \cdots & \frac{\partial f_n}{\partial x_n} \end{pmatrix} \tag{8.10}$$

$$\left(\frac{\partial \boldsymbol{f}}{\partial \boldsymbol{u}}\right) = \begin{pmatrix} \frac{\partial f_1}{\partial u_1} & \frac{\partial f_1}{\partial u_2} & \cdots & \frac{\partial f_1}{\partial u_m} \\ \frac{\partial f_2}{\partial u_1} & \frac{\partial f_2}{\partial u_2} & \cdots & \frac{\partial f_2}{\partial u_m} \\ \vdots & \vdots & \cdots & \vdots \\ \frac{\partial f_n}{\partial u_1} & \frac{\partial f_n}{\partial u_2} & \cdots & \frac{\partial f_n}{\partial u_m} \end{pmatrix} \tag{8.11}$$

are Jacobi matrices. Superscripts $(\cdot)^*$ denote optimal trajectories of $\boldsymbol{u}^*$, $\boldsymbol{x}^*$. Equation (8.9) can be rewritten when considering (8.1) and (8.8) as

$$\frac{\mathrm{d}\,(\delta\boldsymbol{x})}{\mathrm{d}t} = \left(\frac{\partial \boldsymbol{f}}{\partial \boldsymbol{x}}\right) \delta\boldsymbol{x} + \left(\frac{\partial \boldsymbol{f}}{\partial \boldsymbol{u}}\right) \delta\boldsymbol{u} \tag{8.12}$$

The cost function $I(\boldsymbol{u})$ attains the minimum for the function $\boldsymbol{u}^* = \boldsymbol{u}^*(t)$ from a class of permitted functions if an arbitrary function $\boldsymbol{u}(t)$ satisfies inequality (8.3).

In general, an increment of the cost function $I(\boldsymbol{x})$ can be defined as

$$\Delta I = I(\boldsymbol{x} + \delta\boldsymbol{x}) - I(\boldsymbol{x}) \tag{8.13}$$

and thus it can be written as

$$\Delta I = \Delta I(\boldsymbol{x}, \delta\boldsymbol{x}) \tag{8.14}$$

The first variation δI is the part of ΔI that is linear in the variation $\delta\boldsymbol{x}$. The fundamental theorem of the variational calculus says that if $\boldsymbol{x}^*$ is on the extreme then necessary condition for extremum of the cost function is

$$\delta I(\boldsymbol{x}^*, \delta\boldsymbol{x}) = 0 \tag{8.15}$$

for all possible $\delta\boldsymbol{x}$.

It can be noted that there exists a formal similarity between extremes of a function $y = y(x)$ of the form $(\mathrm{d}y/\mathrm{d}x) = 0$ and relation (8.15).

It can be shown that if the increments ΔI and its variation δI are continuous functions and the variation $\delta \boldsymbol{u}$ can be arbitrary, then the equality

$$\delta I = 0 \tag{8.16}$$

is the necessary condition of extremum. If only single-sided variations (for example $\delta \boldsymbol{u} \geq 0$) are permitted then $\delta I = 0$ does not constitute the necessary condition.

The variation of the cost (8.2) can be written as

$$\delta I = \left(\frac{\partial G_{tf}}{\partial \boldsymbol{x}(t_f)}\right)^T \delta x(t_f) + \int_{t_0}^{t_f} \left[\left(\frac{\partial F}{\partial \boldsymbol{x}}\right)^T \delta \boldsymbol{x} + \left(\frac{\partial F}{\partial \boldsymbol{u}}\right)^T \delta \boldsymbol{u}\right] \mathrm{d}t \tag{8.17}$$

Partial derivatives in (8.17) are calculated in optimal trajectories.

Equation (8.17) holds if the final time t_f is fixed. The final state of the system $\boldsymbol{x}(t_f)$ is considered to be free.

Let us define an adjoint vector $\boldsymbol{\lambda}(t)$ and manipulate equation (8.12) as follows

$$\boldsymbol{\lambda}^T \frac{\mathrm{d}\,(\delta \boldsymbol{x})}{\mathrm{d}t} = \boldsymbol{\lambda}^T \left(\frac{\partial \boldsymbol{f}}{\partial \boldsymbol{x}}\right) \delta \boldsymbol{x} + \boldsymbol{\lambda}^T \left(\frac{\partial \boldsymbol{f}}{\partial \boldsymbol{u}}\right) \delta \boldsymbol{u} \tag{8.18}$$

Integrating (8.18) from $t = t_0$ to $t = t_f$ yields

$$\int_{t_0}^{t_f} \left[\boldsymbol{\lambda}^T \frac{\mathrm{d}\,(\delta \boldsymbol{x})}{\mathrm{d}t} - \boldsymbol{\lambda}^T \left(\frac{\partial \boldsymbol{f}}{\partial \boldsymbol{x}}\right) \delta \boldsymbol{x} - \boldsymbol{\lambda}^T \left(\frac{\partial \boldsymbol{f}}{\partial \boldsymbol{u}}\right) \delta \boldsymbol{u}\right] \mathrm{d}t = 0 \tag{8.19}$$

Adding (8.17) to (8.19) gives for δI

$$\begin{aligned}
\delta I = {} & \left(\frac{\partial G_{tf}}{\partial \boldsymbol{x}\,(t_f)}\right)^T \delta \boldsymbol{x}(t_f) \\
& + \int_{t_0}^{t_f} \left\{\left[\left(\frac{\partial F}{\partial \boldsymbol{x}}\right)^T + \boldsymbol{\lambda}^T \left(\frac{\partial \boldsymbol{f}}{\partial \boldsymbol{x}}\right)\right] \delta \boldsymbol{x} + \left[\left(\frac{\partial F}{\partial \boldsymbol{u}}\right)^T + \boldsymbol{\lambda}^T \left(\frac{\partial \boldsymbol{f}}{\partial \boldsymbol{u}}\right)\right] \delta \boldsymbol{u}\right\} \mathrm{d}t \\
& - \int_{t_0}^{t_f} \boldsymbol{\lambda}^T \frac{\mathrm{d}\,(\delta \boldsymbol{x})}{\mathrm{d}t} \mathrm{d}t
\end{aligned} \tag{8.20}$$

Derivative of the term $\boldsymbol{\lambda}^T \delta \boldsymbol{x}$ gives

$$\boldsymbol{\lambda}^T \frac{\mathrm{d}(\delta \boldsymbol{x})}{\mathrm{d}t} = \frac{\mathrm{d}}{\mathrm{d}t} \left(\boldsymbol{\lambda}^T \delta \boldsymbol{x}\right) - \frac{\mathrm{d}\boldsymbol{\lambda}^T}{\mathrm{d}t} \delta \boldsymbol{x} \tag{8.21}$$

and thus equation (8.20) is of the form

$$\begin{aligned}
\delta I = {} & \left(\frac{\partial G_{tf}}{\partial \boldsymbol{x}\,(t_f)}\right)^T \delta \boldsymbol{x}\,(t_f) + \int_{t_0}^{t_f} \left[\left(\frac{\partial F}{\partial \boldsymbol{x}}\right)^T + \boldsymbol{\lambda}^T \left(\frac{\partial \boldsymbol{f}}{\partial \boldsymbol{x}}\right) + \frac{\mathrm{d}\boldsymbol{\lambda}^T}{\mathrm{d}t}\right] \delta \boldsymbol{x} \\
& + \int_{t_0}^{t_f} \left[\left(\frac{\partial F}{\partial \boldsymbol{u}}\right)^T + \boldsymbol{\lambda}^T \left(\frac{\partial \boldsymbol{f}}{\partial \boldsymbol{u}}\right)\right] \delta \boldsymbol{u} \mathrm{d}t + \boldsymbol{\lambda}^T \delta \boldsymbol{x}|_{t=t_0} - \boldsymbol{\lambda}^T \delta \boldsymbol{x}|_{t=t_f}
\end{aligned} \tag{8.22}$$

Let us now introduce the Hamilton function or Hamiltonian

$$H = F + \boldsymbol{\lambda}^T \boldsymbol{f}(\boldsymbol{x}, \boldsymbol{u}) \tag{8.23}$$

If the adjoint vector $\boldsymbol{\lambda}(t)$ satisfies the differential equation

$$\frac{\mathrm{d}\boldsymbol{\lambda}}{\mathrm{d}t} = -\frac{\partial H}{\partial \boldsymbol{x}} \tag{8.24}$$

then for a fixed initial state $\boldsymbol{x}(t_0) = \boldsymbol{x}_0$ (where $\delta\boldsymbol{x}(t_0) = \boldsymbol{0}$) the necessary condition for existence of optimal control is given as

$$\delta I = \int_{t_0}^{t_f} \left(\frac{\partial H}{\partial \boldsymbol{u}}\right)^T \delta\boldsymbol{u}\mathrm{d}t = 0 \tag{8.25}$$

with the terminal condition for the adjoint vector $\boldsymbol{\lambda}(t)$

$$\boldsymbol{\lambda}(t_f) = \left(\frac{\partial G_{tf}}{\partial \boldsymbol{x}}\right)_{t=t_f} \tag{8.26}$$

Equation (8.25) specifies the relation between variation of the cost function and variation of the control trajectory. If some of the elements of the vector $\boldsymbol{x}(t_f)$ are fixed at the terminal time then the variation of control at this point is not arbitrary. However, it can be shown that an equivalent result can be obtained in unspecified or fixed terminal points.

If we suppose that the variation of $\delta\boldsymbol{u}(t)$ is arbitrary ($\boldsymbol{u}(t)$ is unbounded) then the following equation

$$\frac{\partial H}{\partial \boldsymbol{u}} = \boldsymbol{0} \tag{8.27}$$

is the necessary condition for extremum (minimum) of the cost function.

If there are constraints on control variables of the form

$$-\alpha_j \le u_j(t) \le \beta_j, \quad j = 1, 2, \ldots, \tag{8.28}$$

where α_j and β_j are constants for minimum and maximum values of elements u_j of the vector $\boldsymbol{u}$ then $\delta\boldsymbol{u}(t)$ cannot be arbitrary and (8.27) does not guarantee the necessary condition for existence of extremum. If the control variable is on the lower constraint, the only variation allowed is $\delta u_j > 0$. Equation (8.25) then requires that

$$u_j^* = -\alpha_j, \quad \text{if} \quad \frac{\partial H}{\partial u_j} > 0 \tag{8.29}$$

Similarly, if the control variable u_j is on the upper constraint, the only variation allowed is $\delta u_j < 0$. Equation (8.25) then requires that

$$u_j^* = \beta_j, \quad \text{if} \quad \frac{\partial H}{\partial u_j} < 0 \tag{8.30}$$

It is clear in both cases that the control variable minimises H if it is on the constraint.

Necessary conditions for optimal control $\boldsymbol{u}^*(t)$ if the variation $\delta\boldsymbol{u}$ is arbitrary follow from (8.27).

The exact formulation of the principle of minimum will not be given here. Based on the derivation presented above, important relations from (8.23), (8.24) are

$$\frac{\partial H}{\partial \boldsymbol{\lambda}} = \boldsymbol{f}(\boldsymbol{x}, \boldsymbol{u}) \tag{8.31}$$

$$\frac{\partial H}{\partial \boldsymbol{\lambda}} = \frac{\mathrm{d}\boldsymbol{x}}{\mathrm{d}t} \tag{8.32}$$

$$\frac{\partial H}{\partial \boldsymbol{x}} = \frac{\partial F}{\partial \boldsymbol{x}} + \left(\boldsymbol{\lambda}^T \frac{\partial \boldsymbol{f}}{\partial \boldsymbol{x}}\right)^T \tag{8.33}$$

$$\frac{\mathrm{d}\boldsymbol{\lambda}}{\mathrm{d}t} = -\frac{\partial F}{\partial \boldsymbol{x}} - \left(\boldsymbol{\lambda}^T \frac{\partial \boldsymbol{f}}{\partial \boldsymbol{x}}\right)^T \tag{8.34}$$

$$\frac{\mathrm{d}\boldsymbol{\lambda}}{\mathrm{d}t} = -\frac{\partial H}{\partial \boldsymbol{x}} \tag{8.35}$$

Derivative of H with respect to time is given as

$$\frac{\mathrm{d}H}{\mathrm{d}t} = \left(\frac{\partial H}{\partial \boldsymbol{x}}\right)^T \frac{\mathrm{d}\boldsymbol{x}}{\mathrm{d}t} + \left(\frac{\partial H}{\partial \boldsymbol{u}}\right)^T \frac{\mathrm{d}\boldsymbol{u}}{\mathrm{d}t} + \left(\frac{\partial H}{\partial \boldsymbol{\lambda}}\right)^T \frac{\mathrm{d}\boldsymbol{\lambda}}{\mathrm{d}t} \tag{8.36}$$

From (8.32), (8.24) follows

$$\left(\frac{\partial H}{\partial \boldsymbol{x}}\right)^T \frac{\mathrm{d}\boldsymbol{x}}{\mathrm{d}t} + \left(\frac{\partial H}{\partial \boldsymbol{\lambda}}\right)^T \frac{\mathrm{d}\lambda}{\mathrm{d}t} = 0 \tag{8.37}$$

Due to (8.27) is the right hand of equation (8.37) equal to zero

$$\frac{\mathrm{d}H}{\mathrm{d}t} = 0 \tag{8.38}$$

if unconstrained control is assumed (or if control never hits constraints). From (8.38) follows that Hamiltonian is constant when optimal control is applied.

www **Example 8.1:** *Optimal control of a heat exchanger*

Consider a heat exchanger (Fig. 1.1, page 3) for heating a liquid. We assume ideal mixing, negligible heat losses, constant holdup, inlet and outlet flowrates. Mathematical model of the exchanger is given as

$$\frac{V}{q}\frac{\mathrm{d}\vartheta}{\mathrm{d}t'} + \vartheta = \vartheta_v + \frac{\omega}{q\rho c_p}$$

where ϑ is the outlet temperature, ϑ_v – inlet temperature, t' - time, ω – heat input, ρ – liquid density, V – liquid volume in the exchanger, q – volumetric liquid flowrate, c_p – specific heat capacity.
Denote

$$\vartheta_u = \frac{\omega}{q\rho c_p}$$

As $\omega/q\rho c_p$ is in temperature units, ϑ_u can be considered a manipulated variable. Assume that the temperature ϑ_u is constant for a sufficiently long time ($\vartheta_u = \vartheta_{u0}$). If the inlet temperature is constant $\vartheta_v = \vartheta_{v0}$ steady-state of the exchanger is given as

$$\vartheta_0 = \vartheta_{v0} + \vartheta_{u0}$$

Let us now consider another steady-state ϑ_1 given by the manipulated variable ϑ_{u1}

$$\vartheta_1 = \vartheta_{v0} + \vartheta_{u1}$$

The optimal control problem consists in determination of a trajectory $\vartheta_u(t')$ from ϑ_{u0} to ϑ_{u1} in such a way that a given cost function be minimal. This cost function will be defined later. Before it, let us define dimensionless deviation variables from the final steady state.
The new state variable is defined as

$$x(t') = \frac{\vartheta(t') - \vartheta_1}{\vartheta_{u1} - \vartheta_{v0}}$$

The new manipulated variable is defined as

$$u(t') = \frac{\vartheta_u(t') - \vartheta_{u1}}{\vartheta_{u1} - \vartheta_{v0}}$$

Further, we define a new dimensionless time variable

$$t = \frac{q}{V}t'$$

Mathematical model of the heat exchanger is now given by the first order differential equation

$$\frac{\mathrm{d}x(t)}{\mathrm{d}t} + x(t) = u(t)$$

with initial condition

$$x(0) = x_0 = \frac{\vartheta_0 - \vartheta_1}{\vartheta_{u1} - \vartheta_{v0}}$$

Consider now a controlled system described by the first order differential equation $\mathrm{d}x/\mathrm{d}t + x = u$. Let us find a trajectory $u(t)$ such that minimum is attained of the cost function

$$I = \int_{t_0}^{t_f} \left[x^2(t) + ru^2(t)\right] \mathrm{d}t$$

where $t_0 = 0$ is the initial time of optimisation that specifies initial state value $x(0) = x_0$, t_f – final time of optimisation, $r > 0$ – weighting coefficient.
Final time $x(t_f)$ is not specified. Therefore, this is the dynamic optimisation problem with free final time. As the final value of $x(t_f)$ is given, the cost function I depends only on $u(t)$ as $u(t)$ determines $x(t)$ from the differential equation of the controlled process.
Hamiltonian is in our case given as

$$H = x^2 + ru^2 + \lambda(-x + u)$$

From the optimality conditions further holds

$$\frac{\mathrm{d}\lambda}{\mathrm{d}t} = -\frac{\partial H}{\partial x} = -2x + \lambda$$

The final value of the adjoint variable $\lambda(t)$ is

$$\lambda(t_f) = 0$$

Optimality condition of partial derivative of the Hamiltonian H with respect to u determines the optimal control trajectory

$$\frac{\partial H}{\partial u} = 0 \Leftrightarrow 2ru + \lambda = 0$$

Optimal control is thus given as

$$u^*(t) = -\frac{1}{2r}\lambda^*(t)$$

To check whether this extremum is really minimum, the second partial derivative of H with respect to u should be positive

$$\frac{\partial^2 H}{\partial u^2} = 2r > 0$$

which confirms this fact.
Next, the system of equations

$$\frac{\mathrm{d}\lambda^*}{\mathrm{d}t} = -2x^* + \lambda^*$$
$$\frac{\mathrm{d}x^*}{\mathrm{d}t} = -x^* - \frac{1}{2r}\lambda^*$$

with initial and terminal conditions

$$\lambda^*(t_f) = 0$$

$$x^*(0) = x_0$$

can be solved to find the optimal control trajectory $u^*(t) = -(1/2r)\lambda^*(t)$. The system of equations possesses a unique solution

$$\lambda^*(t) = \frac{2x_0}{\beta} \sinh \gamma(t_f - t)$$

$$x^*(t) = \frac{x_0}{\beta} \left[\gamma \cosh \gamma(t_f - t) + \sinh \gamma(t_f - t)\right]$$

where $\beta = \gamma \cosh \gamma t_f + \sinh \gamma t_f$, $\gamma = \sqrt{1 + 1/r}$. The proof that $\lambda^*(t)$ and $x^*(t)$ are solution of the system of differential equations is simple – by substitution. It can be shown that the solution is unique. The existence of the optimal control trajectory can be proved mathematically. However, in applications is this existence confirmed by physical analysis of the problem. $u^*(t)$ specifies the optimal control and $x^*(t)$ specifies the response of the controlled system to it. In our case is the optimal control trajectory given as

$$u^*(t) = \frac{x_0}{r'\beta} \sinh \gamma(t_f - t)$$

In may applications is the final time given as $t = \infty$. In that case the optimal state and control trajectories are given as

$$x^*(t) = x_0 \mathrm{e}^{-\gamma t}$$

$$u^*(t) = -\frac{x_0}{r(1+\gamma)} \mathrm{e}^{-\gamma t}$$

Fig. 8.1 shows optimal trajectories of $u(t)$ and $x(t)$ for $t_f = \infty$, $x_0 = 1$, $r = 1$.

8.2 Feedback Optimal Control

Optimal control of processes presented in Section 8.1 can also be implemented using the feedback configuration.

The problem of optimal feedback control of a linear time-invariant dynamic system can be formulated as follows.

Consider a completely controllable system

$$\dot{\boldsymbol{x}}(t) = \boldsymbol{A}\boldsymbol{x}(t) + \boldsymbol{B}\boldsymbol{u}(t) \tag{8.39}$$

with initial condition at time $t_0 = 0$

$$\boldsymbol{x}(0) = \boldsymbol{x}_0 \tag{8.40}$$

and terminal condition $\boldsymbol{x}(t_f)$, t_f. We assume that t_f is fixed.

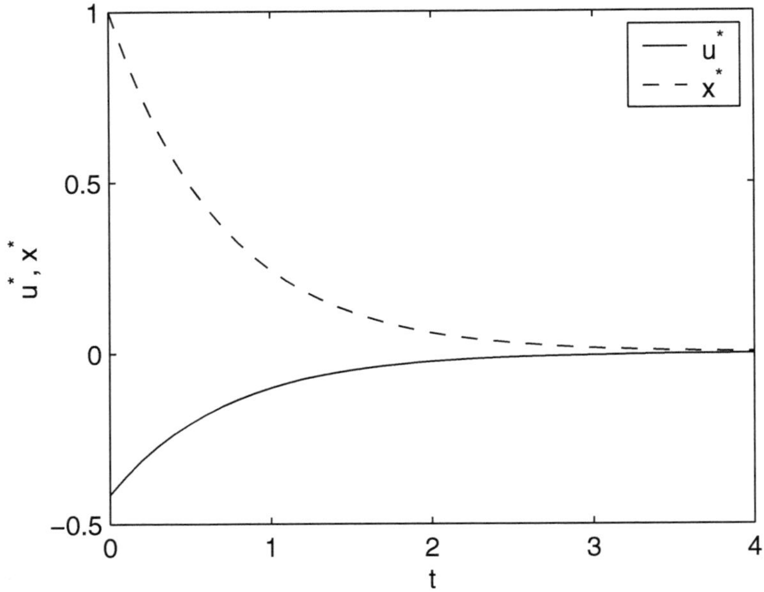

Fig. 8.1. Optimal trajectories of input and state variables of the heat exchanger

The cost function is defined as

$$I = \frac{1}{2}\boldsymbol{x}^T(t_f)\boldsymbol{Q}_{t_f}\boldsymbol{x}(t_f) + \frac{1}{2}\int_0^{t_f} \left(\boldsymbol{x}^T(t)\boldsymbol{Q}\boldsymbol{x}(t) + \boldsymbol{u}^T(t)\boldsymbol{R}\boldsymbol{u}(t)\right) \mathrm{d}t \tag{8.41}$$

where $\boldsymbol{Q}_{t_f}$ and $\boldsymbol{Q}$ are real symmetric positive semidefinite weighting matrices and $\boldsymbol{R}$ is a real symmetric positive definite weighting matrix.

Our aim is to find a feedback control law of the form

$$\boldsymbol{u} = \text{function}(\boldsymbol{x}) \tag{8.42}$$

such that I be minimal for any initial condition $\boldsymbol{x}_0$.

Optimal feedback control guarantees the change of the operating point within $t \in [0, t_f]$ from state $\boldsymbol{x}_0$ to a neighbourhood of the origin $\boldsymbol{x} = \boldsymbol{0}$.

Hamiltonian is for our case defined as

$$H = \frac{1}{2}(\boldsymbol{x}^T\boldsymbol{Q}\boldsymbol{x} + u^T\boldsymbol{R}\boldsymbol{u}) + \boldsymbol{\lambda}^T(\boldsymbol{A}\boldsymbol{x} + \boldsymbol{B}\boldsymbol{u}) \tag{8.43}$$

where the adjoint vector $\boldsymbol{\lambda}(t)$ satisfies the differential equation

$$\dot{\boldsymbol{\lambda}} = -\boldsymbol{Q}\boldsymbol{x} - \boldsymbol{A}^T\boldsymbol{\lambda} \tag{8.44}$$

with terminal condition

$$\boldsymbol{\lambda}(t_f) = \boldsymbol{Q}_{t_f}\boldsymbol{x}(t_f) \tag{8.45}$$

If there are no constraints on $\boldsymbol{u}(t)$ then the optimal control condition can be written as

$$\boldsymbol{R}\boldsymbol{u} + \boldsymbol{B}^T\boldsymbol{\lambda} = \boldsymbol{0} \tag{8.46}$$

Optimal control $\boldsymbol{u}$ is then given as

$$\boldsymbol{u}(t) = -\boldsymbol{R}^{-1}\boldsymbol{B}^T\boldsymbol{\lambda}(t) \tag{8.47}$$

The matrix $\boldsymbol{R}^{-1}$ exists as $\boldsymbol{R}$ is square symmetric positive definite matrix.

Optimal control has to minimise the Hamiltonian. The necessary condition $\partial H/\partial \boldsymbol{u}$ only specifies an extremum. To assure minimum for $\boldsymbol{u}$, the second partial derivative with respect to control $\partial^2 H/\partial \boldsymbol{u}^2$ with dimensions $m \times m$ needs to be positive definite. From (8.46) follows

$$\frac{\partial^2 H}{\partial \boldsymbol{u}^2} = \boldsymbol{R} \tag{8.48}$$

As $\boldsymbol{R}$ is assumed to be positive definite, the optimal control (8.47) indeed minimises the Hamiltonian.

Substitute now $\boldsymbol{u}(t)$ from (8.47) to (8.39)

$$\dot{\boldsymbol{x}}(t) = \boldsymbol{A}\boldsymbol{x}(t) - \boldsymbol{B}\boldsymbol{R}^{-1}\boldsymbol{B}^T\boldsymbol{\lambda}(t) \tag{8.49}$$

Let us denote

$$\boldsymbol{S} = \boldsymbol{B}\boldsymbol{R}^{-1}\boldsymbol{B}^T \tag{8.50}$$

Matrix $\boldsymbol{S}$ has dimensions $n \times n$. Equations (8.49), (8.44) can now be written as

$$\begin{pmatrix} \dot{\boldsymbol{x}}(t) \\ -- \\ \dot{\boldsymbol{\lambda}}(t) \end{pmatrix} = \left(\begin{array}{c|c} \boldsymbol{A} & -\boldsymbol{S} \\ \hline -\boldsymbol{Q} & -\boldsymbol{A}^T \end{array}\right) \begin{pmatrix} \boldsymbol{x}(t) \\ -- \\ \boldsymbol{\lambda}(t) \end{pmatrix} \tag{8.51}$$

This represents $2n$ of linear differential equations with constant coefficients. Its solution can be found if $2n$ initial conditions are known. In our case there are n initial conditions $\boldsymbol{x}(0) = \boldsymbol{x}_0$ and n terminal conditions on the adjoint vector given by (8.45).

Let $\boldsymbol{\Phi}(t, t_0)$ be the state transition matrix of the system (8.51) with dimensions $2n \times 2n$ and $t_0 = 0$. If $\boldsymbol{\lambda}(0)$ denotes the unknown initial value of the adjoint vector, the solution of (8.51) is of the form

$$\begin{pmatrix} \boldsymbol{x}(t) \\ \boldsymbol{\lambda}(t) \end{pmatrix} = \boldsymbol{\Phi}(t, t_0) \begin{pmatrix} \boldsymbol{x}(t_0) \\ \boldsymbol{\lambda}(t_0) \end{pmatrix} \tag{8.52}$$

At $t = t_f$ it holds

$$\begin{pmatrix} \boldsymbol{x}(t_f) \\ \boldsymbol{\lambda}(t_f) \end{pmatrix} = \boldsymbol{\Phi}(t_f, t) \begin{pmatrix} \boldsymbol{x}(t) \\ \boldsymbol{\lambda}(t) \end{pmatrix} \tag{8.53}$$

Let us divide the matrix $\boldsymbol{\Phi}(t_f, t)$ to four matrices of dimensions $n \times n$

$$\boldsymbol{\Phi}(t_f, t) = \left(\begin{array}{c|c} \boldsymbol{\Phi}_{11}(t_f, t) & \boldsymbol{\Phi}_{12}(t_f, t) \\ \hline \boldsymbol{\Phi}_{21}(t_f, t) & \boldsymbol{\Phi}_{22}(t_f, t) \end{array}\right) \tag{8.54}$$

Equation (8.53) can now be written as

$$\boldsymbol{x}(t_f) = \boldsymbol{\Phi}_{11}(t_f, t)\boldsymbol{x}(t) + \boldsymbol{\Phi}_{12}(t_f, t)\boldsymbol{\lambda}(t) \tag{8.55}$$

$$\boldsymbol{\lambda}(t_f) = \boldsymbol{\Phi}_{21}(t_f, t)\boldsymbol{x}(t) + \boldsymbol{\Phi}_{22}(t_f, t)\boldsymbol{\lambda}(t) = \boldsymbol{Q}_{t_f}\boldsymbol{x}(t_f) \tag{8.56}$$

After some manipulations from (8.55) and (8.56) follows

$$\boldsymbol{\lambda}(t) = [\boldsymbol{\Phi}_{22}(t_f, t) - \boldsymbol{Q}_{t_f}\boldsymbol{\Phi}_{12}(t_f, t)]^{-1}[\boldsymbol{Q}_{t_f}\boldsymbol{\Phi}_{11}(t_f, t) - \boldsymbol{\Phi}_{21}(t_f, t)]\boldsymbol{x}(t) \tag{8.57}$$

provided that the inverse matrix in (8.57) exists. Equation (8.57) says that the adjoint vector $\boldsymbol{\lambda}(t)$ and the state vector $\boldsymbol{x}(t)$ are related as

$$\boldsymbol{\lambda}(t) = \boldsymbol{P}(t)\boldsymbol{x}(t) \tag{8.58}$$

where

$$\boldsymbol{P}(t) = [\boldsymbol{\Phi}_{22}(t_f, t) - \boldsymbol{Q}_{t_f}\boldsymbol{\Phi}_{12}(t_f, t)]^{-1}[\boldsymbol{Q}_{t_f}\boldsymbol{\Phi}_{11}(t_f, t) - \boldsymbol{\Phi}_{21}(t_f, t)] \tag{8.59}$$

Equation (8.58) determines the function (8.42) of the optimal feedback control law minimising the cost function I

$$\boldsymbol{u}(t) = -\boldsymbol{R}^{-1}\boldsymbol{B}^T\boldsymbol{P}(t)\boldsymbol{x}(t) \tag{8.60}$$

It can be proved that $\boldsymbol{P}(t)$ exists for any time t where $t_0 \leq t \leq t_f$. However, to determine $\boldsymbol{P}(t)$ from (8.59) is rather difficult. Another possibility of finding it is to derive a differential equation with solution $\boldsymbol{P}(t)$. This equation will be derived as follows.

Suppose that solutions of (8.51) are tied together by equation (8.58) for $t \in \langle t_0, t_f \rangle$.

Differentiating (8.58) with respect to time yields

$$\dot{\boldsymbol{\lambda}}(t) = \dot{\boldsymbol{P}}(t)\boldsymbol{x}(t) + \boldsymbol{P}(t)\dot{\boldsymbol{x}}(t) \tag{8.61}$$

Substituting $\boldsymbol{u}(t)$ from (8.47) and $\boldsymbol{\lambda}(t)$ from (8.58) into (8.39) gives

$$\dot{\boldsymbol{x}}(t) = \boldsymbol{A}\boldsymbol{x}(t) - \boldsymbol{B}\boldsymbol{R}^{-1}\boldsymbol{B}^T\boldsymbol{P}(t)\boldsymbol{x}(t) \tag{8.62}$$

Equations (8.61), (8.62) give

$$\dot{\boldsymbol{\lambda}}(t) = [\dot{\boldsymbol{P}}(t) + \boldsymbol{P}(t)\boldsymbol{A} - \boldsymbol{P}(t)\boldsymbol{B}\boldsymbol{R}^{-1}\boldsymbol{B}^T\boldsymbol{P}(t)]\boldsymbol{x}(t) \tag{8.63}$$

From (8.44) and (8.58) yields

$$\dot{\boldsymbol{\lambda}}(t) = [-\boldsymbol{Q} - \boldsymbol{A}^T \boldsymbol{P}(t)]\boldsymbol{x}(t) \tag{8.64}$$

Equating the right-hand sides of (8.63) and (8.64) gives the relation

$$[\dot{\boldsymbol{P}}(t) + \boldsymbol{P}(t)\boldsymbol{A} - \boldsymbol{P}(t)\boldsymbol{B}\boldsymbol{R}^{-1}\boldsymbol{B}^T\boldsymbol{P}(t) + \boldsymbol{A}^T\boldsymbol{P}(t) + \boldsymbol{Q}]\boldsymbol{x}(t) = \boldsymbol{0}, \quad t \in \langle t_0, t_f \rangle \tag{8.65}$$

As the vector $\boldsymbol{x}(t)$ is a solution of the homogeneous equation (8.62), matrix $\boldsymbol{P}(t)$ obeys the following differential equation

$$\frac{\mathrm{d}\boldsymbol{P}(t)}{\mathrm{d}t} + \boldsymbol{P}(t)\boldsymbol{A} + \boldsymbol{A}^T\boldsymbol{P}(t) - \boldsymbol{P}(t)\boldsymbol{B}\boldsymbol{R}^{-1}\boldsymbol{B}^T\boldsymbol{P}(t) = -\boldsymbol{Q} \tag{8.66}$$

Comparing the expression

$$\boldsymbol{\lambda}(t_f) = \boldsymbol{P}(t_f)\boldsymbol{x}(t_f) \tag{8.67}$$

and expression (8.45) yields a terminal condition needed to solve equation (8.66) of the form

$$\boldsymbol{P}(t_f) = \boldsymbol{Q}_{t_f} \tag{8.68}$$

Equation (8.66) is a Riccati matrix differential equation. Its solution exists and is unique. It can be shown that the matrix $\boldsymbol{P}(t)$ is positive definite and symmetric

$$\boldsymbol{P}(t) = \boldsymbol{P}^T(t) \tag{8.69}$$

Further, optimal control for the system (8.39) exists and is unique for the cost function (8.41). It is given by the equation

$$\boldsymbol{u}(t) = -\boldsymbol{K}(t)\boldsymbol{x}(t) \tag{8.70}$$

where

$$\boldsymbol{K}(t) = \boldsymbol{R}^{-1}\boldsymbol{B}^T\boldsymbol{P}(t) \tag{8.71}$$

It is important to realise that the matrix $\boldsymbol{P}(t)$ can be calculated beforehand.

Controllability of the system (8.39) is not the necessary condition for the fact that the optimal control is given by the control law (8.70). This is because the influence of uncontrollable elements in the cost function I is always finite if the time interval of optimisation is finite. If, on the other hand, the final time t_f goes to infinity, controllability is required to guarantee the finite value of I.

Example 8.2: *Optimal feedback control of a heat exchanger* www

Consider the heat exchanger from Example 8.1 and the problem of its optimal feedback control. The system state-space matrices are $\boldsymbol{A} = -1$, $\boldsymbol{B} = 1$ and the cost function is defined by $\boldsymbol{Q}_{t_f} = 0$, $\boldsymbol{Q} = 2$, $\boldsymbol{R} = 2r' = r$. Optimal feedback control is given as

$$u(t) = -\frac{1}{r}P(t)x(t)$$

where $P(t)$ is the solution of the differential equation of the form

$$\frac{\mathrm{d}P(t)}{\mathrm{d}t} - 2P(t) - \frac{P^2(t)}{r} = -2$$

with terminal condition

$$P(t_f) = 0$$

Fig. 8.2 shows trajectories of P for various values of r if $t_f = 1$.

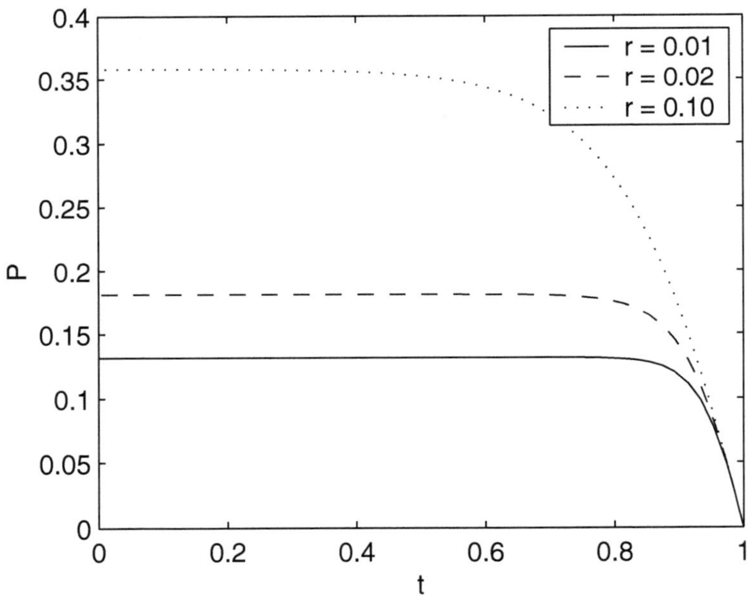

Fig. 8.2. Trajectories of $P(t)$ for various values of r

Example 8.2 shows a very important fact. If the matrix Riccati equation starts at time $t_f = \infty$ then its solutions given by elements of matrix $\boldsymbol{P}(t)$ will be constant. Such a matrix $\boldsymbol{P}$ containing steady-state solution of the differential matrix Riccati equation can be used in optimal control for all finite times with initial time equal to $t = 0$.

Therefore, optimal control design can be simplified considerably in the case of infinite final time as it is shown below.

Consider the system (8.39) with the initial condition (8.40) and the cost function

$$I = \frac{1}{2}\int_0^\infty \left(\boldsymbol{x}^T(t)\boldsymbol{Q}\boldsymbol{x}(t) + \boldsymbol{u}^T(t)\boldsymbol{R}\boldsymbol{u}(t)\right) \mathrm{d}t \tag{8.72}$$

where $\boldsymbol{Q}$ is a real symmetric positive semidefinite weighting matrix and $\boldsymbol{R}$ is a real symmetric positive definite weighting matrix.

Solution of the optimisation problem, i. e. minimisation of I for any $\boldsymbol{x}_0$ satisfies the feedback control law

$$\boldsymbol{u}(t) = -\boldsymbol{K}\boldsymbol{x}(t) \tag{8.73}$$

where

$$\boldsymbol{K} = \boldsymbol{R}^{-1}\boldsymbol{B}^T\boldsymbol{P} \tag{8.74}$$

and where $\boldsymbol{P}$ is a symmetric positive semidefinite solution of the matrix Riccati equation

$$\boldsymbol{P}\boldsymbol{A} + \boldsymbol{A}^T\boldsymbol{P} - \boldsymbol{P}\boldsymbol{B}\boldsymbol{R}^{-1}\boldsymbol{B}^T\boldsymbol{P} = -\boldsymbol{Q} \tag{8.75}$$

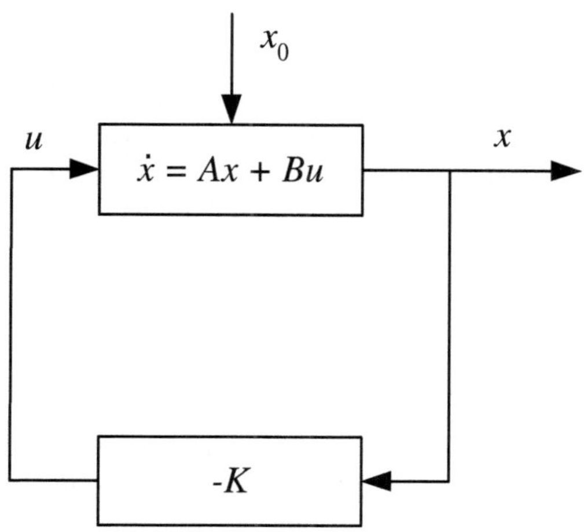

Fig. 8.3. Optimal feedback control

Equation (8.73) is easily implementable. It is a feedback control law. As $\boldsymbol{K}$ is a constant matrix, the controller implements proportional action. This controller is called Linear Quadratic Regulator (LQR). This says that the controlled system is linear, the cost function is quadratic and the controller regulates the system in the neighbourhood of the setpoint $\boldsymbol{x}_w = \mathbf{0}$. Block scheme of feedback LQR control is shown in Fig. 8.3.

Example 8.3: *Optimal feedback control of two heat exchangers in series* WW

Consider two heat exchangers (Fig. 8.4) where a liquid is heated.

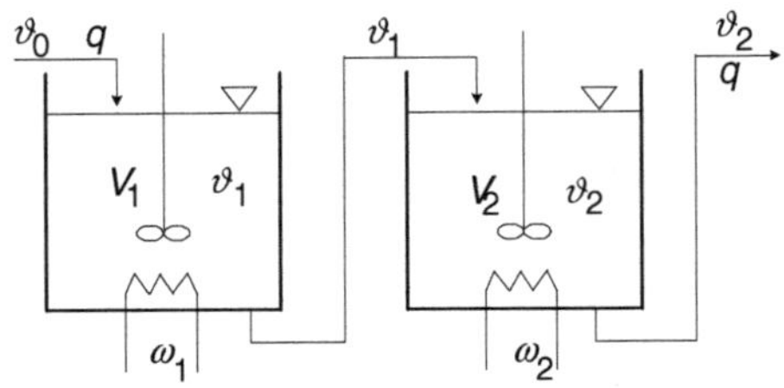

Fig. 8.4. Two heat exchangers in series

Assume that heat flows from heat sources into liquid are independent from liquid temperature. Further assume an ideal liquid mixing and zero heat losses. We neglect accumulation ability of exchangers walls. Hold-ups of exchangers, as well as flow rates and liquid specific heat capacity are constant. Under these assumptions the mathematical model of the exchangers is given as

$$\frac{V_1}{q}\frac{\mathrm{d}\vartheta_1}{\mathrm{d}t'} + \vartheta_1 = \vartheta_0 + \frac{\omega_1}{q\rho c_p}$$

$$\frac{V_2}{q}\frac{\mathrm{d}\vartheta_2}{\mathrm{d}t'} + \vartheta_2 = \vartheta_1 + \frac{\omega_2}{q\rho c_p}$$

where ϑ_1 is temperature in the first exchanger, ϑ_2 – temperature in the second exchanger, t' – time, ϑ_0 – liquid temperature in the first tank inlet stream, ω_1, ω_2 – heat input, q – volumetric flowrate of liquid, ρ – liquid density, V_1, V_2 – liquid volume, c_p – specific heat capacity.

Denote

$$\vartheta_{u1} = \frac{\omega_1}{q\rho c_p}, \quad \vartheta_{u2} = \frac{\omega_2}{q\rho c_p}, \quad T_1 = \frac{V_1}{q}, \quad T_2 = \frac{V_2}{q}$$

$\vartheta_{u1}, \vartheta_{u2}$ are in temperature units and T_1, T_2 are time constants. The process inputs are temperatures $\vartheta_0, \vartheta_{u1}$, and ϑ_{u2}. The state variables are ϑ_1 and ϑ_2. Temperature ϑ_{u1} will be assumed as the manipulated variable. Assume that the manipulated input is constant for a sufficiently long time and equal to ϑ_{u10}. If the inlet temperature to the first exchanger is constant $\vartheta_0 = \vartheta_{00}$ and the heat input to the second exchanger will be constant as well ($\vartheta_{u2} = \vartheta_{u20}$) then the heat exchangers will be in a steady-state described by equations

$$\vartheta_{10} = \vartheta_{00} + \vartheta_{u10}, \quad \vartheta_{20} = \vartheta_{10} + \vartheta_{u20}$$

Let us now assume a new steady-state characterised by the manipulated variable ϑ_{u11} defined by equations

$$\vartheta_{11} = \vartheta_{00} + \vartheta_{u11}, \quad \vartheta_{21} = \vartheta_{11} + \vartheta_{u20}$$

The problem of optimal control is now to find a way to shift the system in the original steady-state with the manipulated variable ϑ_{u10} to the new steady-state with ϑ_{u11} in such a way that some cost function be minimal. At first we define new state variables as dimensionless deviation variables from the final steady-state

$$x_1(t') = \frac{\vartheta_1(t') - \vartheta_{11}}{\vartheta_{u11} - \vartheta_{00}}$$

$$x_2(t') = \frac{\vartheta_2(t') - \vartheta_{21}}{\vartheta_{u20} - \vartheta_{11}}$$

A new manipulated variable is defined as

$$u_1(t') = \frac{\vartheta_{u1}(t') - \vartheta_{u11}}{\vartheta_{u11} - \vartheta_{00}}$$

Further, a dimensionless time variable is given as

$$t = \frac{q}{V_1} t'$$

This gives for the heat exchanger model

$$\frac{\mathrm{d}x_1(t)}{\mathrm{d}t} + x_1(t) = u_1(t)$$

$$k_1 \frac{\mathrm{d}x_2(t)}{\mathrm{d}t} + x_2(t) = k_2 x_1(t)$$

where $k_1 = V_2/V_1$ a $k_2 = (\vartheta_{u11} - \vartheta_{00})/(\vartheta_{u20} - \vartheta_{11})$.
Transformed initial conditions are given as

$$x_1(0) = x_{10} = \frac{\vartheta_{10} - \vartheta_{11}}{\vartheta_{u11} - \vartheta_{00}}$$

$$x_2(0) = x_{20} = \frac{\vartheta_{20} - \vartheta_{21}}{\vartheta_{u20} - \vartheta_{11}}$$

Let us now design an optimal feedback controller that minimises the cost function

$$I = \frac{1}{2} \int_0^\infty \left[\boldsymbol{x}^T(t) \boldsymbol{Q} \boldsymbol{x}(t) + r u_1(t) \right] \mathrm{d}t$$

where $\boldsymbol{x} = (x_1 \; x_2)^T$

$$\boldsymbol{Q} = \begin{pmatrix} q_{11} & 0 \\ 0 & q_{22} \end{pmatrix}, \quad \boldsymbol{R} = r$$

The controlled system is defined as

$$A = \begin{pmatrix} -1 & 0 \\ \dfrac{k_2}{k_1} & -\dfrac{1}{k_1} \end{pmatrix}, \quad B = \begin{pmatrix} 1 \\ 0 \end{pmatrix}$$

Optimal feedback is thus given as

$$u_1(t) = -\frac{1}{r}\boldsymbol{B}^T \boldsymbol{P}\boldsymbol{x}(t)$$

and $\boldsymbol{P}$ is solution of the equation

$$\boldsymbol{P}\boldsymbol{A} + \boldsymbol{A}^T\boldsymbol{P} - \boldsymbol{P}\boldsymbol{B}\boldsymbol{R}^{-1}\boldsymbol{B}^T\boldsymbol{P} = -\boldsymbol{Q}$$

where

$$\boldsymbol{P} = \begin{pmatrix} p_{11} & p_{12} \\ p_{12} & p_{22} \end{pmatrix}$$

Fig. 8.6 shows optimal trajectories x_1, x_2, and u_1 of the heat exchangers (Fig. 8.4) for $k_1 = 1$, $k_2 = -1/5$, $q_{11} = q_{22} = 1$, and $r = 1$.
Matrix

$$\boldsymbol{K} = (k_{11}\ k_{12}) = \frac{1}{r}(p_{11}\ p_{12})$$

was calculated using MATLAB (program 8.1) and trajectories x_1, x_2, and u_1 in Fig. 8.6 were obtained from SIMULINK program shown in Fig. 8.5 with initial conditions $x_{10} = 1$, $x_{20} = -1/3$.

Program 8.1 (Program to calculate LQ gains for Example 8.3)

```
% LQR feedback for the heat exchanger
% program: lqvym2rm.m

x0 = [1;-1/3];

A = [-1 0; -0.2 -1];
B = [1; 0];
C=eye(2);
D=zeros(2,1);

Q=eye(2);
R=1;

[K,S,E]=LQR(A,B,Q,R);
```

Consider now the system shown in Fig. 8.3. It is described by

$$\dot{\boldsymbol{x}}(t) = \boldsymbol{A}\boldsymbol{x}(t) - \boldsymbol{B}\boldsymbol{K}\boldsymbol{x}(t) \tag{8.76}$$

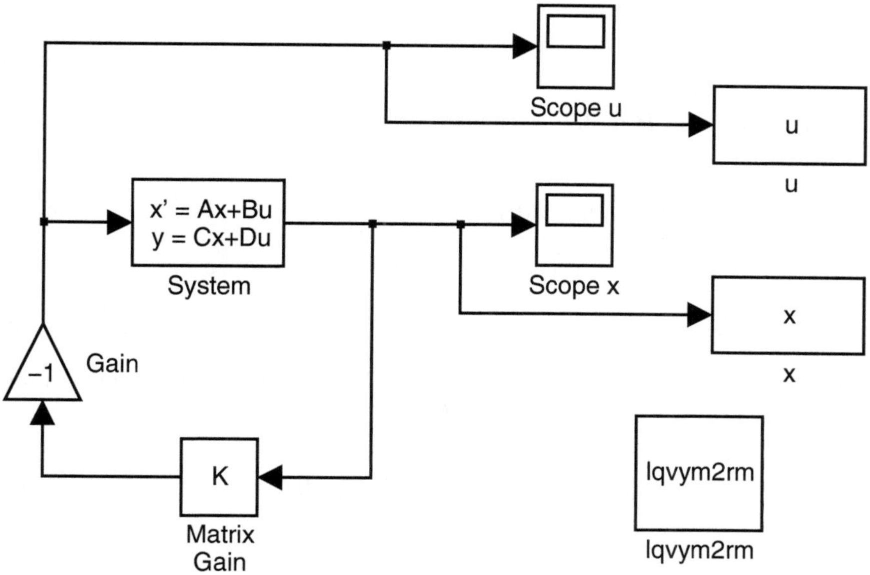

Fig. 8.5. Simulink program to simulate feedback optimal control in Example 8.3

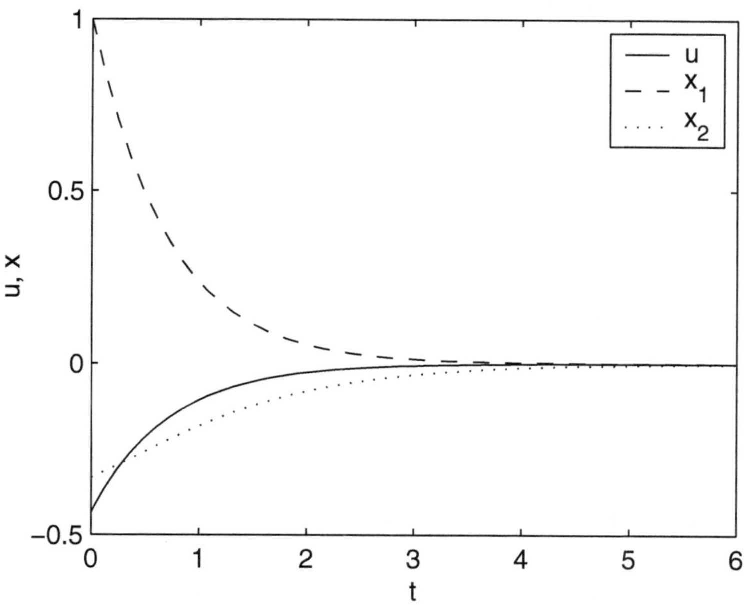

Fig. 8.6. LQ optimal trajectories x_1, x_2, u_1 of heat exchangers

with initial condition at time $t_0 = 0$

$$\boldsymbol{x}(0) = \boldsymbol{x}_0 \tag{8.77}$$

where $\boldsymbol{K} = \boldsymbol{R}^{-1}\boldsymbol{B}^T\boldsymbol{P}$ and $\boldsymbol{P}$ is the solution of (8.75).

Equation (8.76) can also be written as

$$\dot{\boldsymbol{x}}(t) = (\boldsymbol{A} - \boldsymbol{B}\boldsymbol{K})\,\boldsymbol{x}(t), \quad \boldsymbol{x}(0) = \boldsymbol{x}_0 \tag{8.78}$$

The transition matrix of the closed-loop system is given as

$$\bar{\boldsymbol{A}} = \boldsymbol{A} - \boldsymbol{B}\boldsymbol{K} \tag{8.79}$$

The system (8.78) is solution of the optimisation problem based on minimisation of the cost function (8.72). This cost function can be written as

$$I = \frac{1}{2}\int_0^\infty \left(\boldsymbol{x}^T(t)\boldsymbol{Q}\boldsymbol{x}(t) + \boldsymbol{x}^T(t)\boldsymbol{K}^T\boldsymbol{R}\boldsymbol{K}\boldsymbol{x}(t)\right)\mathrm{d}t \tag{8.80}$$

or

$$I = \frac{1}{2}\int_0^\infty \boldsymbol{x}^T(t)\left(\boldsymbol{Q} + \boldsymbol{K}^T\boldsymbol{R}\boldsymbol{K}\right)\boldsymbol{x}(t)\mathrm{d}t \tag{8.81}$$

and can also be formulated as follows

$$I = \frac{1}{2}\int_0^\infty \boldsymbol{x}_0^T \mathrm{e}^{\bar{\boldsymbol{A}}^T t}\bar{\boldsymbol{Q}}\mathrm{e}^{\bar{\boldsymbol{A}}t}\boldsymbol{x}_0\mathrm{d}t \tag{8.82}$$

or

$$I = \frac{1}{2}\boldsymbol{x}_0^T\boldsymbol{P}\boldsymbol{x}_0 \tag{8.83}$$

where

$$\bar{\boldsymbol{Q}} = \boldsymbol{Q} + \boldsymbol{K}^T\boldsymbol{R}\boldsymbol{K} \tag{8.84}$$

and

$$\boldsymbol{P} = \int_0^\infty \mathrm{e}^{\bar{\boldsymbol{A}}^T t}\bar{\boldsymbol{Q}}\mathrm{e}^{\bar{\boldsymbol{A}}t}\mathrm{d}t \tag{8.85}$$

Matrix $\boldsymbol{P}$ can be integrated by parts yielding

$$\boldsymbol{P} = \mathrm{e}^{\bar{\boldsymbol{A}}^T t}\bar{\boldsymbol{Q}}\bar{\boldsymbol{A}}^{-1}\mathrm{e}^{\bar{\boldsymbol{A}}t}\Big|_0^\infty - \int_0^\infty \bar{\boldsymbol{A}}^T\mathrm{e}^{\bar{\boldsymbol{A}}^T t}\bar{\boldsymbol{Q}}\bar{\boldsymbol{A}}^{-1}\mathrm{e}^{\bar{\boldsymbol{A}}^T t}\mathrm{d}t \tag{8.86}$$

If matrix $\bar{\boldsymbol{A}}$ is stable then

$$\boldsymbol{P} = -\bar{\boldsymbol{Q}}\bar{\boldsymbol{A}}^{-1} - \bar{\boldsymbol{A}}^T\int_0^\infty \mathrm{e}^{\bar{\boldsymbol{A}}^T t}\bar{\boldsymbol{Q}}\mathrm{e}^{\bar{\boldsymbol{A}}t}\mathrm{d}t\bar{\boldsymbol{A}}^{-1} \tag{8.87}$$

and after some manipulations

$$\bar{A}^T P + P\bar{A} = -\bar{Q} \tag{8.88}$$

This is the Lyapunov equation. It is known that for an arbitrary symmetric positive definite matrix $\bar{Q}$ there a is symmetric positive definite matrix P that is solution of this equation if matrix $\bar{A}$ is asymptotically stable.

When $\bar{A}$ is back-substituted into (8.88) from (8.79) and $\bar{Q}$ from (8.84), then we obtain

$$(A - BK)^T P + P(A - BK) = -Q - \left(R^{-1}B^T P\right)^T R\left(R^{-1}B^T P\right) \tag{8.89}$$

This equation can be easily rewritten to the form (8.75).

It is important to observe that the optimal feedback control shown in Fig. 8.3 places optimally poles of the closed-loop system.

Pole Placement (PP) represents such feedback control design where the matrix K in (8.79) is chosen so that the matrix $\bar{A}$ is asymptotically stable. LQR control is in this aspect only a special case of the PP control design.

Consider now a system

$$\dot{x}(t) = Ax(t) + Bu(t), \quad x(0) = x_0 \tag{8.90}$$

$$y(t) = Cx(t) \tag{8.91}$$

and cost function

$$I = \frac{1}{2}\int_0^\infty \left(y^T(t)Q_y y(t) + u^T(t)Ru(t)\right) \mathrm{d}t \tag{8.92}$$

Substituting equation (8.91) into (8.92) yields

$$I = \frac{1}{2}\int_0^\infty \left(x^T(t)C^T Q_y Cx(t) + u^T(t)Ru(t)\right) \mathrm{d}t \tag{8.93}$$

Denote as

$$Q = C^T Q_y C \tag{8.94}$$

then the cost function (8.93) is of the same form as (8.72) and the output regulation problem has been transformed into the classical LQR problem. Solution of the optimal problem, i. e. minimisation of I from (8.93) for arbitrary x_0 guarantees the state feedback given by equation (8.73). The block diagram of output LQR is shown in Fig. 8.7.

8.3 Optimal Tracking, Servo Problem, and Disturbance Rejection

Previous section was devoted to the problem when a system was to be steered from some initial state to a new state in some optimal way. In fact, this is only a special case of a general problem of optimal setpoint tracking.

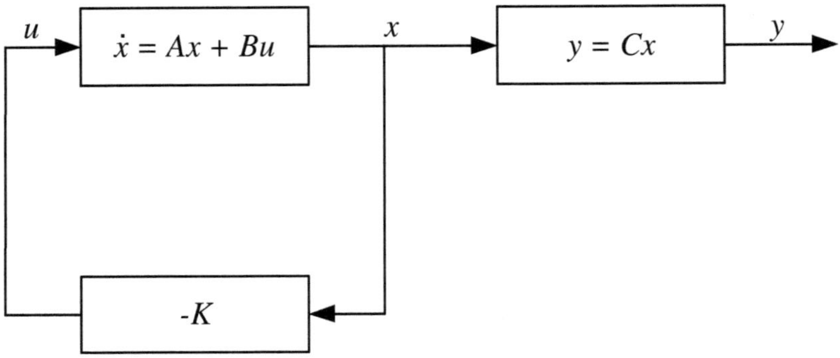

Fig. 8.7. Optimal LQ output regulation

The tracking problem of some desired trajectories can be divided into two subproblems. If desired trajectories are prescribed functions of time then we speak about the *tracking problem.* If the process outputs should follow some class of desired trajectories then we speak about the *servo problem.*

LQ design leads to a proportional feedback control. However, such design results in the steady-state control error if setpoint changes or disturbances occur. If the zero steady-state control error is desired, it is necessary to modify the controller with *integral action.*

8.3.1 Tracking Problem

Consider a controllable and observable linear system

$$\dot{\boldsymbol{x}}(t) = \boldsymbol{A}\boldsymbol{x}(t) + \boldsymbol{B}\boldsymbol{u}(t), \quad \boldsymbol{x}(0) = \boldsymbol{x}_0 \tag{8.95}$$

$$\boldsymbol{y}(t) = \boldsymbol{C}\boldsymbol{x}(t) \tag{8.96}$$

Let $\boldsymbol{w}(t)$ be a vector of setpoint variables of dimensions r. Our aim is to control the system in such a way that the control error vector

$$\boldsymbol{e}(t) = \boldsymbol{w}(t) - \boldsymbol{y}(t) \tag{8.97}$$

should be "close" to zero with a minimum control effort.

The cost function that is to be minimised is of the form

$$I = \frac{1}{2}\boldsymbol{e}^T(t_f)\boldsymbol{Q}_{yt_f}\boldsymbol{e}(t_f) + \frac{1}{2}\int_0^{t_f} \left(\boldsymbol{e}^T(t)\boldsymbol{Q}_y\boldsymbol{e}(t) + \boldsymbol{u}^T(t)\boldsymbol{R}\boldsymbol{u}(t)\right) \mathrm{d}t \tag{8.98}$$

Assume that t_f is fixed, $\boldsymbol{Q}_{yt_f}$ and $\boldsymbol{Q}_y$ are real symmetric positive semidefinite matrices, $\boldsymbol{R}$ is a real symmetric positive definite matrix.

Hamiltonian is of the form

$$H = \frac{1}{2}(\boldsymbol{w} - \boldsymbol{C}\boldsymbol{x})^T \boldsymbol{Q}_y (\boldsymbol{w} - \boldsymbol{C}\boldsymbol{x}) + \frac{1}{2}\boldsymbol{u}^T\boldsymbol{R}\boldsymbol{u} + \boldsymbol{\lambda}^T(\boldsymbol{A}\boldsymbol{x} + \boldsymbol{B}\boldsymbol{u}) \tag{8.99}$$

Optimality condition

$$\frac{\partial H}{\partial \boldsymbol{u}} = \boldsymbol{0}$$

gives

$$\boldsymbol{u}(t) = -\boldsymbol{R}^{-1}\boldsymbol{B}^T\boldsymbol{\lambda}(t) \tag{8.100}$$

Adjoint vector $\boldsymbol{\lambda}(t)$ is the solution of the differential equation

$$\dot{\boldsymbol{\lambda}} = -\frac{\partial H}{\partial \boldsymbol{x}}$$

or

$$\frac{\mathrm{d}\boldsymbol{\lambda}(t)}{\mathrm{d}t} = \boldsymbol{C}^T\boldsymbol{Q}_y\left(\boldsymbol{w}(t) - \boldsymbol{C}\boldsymbol{x}(t)\right) - \boldsymbol{A}^T\boldsymbol{\lambda}(t) \tag{8.101}$$

Similarly, as (8.58) was derived, we can write

$$\boldsymbol{\lambda}(t) = \boldsymbol{P}(t)\boldsymbol{x}(t) - \gamma(t) \tag{8.102}$$

Derivative of (8.102) with respect to time gives

$$\frac{\mathrm{d}\boldsymbol{\lambda}(t)}{\mathrm{d}t} = \frac{\mathrm{d}\boldsymbol{P}(t)}{\mathrm{d}t}\boldsymbol{x}(t) + \boldsymbol{P}(t)\frac{\mathrm{d}\boldsymbol{x}(t)}{\mathrm{d}t} - \frac{\mathrm{d}\gamma(t)}{\mathrm{d}t} \tag{8.103}$$

From (8.95) and (8.100) follows

$$\dot{\boldsymbol{x}}(t) = \boldsymbol{A}\boldsymbol{x}(t) - \boldsymbol{B}\boldsymbol{R}^{-1}\boldsymbol{B}^T\boldsymbol{\lambda}(t) \tag{8.104}$$

Substituting for $\boldsymbol{\lambda}(t)$ from (8.102) gives

$$\dot{\boldsymbol{x}}(t) = \boldsymbol{A}\boldsymbol{x}(t) - \boldsymbol{B}\boldsymbol{R}^{-1}\boldsymbol{B}^T\boldsymbol{P}(t)\boldsymbol{x}(t) + \boldsymbol{B}\boldsymbol{R}^{-1}\boldsymbol{B}^T\gamma(t) \tag{8.105}$$

Differential equation (8.103) can now be written as

$$\dot{\boldsymbol{\lambda}}(t) = \left[\dot{\boldsymbol{P}}(t) + \boldsymbol{P}(t)\boldsymbol{A} - \boldsymbol{P}(t)\boldsymbol{B}\boldsymbol{R}^{-1}\boldsymbol{B}^T\boldsymbol{P}(t)\right]\boldsymbol{x}(t) + \boldsymbol{P}\boldsymbol{B}\boldsymbol{R}^{-1}\boldsymbol{B}^T\gamma(t) - \dot{\gamma}(t) \tag{8.106}$$

This can be simplified using (8.101) and (8.102) as

$$\dot{\boldsymbol{\lambda}}(t) = \left[-\boldsymbol{C}^T\boldsymbol{Q}_y\boldsymbol{C} - \boldsymbol{A}^T\boldsymbol{P}(t)\right]\boldsymbol{x}(t) + \boldsymbol{A}^T\gamma(t) + \boldsymbol{C}^T\boldsymbol{Q}_y\boldsymbol{w}(t) \tag{8.107}$$

As the optimal solution exists, equations (8.106) and (8.107) hold for any $\boldsymbol{x}(t)$ and $\boldsymbol{w}(t)$. From this follows that matrix $\boldsymbol{P}(t)$ with dimensions $n \times n$ has to satisfy the equation

$$\dot{\boldsymbol{P}}(t) = -\boldsymbol{P}(t)\boldsymbol{A} - \boldsymbol{A}^T\boldsymbol{P}(t) + \boldsymbol{P}(t)\boldsymbol{B}\boldsymbol{R}^{-1}\boldsymbol{B}^T\boldsymbol{P}(t) - \boldsymbol{C}^T\boldsymbol{Q}_y\boldsymbol{C} \tag{8.108}$$

Further, for vector $\gamma(t)$ with dimension n holds

$$\dot{\gamma}(t) = \left[\boldsymbol{P}(t)\boldsymbol{B}\boldsymbol{R}^{-1}\boldsymbol{B}^T - \boldsymbol{A}^T\right]\gamma(t) - \boldsymbol{C}^T\boldsymbol{Q}_y\boldsymbol{w}(t) \tag{8.109}$$

Final conditions for the adjoint variables are specified by (8.102)

$$\boldsymbol{\lambda}(t_f) = \boldsymbol{P}(t_f)\boldsymbol{x}(t_f) - \gamma(t_f) \tag{8.110}$$

and the principle of minimum gives

$$\boldsymbol{\lambda}(t_f) = \frac{\partial}{\partial \boldsymbol{x}(t_f)}\left[\frac{1}{2}\boldsymbol{e}^T(t_f)\boldsymbol{Q}_{yt_f}\boldsymbol{e}(t_f)\right] \tag{8.111}$$

$$= \boldsymbol{C}^T\boldsymbol{Q}_{yt_f}\boldsymbol{C}\boldsymbol{x}(t_f) - \boldsymbol{C}^T\boldsymbol{Q}_{yt_f}\boldsymbol{w}(t_f) \tag{8.112}$$

Equations (8.110) and (8.112) hold for any $\boldsymbol{x}(t_f)$ and $\boldsymbol{w}(t_f)$. Thus

$$\boldsymbol{P}(t_f) = \boldsymbol{C}^T\boldsymbol{Q}_{yt_f}\boldsymbol{C} \tag{8.113}$$

$$\gamma(t_f) = \boldsymbol{C}^T\boldsymbol{Q}_{yt_f}\boldsymbol{w}(t_f) \tag{8.114}$$

The optimal state trajectory is defined by solution of linear differential equation (8.105).

The control law can be written as

$$\boldsymbol{u}(t) = \boldsymbol{R}^{-1}\boldsymbol{B}^T\left[\gamma(t) - \boldsymbol{P}(t)\boldsymbol{x}(t)\right] \tag{8.115}$$

Symmetric positive definite matrix $\boldsymbol{P}(t)$ with dimension $n \times n$ is solution of (8.108) with terminal condition (8.113). Vector $\gamma(t)$ with dimension n is solution of (8.109) with terminal condition (8.114).

To calculate $\gamma(t)$, vector $\boldsymbol{w}(t)$ is needed in the entire interval $[0, t_f]$. This is a rather impractical condition. However, if $\boldsymbol{w}(t) = 0$ and $t_f \to \infty$ optimal control law is described by (8.73) where matrix $\boldsymbol{P}$ is given by (8.75) where $\boldsymbol{Q} = \boldsymbol{C}^T\boldsymbol{Q}\boldsymbol{C}$.

8.3.2 Servo Problem

Consider now an optimal feedback control where the desired vector of output variables is generated as

$$\dot{\boldsymbol{x}}_w(t) = \boldsymbol{A}_w\boldsymbol{x}_w(t), \quad \boldsymbol{x}_w(0) = \boldsymbol{x}_{w0} \tag{8.116}$$

$$\boldsymbol{w}(t) = \boldsymbol{C}_w\boldsymbol{x}_w(t) \tag{8.117}$$

We want to ensure that $\boldsymbol{y}(t)$ is "close" to $\boldsymbol{w}(t)$ with minimal control effort. This problem can mathematically be formulated as minimisation of the cost function

$$I = \int_0^\infty \left((\boldsymbol{y}(t) - \boldsymbol{w}(t))^T\boldsymbol{Q}_y(\boldsymbol{y}(t) - \boldsymbol{w}(t)) + \boldsymbol{u}^T(t)\boldsymbol{R}\boldsymbol{u}(t)\right)\mathrm{d}t \tag{8.118}$$

Substituting $\boldsymbol{y}(t)$ from (8.96) and $\boldsymbol{w}(t)$ from (8.117) into the cost gives

$$I = \int_0^\infty \left((\boldsymbol{C}\boldsymbol{x}(t) - \boldsymbol{C}_w\boldsymbol{x}_w(t))^T \boldsymbol{Q}_y (\boldsymbol{C}\boldsymbol{x}(t) - \boldsymbol{C}_w\boldsymbol{x}_w(t)) + \boldsymbol{u}^T(t)\boldsymbol{R}\boldsymbol{u}(t) \right) \mathrm{d}t \tag{8.119}$$

or, after some manipulations

$$I = \int_0^\infty \left(\begin{pmatrix} \boldsymbol{x}^T(t) & \boldsymbol{x}_w^T(t) \end{pmatrix} \begin{pmatrix} \boldsymbol{C}^T\boldsymbol{Q}_y\boldsymbol{C} & -\boldsymbol{C}^T\boldsymbol{Q}_y\boldsymbol{C}_w \\ -\boldsymbol{C}_w^T\boldsymbol{Q}_y\boldsymbol{C} & \boldsymbol{C}_w^T\boldsymbol{Q}_y\boldsymbol{C}_w \end{pmatrix} \begin{pmatrix} \boldsymbol{x}(t) \\ \boldsymbol{x}_w(t) \end{pmatrix} + \boldsymbol{u}^T(t)\boldsymbol{R}\boldsymbol{u}(t) \right) \mathrm{d}t \tag{8.120}$$

To solve the optimal control problem, an expanded controlled system is assumed

$$\begin{pmatrix} \dot{\boldsymbol{x}}(t) \\ \dot{\boldsymbol{x}}_w(t) \end{pmatrix} = \begin{pmatrix} \boldsymbol{A} & \boldsymbol{0} \\ \boldsymbol{0} & \boldsymbol{A}_w \end{pmatrix} \begin{pmatrix} \boldsymbol{x}(t) \\ \boldsymbol{x}_w(t) \end{pmatrix} + \begin{pmatrix} \boldsymbol{B} \\ \boldsymbol{0} \end{pmatrix} \boldsymbol{u}(t) \tag{8.121}$$

From previous derivations follows that the solution of this problem leads to the Riccati equation of the form

$$\begin{pmatrix} \boldsymbol{A} & \boldsymbol{0} \\ \boldsymbol{0} & \boldsymbol{A}_w \end{pmatrix} \boldsymbol{P} + \boldsymbol{P} \begin{pmatrix} \boldsymbol{A} & \boldsymbol{0} \\ \boldsymbol{0} & \boldsymbol{A}_w \end{pmatrix} - \boldsymbol{P} \begin{pmatrix} \boldsymbol{B} \\ \boldsymbol{0} \end{pmatrix} \boldsymbol{R}^{-1} \begin{pmatrix} \boldsymbol{B} \\ \boldsymbol{0} \end{pmatrix}^T \boldsymbol{P} = - \begin{pmatrix} \boldsymbol{C}^T\boldsymbol{Q}_y\boldsymbol{C} & -\boldsymbol{C}^T\boldsymbol{Q}_y\boldsymbol{C}_w \\ -\boldsymbol{C}_w^T\boldsymbol{Q}_y\boldsymbol{C} & \boldsymbol{C}_w\boldsymbol{Q}_y\boldsymbol{C}_w \end{pmatrix} \tag{8.122}$$

If matrix $\boldsymbol{P}$ is decomposed into

$$\boldsymbol{P} = \begin{pmatrix} \boldsymbol{P}_{11} & \boldsymbol{P}_{12} \\ \boldsymbol{P}_{21} & \boldsymbol{P}_{22} \end{pmatrix} \tag{8.123}$$

then the optimal control law is given as

$$\boldsymbol{u}(t) = -\boldsymbol{R}^{-1}\boldsymbol{B}^T \left[\boldsymbol{P}_{11}\boldsymbol{x}(t) + \boldsymbol{P}_{12}\boldsymbol{x}_w(t) \right]$$

8.3.3 LQ Control with Integral Action

LQ control results in proportional feedback controller. This leads to the steady-state control error in case of setpoint changes or disturbances. There are two possible ways to add integral action to the closed-loop system and to remove the steady-state error.

In the first case the LQ cost function is modified and vector $\boldsymbol{u}(t)$ is replaced by its time derivative $\dot{\boldsymbol{u}}(t)$.

Another possibility is to add new states acting as integrators to the closed-loop system. The number of integrators is equal to the dimension of the control error.

8.4 Dynamic Programming

8.4.1 Continuous-Time Systems

Bellman with his coworkers developed around 1950 a new approach to system optimisation – *dynamic programming.* This method is often used in analysis and design of automatic control systems.

Consider the vector differential equation

$$\frac{\mathrm{d}\boldsymbol{x}(t)}{\mathrm{d}t} = \boldsymbol{f}(\boldsymbol{x}(t), \boldsymbol{u}(t), t) \tag{8.124}$$

where $\boldsymbol{f}$ and $\boldsymbol{x}$ are of dimension n, vector $\boldsymbol{u}$ is of dimension m. The control vector $\boldsymbol{u}$ is within a region

$$\boldsymbol{u} \in \boldsymbol{U} \tag{8.125}$$

where $\boldsymbol{U}$ is a closed part of the Euclidean space E_m.

The cost function is given as

$$I = \int_{t_0}^{t_f} F\left(\boldsymbol{x}(t), \boldsymbol{u}(t), t\right)\mathrm{d}t \tag{8.126}$$

The terminal time $t_f > t_0$ can be free of fixed. We will consider the case with fixed time.

Let us define a new state variable

$$x_{n+1}(t) = \int_{t_0}^{t} F\left(\boldsymbol{x}(\tau), \boldsymbol{u}(\tau), \tau\right)\mathrm{d}\tau \tag{8.127}$$

The problem of minimisation of I is equivalent to minimisation of the state $x_{n+1}(t_f)$ of the system described by equations

$$\frac{\mathrm{d}\tilde{\boldsymbol{x}}(t)}{\mathrm{d}t} = \tilde{\boldsymbol{f}}\left(\tilde{\boldsymbol{x}}(t), \boldsymbol{u}(t), t\right) \tag{8.128}$$

where

$$\begin{aligned}\tilde{\boldsymbol{x}}^T &= \left(\boldsymbol{x}^T, x_{n+1}\right)\\ \tilde{\boldsymbol{f}}^T &= \left(\boldsymbol{f}^T, F\right)\end{aligned}$$

and with initial conditions

$$\tilde{\boldsymbol{x}}^T(t_0) = \left(\boldsymbol{x}^T(t_0), 0\right) \tag{8.129}$$

This includes the time optimal control with $F = 1$.

If a direct presence of time t in (8.124) is undesired, it is possible to introduce a new state variable $x_0(t) = t$ and a new differential equation

$$\frac{\mathrm{d}x_0(t)}{\mathrm{d}t} = 1 \tag{8.130}$$

with initial condition $x_0(t_0) = 0$. The new system of differential equations is then defined as

$$\frac{\mathrm{d}\boldsymbol{x}(t)}{\mathrm{d}t} = \boldsymbol{f}(\boldsymbol{x}(t), \boldsymbol{u}(t)) \tag{8.131}$$

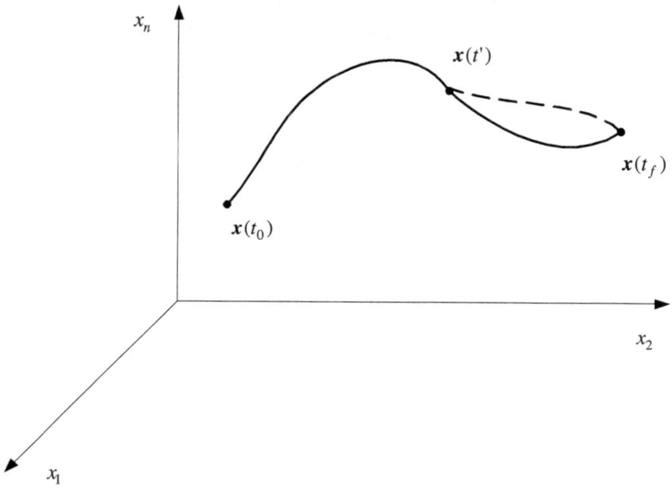

Fig. 8.8. Optimal trajectory in n-dimensional state space

Dynamic programming is based on the principle of optimality first formulated by Bellman. Let us consider the optimal trajectory in n-dimensional state space (Fig. 8.8). Position of moving point in time $t = t'(t_0 < t' < t_f)$ will be denoted by $\boldsymbol{x}(t')$. This point divides the trajectory into two parts.

The principle of optimality states that any part of the optimal trajectory from point $\boldsymbol{x}(t')$ to point $\boldsymbol{x}(t_f)$ is optimal. From this follows that for the initial point $\boldsymbol{x}(t')$ the second part of the trajectory between points $\boldsymbol{x}(t')$ and $\boldsymbol{x}(t_f)$ is optimal and does not depend on the system history, i. e. the way how the point $\boldsymbol{x}(t')$ was reached. Suppose now that this does not hold, i. e. we find a trajectory (shown in Fig. 8.8 by dashed line), where the cost function value is smaller. The overall cost function value is given as the sum of values in the both trajectory parts. This means that it is possible to find a better trajectory than the original. To do so, it is necessary to find a control vector trajectory in such a way that the first part of the state trajectory remains the same and the second coincides with the dashed line. This is in contradiction with the assumption that the original trajectory is optimal. Therefore, it is not possible to improve the second part of the trajectory and it is optimal as well.

The principle of optimality makes it possible to obtain general conditions of optimisation that are valid for both continuous-time and discrete-time systems. Although its formulation is very simple, we can derive the necessary optimisation conditions.

The principle of optimality can alternatively be formulated as follows: optimal trajectory does not depend on previous trajectory and is determined only by the initial condition and the terminal state.

Based on the principle of optimality, determination of the optimal trajectory is started at the terminal state $\boldsymbol{x}(t_f)$. We stress again that the trajectory that ends in the optimal terminal state is optimal. In other words, optimality of one part depends on the optimality of the entire trajectory.

We can check optimality of the last part of the trajectory and after that the optimality of preceding parts. Optimality of partial trajectories depends on the optimality of the entire trajectory. The opposite statement does not hold, i. e. the optimality of the overall trajectory does not follow from the optimality of partial trajectories.

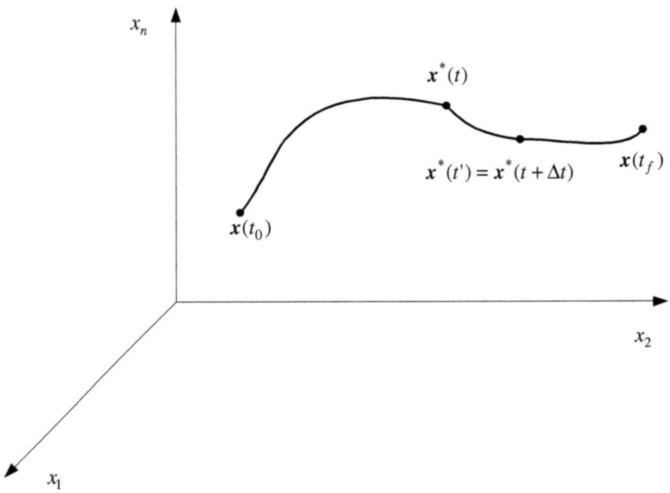

Fig. 8.9. Optimal trajectory

Consider a vector differential equation of a non-linear controlled object of the form

$$\frac{\mathrm{d}\boldsymbol{x}(t)}{\mathrm{d}t} = \boldsymbol{f}\left(\boldsymbol{x}(t), \boldsymbol{u}(t), t\right) \tag{8.132}$$

The cost function to be minimised is given as

$$I = \int_{t_0}^{t_f} F\left(\boldsymbol{x}(t), \boldsymbol{u}(t), t\right) \mathrm{d}t \tag{8.133}$$

Suppose that $\boldsymbol{x}^*(t)$ is the optimal trajectory with the initial state $\boldsymbol{x}(t_0)$ and terminal state $\boldsymbol{x}(t_f)$ (Fig. 8.9). Denote by $I^*(\boldsymbol{x}^*(t_0), t_0)$ minimum of the cost function I. It follows from the principle of optimality that the part of the trajectory with the initial state $\boldsymbol{x}^*(t)$, terminal state $\boldsymbol{x}(t_f)$ and that satisfies equation (8.132) is optimal as well.

From this follows that

$$I^*\left(\boldsymbol{x}^*(t), t\right) = \min_{\boldsymbol{u}\in U} \int_t^{t_f} F\left(\boldsymbol{x}^*(\tau), \boldsymbol{u}(\tau), \tau\right) \mathrm{d}\tau \tag{8.134}$$

This is *Bellman's functional equation.* For a sufficiently small interval Δt and time $t' = t + \Delta t$, the minimum value of the cost function on the interval $\boldsymbol{x}^*(t') = \boldsymbol{x}^*(t + \Delta t)$, $\boldsymbol{x}(t_f)$ is given as

$$I^*\left(\boldsymbol{x}^*(t'), t'\right) = \min_{\boldsymbol{u}\in U} \int_{t'}^{t_f} F\left(\boldsymbol{x}^*(\tau), \boldsymbol{u}(\tau), \tau\right) \mathrm{d}\tau \tag{8.135}$$

The arguments $\boldsymbol{x}^*(t')$, t' of the left hand side of equation (8.135) denote initial state of the given trajectory. Comparing integrals (8.134) and (8.135) gives

$$\begin{aligned} I^*\left(\boldsymbol{x}^*(t), t\right) &= \min_{\boldsymbol{u}\in U} \left\{ \int_{t'}^{t_f} F\left(\boldsymbol{x}^*(\tau), \boldsymbol{u}(\tau), \tau\right) \mathrm{d}\tau + F\left(\boldsymbol{x}^*(t), \boldsymbol{u}(t), t\right) \Delta t \right\} \\ &\quad + o_1\left(\Delta t\right) & (8.136) \\ &= \min_{\boldsymbol{u}\in U} \left\{ I^*\left(\boldsymbol{x}^*(t'), t'\right) + F\left(\boldsymbol{x}^*(t), \boldsymbol{u}(t), t\right) \Delta t \right\} + o_1\left(\Delta t\right) & (8.137) \end{aligned}$$

where $o_1(\Delta t)$ is a remainder satisfying

$$\lim_{\Delta t \to 0} \frac{o_1\left(\Delta t\right)}{\Delta t} = 0 \tag{8.138}$$

The remainder is nonzero because the interval Δt has a finite value in the right hand parts of equations (8.136), (8.137).

The Taylor expansion gives

$$\boldsymbol{x}(t') = \boldsymbol{x}\left(t + \Delta t\right) = \boldsymbol{x}(t) + \dot{\boldsymbol{x}}(t)\Delta t + o_2\left(\Delta t\right) \tag{8.139}$$

where $o_2(\Delta t)$ is a remainder. Taking (8.132) into account gives

$$\boldsymbol{x}(t') = \boldsymbol{x}(t) + \boldsymbol{f}\left(\boldsymbol{x}(t), \boldsymbol{u}(t), t\right) \Delta t + o_2\left(\Delta t\right) \tag{8.140}$$

Substituting $\boldsymbol{x}(t')$ from (8.140) into $I^*(\boldsymbol{x}(t'), t')$ and applying the Taylor expansion (assuming that partial derivatives $\partial I^*/\partial x_i$, $(i = 1, 2, \ldots, n)$, $\partial I^*/\partial t$ exist) gives

$$I^*(\boldsymbol{x}(t'),t') = I^*(\boldsymbol{x}(t+\Delta t), t+\Delta t) \tag{8.141}$$
$$= I^*\{\boldsymbol{x}(t) + \boldsymbol{f}(\boldsymbol{x}(t),\boldsymbol{u}(t),t)\Delta t + o_2(\Delta t), t+\Delta t\} \tag{8.142}$$
$$= I^*(\boldsymbol{x}(t),t) + \sum_{i=1}^{n} \frac{\partial I^*(\boldsymbol{x}(t),t)}{\partial x_i} f_i(\boldsymbol{x}(t),\boldsymbol{u}(t),t)\Delta t + \frac{\partial I^*(\boldsymbol{x}(t),t)}{\partial t}\Delta t + o_3(\Delta t) \tag{8.143}$$

where $o_3(\Delta t)$ is a remainder. Denote as

$$\frac{\partial I^*}{\partial \boldsymbol{x}} = \left(\frac{\partial I^*}{\partial x_1}, \frac{\partial I^*}{\partial x_2}, \cdots, \frac{\partial I^*}{\partial x_n}\right)^T \tag{8.144}$$

then equation (8.143) can be rewritten as

$$I^*(\boldsymbol{x}(t'),t') = I^*(\boldsymbol{x}(t),t) + \frac{\partial I^*(\boldsymbol{x}(t),t)}{\partial \boldsymbol{x}^T} \boldsymbol{f}(\boldsymbol{x}(t),\boldsymbol{u}(t),t)\Delta t + \frac{\partial I^*(\boldsymbol{x}(t),t)}{\partial t}\Delta t + o_3(\Delta t) \tag{8.145}$$

This expression is substituted into the right hand side of equation (8.137) assuming that $\boldsymbol{x}(t) = \boldsymbol{x}^*(t)$. As expressions $I^*(\boldsymbol{x}(t),t)$, $\partial I^*/\partial t$ do not depend on $\boldsymbol{u}(t)$ they can be taken out before the operator min. Simplifying the expression and dividing the equation by Δt yields

$$-\frac{\partial I^*(\boldsymbol{x}^*(t),t)}{\partial t} = \min_{\boldsymbol{u}\in U}\left\{\frac{\partial I^*(\boldsymbol{x}^*(t),t)}{\partial \boldsymbol{x}^{*T}} \boldsymbol{f}(\boldsymbol{x}^*(t),\boldsymbol{u}(t),t) + F(\boldsymbol{x}^*(t),\boldsymbol{u}(t),t)\right\} + \frac{o_4(\Delta t)}{\Delta t} \tag{8.146}$$

where $o_4(\Delta t)$ is a remainder. If $\Delta t \to 0$, the last term on the right hand of the equation vanishes. This gives the *Hamilton-Jacobi equation* (or Hamilton-Jacobi-Bellman (HJB) equation)

$$-\frac{\partial I^*(\boldsymbol{x}^*(t),\, t)}{\partial t} = \min_{\boldsymbol{u}\in U}\left\{\frac{\partial I^*(\boldsymbol{x}^*(t),\, t)}{\partial \boldsymbol{x}^{*T}} \boldsymbol{f}(\boldsymbol{x}^*(t),\, \boldsymbol{u}(t),\, t) + F(\boldsymbol{x}^*(t),\, \boldsymbol{u}(t),\, t)\right\} \tag{8.147}$$

This is the *basic equation of dynamic programming for continuous-time systems*. The first part is independent on control $\boldsymbol{u}(t)$ after minimisation. The equation holds only for optimal control $\boldsymbol{u}^*(t)$. HJB equation is often used in the form

$$-\frac{\partial I^*(\boldsymbol{x}^*(t),t)}{\partial t} = \frac{\partial I^*(\boldsymbol{x}^*(t),t)}{\partial \boldsymbol{x}^{*T}} \boldsymbol{f}(\boldsymbol{x}^*(t),\boldsymbol{u}^*(t),t) + F(\boldsymbol{x}^*(t),\boldsymbol{u}^*(t),t) \tag{8.148}$$

As $(\partial I^*/\partial t + (\partial I^*/\partial \boldsymbol{x}^T)\boldsymbol{f})$ is equal to $\mathrm{d}I^*/\mathrm{d}t$, the equation can also be written as

$$\frac{\mathrm{d}I^*\left(\boldsymbol{x}^*(t), t\right)}{\mathrm{d}t} + F\left(\boldsymbol{x}^*(t), \boldsymbol{u}^*(t), t\right) = 0 \tag{8.149}$$

Similarly, the basic HJB equation (8.147) can be written as

$$\min_{\boldsymbol{u}\in U}\left\{\frac{\mathrm{d}I^*\left(\boldsymbol{x}^*(t), t\right)}{\mathrm{d}t} + F\left(\boldsymbol{x}^*(t), \boldsymbol{u}^*(t), t\right)\right\} = 0 \tag{8.150}$$

The Hamilton-Jacobi-Bellman equation (8.147) or (8.150) specifies necessary optimality conditions. Under some assumptions can sufficent conditions be formulated similarly.

We will now show that optimal control described by the equation (8.70) is unique.

The proof is based on the fact that the solution of the Riccati equation $\boldsymbol{P}(t)$ with terminal condition $\boldsymbol{P}(t) = \boldsymbol{Q}_{t_f}$ is unique.

If the optimal control exists then it is unique and determined by equation (8.70).

If the matrix $\boldsymbol{P}(t)$ is unique then the optimal control is the unambiguos function of the state. Suppose that there are two optimal control trajectories $\boldsymbol{u}_1(t)$, $\boldsymbol{u}_2(t)$ and two optimal state trajectories $\boldsymbol{x}_1(t)$, $\boldsymbol{x}_2(t)$ with $\boldsymbol{x}_1(t_0) = \boldsymbol{x}_2(t_0)$. Vectors $\boldsymbol{x}_1(t)$, $\boldsymbol{x}_2(t)$ are two different solutions of the equation

$$\frac{\mathrm{d}\boldsymbol{x}(t)}{\mathrm{d}t} = \left(\boldsymbol{A} - \boldsymbol{B}\boldsymbol{R}^{-1}\boldsymbol{B}^T\boldsymbol{P}(t)\right)\boldsymbol{x}(t) \tag{8.151}$$

and $\boldsymbol{x}_1(t_0) = \boldsymbol{x}_2(t_0)$. As the matrix $\boldsymbol{P}(t)$ is unique, the same holds for matrix $(\boldsymbol{A} - \boldsymbol{B}\boldsymbol{R}^{-1}\boldsymbol{B}^T\boldsymbol{P})$. Equation (8.151) is linear and homogenous, therefore its solution is unique. It follows that $\boldsymbol{x}_1(t) = \boldsymbol{x}_2(t)$, $\boldsymbol{u}_1(t) = \boldsymbol{u}_2(t)$ in any time $t \in \langle t_0, t_f\rangle$.

The next theorem can be used in proof of existence of optimal control and calculation of value of $I^*(\boldsymbol{x}^*(t), t)$.

Theorem 8.1. *Consider a linear system* (8.39) *and the cost function given by* (8.41). *Denote*

$$I^*\left(\boldsymbol{x}(t), t\right) = \frac{1}{2}\boldsymbol{x}^T(t)\boldsymbol{P}(t)\boldsymbol{x}(t) \tag{8.152}$$

where $\boldsymbol{P}(t)$ is a symmetric matrix of dimensions $n \times n$ that is a solution of the Riccati equation (8.66) *with terminal condition* (8.68). *If for optimal control holds $\boldsymbol{u}(t) \neq \boldsymbol{0}$ for all states then $\boldsymbol{P}(t)$ is positive definite in any time $t, t_0 \leq t \leq t_f$ and $\boldsymbol{P}(t_f) = \boldsymbol{Q}_{t_f}$ is positive semidefinite. As the equation*

$$I^* = \frac{1}{2}\boldsymbol{x}^T(t)\boldsymbol{P}(t)\boldsymbol{x}(t)$$

is determined for all $\boldsymbol{x}(t)$ and t, optimal control exists and the minimum value of the cost function is equal to I^*.

Proof. We show that I^* is the solution of the HJB equation and the corresponding boundary conditions.

At time $t = t_f$ can equation (8.152) be written as

$$I^*\left(\boldsymbol{x}^*(t_f), t_f\right) = \frac{1}{2}\boldsymbol{x}^T(t_f)\boldsymbol{Q}_{t_f}\boldsymbol{x}(t_f) \tag{8.153}$$

and thus it defines the terminal state.

The HJB equation for the system (8.39) and for the cost function (8.41) is of the form

$$\begin{aligned} 0 &= \frac{\partial}{\partial t} I^*\left(\boldsymbol{x}(t), t\right) \\ &+ \min_{\boldsymbol{u}(t)} \left\{ \frac{1}{2}\boldsymbol{x}^T(t)\boldsymbol{Q}\boldsymbol{x}(t) + \frac{1}{2}\boldsymbol{u}^T(t)\boldsymbol{R}\boldsymbol{u}(t) + \frac{\partial I^*\left(\boldsymbol{x}(t), t\right)}{\partial \boldsymbol{x}^T(t)} \left(\boldsymbol{A}\boldsymbol{x}(t) + \boldsymbol{B}\boldsymbol{u}(t)\right) \right\} \end{aligned} \tag{8.154}$$

The term in brackets is minimised for

$$\boldsymbol{u}(t) = -\boldsymbol{R}^{-1}\boldsymbol{B}^T \frac{\partial I^*\left(\boldsymbol{x}(t), t\right)}{\partial \boldsymbol{x}(t)} \tag{8.155}$$

Substituting equation (8.155) into (8.154) yields

$$\begin{aligned} &\frac{\partial I^*}{\partial t} + \frac{1}{2}\boldsymbol{x}^T(t)\boldsymbol{Q}\boldsymbol{x}(t) + \frac{1}{2}\left(\boldsymbol{R}^{-1}\boldsymbol{B}^T\frac{\partial I^*}{\partial \boldsymbol{x}(t)}\right)^T \boldsymbol{B}^T\frac{\partial I^*}{\partial \boldsymbol{x}(t)} \\ &\qquad + \left(\boldsymbol{A}\boldsymbol{x}(t)\right)^T \frac{\partial I^*}{\partial \boldsymbol{x}(t)} - \left(\boldsymbol{B}\boldsymbol{R}^{-1}\boldsymbol{B}^T\frac{\partial I^*}{\partial \boldsymbol{x}(t)}\right)^T \frac{\partial I^*}{\partial \boldsymbol{x}(t)} = 0 \end{aligned} \tag{8.156}$$

The following relations hold

$$\frac{\partial I^*}{\partial t} = \frac{1}{2}\boldsymbol{x}^T(t)\frac{\mathrm{d}\boldsymbol{P}(t)}{\mathrm{d}t}\boldsymbol{x}(t) \tag{8.157}$$

$$\frac{\partial I^*}{\partial \boldsymbol{x}(t)} = \boldsymbol{P}(t)\boldsymbol{x}(t) \tag{8.158}$$

Substituting (8.157) and (8.158) into (8.156) gives

$$\begin{aligned} &\frac{1}{2}\boldsymbol{x}^T(t)\frac{\mathrm{d}\boldsymbol{P}(t)}{\mathrm{d}t}\boldsymbol{x}(t) + \frac{1}{2}\boldsymbol{x}^T(t)\boldsymbol{Q}\boldsymbol{x}(t) + \left(\boldsymbol{A}\boldsymbol{x}(t)\right)^T\boldsymbol{P}(t)\boldsymbol{x}(t) \\ &\qquad - \frac{1}{2}\left(\boldsymbol{B}\boldsymbol{R}^{-1}\boldsymbol{B}^T\boldsymbol{P}(t)\boldsymbol{x}(t)\right)^T \boldsymbol{P}(t)\boldsymbol{x}(t) = 0 \end{aligned} \tag{8.159}$$

As the matrix $\boldsymbol{P}(t)$ is symmetric, the following holds

$$(\boldsymbol{A}\boldsymbol{x}(t))^T \boldsymbol{P}(t)\boldsymbol{x}(t) = \frac{1}{2}(\boldsymbol{A}\boldsymbol{x}(t))^T \boldsymbol{P}(t)\boldsymbol{x}(t) + \frac{1}{2}(\boldsymbol{A}\boldsymbol{x}(t))^T \boldsymbol{P}(t)\boldsymbol{x}(t) \quad (8.160)$$

or

$$(\boldsymbol{A}\boldsymbol{x}(t))^T \boldsymbol{P}(t)\boldsymbol{x}(t) = \frac{1}{2}\boldsymbol{x}^T(t)\boldsymbol{A}^T\boldsymbol{P}(t)\boldsymbol{x}(t) + \frac{1}{2}\boldsymbol{x}^T(t)\boldsymbol{P}(t)\boldsymbol{A}\boldsymbol{x}(t) \quad (8.161)$$

Equation (8.161) can be substituted into (8.159). This gives after some algebraic manipulations

$$\frac{1}{2}\left\{\boldsymbol{x}^T(t)\left(\frac{\mathrm{d}\boldsymbol{P}(t)}{\mathrm{d}t} + \boldsymbol{P}(t)\boldsymbol{A} + \boldsymbol{A}^T\boldsymbol{P}(t) - \boldsymbol{P}(t)\boldsymbol{B}\boldsymbol{R}^{-1}\boldsymbol{B}^T\boldsymbol{P}(t) + \boldsymbol{Q}(t)\right)\boldsymbol{x}(t)\right\} = 0 \quad (8.162)$$

If $\boldsymbol{P}(t)$ is a solution of the Riccati equation (8.66) then the matrix in brackets is zero and equation (8.162) holds. The opposite statement is also true.

Positive definitness of $\boldsymbol{P}$ can be also shown. Suppose that $\boldsymbol{P}(t)$ is not positive definite for $t = t' < t_f$. In this case $\boldsymbol{x}(t')$ can exist such that $\frac{1}{2}\left(\boldsymbol{x}^T(t')\boldsymbol{P}(t')\boldsymbol{x}(t')\right) \leq 0$. This contradicts the fact that $I(\boldsymbol{u})$ is positive for $\boldsymbol{u} \neq \boldsymbol{0}$. Therefore, $\boldsymbol{P}(t)$ is positive definite for any t, $t_0 \leq t \leq t_f$. □

For the optimal system holds

$$F + \frac{\mathrm{d}I^*}{\mathrm{d}t} = 0 \quad (8.163)$$

Hence

$$\frac{\mathrm{d}I^*}{\mathrm{d}t} = -F \quad (8.164)$$

Let the function I^* be the Lyapunov function. As $-F$ is negative definite then positive definite I^* implies asymptotic stability of the optimal system.

8.4.2 Dynamic Programming for Discrete-Time Systems

Let us consider optimal control of discrete-time systems described by the state-space representation

$$\boldsymbol{x}(k+1) = \boldsymbol{f}\left[\boldsymbol{x}(k), \boldsymbol{u}(k)\right] \quad (8.165)$$

where $\boldsymbol{x}(k)$ is the state vector of dimension n, $\boldsymbol{u}(k)$ – vector of input variables of dimension m, and $\boldsymbol{f}$ – vector of functions of dimension n.

As it was shown for the case of continuous-time systems, dynamic programming is based on the principle of optimality. We will present the basic facts.

The cost function is of the form

$$I_N = G_1\left[\boldsymbol{x}(N)\right] + \sum_{k=0}^{N-1} F\left[\boldsymbol{x}(k), \boldsymbol{u}(k)\right] \tag{8.166}$$

The variable N in the cost function (8.166) shows that the system is to be optimised at N intervals (Fig. 8.10). The variable k will denote time in the sequel.

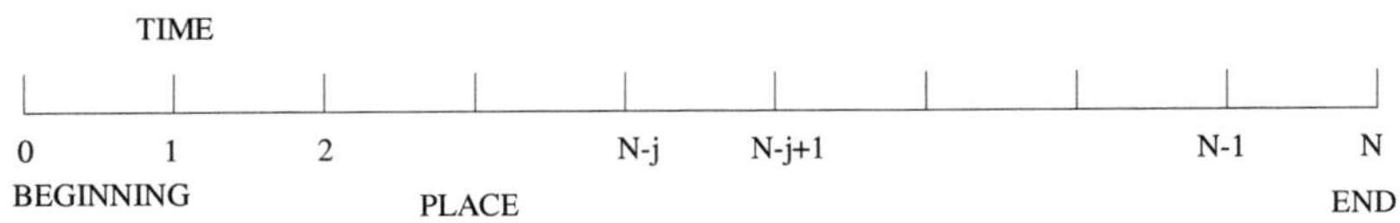

Fig. 8.10. Time scale with time intervals

Denote by I_N^* minimum of the cost function I_N, thus

$$I_N^*\left[\boldsymbol{x}(0)\right] = \min_{\substack{\boldsymbol{u}(k)\\ k=0,\ldots,N-1}} \left\{ G_1\left[\boldsymbol{x}(N)\right] + \sum_{k=0}^{N-1} F\left[\boldsymbol{x}(k), \boldsymbol{u}(k)\right] \right\} \tag{8.167}$$

Further let us define

$$I_{N_j}\left[\boldsymbol{u}, \boldsymbol{x}(N-j)\right] = G_1\left[\boldsymbol{x}(N)\right] + \sum_{k=N-j}^{N-1} F\left[\boldsymbol{x}(k), \boldsymbol{u}(k)\right], \quad j = 1, 2, \ldots, N \tag{8.168}$$

$$I_{N_j}^*\left[\boldsymbol{x}(N-j)\right] = \min_{\substack{\boldsymbol{u}(k)\\ k=N-j,\ldots,N-1}} \left\{ G_1\left[\boldsymbol{x}(N)\right] + \sum_{k=N-j}^{N-1} F\left[\boldsymbol{x}(k), \boldsymbol{u}(k)\right] \right\} \tag{8.169}$$

From the principle of optimality follows that for $j = 1$ holds

$$I_{N_1}^*\left[\boldsymbol{x}(N-1)\right] = \min_{\boldsymbol{u}(N-1)} G_1\left[\boldsymbol{x}(N)\right] + F\left[\boldsymbol{x}(N-1), \boldsymbol{u}(N-1)\right] \tag{8.170}$$

From equation

$$\boldsymbol{x}(N) = \boldsymbol{f}\left[\boldsymbol{x}(N-1), \boldsymbol{u}(N-1)\right] \tag{8.171}$$

follows that $I_{N_1}^*$ depends only on $\boldsymbol{x}(N-1)$.

If $j = 2$

$$I_{N_2}^*\left[\boldsymbol{x}(N-2)\right] = \min_{\boldsymbol{u}(N-2)} \left\{ F\left[\boldsymbol{x}(N-2), \boldsymbol{u}(N-2)\right] + I_{N_1}^*\left[\boldsymbol{x}(N-1)\right] \right\} \tag{8.172}$$

From

$$\boldsymbol{x}(N-1) = \boldsymbol{f}\left[\boldsymbol{x}(N-2), \boldsymbol{u}(N-2)\right] \tag{8.173}$$

is evident that $I^*_{N_2}$ depends only on $\boldsymbol{x}(N-2)$. Continuing with the same line of reasoning, we can write

$$I^*_{N_j}\left[\boldsymbol{x}(N-j)\right] = \min_{\boldsymbol{u}(N-j)} \left\{ F\left[\boldsymbol{x}(N-j), \boldsymbol{u}(N-j)\right] + I^*_{N(j-1)}\left[\boldsymbol{x}(N-j+1)\right] \right\} \tag{8.174}$$

From equation

$$\boldsymbol{x}(N-j+1) = \boldsymbol{f}\left[\boldsymbol{x}(N-j), \boldsymbol{u}(N-j)\right] \tag{8.175}$$

follows that $I^*_{N_j}$ depends only on $\boldsymbol{x}(N-j)$.

Equations (8.170), (8.172)–(8.174), ... make it possible to calculate recursively the optimal control input $\boldsymbol{u}(N-1), \boldsymbol{u}(N-2), \ldots, \boldsymbol{u}(N-j), \ldots$

We give the discrete equivalent of the principle of minimum without details at this place. It can be derived analogically as its continuous-time counterpart.

Consider the system (8.165) with the initial state $\boldsymbol{x}(0)$ and the cost function (8.166). The Hamiltonian of the system is defined as

$$H(k) = F\left[\boldsymbol{x}(k), \boldsymbol{u}(k)\right] + \boldsymbol{\lambda}^T(k+1)\boldsymbol{f}\left[\boldsymbol{x}(k), \boldsymbol{u}(k)\right] \tag{8.176}$$

For $\boldsymbol{\lambda}(k)$ holds

$$\boldsymbol{\lambda}(k) = \frac{\partial H(k)}{\partial \boldsymbol{x}(k)}, \quad k = 0, 1, \ldots, N-1 \tag{8.177}$$

$$\boldsymbol{\lambda}(N) = \frac{\partial G_1}{\partial \boldsymbol{x}(N)} \tag{8.178}$$

The necessary condition for the existence of minimum (8.166) is

$$\frac{\partial H(k)}{\partial \boldsymbol{u}(k)} = \boldsymbol{0}, \quad k = 0, 1, \ldots, N-1 \tag{8.179}$$

8.4.3 Optimal Feedback

Consider the system described by the equation

$$\boldsymbol{x}(k) = \boldsymbol{A}\boldsymbol{x}(k) + \boldsymbol{B}\boldsymbol{u}(k) \tag{8.180}$$

with initial condition $\boldsymbol{x}(0)$. We would like to find $\boldsymbol{u}(0), \boldsymbol{u}(1), \ldots, \boldsymbol{u}(N-1)$ such that the cost function

$$I_N\left[\boldsymbol{u}, \boldsymbol{x}(0)\right] = \boldsymbol{x}^T(N)\boldsymbol{Q}_N\boldsymbol{x}(N) + \sum_{k=0}^{N-1}\left[\boldsymbol{x}^T(k)\boldsymbol{Q}\boldsymbol{x}(k) + \boldsymbol{u}^T(k)\boldsymbol{R}\boldsymbol{u}(k)\right] \tag{8.181}$$

is minimised. Suppose that the constant matrices $\boldsymbol{Q}_N, \boldsymbol{Q}, \boldsymbol{R}$ are symmetric and positive definite.

Equation (8.170) is in our case given as

$$\begin{aligned} I_{N_1}^*\left[\boldsymbol{x}(N-1)\right] = \min_{\boldsymbol{u}(N-1)} &\left\{\boldsymbol{x}^T(N)\boldsymbol{Q}_N\boldsymbol{x}(N)\right. \\ &\left. + \left[\boldsymbol{x}^T(N-1)\boldsymbol{Q}\boldsymbol{x}(N-1) + \boldsymbol{u}^T(N-1)\boldsymbol{R}\boldsymbol{u}(N-1)\right]\right\} \end{aligned} \tag{8.182}$$

We substitute for $\boldsymbol{x}(N)$ from equation

$$\boldsymbol{x}(N) = \boldsymbol{A}\boldsymbol{x}(N-1) + \boldsymbol{B}\boldsymbol{u}(N-1) \tag{8.183}$$

into (8.182) and get

$$\begin{aligned} &I_{N_1}^*\left[\boldsymbol{x}(N-1)\right] = \\ &\min_{\boldsymbol{u}(N-1)}\left\{\left[\boldsymbol{A}\boldsymbol{x}(N-1) + \boldsymbol{B}\boldsymbol{u}(N-1)\right]^T\boldsymbol{Q}_N\left[\boldsymbol{A}\boldsymbol{x}(N-1) + \boldsymbol{B}\boldsymbol{u}(N-1)\right]\right. \\ &\qquad\left. + \left[\boldsymbol{x}^T(N-1)\boldsymbol{Q}\boldsymbol{x}(N-1) + \boldsymbol{u}^T(N-1)\boldsymbol{R}\boldsymbol{u}(N-1)\right]\right\} \end{aligned} \tag{8.184}$$

Optimal control $\boldsymbol{u}(N-1)$ can be derived if the right hand of equation (8.184) is differentiated with respect to $\boldsymbol{u}(N-1)$, equated to zero to yield

$$2\boldsymbol{B}^T\boldsymbol{Q}_N\boldsymbol{A}\boldsymbol{x}(N-1) + 2\boldsymbol{B}^T\boldsymbol{Q}_N\boldsymbol{B}\boldsymbol{u}(N-1) + 2\boldsymbol{R}\boldsymbol{u}(N-1) = \boldsymbol{0} \tag{8.185}$$

$$\boldsymbol{B}^T\boldsymbol{Q}_N\boldsymbol{A}\boldsymbol{x}(N-1) + \left(\boldsymbol{B}^T\boldsymbol{Q}_N\boldsymbol{B} + \boldsymbol{R}\right)\boldsymbol{u}(N-1) = \boldsymbol{0} \tag{8.186}$$

Optimal control is then given as

$$\boldsymbol{u}(N-1) = -\left(\boldsymbol{B}^T\boldsymbol{Q}_N\boldsymbol{B} + \boldsymbol{R}\right)^{-1}\boldsymbol{B}^T\boldsymbol{Q}_N\boldsymbol{A}\boldsymbol{x}(N-1) \tag{8.187}$$

Second partial derivative of equation (8.186) with respect to $\boldsymbol{u}(N-1)$ gives

$$\boldsymbol{B}^T\boldsymbol{Q}_N\boldsymbol{B} + \boldsymbol{R} > \boldsymbol{0} \tag{8.188}$$

This confirms that minimum is attained for $\boldsymbol{u}(N-1)$.

For $\boldsymbol{u}(N-1)$ holds

$$\boldsymbol{u}(N-1) = -\boldsymbol{K}(N-1)\boldsymbol{x}(N-1) \tag{8.189}$$

where

$$\boldsymbol{K}(N-1) = \left(\boldsymbol{B}^T\boldsymbol{Q}_N\boldsymbol{B} + \boldsymbol{R}\right)^{-1}\boldsymbol{B}^T\boldsymbol{Q}_N\boldsymbol{A} \tag{8.190}$$

Denote by

$$\boldsymbol{Q}_N = \boldsymbol{P}(N) \tag{8.191}$$

Then

$$\boldsymbol{K}(N-1) = \left[\boldsymbol{B}^T\boldsymbol{P}(N)\boldsymbol{B} + \boldsymbol{R}\right]^{-1}\boldsymbol{B}^T\boldsymbol{P}(N)\boldsymbol{A} \tag{8.192}$$

Substituting (8.189) into (8.184) gives

$$I_{N_1}^*\left[\boldsymbol{x}(N-1)\right] = \boldsymbol{x}^T(N-1)\,\boldsymbol{P}(N-1)\,\boldsymbol{x}(N-1) \tag{8.193}$$

where

$$\boldsymbol{P}(N-1) = \boldsymbol{A}^T\boldsymbol{P}(N)\boldsymbol{A} - \boldsymbol{A}^T\boldsymbol{P}(N)\boldsymbol{B}\left[\boldsymbol{B}^T\boldsymbol{P}(N)\boldsymbol{B} + \boldsymbol{R}\right]^{-1}\boldsymbol{B}^T\boldsymbol{P}(N)\boldsymbol{A} + \boldsymbol{Q} \tag{8.194}$$

The following holds

$$I_{N_2}^*\left[\boldsymbol{x}(N-2)\right] = \min_{\boldsymbol{u}(N-2)}\Big\{\boldsymbol{x}^T(N-2)\boldsymbol{Q}\boldsymbol{x}(N-2) + \boldsymbol{u}^T(N-2)\boldsymbol{R}\boldsymbol{u}(N-2) + I_{N_1}^*\left[\boldsymbol{x}(N-1)\right]\Big\} \tag{8.195}$$

We substitute for $I_{N_1}^*\left[\boldsymbol{x}(N-1)\right]$ from (8.193) into (8.195) and then subtitute for $\boldsymbol{x}(N-1)$ from the equation

$$\boldsymbol{x}(N-1) = \boldsymbol{A}\boldsymbol{x}(N-2) + \boldsymbol{B}\boldsymbol{u}(N-2) \tag{8.196}$$

This gives

$$I_{N_2}^*\left[\boldsymbol{x}(N-2)\right] = \min_{\boldsymbol{u}(N-2)}\Big\{\Big[\boldsymbol{A}\boldsymbol{x}(N-2) + \boldsymbol{B}\boldsymbol{u}(N-2)\Big]^T \times \boldsymbol{P}(N-1)\left[\boldsymbol{A}\boldsymbol{x}(N-2) + \boldsymbol{B}\boldsymbol{u}(N-2)\right] + \boldsymbol{x}^T(N-2)\,\boldsymbol{Q}\boldsymbol{x}(N-2) + \boldsymbol{u}^T(N-2)\,\boldsymbol{R}\boldsymbol{u}(N-2)\Big\} \tag{8.197}$$

Optimal control $\boldsymbol{u}(N-2)$ can be derived if the right hand of equation (8.197) is differentiated with respect to $\boldsymbol{u}(N-1)$, equated to zero to yield

$$2\boldsymbol{B}^T\boldsymbol{P}(N-1)\,\boldsymbol{A}\boldsymbol{x}(N-2) + 2\boldsymbol{B}^T\boldsymbol{P}(N-1)\,\boldsymbol{B}\boldsymbol{u}(N-2) + 2\boldsymbol{R}\boldsymbol{u}(N-2) = \boldsymbol{0} \tag{8.198}$$

Optimal control is then given as

$$\boldsymbol{u}(N-2) = -\boldsymbol{K}(N-2)\boldsymbol{x}(N-2) \tag{8.199}$$

where

$$\boldsymbol{K}(N-2) = \left[\boldsymbol{B}^T\boldsymbol{P}(N-1)\boldsymbol{B} + \boldsymbol{R}\right]^{-1}\boldsymbol{B}^T\boldsymbol{P}(N-1)\boldsymbol{A} \tag{8.200}$$

Vector $\boldsymbol{u}(N-2)$ guarantees the minimum.

Substituting (8.199) into (8.197) gives

$$I^*_{N_2}\left[\boldsymbol{x}(N-2)\right] = \boldsymbol{x}^T(N-2)\boldsymbol{P}(N-2)\boldsymbol{x}(N-2) \tag{8.201}$$

where

$$\begin{aligned}\boldsymbol{P}(N-2) = {} & \boldsymbol{A}^T\boldsymbol{P}(N-1)\boldsymbol{A} \\ & - \boldsymbol{A}^T\boldsymbol{P}(N-1)\boldsymbol{B}\left[\boldsymbol{B}^T\boldsymbol{P}(N-1)\boldsymbol{B} + \boldsymbol{R}\right]^{-1}\boldsymbol{B}^T\boldsymbol{P}(N-1)\boldsymbol{A} + \boldsymbol{Q}\end{aligned} \tag{8.202}$$

Remaining stages $\boldsymbol{u}(N-3)$, $\boldsymbol{u}(N-4)$,..., or $I^*_{N_3}$, $I^*_{N_4}$, ... can be derived analogously to yield

$$\boldsymbol{u}(N-j) = -\boldsymbol{K}(N-j)\boldsymbol{x}(N-j) \tag{8.203}$$

where

$$\boldsymbol{K}(N-j) = \left[\boldsymbol{B}^T\boldsymbol{P}(N-j)\boldsymbol{B} + \boldsymbol{R}\right]^{-1}\boldsymbol{B}^T\boldsymbol{P}(N-j)\boldsymbol{A} \tag{8.204}$$

$$I^*_{N_j}\left[\boldsymbol{x}(N-j)\right] = \boldsymbol{x}^T(N-j)\boldsymbol{P}(N-j)\boldsymbol{x}(N-j) \tag{8.205}$$

where

$$\begin{aligned}\boldsymbol{P}(N-j) = {} & \boldsymbol{Q} + \boldsymbol{A}^T\boldsymbol{P}(N-j+1)\boldsymbol{A} \\ & -\boldsymbol{A}^T\boldsymbol{P}(N-j+1)\boldsymbol{B}\left[\boldsymbol{B}^T\boldsymbol{P}(N-j+1)\boldsymbol{B} + \boldsymbol{R}\right]^{-1}\boldsymbol{B}^T\boldsymbol{P}(N-j+1)\boldsymbol{A}\end{aligned} \tag{8.206}$$

If $N \to \infty$ then $\boldsymbol{P}(N-j)$ and $\boldsymbol{K}(N-j)$ asymptotically converge to matrices $\boldsymbol{P}$ and $\boldsymbol{K}$ given as solutions of equations

$$\boldsymbol{P} = \boldsymbol{A}^T\boldsymbol{P}\boldsymbol{A} - \boldsymbol{A}^T\boldsymbol{P}\boldsymbol{B}\left[\boldsymbol{B}^T\boldsymbol{P}\boldsymbol{B} + \boldsymbol{R}\right]^{-1}\boldsymbol{B}^T\boldsymbol{P}\boldsymbol{A} + \boldsymbol{Q} \tag{8.207}$$

$$\boldsymbol{K} = \left[\boldsymbol{B}^T\boldsymbol{P}\boldsymbol{B} + \boldsymbol{R}\right]^{-1}\boldsymbol{B}^T\boldsymbol{P}\boldsymbol{A} \tag{8.208}$$

and optimal control is given as

$$\boldsymbol{u}(k) = -\boldsymbol{K}\boldsymbol{x}(k), \quad i = 0, 1, 2, \ldots \tag{8.209}$$

8.5 Observers and State Estimation

LQ control, as well as other control designs supposes that all states $\boldsymbol{x}(t)$ are fully measurable. This cannot be guaranteed in practice. A possible remedy is to estimate states $\boldsymbol{x}(t)$ based on measurement of output variables $\boldsymbol{y}(t)$. Below we give derivations and properties of deterministic and stochastic state estimation techniques.

8.5.1 State Observation

The deterministic state estimation or state observation can be implemented using an observer. An observer is a dynamic system having the property that its state converges to the state of the observed deterministic system i. e. the system without noise and measurement error. Observers use information about the output $\boldsymbol{y}(t)$ and input $\boldsymbol{u}(t)$ vectors to determine the state vector $\boldsymbol{x}(t)$.

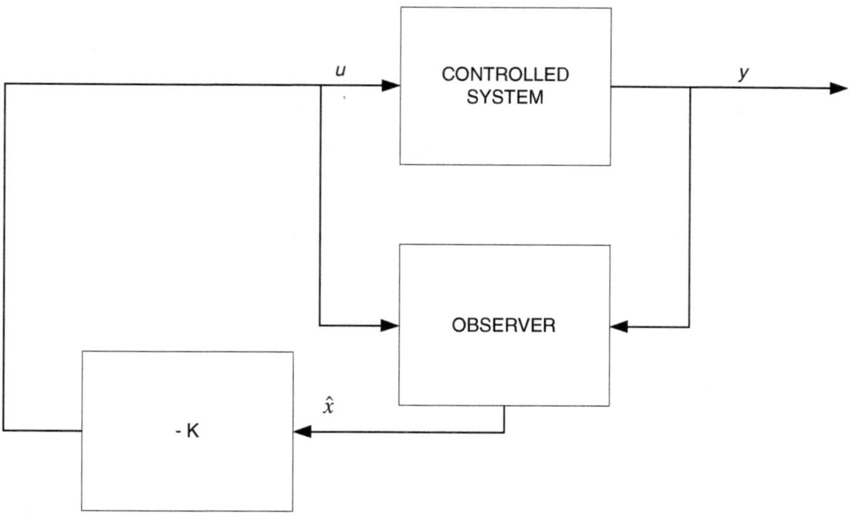

Fig. 8.11. Block diagram of the closed-loop system with a state observer

Nowadays, theory of deterministic state observation is in mature state. Fig. 8.11 shows how the observed state $\hat{\boldsymbol{x}}(t)$ is used in automatic control.

Luenberger introduced a concept of an observer illustrated in scheme in Fig. 8.12. The observer has the form of the estimated linear process extended with a feedback

$$\boldsymbol{u}_L(t) = \boldsymbol{L}\left[\boldsymbol{y}(t) - \hat{\boldsymbol{y}}(t)\right] \tag{8.210}$$

This is a proportional feedback loop designed to minimise the difference $(\boldsymbol{y}(t) - \hat{\boldsymbol{y}}(t))$. The task of constructing a desired observer is thus transformed into the problem of finding a suitable $\boldsymbol{L}$.

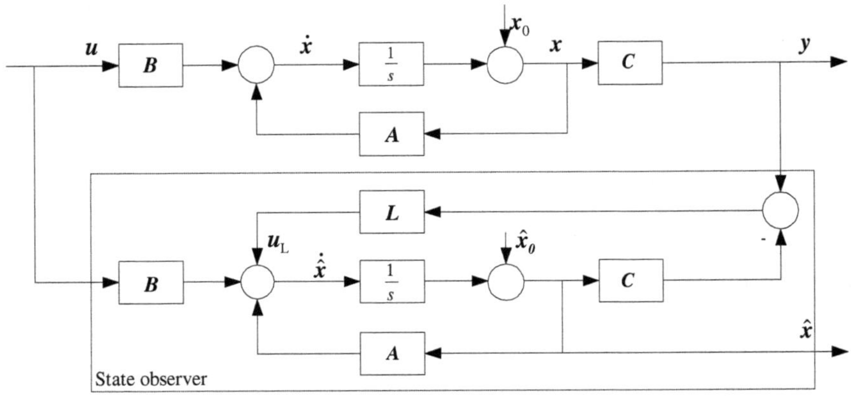

Fig. 8.12. Scheme of an observer

The observer will be designed for an observed system

$$\dot{\boldsymbol{x}}(t) = \boldsymbol{A}\boldsymbol{x}(t) + \boldsymbol{B}\boldsymbol{u}(t), \quad \boldsymbol{x}(0) = \boldsymbol{x}_0 \tag{8.211}$$

$$\boldsymbol{y}(t) = \boldsymbol{C}\boldsymbol{x}(t) \tag{8.212}$$

The observer from Fig. 8.12 is described by equations

$$\dot{\hat{\boldsymbol{x}}}(t) = \boldsymbol{A}\hat{\boldsymbol{x}}(t) + \boldsymbol{B}\boldsymbol{u}(t) + \boldsymbol{L}\left[\boldsymbol{y}(t) - \hat{\boldsymbol{y}}(t)\right], \quad \hat{\boldsymbol{x}}(0) = \hat{\boldsymbol{x}}_0 \tag{8.213}$$

$$\hat{\boldsymbol{y}}(t) = \boldsymbol{C}\hat{\boldsymbol{x}}(t) \tag{8.214}$$

Let us now find a matrix $\boldsymbol{L}$ such that the estimation error

$$\boldsymbol{e}(t) = \boldsymbol{x}(t) - \hat{\boldsymbol{x}}(t) \tag{8.215}$$

with initial value

$$\boldsymbol{e}(0) = \boldsymbol{x}_0 - \hat{\boldsymbol{x}}_0 \tag{8.216}$$

converges asymptotically to zero.

From (8.215) follows

$$\dot{\boldsymbol{e}}(t) = \dot{\boldsymbol{x}}(t) - \dot{\hat{\boldsymbol{x}}}(t) \tag{8.217}$$

Substituting $\dot{\boldsymbol{x}}$ and $\dot{\hat{\boldsymbol{x}}}$ form (8.211) and (8.213) yields

$$\dot{\boldsymbol{e}}(t) = \boldsymbol{A}\boldsymbol{x}(t) - \boldsymbol{A}\hat{\boldsymbol{x}}(t) - \boldsymbol{L}\left[\boldsymbol{y}(t) - \hat{\boldsymbol{y}}(t)\right] \tag{8.218}$$

Finally, using (8.212) and (8.214) gives

$$\dot{\boldsymbol{e}}(t) = (\boldsymbol{A} - \boldsymbol{L}\boldsymbol{C})\,\boldsymbol{e}(t), \quad \boldsymbol{e}(0) = \boldsymbol{e}_0 \tag{8.219}$$

The system (8.219) meets the requirements on observer design if it is asymptotically stable.

Asymptotic stability of the system (8.219) can be guaranteed for arbitrary initial conditions of the observed system and the observer if and only if eigenvalues of the matrix $(\boldsymbol{A} - \boldsymbol{L}\boldsymbol{C})$ lie in the left half-plane. Thus, matrix $\boldsymbol{L}$ has to be chosen in such a way that the matrix $(\boldsymbol{A} - \boldsymbol{L}\boldsymbol{C})$ be stable.

8.5.2 Kalman Filter

The Kalman filter is a special case of the Luenberger observer that is optimised for the observation and input noises. We consider an optimal state estimation of a linear system

$$\dot{\boldsymbol{x}}(t) = \boldsymbol{A}\boldsymbol{x}(t) + \boldsymbol{\xi}_x(t) \tag{8.220}$$

where $\boldsymbol{\xi}_x(t)$ is n-dimensional stochastic process vector. We assume that the processes have properties of a Gaussian noise

$$E\{\boldsymbol{\xi}_x(t)\} = \boldsymbol{0} \tag{8.221}$$

$$\text{Cov}(\boldsymbol{\xi}_x(t), \boldsymbol{\xi}_x(\tau)) = E\left\{\boldsymbol{\xi}_x(t)\boldsymbol{\xi}_x^T(\tau)\right\} = \boldsymbol{V}\delta(t-\tau) \tag{8.222}$$

Initial condition is given as

$$\boldsymbol{x}(0) = \bar{\boldsymbol{x}}_0 + \boldsymbol{\xi}_{x0} \tag{8.223}$$

where

$$E\{\boldsymbol{x}(0)\} = \bar{\boldsymbol{x}}_0 \tag{8.224}$$

$$\text{Cov}(\boldsymbol{x}(0)) = E\left\{[\bar{\boldsymbol{x}}_0 - \boldsymbol{x}(0)]\,[\bar{\boldsymbol{x}}_0 - \boldsymbol{x}(0)]^T\right\} = \boldsymbol{N}_0 \tag{8.225}$$

The mathematical model of the measurement process is of the form

$$\boldsymbol{y}(t) = \boldsymbol{C}\boldsymbol{x}(t) + \boldsymbol{\xi}(t) \tag{8.226}$$

where

$$E\{\boldsymbol{\xi}(t)\} = \boldsymbol{0} \tag{8.227}$$

$$\text{Cov}(\boldsymbol{\xi}(t), \boldsymbol{\xi}(\tau)) = E\left\{\boldsymbol{\xi}(t)\boldsymbol{\xi}^T(\tau)\right\} = \boldsymbol{S}\delta(t-\tau) \tag{8.228}$$

$\boldsymbol{\xi}_0(t)$ and $\boldsymbol{\xi}(t)$ are uncorrelated in time and with the initial state.

The problem is now to find a state estimate so that the cost function

$$\begin{aligned} I = {} & \frac{1}{2}\left[\boldsymbol{x}(0) - \bar{\boldsymbol{x}}_0\right]^T \boldsymbol{N}_0^{-1}\left[\boldsymbol{x}(0) - \bar{\boldsymbol{x}}_0\right] \\ & + \frac{1}{2}\int_0^{t_f} \left(\left[\dot{\boldsymbol{x}}(t) - \boldsymbol{A}\boldsymbol{x}(t)\right]^T \boldsymbol{V}^{-1}\left[\dot{\boldsymbol{x}}(t) - \boldsymbol{A}\boldsymbol{x}(t)\right]\right) \mathrm{d}t \\ & + \frac{1}{2}\int_0^{t_f} \left(\left[\boldsymbol{y}(t) - \boldsymbol{C}\boldsymbol{x}(t)\right]^T \boldsymbol{S}^{-1}\left[\boldsymbol{y}(t) - \boldsymbol{C}\boldsymbol{x}(t)\right]\right) \mathrm{d}t \end{aligned} \tag{8.229}$$

is minimised. The first right-hand term in (8.229) minimises the squared estimation error in initial conditions. The second term minimises the integral containing squared model error. Finally, the third term penalises the squared measurement error.

Let us define a fictitious control vector

$$\boldsymbol{u}(t) = \dot{\boldsymbol{x}}(t) - \boldsymbol{A}\boldsymbol{x}(t) \tag{8.230}$$

The cost function I is then of the form

$$I = \frac{1}{2}\left[\boldsymbol{x}(0) - \bar{\boldsymbol{x}}_0\right]^T \boldsymbol{N}_0^{-1}\left[\boldsymbol{x}(0) - \bar{\boldsymbol{x}}_0\right] + \frac{1}{2}\int_0^{t_f}\left[\boldsymbol{u}^T(t)\boldsymbol{V}^{-1}\boldsymbol{u}(t)\right] + \left[\boldsymbol{y}(t) - \boldsymbol{C}\boldsymbol{x}(t)\right]^T \boldsymbol{S}^{-1}\left[\boldsymbol{y}(t) - \boldsymbol{C}\boldsymbol{x}(t)\right]\mathrm{d}t \tag{8.231}$$

and the optimal state estimation problem is then transformed into the deterministic LQ control problem. The aim of this *deterministic optimal control* is to find the control $\boldsymbol{u}(t)$ that minimises I of the form (8.231) subject to constraint

$$\dot{\boldsymbol{x}} = \boldsymbol{A}\boldsymbol{x} + \boldsymbol{u} \tag{8.232}$$

Hamiltonian of this problem is defined as

$$H = \frac{1}{2}\left[\boldsymbol{u}^T\boldsymbol{V}^{-1}\boldsymbol{u} + (\boldsymbol{y} - \boldsymbol{C}\boldsymbol{x})^T \boldsymbol{S}^{-1}(\boldsymbol{y} - \boldsymbol{C}\boldsymbol{x})\right] + \boldsymbol{\lambda}^T(\boldsymbol{A}\boldsymbol{x} + \boldsymbol{u}) \tag{8.233}$$

The adjoint vector $\boldsymbol{\lambda}(t)$ is given as

$$\begin{aligned}\dot{\boldsymbol{\lambda}} &= -\frac{\partial H}{\partial \boldsymbol{x}} \\ &= \boldsymbol{C}^T\boldsymbol{S}^{-1}\boldsymbol{y} - \boldsymbol{C}^T\boldsymbol{S}^{-1}\boldsymbol{C}\boldsymbol{x} - \boldsymbol{A}^T\boldsymbol{\lambda}\end{aligned} \tag{8.234}$$

Final state is not fixed and gives the terminal condition for the adjoint vector

$$\boldsymbol{\lambda}(t_f) = \boldsymbol{0} \tag{8.235}$$

$\boldsymbol{x}(0)$ is free as well and thus

$$\boldsymbol{x}(0) = \bar{\boldsymbol{x}}_0 + \boldsymbol{N}_0\boldsymbol{\lambda}(0) \tag{8.236}$$

Optimal control follows from the optimality condition

$$\frac{\partial H}{\partial \boldsymbol{u}} = \boldsymbol{0} \tag{8.237}$$

This gives

$$\boldsymbol{V}^{-1}\boldsymbol{u} + \boldsymbol{\lambda} = 0 \tag{8.238}$$

$$\boldsymbol{u}(t) = -\boldsymbol{V}\boldsymbol{\lambda}(t) \tag{8.239}$$

We note that our original problem is the state estimation of the system with random signals. This problem is commonly denoted as filtration and belongs to a broader class of interpolation, filtration, and prediction. All three problems are closely tied together and the same mathematical apparatus can be used to solve each of them. In the sequel only the filtration problem in the Kalman sense will be investigated.

Filtration can be thought of as the state estimation at time t' where all information up to time t' is used.

Let $\hat{\boldsymbol{x}}\,(t|t_f)$ be the optimal estimate at time t_f with known data $\boldsymbol{y}(t)$ up to time t_f. The optimal estimate can be determined as

$$\dot{\hat{\boldsymbol{x}}}\,(t|t_f) = \boldsymbol{A}\hat{\boldsymbol{x}}\,(t|t_f) - \boldsymbol{V}\boldsymbol{\lambda}(t) \tag{8.240}$$

Let us introduce a transformation

$$\hat{\boldsymbol{x}}\,(t|t_f) = \boldsymbol{z}(t) - \boldsymbol{N}(t)\boldsymbol{\lambda}(t) \tag{8.241}$$

The problem of finding the optimal state estimate can be solved if $\boldsymbol{z}(t)$ and $\boldsymbol{N}(t)$ will be found. Equation (8.240) can be rewritten using the transformation (8.241) as

$$\dot{\boldsymbol{z}}(t) - \dot{\boldsymbol{N}}(t)\boldsymbol{\lambda}(t) - \boldsymbol{N}(t)\dot{\boldsymbol{\lambda}}(t) = \boldsymbol{A}\left[\boldsymbol{z}(t) - \boldsymbol{N}(t)\boldsymbol{\lambda}(t)\right] - \boldsymbol{V}\boldsymbol{\lambda}(t) \tag{8.242}$$

Substituting $\dot{\boldsymbol{\lambda}}(t)$ from (8.234) and using (8.241) gives

$$\begin{aligned}
&\dot{\boldsymbol{z}}(t) - \dot{\boldsymbol{N}}(t)\boldsymbol{\lambda}(t) \\
&- \boldsymbol{N}(t)\left[\boldsymbol{C}^T\boldsymbol{S}^{-1}\boldsymbol{y}(t) - \boldsymbol{C}\boldsymbol{S}^{-1}\boldsymbol{C}\left(\boldsymbol{z}(t) - \boldsymbol{N}(t)\boldsymbol{\lambda}(t)\right) - \boldsymbol{A}^T\boldsymbol{\lambda}(t)\right] \\
&\qquad = \boldsymbol{A}\left[\boldsymbol{z}(t) - \boldsymbol{N}(t)\boldsymbol{\lambda}(t)\right] - \boldsymbol{V}\boldsymbol{\lambda}(t)
\end{aligned} \tag{8.243}$$

and after some manipulations

$$\begin{aligned}
&\dot{\boldsymbol{z}}(t) - \boldsymbol{N}(t)\boldsymbol{C}^T\boldsymbol{S}^{-1}\left(\boldsymbol{y}(t) - \boldsymbol{C}\boldsymbol{z}(t)\right) - \boldsymbol{A}\boldsymbol{z}(t) \\
&\quad = \left[\dot{\boldsymbol{N}}(t) - \boldsymbol{N}(t)\boldsymbol{A}^T - \boldsymbol{A}\boldsymbol{N}(t) + \boldsymbol{N}(t)\boldsymbol{C}\boldsymbol{S}^{-1}\boldsymbol{C}\boldsymbol{N}(t) - \boldsymbol{V}\right]\boldsymbol{\lambda}(t)
\end{aligned} \tag{8.244}$$

We can choose $\boldsymbol{z}(t)$ and $\boldsymbol{N}(t)$ such that

$$\dot{\boldsymbol{z}}(t) = \boldsymbol{A}\boldsymbol{z}(t) + \boldsymbol{N}(t)\boldsymbol{C}^T\boldsymbol{S}^{-1}[\boldsymbol{y}(t) - \boldsymbol{C}\boldsymbol{z}(t)] \tag{8.245}$$

$$\boldsymbol{z}(0) = \bar{\boldsymbol{x}}_0 \tag{8.246}$$

$$\boldsymbol{V} = \dot{\boldsymbol{N}}(t) - \boldsymbol{N}(t)\boldsymbol{A}^T - \boldsymbol{A}\boldsymbol{N}(t) + \boldsymbol{N}(t)\boldsymbol{C}\boldsymbol{S}^{-1}\boldsymbol{C}\boldsymbol{N}(t) \tag{8.247}$$

$$\boldsymbol{N}(0) = \boldsymbol{N}_0 \tag{8.248}$$

Note that both equations (8.245) and (8.247) are solved forward in time from $t = 0$.

The state estimate $\hat{\boldsymbol{x}}\,(t_f|t_f)$ is based on available information up to time t_f. If $t = t_f$, the transformation (8.241) is given as

$$\hat{\boldsymbol{x}}\,(t_f|t_f) = \boldsymbol{z}(t_f) - \boldsymbol{N}(t_f)\boldsymbol{\lambda}(t_f) \tag{8.249}$$

Using (8.235) gives

$$\hat{\boldsymbol{x}}\left(t_f|t_f\right) = \boldsymbol{z}(t_f) \tag{8.250}$$

State estimates are thus given continuously at time t as

$$\dot{\hat{\boldsymbol{x}}}\left(t|t\right) = \boldsymbol{A}\hat{\boldsymbol{x}}\left(t|t\right) + \boldsymbol{N}(t)\boldsymbol{C}^T\boldsymbol{S}^{-1}\left[\boldsymbol{y}(t) - \boldsymbol{C}\hat{\boldsymbol{x}}\left(t|t\right)\right] \tag{8.251}$$

$$\hat{\boldsymbol{x}}(0) = \bar{\boldsymbol{x}}_0 \tag{8.252}$$

where $\boldsymbol{N}(t)$ is solution of (8.247). Equations (8.251) and (8.247) describe the Kalman filter.

Its gain is given for steady-state solution of (8.247) as

$$\boldsymbol{L} = \boldsymbol{N}\boldsymbol{C}^T\boldsymbol{S}^{-1} \tag{8.253}$$

Define the estimation error

$$\boldsymbol{e}(t) = \boldsymbol{x}(t) - \hat{\boldsymbol{x}}\left(t|t\right) \tag{8.254}$$

then we can write

$$\dot{\boldsymbol{e}}(t) = \left(\boldsymbol{A} - \boldsymbol{N}(t)\boldsymbol{C}^T\boldsymbol{S}^{-1}\boldsymbol{C}\right)\boldsymbol{e}(t) + \boldsymbol{\xi}_x(t) - \boldsymbol{N}(t)\boldsymbol{C}^T\boldsymbol{S}^{-1}\boldsymbol{\xi}(t) \tag{8.255}$$

where

$$\boldsymbol{e}(0) = \boldsymbol{\xi}_{x0} \tag{8.256}$$

From (8.255) and (8.256) follows

$$\frac{\mathrm{d}}{\mathrm{d}t}\left(E\left\{\boldsymbol{e}(t)\right\}\right) = \left(\boldsymbol{A} - \boldsymbol{N}\boldsymbol{C}^T\boldsymbol{S}^{-1}\boldsymbol{C}\right)E\left\{\boldsymbol{e}(t)\right\}, \quad E\left\{\boldsymbol{e}(0)\right\} = \boldsymbol{0} \tag{8.257}$$

Equation (8.257) gives

$$E\left\{\boldsymbol{e}(t)\right\} = 0 \tag{8.258}$$

and thus the estimate is unbiased. It can be shown that the covariance matrix of the estimate is

$$\mathrm{Cov}\left(\boldsymbol{e}(t)\right) = \boldsymbol{N}(t) \tag{8.259}$$

The covariance matrix $\boldsymbol{N}$ is symmetric and positive semidefinite and it is a solution of the Riccati equation similar to the LQ Riccati equation.

8.6 Analysis of State Feedback with Observer and Polynomial Pole Placement

8.6.1 Properties of State Feedback with Observer

We have seen in previous chapters that the optimal LQ control design leads to state feedback

$$\boldsymbol{u}(t) = -\boldsymbol{K}\boldsymbol{x}(t) \tag{8.260}$$

where $\boldsymbol{K}$ is a constant matrix. This controller implements proportional action and it is not able to guarantee the zero steady-state control error. This drawback can be circumvented by introducing a state feedback of the form

$$\boldsymbol{u}(t) = -\boldsymbol{K}\boldsymbol{x}(t) + \tilde{\boldsymbol{w}}(t) \tag{8.261}$$

However, this state feedback control law cannot usually be implemented as not all states are measurable. A possible remedy is to implement the state feedback as

$$\boldsymbol{u}(t) = -\boldsymbol{K}\hat{\boldsymbol{x}}(t) + \tilde{\boldsymbol{w}}(t) \tag{8.262}$$

where $\hat{\boldsymbol{x}}$ is the deterministic state estimate. The realisation of the feedback (8.262), i. e. state feedback with state observer is shown in Fig. 8.13. It is important to observe that even if the control law of the form (8.262) is realised and $\boldsymbol{x}(t)$ and $\hat{\boldsymbol{x}}(t)$ are different, the stability of the closed-loop system is still assured. Combined state feedback with state observer can be separated into two independent designs as the following theorem states.

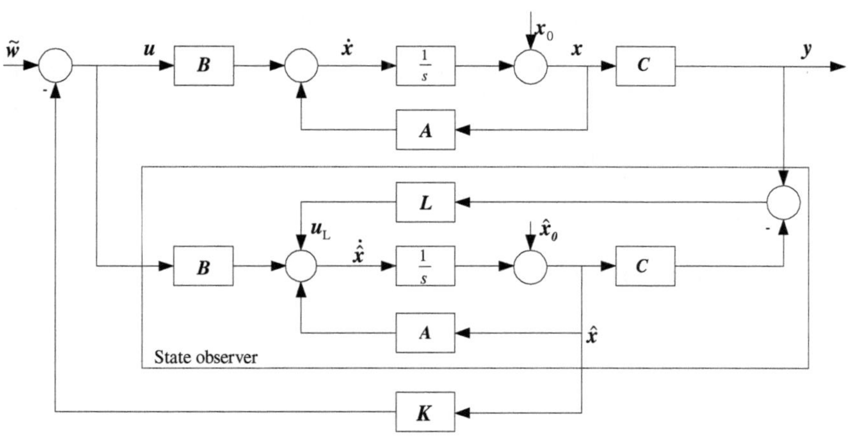

Fig. 8.13. State feedback with observer

Theorem 8.2 (Separation principle). *Consider state the feedback control with observer for a completely controllable and observable system (8.211), (8.212), with observer (8.213) and state feedback (8.262). Closed-loop eigenvalues consist of eigenvalues of matrix* $\boldsymbol{A} - \boldsymbol{BK}$ *(state feedback without observer) and of eigenvalues of matrix* $\boldsymbol{A} - \boldsymbol{LC}$ *(observer dynamics).*

Proof. Substituting (8.262) and (8.215) into (8.211) yields

$$\dot{\boldsymbol{x}}(t) = (\boldsymbol{A} - \boldsymbol{B}\boldsymbol{K})\,\boldsymbol{x}(t) + \boldsymbol{B}\boldsymbol{K}\boldsymbol{e}(t) + \boldsymbol{B}\tilde{\boldsymbol{w}}(t), \quad \boldsymbol{x}(0) = \boldsymbol{x}_0 \tag{8.263}$$

Equations (8.263), (8.219) give the complete closed-loop dynamics

$$\begin{pmatrix} \dot{\boldsymbol{x}}(t) \\ \dot{\boldsymbol{e}}(t) \end{pmatrix} = \begin{pmatrix} \boldsymbol{A} - \boldsymbol{B}\boldsymbol{K} & \boldsymbol{B}\boldsymbol{K} \\ \boldsymbol{0} & \boldsymbol{A} - \boldsymbol{C}\boldsymbol{L} \end{pmatrix} \begin{pmatrix} \boldsymbol{x}(t) \\ \boldsymbol{e}(t) \end{pmatrix} + \begin{pmatrix} \boldsymbol{B} \\ \boldsymbol{0} \end{pmatrix} \tilde{\boldsymbol{w}}(t), \tag{8.264}$$

$$\begin{pmatrix} \boldsymbol{x}(0) \\ \boldsymbol{e}(0) \end{pmatrix} = \begin{pmatrix} \boldsymbol{x}_0 \\ \boldsymbol{e}_0 \end{pmatrix}$$

$$\boldsymbol{y}(t) = (\boldsymbol{C}\ \boldsymbol{0}) \begin{pmatrix} \boldsymbol{x}(t) \\ \boldsymbol{e}(t) \end{pmatrix} \tag{8.265}$$

Eigenvalues of this upper triangular matrix are given by the eigenvalues at the main diagonal. As the closed-loop dynamics is governed by such a matrix, the proof is complete. □

We note that the estimation error is not reachable from the reference input $\tilde{\boldsymbol{w}}$.

The separation principle thus makes it possible to design the state feedback independently on the fact whether all states are measurable or not. If the controlled system is fully controllable, eigenvalues of the closed-loop system can be placed arbitrarily. We illustrate this fact on a singlevariable system.

Consider a singlevariable fully controllable system

$$\dot{\boldsymbol{x}}(t) = \boldsymbol{A}\boldsymbol{x}(t) + \boldsymbol{B}u(t), \quad \boldsymbol{x}(0) = \boldsymbol{x}_0 \tag{8.266}$$

$$y(t) = \boldsymbol{C}\boldsymbol{x}(t) \tag{8.267}$$

It can be transformed using

$$\boldsymbol{x} = \boldsymbol{T}\boldsymbol{x}_R \tag{8.268}$$

to the controllable canonical form

$$\dot{\boldsymbol{x}}_R(t) = \boldsymbol{A}_R\boldsymbol{x}_R(t) + \boldsymbol{B}_R u(t) \tag{8.269}$$

$$y(t) = \boldsymbol{C}_R\boldsymbol{x}_R(t) \tag{8.270}$$

where

$$\boldsymbol{A}_R = \boldsymbol{T}^{-1}\boldsymbol{A}T \tag{8.271}$$

$$\boldsymbol{B}_R = \boldsymbol{T}^{-1}B \tag{8.272}$$

$$\boldsymbol{C}_R = \boldsymbol{C}\boldsymbol{T} \tag{8.273}$$

Matrix $\boldsymbol{T}$ is non-singular and we will not go into details of its construction here. $\boldsymbol{x}_R$ is a new state vector with dimension n. Matrices $\boldsymbol{A}_R$, $\boldsymbol{B}_R$, $\boldsymbol{C}_R$ of the controllable canonical form are

$$\boldsymbol{A}_R = \begin{pmatrix} 0 & 1 & 0 & \dots & 0 \\ 0 & 0 & 1 & \dots & 0 \\ \vdots & & & \ddots & \vdots \\ 0 & 0 & 0 & \cdots & 1 \\ -a_0 & -a_1 & -a_2 & \cdots & -a_{n-1} \end{pmatrix} \tag{8.274}$$

$$\boldsymbol{B}_R = \begin{pmatrix} 0 \\ 0 \\ \vdots \\ 0 \\ 1 \end{pmatrix} \tag{8.275}$$

$$\boldsymbol{C}_R = \begin{pmatrix} b_0 \; b_1 \; \cdots \; b_{n-1} \end{pmatrix} \tag{8.276}$$

The state feedback applied to the system (8.266)

$$u(t) = -\boldsymbol{K}_R \boldsymbol{T}^{-1} \boldsymbol{x}(t) + \tilde{w}(t) \tag{8.277}$$

corresponds to the state feedback (cf. (8.268))

$$u(t) = -\boldsymbol{K}_R \boldsymbol{x}_R(t) + \tilde{w}(t) \tag{8.278}$$

If

$$\boldsymbol{K}_R = \begin{pmatrix} k_{R1} \; k_{R2} \; \cdots \; k_{Rn} \end{pmatrix} \tag{8.279}$$

and the state feedback (8.278) is applied to the system (8.269) then the closed-loop state representation is given by

$$\dot{\boldsymbol{x}}_R(t) = \begin{pmatrix} 0 & 1 & \cdots & 0 \\ 0 & 0 & \ddots & 0 \\ \vdots & & & \vdots \\ 0 & 0 & \cdots & 1 \\ -a_0 - k_{R1} & -a_1 - k_{R2} & \cdots & -a_{n-1} - k_{Rn} \end{pmatrix} \boldsymbol{x}_R(t) + \begin{pmatrix} 0 \\ 0 \\ \vdots \\ 0 \\ 1 \end{pmatrix} \tilde{w}(t) \tag{8.280}$$

The characteristic polynomial of the controlled system is

$$a_R(s) = \det(s\boldsymbol{I} - \boldsymbol{A}_R) = s^n + a_{n-1}s^{n-1} + \cdots + a_1 s + a_0 \tag{8.281}$$

and the characteristic polynomial of the closed-loop system is given as

$$c_{Rcl}(s) = \det(s\boldsymbol{I} - \boldsymbol{A}_{Rcl}) \tag{8.282}$$

where

$$A_{Rcl} = \begin{pmatrix} 0 & 1 & 0 & \cdots & 0 \\ 0 & 0 & 1 & \cdots & 0 \\ \vdots & & & \ddots & \vdots \\ 0 & 0 & 0 & \cdots & 1 \\ -a_0 - k_{R1} & -a_1 - k_{R2} & -a_2 - k_{R3} & \cdots & -a_{n-1} - k_{Rn} \end{pmatrix} \tag{8.283}$$

and after a manipulation

$$c_{Rcl}(s) = s^n + a_{Rn-1}s^{n-1} + \cdots a_{R1}s + a_{R0} \tag{8.284}$$

where

$$\begin{aligned} a_{R0} &= a_0 + k_{R1} \\ a_{R1} &= a_1 + k_{R2} \\ &\vdots \\ a_{Rn-1} &= a_{n-1} + k_{Rn} \end{aligned} \tag{8.285}$$

The last row of the matrix $\boldsymbol{A}_R$ contains coefficients of the characteristic polynomial of the closed-loop system $c_{Rcl}(s)$. Therefore, the choice of elements of matrix $\boldsymbol{K}_R$ places arbitrarily eigenvalues of the closed-loop system.

If $a_{R0}, a_{R1}, \ldots, a_{Rn-1}$ are prescribed then the feedback is given as

$$\boldsymbol{K}_R = \left(a_{R0}\ a_{R1}\ \cdots\ a_{Rn-1}\right) - \left(a_0\ a_1\ \cdots\ a_{n-1}\right) \tag{8.286}$$

We can notice that parameters of a state feedback controller applied to a system in controllable canonical form are given as the difference between desired coefficients of the closed-loop characteristic polynomial and coefficients of the characteristic polynomial of the controlled system.

Of course, the same holds for the closed-loop system consisting of the original system (8.266) and feedback controller (8.277) where

$$\boldsymbol{K} = \boldsymbol{K}_R \boldsymbol{T}^{-1} \tag{8.287}$$

Therefore, the first step in combined state feedback control with observer is to determine matrix $\boldsymbol{K}$. This specifies eigenvalues of the closed-loop system $\boldsymbol{A} - \boldsymbol{BK}$.

The second step is to design an observer. The design consists in placing its eigenvalues provided the controlled system is fully observable. The eigenvalues of the observer are eigenvalues of matrix $\boldsymbol{A} - \boldsymbol{LC}$ and are chosen so that the estimation error $\boldsymbol{e}(t)$ is removed as fast as possible. Consequence of this is high gain matrix $\boldsymbol{L}$ implying that noise $\boldsymbol{\xi}(t)$ has a great influence on the estimation error $\boldsymbol{e}(t)$. This can be seen in

$$\dot{\boldsymbol{e}}(t) = (\boldsymbol{A} - \boldsymbol{LC})\,\boldsymbol{e}(t) - \boldsymbol{L}\boldsymbol{\xi}(t) \tag{8.288}$$

where

$$\boldsymbol{\xi}(t) = \boldsymbol{y}(t) - \boldsymbol{C}\boldsymbol{x}(t) \tag{8.289}$$

8.6.2 Input-Output Interpretation of State Feedback with Observer

We analyse the same problem as before using the input-output representation of the controller and the controlled process. Mainly singlevariable systems will be investigated but some results will also be presented for multivariable systems.

Polynomial interpretation of the state feedback with observer is governed in the following theorem.

Theorem 8.3 (Pole placement design). *Consider a state feedback*

$$u(t) = -\boldsymbol{K}\hat{\boldsymbol{x}}(t) + \tilde{w}(t) \tag{8.290}$$

and an observer

$$\dot{\hat{\boldsymbol{x}}}(t) = \boldsymbol{A}\hat{\boldsymbol{x}}(t) + \boldsymbol{B}u + \boldsymbol{L}\left(y - \boldsymbol{C}\hat{\boldsymbol{x}}\right) \tag{8.291}$$

with a completely reachable and observable system

$$\dot{\boldsymbol{x}}(t) = \boldsymbol{A}\boldsymbol{x}(t) + \boldsymbol{B}u \tag{8.292}$$

$$y(t) = \boldsymbol{C}\boldsymbol{x}(t) \tag{8.293}$$

The state feedback (8.290) with observer can be written as

$$\frac{p(s)}{o(s)}u = -\frac{q(s)}{o(s)}y + \tilde{w} \tag{8.294}$$

where $p(s)$ is a polynomial with degree n and $q(s)$ is a polynomial with degree $n-1$. Polynomials $p(s)$ and $q(s)$ satisfy the equation

$$a(s)p(s) + b(s)q(s) = o(s)f(s) \tag{8.295}$$

and $a(s)$, $b(s)$ are defined as

$$\boldsymbol{C}\left(s\boldsymbol{I} - \boldsymbol{A}\right)^{-1}\boldsymbol{B} = \frac{b(s)}{a(s)} \tag{8.296}$$

$f(s)$ is the characteristic polynomial of the matrix $\boldsymbol{A} - \boldsymbol{B}\boldsymbol{K}$ and $o(s)$ is characteristic polynomial of the matrix $\boldsymbol{A} - \boldsymbol{L}\boldsymbol{C}$.

Proof. To prove the theorem we make use of the following properties of polynomials.

$a(s)$ is the characteristic polynomial of matrix $\boldsymbol{A}$. If the controlled system is reachable and observable then polynomials $a(s)$ and $b(s)$ are coprime. This is an important condition related to the solution of equation (8.295).

If the controlled system is reachable then $a(s)$ and $\boldsymbol{B}_{Rs}(s)$ given by

$$(s\boldsymbol{I} - \boldsymbol{A})^{-1} \boldsymbol{B} = \frac{\boldsymbol{B}_{Rs}(s)}{a(s)} \tag{8.297}$$

are right coprime. Equation (8.297) can be written as

$$(s\boldsymbol{I} - \boldsymbol{A})\boldsymbol{B}_{Rs}(s) = \boldsymbol{B}a(s) \tag{8.298}$$

Adding $\boldsymbol{BKB}_{Rs}(s)$ to both sides of equation gives

$$(s\boldsymbol{I} - \boldsymbol{A})\, \boldsymbol{B}_{Rs}(s) + \boldsymbol{BKB}_{Rs}(s) = \boldsymbol{B}a(s) + \boldsymbol{BKB}_{Rs}(s) \tag{8.299}$$

$$(s\boldsymbol{I} - (\boldsymbol{A} - \boldsymbol{BK}))\, \boldsymbol{B}_{Rs}(s) = \boldsymbol{B}\,(a(s) + \boldsymbol{KB}_{Rs}(s)) \tag{8.300}$$

$$\frac{\boldsymbol{B}_{Rs}(s)}{a(s) + \boldsymbol{KB}_{Rs}(s)} = \frac{\operatorname{adj}(s\boldsymbol{I} - (\boldsymbol{A} - \boldsymbol{BK}))}{\det(s\boldsymbol{I} - (\boldsymbol{A} - \boldsymbol{BK}))}\boldsymbol{B} \tag{8.301}$$

From (8.301) follows that the feedback gain can be calculated from equation

$$a(s) + \boldsymbol{KB}_{Rs}(s) = f(s) \tag{8.302}$$

where $f(s)$ is the characteristic polynomial of the closed-loop matrix $\boldsymbol{A} - \boldsymbol{BK}$. From stability of the closed-loop system follows that the polynomial $a(s) + \boldsymbol{KB}_{Rs}(s)$ is stable as well.

The observer equations can be written as

$$s\hat{\boldsymbol{x}} = (\boldsymbol{A} - \boldsymbol{LC})\,\hat{\boldsymbol{x}} + \boldsymbol{B}u + \boldsymbol{L}y \tag{8.303}$$

$$(s\boldsymbol{I} - (\boldsymbol{A} - \boldsymbol{LC}))\,\hat{\boldsymbol{x}} = \boldsymbol{B}u + \boldsymbol{L}y \tag{8.304}$$

$$\hat{\boldsymbol{x}} = (s\boldsymbol{I} - (\boldsymbol{A} - \boldsymbol{LC}))^{-1}\,\boldsymbol{B}u + (s\boldsymbol{I} - (\boldsymbol{A} - \boldsymbol{LC}))^{-1}\,\boldsymbol{L}y \tag{8.305}$$

$$\hat{\boldsymbol{x}} = \frac{\boldsymbol{T}_u(s)}{o(s)}u + \frac{\boldsymbol{T}_y(s)}{o(s)}y \tag{8.306}$$

where

$$\boldsymbol{T}_u(s) = \operatorname{adj}(s\boldsymbol{I} - (\boldsymbol{A} - \boldsymbol{LC}))\,\boldsymbol{B} \tag{8.307}$$

$$\boldsymbol{T}_y(s) = \operatorname{adj}(s\boldsymbol{I} - (\boldsymbol{A} - \boldsymbol{LC}))\,\boldsymbol{L} \tag{8.308}$$

$$o(s) = \det(sI - (\boldsymbol{A} - \boldsymbol{LC})) \tag{8.309}$$

Equation (8.306) demonstrates that the observer can be defined as in Fig. 8.14. If the output of the system in Fig. 8.14 would be the variable $\tilde{u} = \boldsymbol{K}\hat{\boldsymbol{x}}$ then

$$\boldsymbol{K}\hat{\boldsymbol{x}} = \frac{r(s)}{o(s)}u + \frac{q(s)}{o(s)}y \tag{8.310}$$

where

$$r(s) = \boldsymbol{KT}_u(s) \tag{8.311}$$

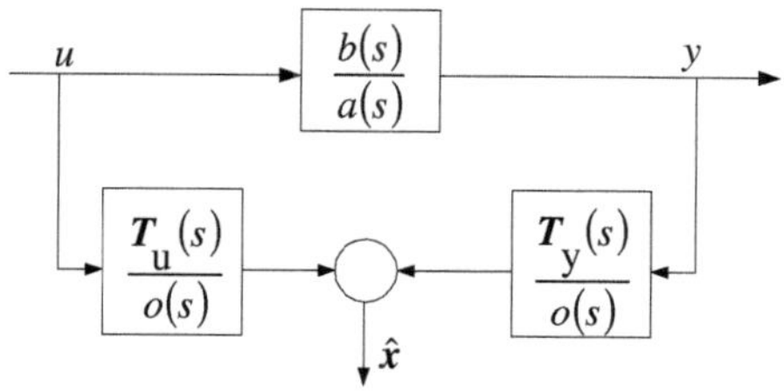

Fig. 8.14. Block diagram of state estimation of a singlevariable system

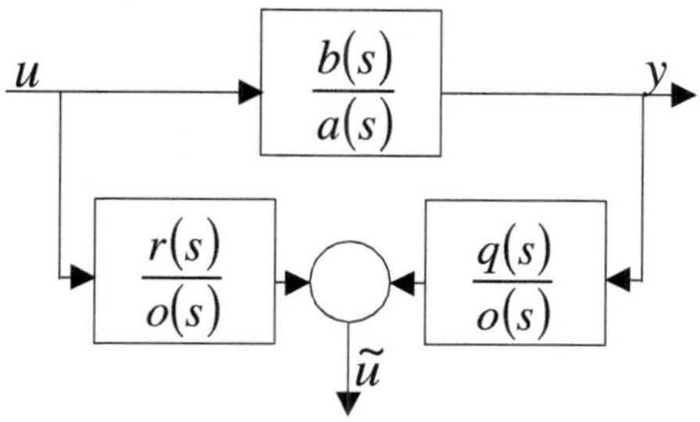

Fig. 8.15. Polynomial state estimation

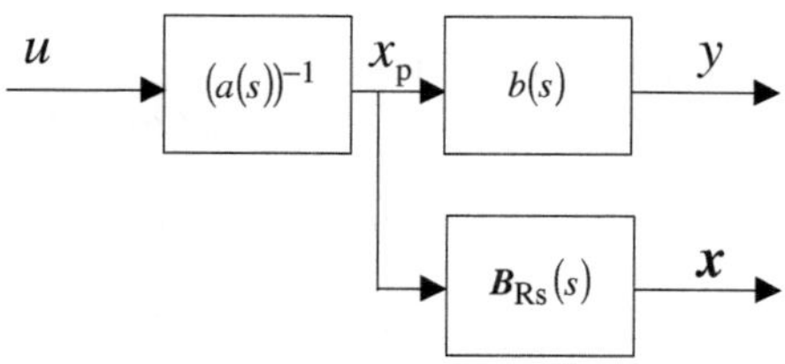

Fig. 8.16. Interpretation of a pseudo-state

$$q(s) = \boldsymbol{K}\boldsymbol{T}_y(s) \tag{8.312}$$

Equation (8.310) is shown in the block diagram in Fig. 8.15.

If we define the pseudo-state as in Fig. 8.16 then

$$y = b(s)x_p \tag{8.313}$$

$$u = a(s)x_p \tag{8.314}$$

$$\boldsymbol{x} = \boldsymbol{B}_{Rs}(s)x_p \tag{8.315}$$

$$\boldsymbol{K}\boldsymbol{x} = k(s)x_p \tag{8.316}$$

where

$$k(s) = \boldsymbol{K}\boldsymbol{B}_{Rs}(s) \tag{8.317}$$

Substitution of (8.313) and (8.314) into (8.310) and using (8.316) gives

$$o(s)\tilde{u} - o(s)\boldsymbol{K}\boldsymbol{x} = r(s)a(s)x_p + q(s)b(s)x_p - o(s)k(s)x_p \tag{8.318}$$

$$o(s)\left(\tilde{u} - \boldsymbol{K}\boldsymbol{x}\right) = \left(r(s)a(s) + q(s)b(s) - o(s)k(s)\right)x_p \tag{8.319}$$

Asymptotic observer is described by (8.310) if and only if $o(s)$ is a stable polynomial and $r(s)$, $q(s)$ satisfy equation

$$r(s)a(s) + q(s)b(s) = o(s)k(s) \tag{8.320}$$

From (8.309) follows that the degree of the polynomial $o(s)$ is n. This corresponds to a full order observer. If the state estimation has been realised by a reduced order observer, polynomial $o(s)$ would be a stable monic polynomial with degree $n-1$.

Let us return to the feedback (8.290). If $\boldsymbol{K}\hat{\boldsymbol{x}}$ from (8.310) is substituted into (8.290) then

$$u = -\frac{r(s)}{o(s)}u - \frac{q(s)}{o(s)}y + \tilde{w} \tag{8.321}$$

If

$$p(s) = o(s) + r(s) \tag{8.322}$$

then equation (8.321) reduces to (8.294).

Adding $a(s)o(s)$ to either side of equation (8.320) gives

$$r(s)a(s) + q(s)b(s) + a(s)o(s) = o(s)k(s) + a(s)o(s) \tag{8.323}$$

$$a(s)\left(r(s) + o(s)\right) + q(s)b(s) = o(s)\left(k(s) + a(s)\right) \tag{8.324}$$

Taking into account equations (8.302), (8.317), (8.322) then (8.324) is of the form (8.295). This concludes the proof. □

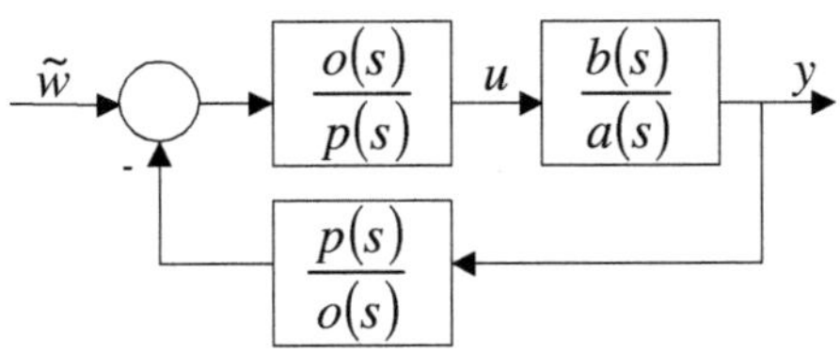

Fig. 8.17. Realisation of a state feedback with an observer

Pole placement theorem thus shows that the state feedback with observer can be implemented as shown in Fig. 8.17. Transfer function between input $\tilde{w}$ and output y is given as

$$y = \frac{b(s)o(s)}{a(s)p(s) + b(s)q(s)}\tilde{w} \tag{8.325}$$

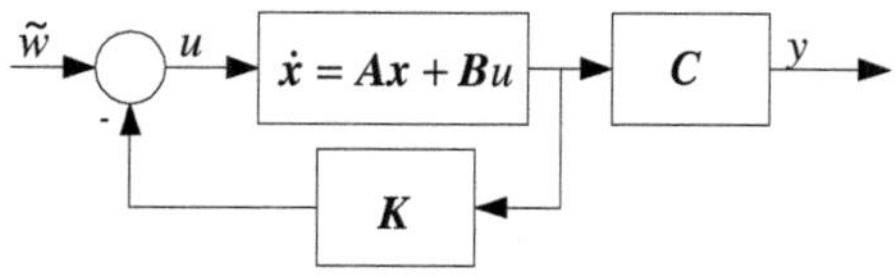

Fig. 8.18. State feedback control

$$y = \frac{b(s)}{f(s)}\tilde{w} \tag{8.326}$$

Based on Fig. 8.18 this further gives

$$y = \boldsymbol{C}\left(s\boldsymbol{I} - (\boldsymbol{A} - \boldsymbol{B}\boldsymbol{K})\right)^{-1}\boldsymbol{B}\tilde{w} \tag{8.327}$$

$$y = \boldsymbol{C}\frac{\operatorname{adj}\left(s\boldsymbol{I} - (\boldsymbol{A} - \boldsymbol{B}\boldsymbol{K})\right)}{\det\left(s\boldsymbol{I} - (\boldsymbol{A} - \boldsymbol{B}\boldsymbol{K})\right)}\boldsymbol{B}\tilde{w} \tag{8.328}$$

From (8.301) follows

$$\boldsymbol{B}_{Rs}(s) = \operatorname{adj}\left(s\boldsymbol{I} - (\boldsymbol{A} - \boldsymbol{B}\boldsymbol{K})\right)\boldsymbol{B} \tag{8.329}$$

and from (8.296), (8.297), (8.301), (8.302) implies

$$\boldsymbol{C}\boldsymbol{B}_{Rs}(s) = b(s) \tag{8.330}$$

$$\det\left(s\boldsymbol{I} - (\boldsymbol{A} - \boldsymbol{B}\boldsymbol{K})\right) = f(s) \tag{8.331}$$

The relation between the closed-loop input $\tilde{w}$ and output y in figures 8.17 and 8.18 is thus the same. Hence, the pole placement design is entirely the same in both state-space and input-output representations. However, the closed-loop structures in figures 8.17, 8.18 are different.

From (8.307) and (8.308) follows that $\boldsymbol{T}_u(s)$ and $\boldsymbol{T}_y(s)$ are n-dimensional vectors with elements being polynomials with degree $n-1$. As $o(s)$ is a polynomial with degree n, the polynomial form of the observer consists of strictly proper transfer functions. The right-hand side polynomial in equation (8.295) has degree $2n$.

The fact that derivation of (8.324) involves cancellation of $o(s)$ implies that estimated states are not controllable by the input $\tilde{w}$.

The realisation of state feedback with observer in Fig. 8.17 takes the form of a closed-loop system with the two-degree-of-freedom controller.

If the closed-loop input in Fig. 8.17 is given as

$$\tilde{w} = \frac{q(s)}{o(s)}w \tag{8.332}$$

then (8.294) can be written as

$$u = -\frac{q(s)}{p(s)}y + \frac{q(s)}{p(s)}w \tag{8.333}$$

or

$$u = -\frac{q(s)}{p(s)}(w - y) \tag{8.334}$$

The choice of $\tilde{w}$ from (8.332) changes the closed-loop system in Fig. 8.17 from a two-degree-of-freedom to an one-degree-of-freedom system (Fig. 8.19).

The transfer function between w and y is given as

$$y = \frac{b(s)q(s)}{o(s)f(s)}w \tag{8.335}$$

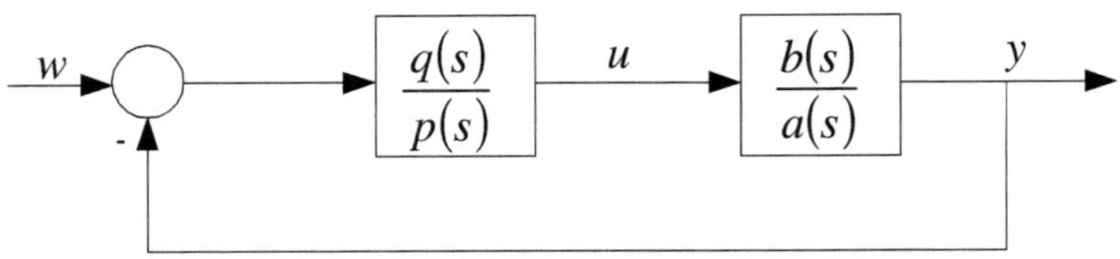

Fig. 8.19. Feedback control with one-degree-of-freedom controller

The design procedure of the polynomial pole placement controller as an alternative to the state feedback controller with observer is as follows:

1. Given are state-space matrices of the controlled process $\boldsymbol{A}$, $\boldsymbol{B}$, $\boldsymbol{C}$. Gain matrices $\boldsymbol{K}$ and $\boldsymbol{L}$ are also given.
2. Polynomials $a(s)$ and $b(s)$ are calculated from

$$\boldsymbol{C}\left(s\boldsymbol{I} - \boldsymbol{A}\right)^{-1}\boldsymbol{B} = \frac{b(s)}{a(s)} \tag{8.336}$$

3. Controller polynomials $p(s)$ and $q(s)$ are found from the equation

$$a(s)p(s) + b(s)q(s) = o(s)f(s) \tag{8.337}$$

where

$$o(s) = \det\left(s\boldsymbol{I} - (\boldsymbol{A} - \boldsymbol{L}\boldsymbol{C})\right) \tag{8.338}$$

$$f(s) = \det\left(s\boldsymbol{I} - (\boldsymbol{A} - \boldsymbol{B}\boldsymbol{K})\right) \tag{8.339}$$

are stable polynomials. Polynomial $c(s) = o(s)f(s)$ has the degree $2n$ (full order observer) or $2n - 1$ (reduced order observer).

www

Example 8.4:

Consider a controlled process of the form

$$G(s) = \frac{s+6}{s^2 + 5s + 6}$$

We choose the closed-loop characteristic polynomial as

$$o(s)f(s) = (s+2)(s+4)(s+5)(s+6)$$

Our aim is to find the transfer function of a controller that places the closed-loop poles to locations specified by zeros of $o(s)f(s)$. This transfer function is of the form

$$R(s) = \frac{q_1 s + q_0}{p_2 s^2 + p_1 s + p_0}$$

and the coefficients p_2, p_1, p_0, q_1, q_0 can be found as the solution of equation

$$\left(s^2+5s+6\right)\left(p_2 s^2 + p_1 s + p_0\right) + (s+6)(q_1 s + q_0) = (s+2)(s+4)(s+5)(s+6)$$

Equating coefficients at either side of this equation gives five equations with five unknowns

$$\begin{aligned} s^4 &: p_2 = 1 \\ s^3 &: 5p_2 + p_1 = 17 \\ s^2 &: 6p_2 + 5p_1 + p_0 + q_1 = 104 \\ s &: 6p_1 + 5p_0 + 6q_1 + q_0 = 268 \\ s^0 &: 6p_0 + 6q_0 = 240 \end{aligned}$$

The controller polynomials are then given as

$$\begin{aligned} p(s) &= s^2 + 12s + 36 \\ q(s) &= 2s + 4 \end{aligned}$$

The derivation of the controller shows that pole placement as the only performance criterium need not be satisfactory. For example, as the controller does not contain integral part, the steady-state control error is non-zero – if unit step change of the setpoint is assumed then is settles at the value of 0.25.

The state feedback with observer interpreted using the input-output models naturally leads to the problem of the polynomial pole placement.

Calculation of $o(s)$ and $f(s)$ from (8.309), (8.331) is frequently not necessary as the closed-loop poles can directly be assigned by the choice of a polynomial $c(s)$ on the right-hand side of the equation

$$a(s)p(s) + b(s)q(s) = c(s) \tag{8.340}$$

Similarly, polynomials $a(s)$, $b(s)$ are often known and it is not necessary to solve equation (8.296). Equation (8.340) plays a fundamental role in polynomial pole placement.

8.6.3 Diophantine Equations

A scalar linear polynomial equation

$$a(s)x(s) + b(s)y(s) = c(s) \tag{8.341}$$

is called the Diophantine equation – after the mathematician Diophantos from Alexandria (275 a. d.).

Polynomials $a(s)$, $b(s)$, and $c(s)$ are given and polynomials $x(s)$, $y(s)$ are unknown. The solvability of a Diophantine equation is characterised by the next theorem.

Theorem 8.4 (Solvability of Diophantine equations). *Let $a(s)$, $b(s)$, $c(s)$ be real valued polynomials. Equation (8.341) has a solution if and only if the greatest common divisor of $a(s)$, $b(s)$ divides $c(s)$.*

It follows from the theorem that if $a(s)$ and $b(s)$ are coprime then the Diophantine equation is solvable for any $c(s)$ including $c(s) = 1$.

If Diophantine equation is solvable then it has infinitely many solutions. If $x'(s)$, $y'(s)$ denote a particular solution then general solution can be parametrised as

$$x(s) = x'(s) + \bar{b}(s)t(s) \tag{8.342}$$

$$y(s) = y'(s) - \bar{a}(s)t(s) \tag{8.343}$$

where $t(s)$ is an arbitrary polynomial, $\bar{a}(s)$, $\bar{b}(s)$ are coprime and

$$\frac{\bar{b}(s)}{\bar{a}(s)} = \frac{b(s)}{a(s)} \tag{8.344}$$

If $a(s)$, $b(s)$ are coprime then $\bar{a}(s) = a(s)$, $\bar{b}(s) = b(s)$.

Corollary: Among all solutions of the Diophantine equation there is one $x(s)$, $y(s)$ where

$$\deg x(s) < \deg \bar{b}(s) \tag{8.345}$$

Similarly, there exists unique solution of the same Diophantine equation where

$$\deg y(s) < \deg \bar{a}(s) \tag{8.346}$$

Both solutions coincide if

$$\deg a(s) + \deg b(s) \geq \deg c(s) \tag{8.347}$$

If equation (8.347) holds then the Diophantine equation has the minimum degree solution and

$$\deg x(s) = \deg y(s) \tag{8.348}$$

If equation (8.347) does not hold, there are solutions for minimal degree of $x(s)$ or for minimal degree of $y(s)$.

A versatile tool for Diophantine equations is *Polynomial Toolbox for MATLAB*. Its use will be illustrated on examples.

W **Example 8.5:**

Consider polynomials (see Example 8.4) $a(s) = s^2 + 5s + 6$, $b(s) = s + 6$. Find polynomials $x(s)$, $y(s)$ satisfying equation

$$a(s)x(s) + b(s)y(s) = c(s)$$

for various choices of polynomial $c(s)$. Polynomial Toolbox for MATLAB will be used – function `[x,y]=axbyc(a,b,c)` where the function arguments are the same as our polynomials.

Case: $c(s) = 1 + s$. Entering the following commands in MATLAB

```
>> a  = 6 + 5*s+s^2;
>> b  = 6 + s;
>> [x, y] = axbyc(a, b, 1+s)
```

gives the solution:

```
x  = -0.42
y  = 0.58 + 0.42 s
```

Case: $c(s) = (1 + s)^2$:

```
>> [x, y] = axbyc(a, b, (1+s)^2)
x  = 2.1
y  = 1.9 - 1.1 s
```

Case: $c(s) = (1 + s)^3$ (note: inequality (8.347) holds)

```
>> [x, y] = axbyc(a, b, (1+s)^3)
x  = -4.4 + s
y  = 4.6 + 2.4 s
```

Case: $c(s) = (1 + s)^4$ (note: inequality (8.347) does not hold)

```
>> [x1, y1] = axbyc(a, b, (1+s)^4)
x1 = 1.3 - 2.5 s + s^2
y1 = -1.2 + 2.2 s + 1.5 s^2
```

This is minimum degree solution where $\deg(x_1) = \deg(y_1) = 2$.

Case: $c(s) = (1 + s)^4$. Minimise the degree of polynomial x.

```
>> [x2, y2] = axbyc(a, b, (1+s)^4, 'minx')
x2 = 52
y2 = -52 - 34s - 2 s^2 + s^3
```

Case: $c(s) = (1 + s)^4$. Minimise the degree of polynomial y.

```
>> [x2, y2] = axbyc(a, b, (1+s)^4, 'miny')
x2 = 10 - s + s^2
y2 = -9.9 - 5.1s
```

Case: $c = 1+s$. General solution of the form $x(s)+f(s)t(s)$, $y(s)+g(s)t(s)$ is calculated where $t(s)$ is arbitrary polynomial.

```
>> [x,y,f,g] = axbyc(a, b, 1+s)
x = -0.42
y = 0.58 + 0.42s
f = -0.6 - 0.1s
```

```
g = 0.6 + 0.5s + 0.1 s^2
```

8.6.4 Polynomial Pole Placement Control Design

This control design is closely related to the solution of the Diophantine equation (8.340).

Consider the feedback system shown in Fig. 8.19. The process transfer function is of the form

$$G(s) = \frac{b(s)}{a(s)} \tag{8.349}$$

where polynomials

$$a(s) = a_n s^n + a_{n-1} s^{n-1} + \ldots + a_0 \tag{8.350}$$

$$b(s) = b_{n-1} s^{n-1} + b_{n-2} s^{n-2} + \ldots + b_0 \tag{8.351}$$

are coprime and the system is strictly proper.

The controller transfer function is

$$R(s) = \frac{q(s)}{p(s)} \tag{8.352}$$

where

$$p(s) = p_{n_p} s^{n_p} + p_{n_p-1} s^{n_p-1} + \ldots + p_0 \tag{8.353}$$

$$q(s) = q_{n_q} s^{n_q} + q_{n_q-1} s^{n_q-1} + \ldots + q_0 \tag{8.354}$$

Desired stable characteristic polynomial of the closed-loop system is specified as

$$c(s) = c_{n_c} s^{n_c} + c_{n_c-1} s^{n_c-1} + \ldots + c_0 \tag{8.355}$$

PP controller, case 1: Let $c(s)$ be an arbitrary polynomial of degree $n_c = 2n - 1$. Then there are polynomials $p(s)$ of degree $n_p = n - 1$ and $q(s)$ of degree $n_q = n - 1$ such that

$$a(s)p(s) + b(s)q(s) = c(s) \tag{8.356}$$

holds (see Example 8.5 – case 3). Controller $R(s)$ is proper.

PP controller, case 2: Let $c(s)$ be an arbitrary polynomial of degree $n_c = 2n$. Then there are polynomials $p(s)$ of degree $n_p = n$ and $q(s)$ of degree $n_q = n - 1$ such that equation (8.356) holds (see Example 8.5 – case 6). Controller $R(s)$ is strictly proper.

We note that the choice of the polynomial degrees is not arbitrary. The simplest controller is that with minimum degree of $q(s)$. From the Diophantine equation (8.356) follows that the minimum degree of $q(s)$ is less than the degree of $a(s)$, thus $n_q = n - 1$. The degree of the polynomial $p(s)$ is them given from realisability of the controller. Its degree has to be equal (PP 1) or larger (PP 2) than the degree of $q(s)$. Finally, the degree of $c(s)$ can be determined from the degree of the left hand side of (8.356).

8.6.5 Integrating Behaviour of Controller

The closed-loop system consisting of the controlled system (8.296) and controller (8.334) calculated from (8.295) cannot guarantee the zero steady-state control error. As we have seen in previous chapters, to solve this issue, pure integrator is needed in the controller as in Fig. 8.20.

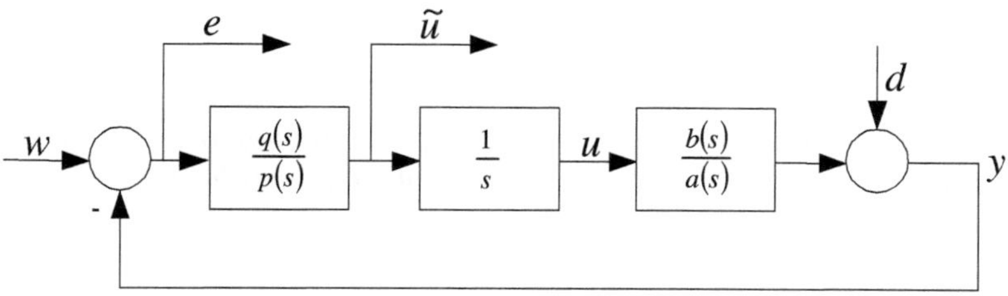

Fig. 8.20. Closed-loop system with a pure integrator

The controller formed by the blocks $q(s)/p(s)$ and $1/s$ assures that for step changes of the setpoint variable w and step changes of disturbance d approach the control error e and variable $\tilde{u}$ in the loop in Fig. 8.20 asymptotically zero if $q(s)$ and $p(s)$ are solution of the equation

$$a(s)sp(s) + b(s)q(s) = c(s) \tag{8.357}$$

and $c(s)$ is a stable polynomial.

This can easily be proved using transfer functions between inputs w, d and outputs e, $\tilde{u}$ of the closed-loop in Fig. 8.20. These are of the form

$$G_{ew}(s) = \frac{a(s)sp(s)}{a(s)sp(s) + b(s)q(s)} \tag{8.358}$$

$$G_{ed}(s) = \frac{-a(s)sp(s)}{a(s)sp(s) + b(s)q(s)} \tag{8.359}$$

$$G_{\tilde{u}w}(s) = \frac{a(s)sq(s)}{a(s)sp(s) + b(s)q(s)} \tag{8.360}$$

$$G_{\tilde{u}d}(s) = \frac{-a(s)sq(s)}{a(s)sp(s) + b(s)q(s)} \tag{8.361}$$

Closed-loop poles are characterised by the polynomial $a(s)sp(s) + b(s)q(s)$, hence $c(s)$ must be stable polynomial.

If $w(t) = 1(t)$ then the final value theorem gives

$$\lim_{t\to\infty} e(t) = \lim_{t\to\infty} \tilde{u}(t) = 0 \tag{8.362}$$

Equation (8.362) also holds for $d(t) = 1(t)$. This proves asymptotic properties of the pole placement controller with integrator.

The design procedure of the controller with integral action is as follows. First the controlled system is formally modified to contain a pure integrator. Polynomials $p(s)$, $q(s)$ are calculated from the Diophantine equation

$$\tilde{a}(s)p(s) + b(s)q(s) = c(s) \tag{8.363}$$

where

$$\tilde{a}(s) = sa(s) \tag{8.364}$$

with degree $n_I = n + 1$.

In the second step after polynomials $p(s)$, $q(s)$ are found from (8.363) the integrator is moved to the controller. The controller transfer function is then given as

$$R(s) = \frac{q(s)}{\tilde{p}(s)} \tag{8.365}$$

where

$$\tilde{p}(s) = sp(s) \tag{8.366}$$

PP controller with integral action, case 1: Let $c(s)$ be an arbitrary polynomial of degree $2n$. Then there are polynomials $p(s)$ of degree $n_p = n - 1$ and $q(s)$ of degree $n_q = n$ such that equation (8.363) holds. The controller transfer function $q(s)\ /\ p(s)$ is not proper. However, the real controller is proper.

PP controller with integral action, case 2: Let $c(s)$ be an arbitrary polynomial of degree $n_c = 2n_I - 1 = 2n + 1$. Then there are polynomials $p(s)$ of degree $n_p = n_I - 1 = n$ and $q(s)$ of degree $n_q = n_I - 1 = n$ such that equation (8.363) holds. The controller transfer function $q(s)/p(s)$ is proper. The real controller corresponds to the transfer function (8.365) and is strictly proper.

PP controller with integral action, case 3: Let $c(s)$ be an arbitrary polynomial of degree $n_c = 2n_I = 2(n + 1)$. Then there are polynomials $p(s)$ of degree $n_p = n_I = n + 1$ and $q(s)$ of degree $n_q = n_I - 1 = n$ such that equation (8.363) holds. The controller transfer function $q(s)\ /\ p(s)$ is strictly proper.

We note again that the choice of the polynomial degrees is not arbitrary and follows from the requirement that the controller numerator degree $q(s)$ should be minimal. In this case it is given by the Diophantine equation (8.363) as $n_q = n$. The controller cases then follow from the fact that the minimum realisable degree of the controller denominator $sp(s)$ is equal to the numerator degree. Finally, when the degrees of $p(s), q(s)$ are known, the degree of the right hand side polynomial $c(s)$ can de determined from the left hand side polynomial of (8.363). The degree of $c(s)$ can also be larger than $2n_I$. However, such controller would be too complicated. In practice, the controller degrees are usually derived from cases 1 and 2.

Example 8.6:

The model of the controlled process is of the form

$$G(s) = \frac{b(s)}{a(s)}, \quad a(s) = s + 2, \quad b(s) = 1$$

The desired controller transfer function is given as

$$R(s) = \frac{q(s)}{sp(s)}$$

where polynomials $p(s)$, $q(s)$ are solution of the equation

$$a(s)sp(s) + b(s)q(s) = c(s)$$

for various cases of the polynomial $c(s)$. We use Polynomial Toolbox for MATLAB.

Case: $c(s) = (s + 3)^2$. Typing in MATLAB Command Window

```
>> ai = s * (s+2);
>> b = 1;
>> c = (s+3)^2;
>> [p,q] = axbyc(ai,b,c)
```

gives

```
p = 1
q = 9+4s
```

The pole placement controller with integral action is of the form

$$R(s) = \frac{4s + 9}{1s} = 4 + \frac{9}{s}$$

Case: $c(s) = (s + 3)^2(s + 4)$.

```
>> ai = s * (s+2);
>> b = 1;
>> c = (s+4) * (s+3)^2;
>> [p,q] = axbyc(ai,b,c)
```

yields

```
p = 8+s
q = 36+17s
```

The controller with integral action is of the form

$$R(s) = \frac{17s + 36}{s\,(s + 8)} = \left(17 + \frac{36}{s}\right)\left(\frac{1}{s + 8}\right)$$

Case: $c(s) = (s + 3)^2(s + 4)^2$.

```
>> ai = s * (s+2);
>> b = 1;
>> c = (s+3)^2 * (s+4)^2;
>> [p,q] = axbyc(ai,b,c)
```

Result:

```
p = 64+12s+s^2
q = 140+40s-15 s^2
```

The controller with integral action is of the form

$$R(s) = \frac{-15s^2 + 40s + 140}{s\,(s^2 + 12s + 64)}$$

WWW

Example 8.7:

The model of the controlled process is of the form

$$G(s) = \frac{b(s)}{a(s)}, \quad a(s) = s^2 + 4s + 4, \quad b(s) = s + 1$$

The task is to design a controller in the same way as in Example 8.6.
Case: $c(s) = (s + 3)^4$. Typing in MATLAB Command Window

```
>> ai = s * (s^2+4s+4);
>> b = s+1;
>> c = (s+3)^4;
>> [p,q] = axbyc(ai,b,c)
```

gives

```
p = -15+s
q = 81+87s+23 s^2
```

The pole placement controller with integral action is of the form

$$R(s) = \frac{23s^2 + 87s + 81}{s\,(s - 15)} = \frac{1}{s - 15}\left(87 + \frac{81}{s} + 23s\right)$$

Case: $c(s) = (s + 3)^5$.

```
>> ai = s * (s^2+4s+4);
>> b = s+1;
>> c = (s+3)^5;
>> [p,q] = axbyc(ai,b,c)
```

yields

```
p = -22+11s+ s^2
q = 240+250s+64s^2
```

The controller with integral action is of the form

$$R(s) = \frac{64s^2 + 250s + 240}{s\,(s^2 + 11s - 22)} = \frac{1}{s^2 + 11s - 22}\left(250 + \frac{240}{s} + 64s\right)$$

We can see that the controller denominator polynomial $p(s)$ is unstable. However, if the roots of the polynomial $c(s)$ have been changed from -3 to -1.5, the denominator would have been stable.

Even if an unstable controller is capable to guarantee stable closed-loop, from practical point of view it is more suitable to have a stable controller. In general, stability of $p(s)$ is influenced by its coefficients that are determined from from the closed-loop roots. Therefore, placement of the poles for strong stability is not trivial.

Example 8.6 shows that the pole placement method can be used for design of PI and PID controllers.

PI Controller for the First Order System

Consider the controlled process model described by the first order transfer function. If we choose the characteristic polynomial of the closed-loop systems $c(s)$ to be of the second degree then the structure of the pole placement controller is with the transfer function (see Example 8.6 – case 1)

$$R(s) = \frac{q_1 s + q_0}{s} \tag{8.367}$$

This transfer function is equivalent to the structure of the PI controller

$$R_{\mathrm{PI}}(s) = K_P + \frac{K_I}{s} \tag{8.368}$$

where

$$Z_R = q_1, \quad T_I = \frac{1}{q_0} \tag{8.369}$$

This controller was also derived in Chapter 7.4.6 (see page 285).

PI Controller with a Filter for the First Order System

Consider the controlled process model described by the first order transfer function. If we choose the characteristic polynomial of the closed-loop systems $c(s)$ to be of the third degree then the structure of the pole placement controller is with the transfer function (see Example 8.6 – case 2)

$$R(s) = \frac{q_1 s + q_0}{s\,(s + p_0)} \tag{8.370}$$

This transfer function is equivalent to the structure of the PI controller with a first order filter

$$R_{\mathrm{PIF}}(s) = \left(Z_R + \frac{1}{T_I s}\right)\left(\frac{1}{T_{\mathrm{PIF}} s + 1}\right) \tag{8.371}$$

where

$$Z_R = \frac{q_1}{p_0}, \quad T_I = \frac{p_0}{q_0}, \quad T_{\mathrm{PIF}} = \frac{1}{p_0} \tag{8.372}$$

PID Controller for the Second Order System

Consider the controlled process model described by a strictly proper second order system. If we choose the characteristic polynomial of the closed-loop systems $c(s)$ to be of the fourth degree then the structure of the pole placement controller is with the transfer function (see Example 8.7 – case 1)

$$R(s) = \frac{q_2 s^2 + q_1 s + q_0}{s\,(s + p_0)} \tag{8.373}$$

This transfer function is equivalent to the structure of the PID controller

$$R_{\text{PIDF}}(s) = \left(Z_R + \frac{1}{T_I s} + T_D s\right)\left(\frac{1}{T_{\text{PIDF}} s + 1}\right) \tag{8.374}$$

where

$$Z_R = \frac{q_1}{p_0}, \quad T_I = \frac{p_0}{q_0}, \quad T_D = \frac{q_2}{p_0}, \quad T_{\text{PIDF}} = \frac{1}{p_0} \tag{8.375}$$

The transfer function (8.373) also corresponds to the PID controller with derivative action filtered by the first order filter

$$R_{\text{PIDZ}}(s) = Z_R + \frac{1}{T_I s} + \frac{T_D s}{T_{\text{PIDZ}} s + 1} \tag{8.376}$$

where

$$Z_R = \frac{p_0 q_1 - q_0}{p_0^2}, \quad T_I = \frac{p_0}{q_0}, \quad T_D = \frac{p_0^2 q_2 - p_0 q_1 + q_0}{p_0^3}, \quad T_{\text{PIDZ}} = \frac{1}{p_0} \tag{8.377}$$

PID Controller with a Filter for the Second Order System

Consider the controlled process model described by a strictly proper second order system. If we choose the characteristic polynomial of the closed-loop systems $c(s)$ to be of the fifth degree then the structure of the pole placement controller is with the transfer function (see Example 8.7 – case 2)

$$R(s) = \frac{q_2 s^2 + q_1 s + q_0}{s\,(s^2 + p_1 s + p_0)} \tag{8.378}$$

This transfer function is equivalent to the structure of the PID controller with a filter

$$R_{\text{PIDF2}}(s) = \left(Z_R + \frac{1}{T_I s} + T_D s\right)\left(\frac{1}{p_{F2} s^2 + p_{F1} s + 1}\right) \tag{8.379}$$

where

$$Z_R = \frac{q_1}{p_0}, \quad T_I = \frac{p_0}{q_0}, \quad T_D = \frac{q_2}{p_0}, \quad p_{F2} = \frac{1}{p_0}, \quad p_{F1} = \frac{p_1}{p_0} \tag{8.380}$$

Non-Proper PID Controller for the Second Order System

If the structure of the classical PID controller is desired, the characteristic polynomial of the closed-loop systems $c(s)$ needs to be of the second degree. The pole placement controller has the structure with the transfer function (see Example 8.7 – case 2)

$$R(s) = \frac{q_2 s^2 + q_1 s + q_0}{p_0 s} \tag{8.381}$$

This transfer function is equivalent to the structure of the PID controller of the form

$$R_{\text{PID}}(s) = Z_R + \frac{1}{T_I s} + T_D s \tag{8.382}$$

where

$$Z_R = \frac{q_2}{p_0}, \quad T_I = \frac{p_0}{q_1}, \quad T_D = \frac{q_0}{p_0} \tag{8.383}$$

This controller was also derived in Chapter 7.4.6 (see page 286).

We can see that controllers R_{PID} from (8.382) and R_{PIDF} from (8.374) have similar behaviour if the filter time constant T_{PIDF} is sufficiently small.

8.6.6 Polynomial Pole Placement Design for Multivariable Systems

The state feedback control with observer for multivariable systems shown in Fig. 8.21 can be interpreted as a closed-loop system with the two-degree-of-freedom controller.

The observer dynamics is given as

$$\hat{\boldsymbol{x}} = (s\boldsymbol{I} - \boldsymbol{A} + \boldsymbol{LC})^{-1} \boldsymbol{Bu} + (s\boldsymbol{I} - \boldsymbol{A} + \boldsymbol{LC})^{-1} \boldsymbol{Ly} \tag{8.384}$$

The controller output is

$$\boldsymbol{u} = -\boldsymbol{K}\hat{\boldsymbol{x}} + \tilde{\boldsymbol{w}} \tag{8.385}$$

and after substituting for $\hat{\boldsymbol{x}}$ from (8.384) yields

$$\boldsymbol{u} = -\boldsymbol{K}(s\boldsymbol{I} - \boldsymbol{A} + \boldsymbol{LC})^{-1} \boldsymbol{Bu} - \boldsymbol{K}(s\boldsymbol{I} - \boldsymbol{A} + \boldsymbol{LC})^{-1} \boldsymbol{Ly} + \tilde{\boldsymbol{w}} \tag{8.386}$$

After some manipulations the controller can be written as

$$\left(\boldsymbol{I} + \boldsymbol{K}(s\boldsymbol{I} - \boldsymbol{A} + \boldsymbol{LC})^{-1} \boldsymbol{B}\right) \boldsymbol{u} = -\boldsymbol{K}(s\boldsymbol{I} - \boldsymbol{A} + \boldsymbol{LC})^{-1} \boldsymbol{Ly} + \tilde{\boldsymbol{w}} \tag{8.387}$$

The feedforward part is given as

$$\boldsymbol{R}_f(s) = \left(\boldsymbol{I} + \boldsymbol{K}(s\boldsymbol{I} - \boldsymbol{A} + \boldsymbol{LC})^{-1} \boldsymbol{B}\right)^{-1} \tag{8.388}$$

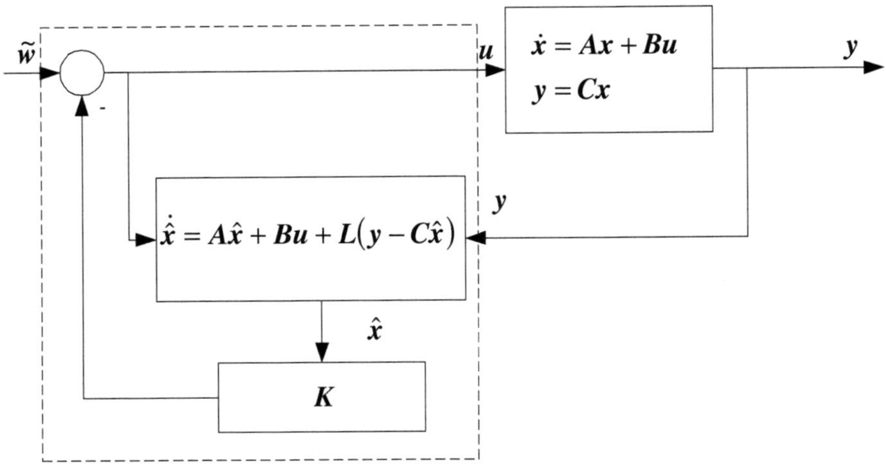

Fig. 8.21. State feedback control with state observation and reference input

The feedback part is given as

$$R(s) = -\left(I + K(sI - A + LC)^{-1}B\right)^{-1} K(sI - A + LC)^{-1} L \tag{8.389}$$

The controller in Fig. 8.21 has two degrees of freedom and its description using matrix transfer functions $R_f(s)$ and $R(s)$ is of the form

$$u = \left(R_f(s)\ R(s)\right)\begin{pmatrix}\tilde{w}\\ y\end{pmatrix} \tag{8.390}$$

Using the matrix inversion lemma (Theorem 6.1)

$$\left(I + K(sI - A + LC)^{-1}B\right)^{-1} = I - K(sI - A + LC + BK)^{-1}B \tag{8.391}$$

and thus

$$R_f(s) = I - K(sI - A + LC + BK)^{-1}B \tag{8.392}$$

$$\begin{aligned} R(s) &= \left(I - K(sI - A + LC + BK)^{-1}B\right)\left(K(sI - A + LC)^{-1}L\right) \\ &= K\Big((sI - A + LC)^{-1} \\ &\quad - (sI - A + LC + BK)^{-1}BK(sI - A + LC)^{-1}\Big)L \\ &= K(sI - A + LC + BK)^{-1}L \end{aligned} \tag{8.393}$$

As it was shown in the singlevariable case, pole locations of the closed-loop system are given by the feedback part of the state controller with observer with the matrix transfer function (8.393). If left or right matrix fraction of the pole placement controller is to be determined (8.393), it suffices to study the case with the zero reference input.

Consider state the feedback

$$\boldsymbol{u}(t) = -\boldsymbol{K}\hat{\boldsymbol{x}}(t) \tag{8.394}$$

with observer

$$\dot{\hat{\boldsymbol{x}}}(t) = \boldsymbol{A}\hat{\boldsymbol{x}}(t) + \boldsymbol{B}\boldsymbol{u}(t) + \boldsymbol{L}\left(\boldsymbol{y}(t) - \boldsymbol{C}\hat{\boldsymbol{x}}(t)\right) \tag{8.395}$$

applied to a controllable and observable system

$$\dot{\boldsymbol{x}}(t) = \boldsymbol{A}\boldsymbol{x}(t) + \boldsymbol{B}\boldsymbol{u}(t) \tag{8.396}$$

$$\boldsymbol{y}(t) = \boldsymbol{C}\boldsymbol{x}(t) \tag{8.397}$$

Matrix transfer functions of the controlled system are given as

$$(s\boldsymbol{I} - \boldsymbol{A})^{-1}\boldsymbol{B} = \boldsymbol{B}_{Rs}(s)\boldsymbol{A}_R^{-1}(s) \tag{8.398}$$

where $\boldsymbol{A}_R$, $\boldsymbol{B}_{Rs}$ are right coprime polynomial matrices and $\boldsymbol{A}_R$ is column reduced.

$$\boldsymbol{C}\left(s\boldsymbol{I} - \boldsymbol{A}\right)^{-1} = \boldsymbol{A}_L^{-1}(s)\boldsymbol{B}_{Ls}(s) \tag{8.399}$$

where $\boldsymbol{A}_L$, $\boldsymbol{B}_{Ls}$ are left coprime polynomial matrices and $\boldsymbol{A}_L$ is row reduced.

$$\boldsymbol{C}\left(s\boldsymbol{I} - \boldsymbol{A}\right)^{-1}\boldsymbol{B} = \boldsymbol{A}_L^{-1}(s)\boldsymbol{B}_L(s) = \boldsymbol{B}_R(s)\boldsymbol{A}_R^{-1}(s) \tag{8.400}$$

where $\boldsymbol{A}_L$, $\boldsymbol{B}_L$ are left coprime polynomial matrices, $\boldsymbol{A}_R$, $\boldsymbol{B}_R$ are right coprime polynomial matrices and

$$\boldsymbol{B}_L(s) = \boldsymbol{B}_{Ls}(s)\boldsymbol{B} \tag{8.401}$$

$$\boldsymbol{B}_R(s) = \boldsymbol{C}\boldsymbol{B}_{Rs}(s) \tag{8.402}$$

Matrix $\boldsymbol{K}$ is chosen such that the closed-loop matrix $\boldsymbol{A} - \boldsymbol{B}\boldsymbol{K}$ reflects desired dynamics. From (8.398) follows

$$\boldsymbol{B}\boldsymbol{A}_R(s) = (s\boldsymbol{I} - \boldsymbol{A})\boldsymbol{B}_{Rs}(s) \tag{8.403}$$

Adding $\boldsymbol{B}\boldsymbol{K}\boldsymbol{B}_{Rs}$ to either side of (8.403) gives

$$\boldsymbol{B}(\boldsymbol{A}_R(s) + \boldsymbol{K}\boldsymbol{B}_{Rs}(s)) = (s\boldsymbol{I} - \boldsymbol{A} + \boldsymbol{B}\boldsymbol{K})\boldsymbol{B}_{Rs}(s) \tag{8.404}$$

$$(s\boldsymbol{I} - \boldsymbol{A} + \boldsymbol{B}\boldsymbol{K})^{-1}\boldsymbol{B} = \boldsymbol{B}_{Rs}(s)(\boldsymbol{A}_R(s) + \boldsymbol{K}\boldsymbol{B}_{Rs}(s))^{-1} \tag{8.405}$$

$(s\boldsymbol{I} - \boldsymbol{A})$ and $\boldsymbol{B}$ are left coprime because the controlled system is controllable. This implies that $(s\boldsymbol{I} - \boldsymbol{A} + \boldsymbol{B}\boldsymbol{K})$, $\boldsymbol{B}$ are left coprime as well. Similarly, it follows from the definition that $\boldsymbol{A}_R$, $\boldsymbol{B}_{Rs}$ are right coprime and thus $(\boldsymbol{A}_R +$

$\boldsymbol{KB}_{Rs}$), $\boldsymbol{B}_{Rs}$ are right coprime polynomial matrices. This also implies that $(s\boldsymbol{I} - \boldsymbol{A} + \boldsymbol{BK})$ and $(\boldsymbol{A}_R(s) + \boldsymbol{KB}_{Rs}(s))$ have the same determinants.

If the gain matrix $\boldsymbol{K}$ exists, it is unique and of the form

$$\boldsymbol{K} = \boldsymbol{X}_L^{-1}\boldsymbol{Y}_L \tag{8.406}$$

Then $\boldsymbol{X}_L$, $\boldsymbol{Y}_L$ are solution of the equation

$$\boldsymbol{X}_L\boldsymbol{A}_R(s) + \boldsymbol{Y}_L\boldsymbol{B}_{Rs}(s) = \boldsymbol{F}_R(s) \tag{8.407}$$

$\boldsymbol{F}_R(s)$ is a stable polynomial matrix with $\det \boldsymbol{F}_R(s) \neq 0$. $\boldsymbol{F}_R(s)$ is a matrix with dimensions $m \times m$, it is column reduced and has the same degree as the matrix $\boldsymbol{A}_R(s)$.

The observer gain matrix $\boldsymbol{L}$ is chosen such that the matrix $\boldsymbol{A} - \boldsymbol{LC}$ corresponds to the desired dynamics of the observer. From (8.399) follows

$$\boldsymbol{A}_L(s)\boldsymbol{C} = \boldsymbol{B}_{Ls}(s)(s\boldsymbol{I} - \boldsymbol{A}) \tag{8.408}$$

Adding $\boldsymbol{B}_{Ls}(s)\boldsymbol{LC}$ to either side of (8.408) gives

$$(\boldsymbol{A}_L(s) + \boldsymbol{B}_{Ls}(s)\boldsymbol{L})\boldsymbol{C} = \boldsymbol{B}_{Ls}(s)(s\boldsymbol{I} - \boldsymbol{A} + \boldsymbol{LC}) \tag{8.409}$$

$$\boldsymbol{C}(s\boldsymbol{I} - \boldsymbol{A} + \boldsymbol{LC})^{-1} = (\boldsymbol{A}_L(s) + \boldsymbol{B}_{Ls}(s)\boldsymbol{L})^{-1}\boldsymbol{B}_{Ls}(s) \tag{8.410}$$

$\boldsymbol{C}$ and $(s\boldsymbol{I} - \boldsymbol{A})$ are right coprime as the controlled system is observable. This implies that $\boldsymbol{C}$ and $(s\boldsymbol{I} - \boldsymbol{A} + \boldsymbol{LC})$ are right coprime. Similarly, it follows from the definition that $\boldsymbol{A}_L$ and $\boldsymbol{B}_{Ls}$ are left coprime and thus $(\boldsymbol{A}_L + \boldsymbol{B}_{Ls}\boldsymbol{L})$ and $\boldsymbol{B}_{Ls}$ are left coprime. This also implies that $(s\boldsymbol{I} - \boldsymbol{A} + \boldsymbol{LC})$ and $(\boldsymbol{A}_L(s) + \boldsymbol{B}_{Ls}(s)\boldsymbol{L})$ have the same determinants.

If the gain matrix $\boldsymbol{L}$ exists, it is unique and of the form

$$\boldsymbol{L} = \boldsymbol{Y}_R\boldsymbol{X}_R^{-1} \tag{8.411}$$

Then $\boldsymbol{X}_R$ a $\boldsymbol{Y}_R$ are solution of the equation

$$\boldsymbol{A}_L(s)\boldsymbol{X}_R + \boldsymbol{B}_{Ls}(s)\boldsymbol{Y}_R = \boldsymbol{O}_L(s) \tag{8.412}$$

$\boldsymbol{O}_L(s)$ is a stable polynomial matrix with $\det \boldsymbol{O}_L(s) \neq 0$. $\boldsymbol{O}_L(s)$ is a matrix with dimensions $r \times r$, it is row reduced and has the same degree as matrix $\boldsymbol{A}_L(s)$.

The matrix transfer function of the feedback controller corresponding to the state feedback controller with observer is given as

$$\boldsymbol{K}(s\boldsymbol{I} - \boldsymbol{A} + \boldsymbol{BK} + \boldsymbol{LC})^{-1}\boldsymbol{L} = \boldsymbol{P}_L^{-1}(s)\boldsymbol{Q}_L(s) = \boldsymbol{Q}_R(s)\boldsymbol{P}_R^{-1}(s) \tag{8.413}$$

Matrices $\boldsymbol{P}_L(s)$ and $\boldsymbol{Q}_L(s)$ of the controller left matrix fraction are solution of the matrix Diophantine equation

$$\boldsymbol{P}_L(s)\boldsymbol{A}_R(s) + \boldsymbol{Q}_L(s)\boldsymbol{B}_R(s) = \tilde{\boldsymbol{O}}_L(s)\boldsymbol{F}_R(s) \tag{8.414}$$

Matrices $\boldsymbol{P}_R(s)$ and $\boldsymbol{Q}_R(s)$ of the controller right matrix fraction are solution of the matrix Diophantine equation

$$\boldsymbol{A}_L(s)\boldsymbol{P}_R(s) + \boldsymbol{B}_L(s)\boldsymbol{Q}_R(s) = \boldsymbol{O}_L(s)\tilde{\boldsymbol{F}}_R(s) \tag{8.415}$$

Matrices $\tilde{\boldsymbol{O}}_L(s)$ and $\tilde{\boldsymbol{F}}_R(s)$ are calculated using auxiliary factorisations of matrices$\boldsymbol{O}_L(s)$ and $\boldsymbol{F}_R(s)$. From the practical point of view it is advantageous to use equations (8.393), (8.407) and (8.412) for calculation of the pole placement controller.

If the singlevariable case is considered i. e. $m = r = 1$, then

$$\boldsymbol{A}_L(s) = \boldsymbol{A}_R(s) = a(s) \tag{8.416}$$

$$\boldsymbol{B}_L(s) = \boldsymbol{B}_R(s) = b(s) \tag{8.417}$$

$$\boldsymbol{O}_L(s) = o(s) \tag{8.418}$$

$$\boldsymbol{F}_R(s) = f(s) \tag{8.419}$$

and equations (8.414) and (8.415) are replaced by (8.295) that determines transfer function of the singlevariable pole placement controller.

8.7 The Youla-Kučera Parametrisation

Previous sections of control design presented the optimal feedback control as the pole placement problem. This section deals with all feedback control laws that guarantee some desired properties of the closed-loop system.

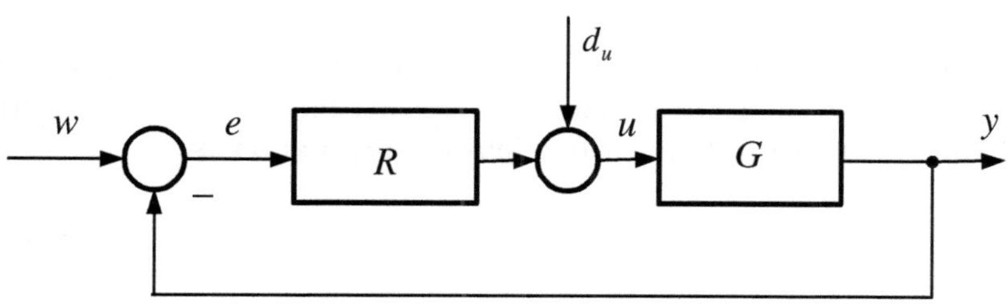

Fig. 8.22. Feedback control system

The control problem shown in Fig. 8.22 consists in findig such a controller $R(s)$ for the controlled system $G(s)$ that the closed-loop system satisfies desired specifications. This problem can be divided into two tasks. The first is to stabilise the closed-loop system and the second is to guarantee additional performance specifications.

The Youla-Kučera parametrisation employed in the second step uses fractional models for all systems. It is possible to parametrise feedback controllers from a set of fractions of stable rational functions. Another use of the Youla-Kučera parametrisation is in the dual parametrisation when a fixed controller is known and a set of plants that can be stabilised by it is searched. The dual parametrisation can be used in identification of processes. Finally, the Youla-Kučera parametrisation can be employed simultaneously for both the controller and the plant.

The Youla-Kučera parametrisation is based on assumption that a stabilising controller is known. For its purposes the bounded-input bounded-output (BIBO) stability definition is suitable.

The closed-loop system is stable if arbitrary bounded input causes results in a bounded output at any place of the closed-loop system. A system with transfer function $F(s)$ is BIBO stable if and only if $F(s)$ is proper and Hurwitz-stable.

From this definition follows that all transfer functions of the closed-loop system have to be stable and proper.

A system is asymptotically stable if its characteristic polynomial is stable.

We will at first investigate singlevariable systems and then the problem will be generalised to multivariable systems. Finally, parametrisation of discrete-time systems will be treated.

8.7.1 Fractional Representation

Fractional representation of systems (plants, controllers) consists in expressing transfer functions as a fraction of two stable transfer functions. Consider for example a controlled process of the form

$$G(s) = \frac{b(s)}{a(s)} \tag{8.420}$$

where $a(s)$, $b(s)$ are coprime polynomials with $\deg b(s) \leq \deg a(s)$, and deg denotes the degree of a polynomial. This transfer function can be rewritten into a Hurwitz-stable and proper fractional representation as

$$G(s) = \frac{B(s)}{A(s)} \tag{8.421}$$

where $B(s)$ and $A(s)$ are stable transfer functions of the form

$$B(s) = \frac{b(s)}{c(s)}, \quad A(s) = \frac{a(s)}{c(s)} \tag{8.422}$$

and $c(s)$ is a monic Hurwitz polynomial, i. e. $1/c(s)$ is asymptotically stable, and $\deg c(s) \geq \deg a(s)$. A monic polynomial contains unit coefficient at the maximum power of s.

Consider for example a transfer function

$$G(s) = \frac{b_1 s + b_0}{s^2 + a_1 s + a_0} \tag{8.423}$$

Its fractional representation can be of the form

$$G(s) = \frac{\dfrac{b_1 s + b_0}{s^2 + c_1 s + c_0}}{\dfrac{s^2 + a_1 s + a_0}{s^2 + c_1 s + c_0}} \tag{8.424}$$

where $s^2 + c_1 s + c_0$ is Hurwitz polynomial.

Let F_{ps} denotes a set of stable proper rational transfer functions in the above defined sense. Two transfer functions $B(s), A(s) \in F_{ps}$ are coprime if there exist transfer functions $X(s), Y(s) \in F_{ps}$ satisfying the Diophantine equation

$$A(s)X(s) + B(s)Y(s) = 1 \tag{8.425}$$

Let us now consider the feedback control system shown in Fig. 8.22 and apply the concept of the BIBO stability. Input variables of the closed-loop system are w, d_u and output variables are u, y. We express the plant and the controller as ratios of proper stable restional functions. The controlled system can be written as

$$G(s) = \frac{B(s)}{A(s)} \tag{8.426}$$

where $B(s), A(s) \in F_{ps}$ are coprime. The controller transfer function is of the form

$$R(s) = \frac{Q(s)}{P(s)} \tag{8.427}$$

where $Q(s), P(s) \in F_{ps}$ are coprime.

The closed-loop system in Fig. 8.22 is BIBO stable if and only if all transfer functions between inputs w, d_u and outputs u a y are proper and Hurwitz-stable.

The transfer functions are given as

$$G_{uw}(s) = \frac{R(s)}{1 + G(s)R(s)} = \frac{A(s)Q(s)}{A(s)P(s) + B(s)Q(s)} \tag{8.428}$$

$$G_{yw}(s) = \frac{G(s)R(s)}{1 + G(s)R(s)} = \frac{B(s)Q(s)}{A(s)P(s) + B(s)P(s)} \tag{8.429}$$

$$G_{ud_u}(s) = \frac{1}{1 + G(s)R(s)} = \frac{A(s)P(s)}{A(s)P(s) + B(s)Q(s)} \tag{8.430}$$

$$G_{yd_u}(s) = \frac{G(s)}{1 + G(s)R(s)} = \frac{B(s)P(s)}{A(s)P(s) + B(s)Q(s)} \tag{8.431}$$

From equations

$$\begin{pmatrix} u(s) \\ y(s) \end{pmatrix} = \frac{1}{C(s)} \begin{pmatrix} A(s)Q(s) & A(s)P(s) \\ B(s)Q(s) & B(s)P(s) \end{pmatrix} \begin{pmatrix} w(s) \\ d_u(s) \end{pmatrix} \tag{8.432}$$

$$A(s)P(s) + B(s)Q(s) = C(s) \tag{8.433}$$

follows that the closed-loop system in Fig. 8.22 is BIBO stable if and only if $1/C(s)$ is proper and Hurwitz-stable.

8.7.2 Parametrisation of Stabilising Controllers

A controller with transfer function $R(s)$ that guarantees the BIBO closed-loop stability for a controlled system $G(s)$ is called a *stabilising* controller for the given controlled system.

Suppose that the controlled system gives rise to the transfer function

$$G(s) = \frac{B(s)}{A(s)} \tag{8.434}$$

and $B(s), A(s) \in F_{ps}$ are coprime. If some controller with $Q'(s), P'(s) \in F_{ps}$

$$R(s) = \frac{Q'(s)}{P'(s)} \tag{8.435}$$

stabilises the controlled system (8.434) then

$$C(s) = A(s)P'(s) + B(s)Q'(s) \tag{8.436}$$

and $1/C(s)$ is proper and Hurwitz-stable.

Equation (8.436) can be divided by a stable factor $C(s)$ and yields

$$A(s)\frac{P'(s)}{C(s)} + B(s)\frac{Q'(s)}{C(s)} = 1 \tag{8.437}$$

Introducing the notation

$$P(s) = \frac{P'(s)}{C(s)}, \quad Q(s) = \frac{Q'(s)}{C(s)} \tag{8.438}$$

follows that $P(s)$, $Q(s)$ are a solution of the equation

$$A(s)P(s) + B(s)Q(s) = 1 \tag{8.439}$$

where $Q(s), P(s) \in F_{ps}$.

Stabilising controllers from (8.436) and (8.439) are the same as

$$R(s) = \frac{Q'(s)}{P'(s)} = \frac{Q(s)}{P(s)} \tag{8.440}$$

where $P(s) \neq 0$. Equation (8.439) is a Diophantine equation that has infinity of solutions. These solutions can be easily parametrised.

We will now show that all controllers that stabilise the controlled system are solutions of equation (8.439) of the form

$$P(s) = X(s) - B(s)T(s), \quad Q(s) = Y(s) + A(s)T(s) \tag{8.441}$$

where $X(s)$ and $Y(s)$ is a particular solution of the equation and $T(s) \in F_{ps}$. This follows from the substitution of $P(s)$ and $Q(s)$ from (8.441) into (8.439)

$$A(s)(X(s) - B(s)T(s)) + B(s)(Y(s) + A(s)T(s)) = 1 \tag{8.442}$$

or

$$A(s)X(s) + B(s)Y(s) = 1 \tag{8.443}$$

This is a very important result as it gives parametrisation of the set of all stabilising controllers for the controlled system $G(s)$ of the form

$$R(s) = \frac{Y(s) + A(s)T(s)}{X(s) - B(s)T(s)} \tag{8.444}$$

where $Y(s), X(s) \in F_{ps}$ is arbitrary particular solution of the equation

$$A(s)X(s) + B(s)Y(s) = 1 \tag{8.445}$$

and $T(s) \in F_{ps}$ under the condition that $X(s) - B(s)T(s) \neq 0$ holds.

Substracting (8.439) from (8.445) gives

$$A(s)(X(s) - P(s)) + B(s)(Y(s) - Q(s)) = 0 \tag{8.446}$$

As $A(s)$, $B(s)$ are coprime, $A(s)$ divides $Y(s) - Q(s)$ and $B(s)$ divides $X(s) - P(s)$. Denote

$$\frac{Q(s) - Y(s)}{A(s)} = T(s) \tag{8.447}$$

Then from (8.446) follows

$$P(s) = X(s) - B(s)T(s) \tag{8.448}$$

Similarly, denote

$$\frac{X(s) - P(s)}{B(s)} = T(s) \tag{8.449}$$

Then from (8.446) follows

$$Q(s) = Y(s) + A(s)T(s) \tag{8.450}$$

This shows that any solution of (8.439) is of the form (8.448) and (8.450) for some $T(s) \in F_{ps}$.

The detailed block diagram of the parametrised controller in the closed-loop system is shown in Fig. 8.23.

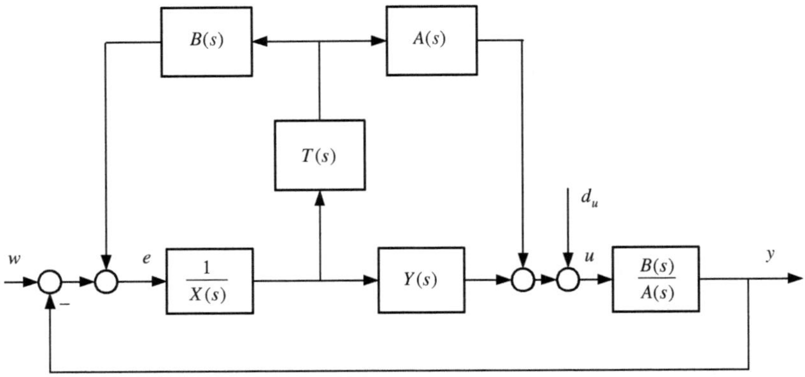

Fig. 8.23. Block diagram of the parametrised controller

8.7.3 Parametrised Controller in the State-Space Representation

Consider a minimal realisation of the controlled system

$$G(s) = \boldsymbol{C}(s\boldsymbol{I} - \boldsymbol{A})^{-1}\boldsymbol{B} \tag{8.451}$$

State feedback controller with observer is then given as

$$\dot{\hat{\boldsymbol{x}}}(t) = (\boldsymbol{A} - \boldsymbol{B}\boldsymbol{K} - \boldsymbol{L}\boldsymbol{C})\hat{\boldsymbol{x}}(t) + \boldsymbol{L}y(t) \tag{8.452}$$

$$u(t) = -\boldsymbol{K}\hat{\boldsymbol{x}}(t) \tag{8.453}$$

This controller is shown in Fig. 8.24. It is not difficult to show that this diagram is the same as the controller in Fig. 8.25.

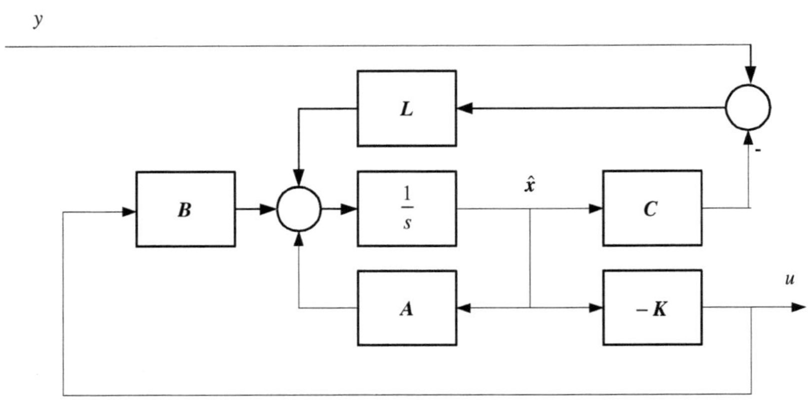

Fig. 8.24. State feedback controller with observer

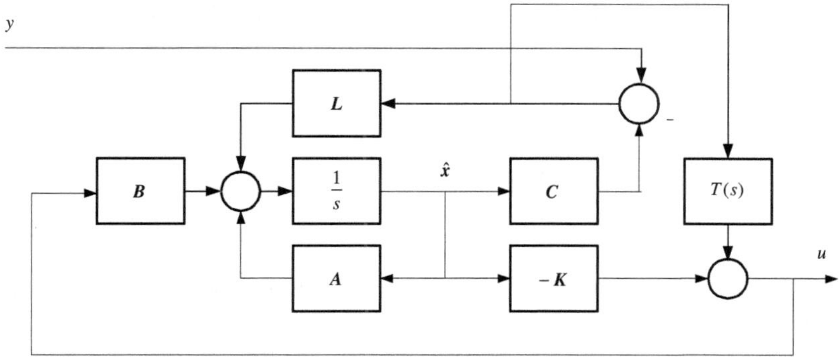

Fig. 8.25. State parametrised controller

8.7.4 Parametrisation of Transfer Functions of the Closed-Loop System

The mathematical model of the closed-loop system in Fig. 8.23 can be written as

$$\begin{pmatrix} u(s) \\ y(s) \end{pmatrix} = \begin{pmatrix} A(s)(Y(s) + A(s)T(s)) & A(s)(X(s) - B(s)T(s)) \\ B(s)(Y(s) + A(s)T(s) & B(s)(X(s) - B(s)T(s)) \end{pmatrix} \begin{pmatrix} w(s) \\ d_u(s) \end{pmatrix} \tag{8.454}$$

Optimal control design for the controlled system with transfer function $G(s)$ can be performed in two steps. In the first a stabilising controller is found and in the second a parameter $T(s)$ is chosen such that the performance of the closed-loop system is optimised. The cost function can be for example LQ. This procedure will be explained in the following example.

Example 8.8:

Consider a first order controlled system with the transfer function

$$G(s) = \frac{1}{s+1}$$

We want to find a stabilising controller for this system. Based on it, the set of all stabilising controllers will be parametrised. Finally one controller will be chosen that rejects asymptotically step disturbance.

Numerator and denominator of the process in fractional representation can for example be given as

$$A(s) = \frac{s+1}{s+c_0}, \quad B(s) = \frac{1}{s+c_0}$$

where $A(s), B(s) \in F_{ps}$. Equation (8.445) is of the form

$$\frac{s+1}{s+c_0}X(s) + \frac{1}{s+c_0}Y(s) = 1$$

One particular solution is $X(s) = 1$, $Y(s) = c_0 - 1$. This specifies one stabilising controller. The set of all stabilising controllers is then of the form

$$R(s) = \frac{c_0 - 1 + \dfrac{s+1}{s+c_0}T(s)}{1 - \dfrac{1}{s+c_0}T(s)}$$

The transfer function of the parametrised closed-loop system for input variable d_u and output variable $e = -y$ is

$$G_{ed_u}(s) = -B(s)(X(s) - B(s)T(s))$$

Substituting for $B(s) = \frac{1}{s+c_0}$ and $X(s) = 1$ gives

$$G_{ed_u}(s) = -\frac{1}{s+c_0}\left(1 - \frac{1}{s+c_0}T(s)\right) = \frac{-s - c_0 + T(s)}{(s+c_0)^2}$$

The final value theorem implies that the steady-state control error is zero if

$$\lim_{s\to 0} G_{ed_u}(s) = 0$$

or if

$$T(s) = c_0$$

If this relation is substituted into the transfer function of the parametrised controller then

$$R(s) = 2c_0 - 1 + \frac{c_0^2}{s}$$

This is a PI controller with one design parameter. If $c_0 = 0.5$ then the response to the unit step change of the disturbance is shown on Fig. 8.26.

8.7.5 Dual Parametrisation

It is not important from the parametrisation point of view which system in the closed-loop system is parametrised. So far, the controller transfer function was parametrised. If the controlled system is parametrised we speak about the *dual parametrisation.*

Consider again the closed-loop system in Fig. 8.22. The controlled system is described by transfer function

$$G(s) = \frac{B(s)}{A(s)} \tag{8.455}$$

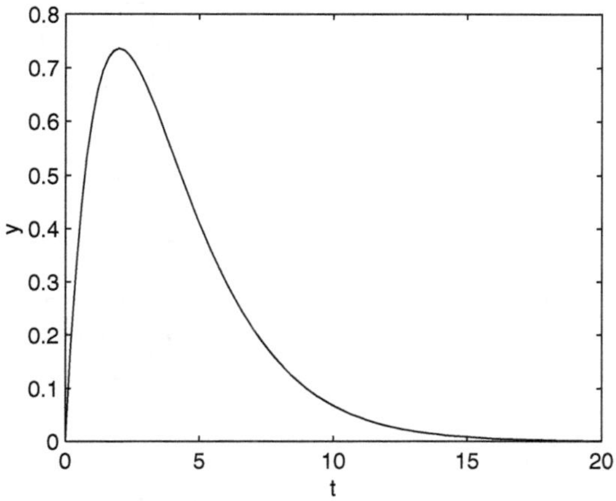

Fig. 8.26. Closed-loop system response to the unit step disturbance on input

where $B(s), A(s) \in F_{ps}$ are coprime. This controlled system can be stabilised by a controller with transfer function

$$R(s) = \frac{Y(s)}{X(s)} \tag{8.456}$$

where $Y(s), X(s) \in F_{ps}$ are coprime. The set of all controlled systems stabilised by the controller (8.456) is given as

$$\frac{B(s) + X(s)S(s)}{A(s) - Y(s)S(s)} \tag{8.457}$$

where $S(s) \in F_{ps}$. The detailed block diagram of the parametrised controlled system is given in Fig. 8.27.

Dual parametrisation is suitable for identification of the controlled system in the closed-loop. Here, not the whole transfer function has to be identified. Instead, only some deviation $S(s)$ from nominal process can be estimated. Moreover, the problem of estimation of $S(s)$ represents only an open-loop identification problem as

$$z(s) = S(s)x(s) \tag{8.458}$$

and $x(t)$, $z(t)$ can be expressed in terms of $u(t)$, $y(t)$ that are measurable.

From Fig. 8.27 follows that $u(s)$, $y(s)$ are given as

$$u(s) = (A(s) - Y(s)S(s))x(s) \tag{8.459}$$

$$y(s) = (B(s) + X(s)S(s))x(s) \tag{8.460}$$

Multiplying (8.459) by $X(s)$, (8.460) by $Y(s)$ and summing them gives

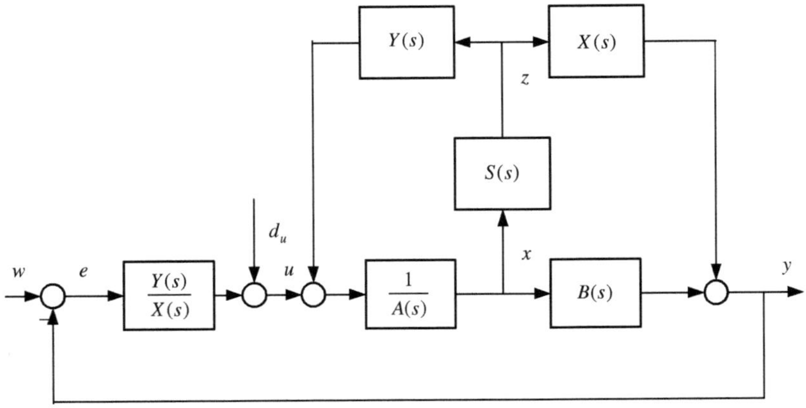

Fig. 8.27. Parametrised controlled system in the closed-loop system

$$(A(s)X(s) + B(s)Y(s))x(s) = X(s)u(s) + Y(s)y(s) \tag{8.461}$$

or

$$x(s) = X(s)u(s) + Y(s)y(s) \tag{8.462}$$

From Fig. 8.27 also follows that

$$A(s)x(s) = u(s) + Y(s)z(s) \tag{8.463}$$
$$B(s)x(s) = y(s) - X(s)z(s) \tag{8.464}$$

Let us multiply (8.463) by $B(s)$, (8.464) by $Y(s)$ and substract (8.464) from (8.463). This yields

$$(A(s)X(s) + B(s)Y(s))z(s) = A(s)y(s) - B(s)y(s) \tag{8.465}$$

or

$$z(s) = A(s)y(s) - B(s)y(s) \tag{8.466}$$

The data needed for identification based on equation (8.458) are obtained from (8.462) and (8.466).

We note that if both controller and plant are parametrised then the relation between Youla-Kučera parameters $T(s)$, $S(s)$ can significantly simplify the control design problem.

8.7.6 Parametrisation of Stabilising Controllers for Multivariable Systems

Matrices $\tilde{\boldsymbol{B}}_R(s)$, $\tilde{\boldsymbol{A}}_R(s)$ with elements in F_{ps} ($\tilde{\boldsymbol{B}}_R(s)$, $\tilde{\boldsymbol{A}}_R(s) \in F_{ps}$) are rigth coprime if

$$\tilde{\boldsymbol{X}}_L(s)\tilde{\boldsymbol{A}}_R(s) + \tilde{\boldsymbol{Y}}_L(s)\tilde{\boldsymbol{B}}_R(s) = \boldsymbol{I} \tag{8.467}$$

where $\tilde{\boldsymbol{Y}}_L(s)$, $\tilde{\boldsymbol{X}}_L(s) \in F_{ps}$. The left coprimeness of matrices $\tilde{\boldsymbol{A}}_L(s)$ and $\tilde{\boldsymbol{B}}_L(s)$ with matrix elements in F_{ps} can be similarly defined.

The same procedure as for singlevariable systems can be performed to show that the multivariable closed-loop system with structure given in Fig. 8.22 can be parametrised to guarantee BIBO stability.

Consider a controlled system with a matrix transfer function

$$\boldsymbol{G}(s) = \tilde{\boldsymbol{A}}_L^{-1}(s)\tilde{\boldsymbol{B}}_L(s) = \tilde{\boldsymbol{B}}_R(s)\tilde{\boldsymbol{A}}_R^{-1}(s) \tag{8.468}$$

where $\tilde{\boldsymbol{A}}_L(s)$, $\tilde{\boldsymbol{B}}_L(s) \in F_{ps}$ are left coprime and $\tilde{\boldsymbol{B}}_R(s)$, $\tilde{\boldsymbol{A}}_R(s) \in F_{ps}$ are right coprime. The set of all stabilising controllers is given as the matrix of transfer functions

$$\boldsymbol{R}(s) = (\tilde{\boldsymbol{Y}}_R(s) + \tilde{\boldsymbol{A}}_R(s)\tilde{\boldsymbol{T}}(s))(\tilde{\boldsymbol{X}}_R(s) - \tilde{\boldsymbol{B}}_R(s)\tilde{\boldsymbol{T}}(s))^{-1} \tag{8.469}$$

or

$$\boldsymbol{R}(s) = (\tilde{\boldsymbol{X}}_L(s) - \tilde{\boldsymbol{T}}(s)\tilde{\boldsymbol{B}}_L(s))^{-1}(\tilde{\boldsymbol{Y}}_L(s) + \tilde{\boldsymbol{T}}(s)\tilde{\boldsymbol{A}}_L(s)) \tag{8.470}$$

where $\tilde{\boldsymbol{Y}}_L(s)$, $\tilde{\boldsymbol{X}}_L(s) \in F_{ps}$ and $\tilde{\boldsymbol{Y}}_R(s)$, $\tilde{\boldsymbol{X}}_R(s) \in F_{ps}$ are solution of the equation

$$\begin{pmatrix} \tilde{\boldsymbol{A}}_L(s) & -\tilde{\boldsymbol{B}}_L(s) \\ \tilde{\boldsymbol{Y}}_L(s) & \tilde{\boldsymbol{X}}_L(s) \end{pmatrix} \begin{pmatrix} \tilde{\boldsymbol{X}}_R(s) & \tilde{\boldsymbol{B}}_R(s) \\ -\tilde{\boldsymbol{Y}}_R(s) & \tilde{\boldsymbol{A}}_R(s) \end{pmatrix} = \boldsymbol{I} \tag{8.471}$$

The Youla-Kučera parameter $\tilde{\boldsymbol{T}}(s) \in F_{ps}$ has to be chosen such that $\tilde{\boldsymbol{X}}_L(s) - \tilde{\boldsymbol{T}}(s)\tilde{\boldsymbol{B}}_L(s)$ and $\tilde{\boldsymbol{X}}_R(s) - \tilde{\boldsymbol{B}}_R(s)\tilde{\boldsymbol{T}}(s)$ be non-singular.

8.7.7 Parametrisation of Stabilising Controllers for Discrete-Time Systems

Parametrisation of controllers in the discrete-time domain has to take into account some specific issues.

A system described by a proper rational transfer function $F(z)$ is BIBO stable if and only if it is Schur-stable. This implies that the transfer function does not contain poles in the region $|z| \geq 1$.

To have a causal discrete-time closed-loop system and assuming proper transfer function of the controller, the transfer function of the controlled system has to be strictly proper. Stable rational functions in the z operator that have all poles in the origin $z = 0$ are in fact polynomials in z^{-1}.

Consider a transfer function of a discrete-time system of the form

$$G(z) = \frac{b_0 z + b_1}{z^2 + a_1 z + a_2} \tag{8.472}$$

If we choose a stable denominator of the fractional representation as z^2 then the transfer function can be written as

$$G(z) = \frac{b_0 z^{-1} + b_1 z^{-2}}{1 + a_1 z^{-1} + a_2 z^{-2}} \tag{8.473}$$

All relations for calculation of the parametrisation of all stabilising controllers are the same for continuous and discrete-time systems. The difference between them lies in the fact that continuous-time systems are represented by stable fractions in s whereas discrete-time systems can be defined by polynomials in z^{-1}.

It can be shown that the optimal feedback control can be interpreted as the pole placement design problem. From the nature of continuous-time systems follows that the response of the closed-loop system to a step change in input settles theoretically in infinity. The discrete-time control however, makes it possible to force the output to follow the reference value in a finite time.

Dead-Beat Control

Consider the feedback closed-loop system shown in Fig. 8.28. The transfer function of the controlled process is assumed to be strictly proper. We want to design such a controller that the control error be zero after a finite time. When it is found, the Youla-Kučera parametrisation makes it possible to find a set of such controllers and to choose among them the one that generates the fastest response. Such a controller is called the dead-beat controller.

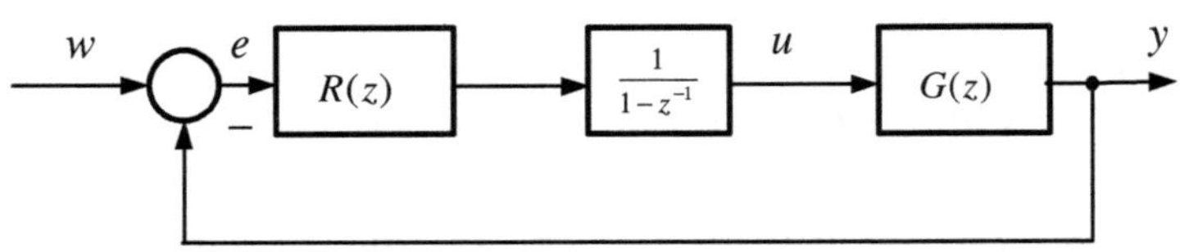

Fig. 8.28. Discrete-time feedback closed-loop system

The controller consists of two parts in series. The first part with the transfer function $R(z)$ will be calculated by the control design, the second part is fixed and guarantees the integrating properties of the controller.

The controlled process can be characterised by the fractional representation

$$G(z^{-1}) = \frac{B(z^{-1})}{A(z^{-1})} \tag{8.474}$$

Fractional representation of the controller is of the form

$$R(z^{-1}) = \frac{Q(z^{-1})}{P(z^{-1})} \tag{8.475}$$

Integral part of the controller has been in the design procedure moved to the controlled system $G_I = G/(1 - z^{-1})$. All poles of the closed-loop system are placed in origin $z = 0$ for dead-beat control. The tracking error is given as

$$e(z^{-1}) = \frac{1}{1 + G_I(z^{-1})R(z^{-1})} w(z^{-1}) \tag{8.476}$$

or

$$e(z^{-1}) = (1 - z^{-1})A(z^{-1})P(z^{-1})w(z^{-1}) \tag{8.477}$$

where

$$(1 - z^{-1})A(z^{-1})P(z^{-1}) + B(z^{-1})Q(z^{-1}) = 1 \tag{8.478}$$

Transfer functions of all stabilising controllers are given as

$$R(z^{-1}) = \frac{Q(z^{-1})}{P(z^{-1})} = \frac{Y(z^{-1}) + (1 - z^{-1})A(z^{-1})T(z^{-1})}{X(z^{-1}) - B(z^{-1})T(z^{-1})} \tag{8.479}$$

If reference sequence is given as a unit step change then

$$w(z^{-1}) = \frac{1}{1 - z^{-1}} \tag{8.480}$$

and the control error that is given as

$$e(z^{-1}) = A(z^{-1})\left(X(z^{-1}) - B(z^{-1})T(z^{-1})\right) \tag{8.481}$$

is a polynomial. Equation (8.479) characterises a set of finite response controllers with a parameter $T(z^{-1})$.

The dead-beat controller that generates zero control error in the least number of steps is obtained when

$$T(z^{-1}) = 0 \tag{8.482}$$

Example 8.9: *The dead-beat controller for a continuous-time second order system* WWW

Consider a continuous-time system with a transfer function

$$G(s) = \frac{Z_2}{(T_1 s + 1)(T_2 + 1)}$$

where Z_2 is the process gain and T_1, T_2 are time constants. The transfer function can be rewritten as

$$G(s) = \frac{b_{s0}}{(a_{s2}s^2 + a_{s1}s + 1)}$$

where $b_{s0} = Z_2$, $a_{s2} = T_1T_2$, $a_{s1} = T_1 + T_2$.

The continuous-time system will be controlled by the feedback controller (Fig.8.28). The controller $R(z^{-1})$ has to guarantee

1. finite time response of the closed-loop system,
2. finite time response of the closed-loop system in the minimum number of steps.

The discretised transfer function of the controlled system is given in the fractional representation as

$$G(z^{-1}) = \frac{b_1 z^{-1} + b_2 z^{-2}}{1 + a_{I1} z^{-1} + a_{I2} z^{-2}}$$

where

$$\begin{aligned}
b_1 &= Z_2 \left(-d_1 - d_2 - \frac{T_1(1+d_2)}{T_2 - T_1} + \frac{T_2(1+d_1)}{T_2 - T_1} \right) \\
b_2 &= Z_2 \left(d_1 + d_2 + \frac{T_1 d_2}{T_2 - T_1} - \frac{T_2 d_1}{T_2 - T_1} \right) \\
d_1 &= \mathrm{e}^{-\frac{T_s}{T_1}} \\
d_2 &= \mathrm{e}^{-\frac{T_s}{T_2}} \\
a_{I1} &= -d_1 - d_2 \\
a_{I2} &= d_1 d_2
\end{aligned}$$

and T_s is the sampling period.

The discrete-time transfer function of the controlled system in series with the discrete-time integrator is given as

$$G_i(z^{-1}) = \frac{b_1 z^{-1} + b_2 z^{-2}}{1 + a_1 z^{-1} + a_2 z^{-2} + a_3 z^{-3}}$$

where $a_1 = a_{I1} - 1$, $a_2 = a_{I2} - a_{I1}$, $a_3 = -a_{I2}$.

The minimum degree controller has the structure of the transfer function given by (8.479) where polynomials

$$Y(z^{-1}) = y_0 + y_1 z^{-1} + y_2 z^{-2}, \quad X(z^{-1}) = x_0 + x_1 z^{-1}$$

are solution of the Diophantine equation

$$(1 + a_1 z^{-1} + a_2 z^{-2} + a_3 z^{-3}) X(z^{-1}) + (b_1 z^{-1} + b_2 z^{-2}) Y(z^{-1}) = 1$$

If $T(z^{-1}) = t_0$ then the controller with the transfer function

$$R(z^{-1}) = \frac{y_0 + t_0 + (y_1 + t_0 a_1) z^{-1} + (y_2 + t_0 a_2) z^{-2} + t_0 a_3 z^{-3}}{x_0 + (x_1 - t_0 b_1) z^{-1} - t_0 b_2 z^{-2}}$$

guarantees the finite-time response in four sampling steps.

If $T(z^{-1}) = 0$, the controller with the transfer function

$$R(z^{-1}) = \frac{y_0 + y_1 z^{-1} + y_2 z^{-2}}{x_0 + x_1 z^{-1}}$$

guarantees finite-time response in a minimum time, i. e. in three sampling steps.

MATLAB Program 8.2 implements the whole finite-time design for parameters $Z_2 = 2$, $T_1 = 1$, $T_2 = 2$, $T_s = 1$ and $T(z^{-1}) = -1$.

Program 8.2 (Calculation of the dead-beat controller)
(requires the Polynomial Toolbox for MATLAB)

```
z = 2; t1 = 1; t2 = 2; ts = 1;
G = tf(z, conv([t1 1],[t2 1]));
gz = c2d(G, ts);
zinteg = tf([1 0],[1 -1], ts);

% convert to Polynomial toolbox format manually
[b,a]=tfdata(gz, 'v');
zii = [1; zi; zi^2];
b = b * [1; zi; zi^2];
a = a * [1; zi; zi^2];
a = a * (1-zi); % add integrator

%Design the DB controller
[x0,y0]=axbyc(a,b,1);
tt = -1;
x = x0 - b * tt;
y = y0 + a * tt;
```

Fig. 8.29 shows the Simulink diagram using Program 8.2 for finite-time control. Fig. 8.30a shows trajectories of the manipulated and controlled variable of the continuous-time second order system with the finite-time control design and for reference $w(t) = 1(t)$.
Fig. 8.30b shows trajectories of the manipulated and controlled variable of the same system with the dead-beat control design.
We can notice that the non-zero parameter $T(z^{-1})$ influences for example magnitude of the manipulated variable. The price paid for it is a longer duration of the closed-loop response. The minimum number of non-zero values of the control error is 3, a constant $T(z^{-1})$ gives one step more. The same situation is for the manipulated variable where there are at least 4 non-zero values and one more for a constant $T(z^{-1})$.
It is still possible for our second order system to find a dead-beat controller that gives a faster response with smaller number of non-zero control error values. This can be obtained using a two-degrees-of-freedom controller. The minimum number of non-zero values of the control error is 2 and the number of non-zero values of control increments is 3.
If the condition on finite number of control increment steps is relaxed to a asymptotically stable sequence then it is possible for stable systems to obtain the zero control error after one sampling step.

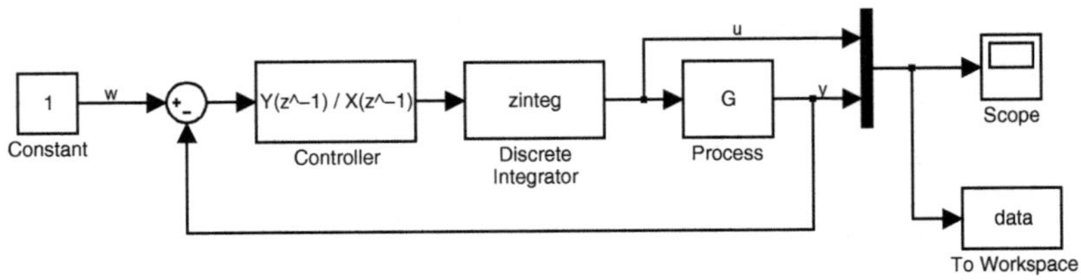

Fig. 8.29. Simulink diagram for finite-time tracking of a continuous-time second order system

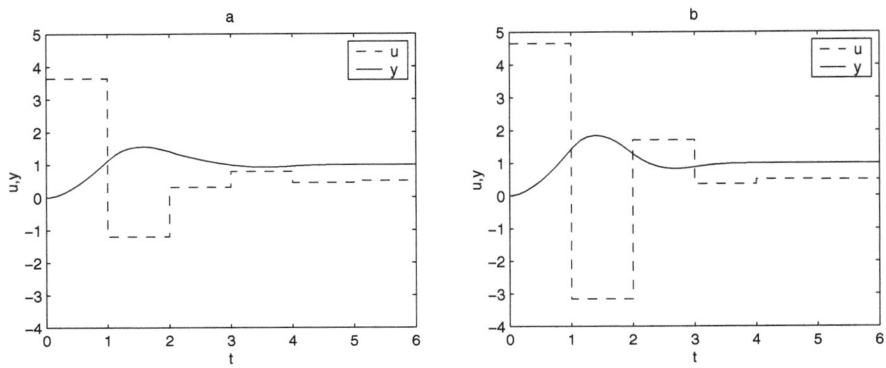

Fig. 8.30. Tracking in (a) finite (b) the least number of steps

8.8 LQ Control with Observer, State-Space and Polynomial Interpretations

In Section 8.6 the state feedback control with observer was analysed and it is shown in Fig. 8.13. One of the results is the separation theorem saying that state feedback can be designed independently on the observer part.

Eigenvalues of the closed-loop system are composed of eigenvalues of the matrix $\boldsymbol{A} - \boldsymbol{BK}$ (state feedback without observer) and of eigenvalues of the matrix $\boldsymbol{A} - \boldsymbol{LC}$. LQ control is then designed as follows:

1. LQ calculation of $\boldsymbol{K}$,
2. design of $\boldsymbol{L}$ to guarantee stability,
3. state feedback control for singlevariable systems is implemented using the control law

$$\boldsymbol{u} = \boldsymbol{K}\hat{\boldsymbol{x}} \tag{8.483}$$

Poles of the closed-loop system are determined by the poles of

$$o(s) = \det\left(s\boldsymbol{I} - (\boldsymbol{A} - \boldsymbol{LC})\right) \tag{8.484}$$

$$f(s) = \det\left(s\boldsymbol{I} - (\boldsymbol{A} - \boldsymbol{BK})\right) \tag{8.485}$$

Based on the results in the state-space domain we will now derive LQ control in the polynomial domain.

8.8.1 Polynomial LQ Control Design with Observer for Singlevariable Systems

Consider a linear reachable and observable singlevariable system

$$\dot{\boldsymbol{x}}(t) = \boldsymbol{A}\boldsymbol{x}(t) + \boldsymbol{B}u(t), \quad \boldsymbol{x}(0) = \boldsymbol{x}_0 \tag{8.486}$$

$$y(t) = \boldsymbol{C}\boldsymbol{x}(t) \tag{8.487}$$

and a cost function

$$I = \frac{1}{2}\int_0^\infty \left(\boldsymbol{x}^T(t)\boldsymbol{Q}\boldsymbol{x}(t) + u^2(t)\right) \mathrm{d}t \tag{8.488}$$

where $\boldsymbol{Q}$ is a real symmetric positive semidefinite matrix. LQ control problem consists in finding a control law ($u = \text{function}(\boldsymbol{x})$) such that guarantees asymptotic stability of the closed-loop system and minimisation of I for any $\boldsymbol{x}_0$.

As it was shown in Section 8.2, the control law is of the form

$$u = -\boldsymbol{K}\boldsymbol{x} \tag{8.489}$$

where

$$\boldsymbol{K} = \boldsymbol{B}^T\boldsymbol{P} \tag{8.490}$$

and $\boldsymbol{P}$ is a symmetric positive semidefinite solution of the matrix Riccati equation

$$\boldsymbol{P}\boldsymbol{A} + \boldsymbol{A}^T\boldsymbol{P} - \boldsymbol{P}\boldsymbol{B}\boldsymbol{B}^T\boldsymbol{P} = -\boldsymbol{Q} \tag{8.491}$$

Deterministic state estimate is given as

$$\dot{\hat{\boldsymbol{x}}}(t) = \boldsymbol{A}\hat{\boldsymbol{x}}(t) + \boldsymbol{B}u(t) + \boldsymbol{L}\left(y(t) - \boldsymbol{C}\hat{\boldsymbol{x}}\right), \quad \hat{\boldsymbol{x}}(0) = \hat{\boldsymbol{x}}_0 \tag{8.492}$$

Matrix $\boldsymbol{L}$ is chosen in such a way that the estimation error

$$\boldsymbol{e}(t) = \boldsymbol{x}(t) - \hat{\boldsymbol{x}}(t) \tag{8.493}$$

with initial condition

$$\boldsymbol{e}(0) = \boldsymbol{x}_0 - \hat{\boldsymbol{x}}_0 \tag{8.494}$$

asymptotically converges to origin.

Control law (8.489) cannot directly be used. It was shown in Section 8.6 that polynomial implementation of the observer gives the control law of the form

$$u = \tilde{u} \tag{8.495}$$

where

$$\tilde{u} = -\boldsymbol{K}\hat{\boldsymbol{x}} \tag{8.496}$$

Polynomial solution of the deterministic control design with controller input being y can be obtained as a combination of polynomial controller and observer.

Theorem 8.5 (LQ control with observer). *Consider a reachable and observable singlevariable system*

$$\dot{\boldsymbol{x}}(t) = \boldsymbol{A}\boldsymbol{x}(t) + \boldsymbol{B}u(t), \quad \boldsymbol{x}(0) = \boldsymbol{x}_0 \tag{8.497}$$

$$y(t) = \boldsymbol{C}\boldsymbol{x}(t) \tag{8.498}$$

where

$$(s\boldsymbol{I} - \boldsymbol{A})^{-1}\boldsymbol{B} = \frac{\boldsymbol{B}_{Rs}(s)}{a(s)} \tag{8.499}$$

$$\boldsymbol{C}(s\boldsymbol{I} - \boldsymbol{A})^{-1}\boldsymbol{B} = \frac{b(s)}{a(s)} \tag{8.500}$$

and where $a(s)$ is the characteristic polynomial of the matrix $\boldsymbol{A}$ and polynomials $a(s)$, $b(s)$ are coprime.

The control law based on state estimation and minimising the cost function I is given as

$$u = -\frac{q(s)}{p(s)}y \tag{8.501}$$

where polynomials $p(s)$, $q(s)$ are a solution of the equation

$$a(s)p(s) + b(s)q(s) = o(s)f(s) \tag{8.502}$$

Polynomial $p(s)$ is of degree n if stable polynomial $o(s)$ is of degree n (full order observer case), or is of degree $n-1$ is stable polynomial $o(s)$ is of degree $n-1$ (reduced order observer case). Polynomial $q(s)$ is of degree $n-1$. Stable monic polynomial $f(s)$ is defined as

$$a(-s)a(s) + \boldsymbol{B}_{Rs}(-s)\boldsymbol{Q}\boldsymbol{B}_{Rs}(s) = f(-s)f(s) \tag{8.503}$$

Proof. The proof of Theorem 8.3 shows that the characteristic polynomial $f(s)$ of the closed-loop matrix $\boldsymbol{A} - \boldsymbol{B}\boldsymbol{K}$ and the controller gain $\boldsymbol{K}$ satisfy the following relation

$$a(s) + \boldsymbol{K}\boldsymbol{B}_{Rs}(s) = f(s) \tag{8.504}$$

where the stable monic polynomial $f(s)$ is solution of spectral factorisation equation (8.503). If $f(s)$ exists, it is unique. Matrix $\boldsymbol{K}$ from (8.504) is also solution of the Riccati equation

$$PA + A^T P - PBB^T P = -Q + sP - sP \tag{8.505}$$

This equation can be manipulated as follows

$$\left(-sI - A^T\right) P + P\left(sI - A\right) = Q - PBB^T P \tag{8.506}$$

$$\begin{aligned} &B^T\left(-sI - A^T\right)^{-1}\left(-sI - A^T\right)P\left(sI - A\right)^{-1}B \\ &+ B^T\left(-sI - A^T\right)^{-1}P\left(sI - A\right)\left(sI - A\right)^{-1}B \\ &\qquad = B^T\left(-sI - A^T\right)^{-1}\left(Q - PBB^T P\right)\left(sI - A\right)^{-1}B \end{aligned} \tag{8.507}$$

$$\begin{aligned} &B^T P\left(sI - A\right)^{-1}B + B^T\left(-sI - A^T\right)^{-1}PB \\ &\qquad = B^T\left(-sI - A^T\right)^{-1}\left(Q - PBB^T P\right)\left(sI - A\right)^{-1}B \end{aligned} \tag{8.508}$$

As (8.499) holds, equation (8.508) can be written as

$$B^T P B_{Rs}(s)a(-s) + B_{Rs}^T(-s)PBa(s) = B_{Rs}^T(-s)\left(Q - PBB^T P\right)B_{Rs} \tag{8.509}$$

$$\begin{aligned} &KB_{Rs}(s)a(-s) + B_{Rs}^T(-s)K^T a(s) \\ &\qquad = B_{Rs}^T(-s)QB_{Rs}(s) - B_{Rs}^T(-s)K^T K B_{Rs}(s) \end{aligned} \tag{8.510}$$

$$\begin{aligned} &a(-s)a(s) + B_{Rs}^T(-s)K^T a(s) + KB_{Rs}(s)a(-s) + B_{Rs}^T(-s)K^T K B_{Rs}(s) \\ &\qquad = a(-s)a(s) + B_{Rs}^T(-s)QB_{Rs}(s) \end{aligned} \tag{8.511}$$

$$\begin{aligned} &\left(a(-s) + B_{Rs}^T(-s)K^T\right)\left(a(s) + KB_{Rs}(s)\right) \\ &\qquad = a(-s)a(s) + B_{Rs}^T(-s)QB_{Rs}(s) \end{aligned} \tag{8.512}$$

Considering (8.504) shows that (8.512) is the same as (8.503) which proves one part of the theorem.

The proof of Theorem 8.3 also gives equation of the asymptotic observer (8.310) where polynomials on this equation are solution of (8.320).

Substituting $K\hat{x}$ from (8.310) into (8.496) yields

$$u = -\frac{r(s)}{o(s)}u - \frac{q(s)}{o(s)}y \tag{8.513}$$

If

$$p(s) = o(s) + r(s) \tag{8.514}$$

then (8.513) is of the form (8.501). As (8.302), (8.317), (8.514) hold, equation (8.310) can be transformed into (8.502). This concludes the proof. □

Example 8.10:

Consider a controlled process with transfer function of the form

$$G(s) = \frac{b_0}{s + a_0}$$

The objective is to find for this system the transfer function of the feedback LQ controller with state observer. It is required that the controller contains an integration term. Therefore, we consider in the cost function (8.488) derivative $\dot{u}(t)$ in place of $u(t)$ and the weighting matrix $\boldsymbol{Q}$ is reduced to $\boldsymbol{Q} = Q_1$.

From Theorem 8.5 follows that the controller transfer function is in our case given as

$$R(s) = \frac{q_1 + q_0}{s(p_1 s + p_0)}$$

where

$$p_0 = o_0 + f_1 - a_0 p_1 \qquad p_1 = 1$$

$$q_0 = \frac{o_0 f_0}{b_0} \qquad q_1 = \frac{o_0 f_1 + f_0 - a_0 p_0}{b_0}$$

are solutions of the Diophantine equation

$$(s + a_0)s(p_1 s + p_0) + b_0(q_1 s + q_0) = o(s)f(s)$$

and

$$o(s) = s + o_0$$

is a stable monic polynomial. Coefficients of the stable monic polynomial

$$f(s) = s^2 + f_1 s + f_0$$

follow from the spectral factorisation equation

$$(-s + a_0)(-s)(s + a_0)s + \begin{pmatrix} b_0 & -b_0 s \end{pmatrix} \begin{pmatrix} Q_1 & 0 \\ 0 & 0 \end{pmatrix} \begin{pmatrix} b_0 \\ b_0 s \end{pmatrix} = f(-s)f(s)$$

and are given as

$$f_0 = b_0 \sqrt{Q_1}, \quad f_1 = \sqrt{a_0^2 + 2f_0}$$

The observer polynomial $o(s)$ can be suitably chosen according to further requirements on control performance.

8.8.2 Polynomial LQ Design with State Estimation for Multivariable Systems

Consider a controllable and observable multivariable system

$$\dot{\boldsymbol{x}}(t) = \boldsymbol{A}\boldsymbol{x}(t) + \boldsymbol{B}\boldsymbol{u}(t), \quad \boldsymbol{x}(0) = \boldsymbol{x}_0 \tag{8.515}$$

$$\boldsymbol{y}(t) = \boldsymbol{C}\boldsymbol{x}(t) \tag{8.516}$$

and a cost function

$$I = \frac{1}{2}\int_0^\infty (\boldsymbol{x}^T(t)\boldsymbol{Q}\boldsymbol{x}(t) + \boldsymbol{u}^T(t)\boldsymbol{u}(t))\mathrm{d}t \tag{8.517}$$

LQ control law is implemented as

$$\boldsymbol{u}(t) = -\boldsymbol{K}\hat{\boldsymbol{x}}(t) \tag{8.518}$$

where

$$\boldsymbol{K} = \boldsymbol{B}^T\boldsymbol{P} \tag{8.519}$$

and $\boldsymbol{P}$ is a symmetric positive semidefinite solution of the algebraic Riccati equation

$$\boldsymbol{P}\boldsymbol{A} + \boldsymbol{A}^T\boldsymbol{P} - \boldsymbol{P}\boldsymbol{B}\boldsymbol{B}^T\boldsymbol{P} = -\boldsymbol{Q} \tag{8.520}$$

Deterministic state estimate is given as

$$\dot{\hat{\boldsymbol{x}}}(t) = \boldsymbol{A}\hat{\boldsymbol{x}}(t) + \boldsymbol{B}\boldsymbol{u}(t) + \boldsymbol{L}(\boldsymbol{y}(t) - \boldsymbol{C}\hat{\boldsymbol{x}}(t)) \quad , \quad \hat{\boldsymbol{x}}(0) = \boldsymbol{x}_0 \tag{8.521}$$

As (8.398) holds then

$$\boldsymbol{K} = \boldsymbol{X}_L^{-1}\boldsymbol{Y}_L \tag{8.522}$$

where $\boldsymbol{X}_L$, $\boldsymbol{Y}_L$ are a solution of the Diophantine equation

$$\boldsymbol{X}_L\boldsymbol{A}_R(s) + \boldsymbol{Y}_L\boldsymbol{B}_{Rs}(s) = \boldsymbol{F}_R(s) \tag{8.523}$$

This was derived in polynomial design of pole placement controllers for multivariable systems. LQ control can be interpreted as the pole placement design. This implies that we need to determine a matrix $\boldsymbol{F}_R(s)$ such that LQ cost function is minimised. To find such a matrix, we will transform the Riccati equation as follows.

Adding $s\boldsymbol{P}$ to either side of equation (8.520) gives

$$\left(-s\boldsymbol{I} - \boldsymbol{A}^T\right)\boldsymbol{P} + \boldsymbol{P}\left(s\boldsymbol{I} - \boldsymbol{A}\right) = \boldsymbol{Q} - \boldsymbol{P}\boldsymbol{B}\boldsymbol{B}^T\boldsymbol{P} \tag{8.524}$$

Multiplying from the left by $\boldsymbol{B}^T(-s\boldsymbol{I}-\boldsymbol{A}^T)^{-1}$, from the right by $(s\boldsymbol{I}-\boldsymbol{A})^{-1}\boldsymbol{B}$ and using (8.522), (8.398) gives

$$\boldsymbol{B}_{Rs}^T(-s)\boldsymbol{K}^T\boldsymbol{A}_R(s) + \boldsymbol{A}_R^T(-s)\boldsymbol{K}\boldsymbol{B}_{Rs}(s) + \boldsymbol{B}_{Rs}^T(-s)\boldsymbol{K}^T\boldsymbol{K}\boldsymbol{B}_{Rs}(s) = \boldsymbol{B}_{Rs}^T(-s)\boldsymbol{Q}\boldsymbol{B}_{Rs}(s) \quad (8.525)$$

Adding $\boldsymbol{A}_R^T(-s)\boldsymbol{A}_R(s)$ to either side of equation yields

$$(\boldsymbol{A}_R^T(-s) + \boldsymbol{B}_{Rs}^T(-s)\boldsymbol{K}^T)(\boldsymbol{A}_R(s) + \boldsymbol{K}\boldsymbol{B}_{Rs}(s)) = \boldsymbol{A}_R^T(-s)\boldsymbol{A}_R(s) + \boldsymbol{B}_{Rs}^T(-s)\boldsymbol{Q}\boldsymbol{B}_{Rs}(s) \quad (8.526)$$

From (8.405) follows that LQ design $\boldsymbol{F}_R(s)$ needed for determination of $\boldsymbol{K}$ from (8.523) is a solution of the spectral factorisation equation

$$\boldsymbol{F}_R^T(-s)\boldsymbol{F}_R(s) = \boldsymbol{A}_R^T(-s)\boldsymbol{A}_R(s) + \boldsymbol{B}_{Rs}^T(-s)\boldsymbol{Q}\boldsymbol{B}_{Rs}(s) \quad (8.527)$$

All subsequent steps of the multivariable LQ control design are the same as in the pole placement design. Thus, $\boldsymbol{L}$ is determined from (8.411) and the controller transfer function matrix from (8.413).

WWW

Example 8.11: *Multivariable LQ control*

Consider a continuous-time multivariable controlled system with the transfer function matrix

$$\boldsymbol{G}(s) = \boldsymbol{A}_L^{-1}(s)\boldsymbol{B}_L(s)$$

where

$$\boldsymbol{A}_L(s) = \begin{pmatrix} 1+0.3s & 0.5s \\ 0.1s & 1+0.7s \end{pmatrix}, \quad \boldsymbol{B}_L(s) = \begin{pmatrix} 0.2 & 0.4 \\ 0.6 & 0.8 \end{pmatrix}$$

The task is to design LQ controller with observer. It is required that the controller contains an integration term.

The problem can be solved using the relations given above and it is shown in Program 8.3. We use the Polynomial toolbox for MATLAB.

Program 8.3 (LQ controller design – `optlq22.m`**)**

```
%PLANT
al = [1+.3*s, .5*s; .1*s 1+0.7*s];
bl = [.2 .4;.6 .8];
bl = pol(bl);
[a,b,c,d] = lmf2ss(bl,al);
G = ss(a,b,c,d);

%CONTROLLER
als = al*s;
[As,Bs,Cs,Ds] = lmf2ss(bl,als);
    [Brs,Ar] = ss2rmf(As,Bs,eye(4));
Q =10*eye(4);
Fr = spf(Ar'*Ar+Brs'*Q*Brs);
```

```
[Xl,Yl] = xaybc(Ar,Brs,Fr);
K=mldivide(Xl,Yl);
K=lcoef(K);
[Bls,Al] = ss2lmf(As,eye(4),Cs);
%V=10*eye(4);
%Ol = spf(Al*Al'+Bls*V*Bls');
Ol =[-3.1-3.4*s-0.18*s^2    -0.5-0.58*s-0.011*s^2;
     -0.5-0.62*s-0.055*s^2   3.1+4*s+0.9*s^2];

[Xr,Yr] = axbyc(Al,Bls,Ol);
L=mrdivide(Yr,Xr);
L=lcoef(L);
acp = As-L*Cs-Bs*K;
bcp = L;
ccp = K;
dcp = zeros(2);

[brp,arp]=ss2rmf(acp,bcp,ccp,dcp);
[ac,bc,cc,dc] = rmf2ss(brp,arp*s)
```

The closed-loop system is implemented in Simulink scheme `optlq22s.mdl` shown in Fig. 8.31.
Fig. 8.32a,b shows trajectories of input and output variables of the controlled process.

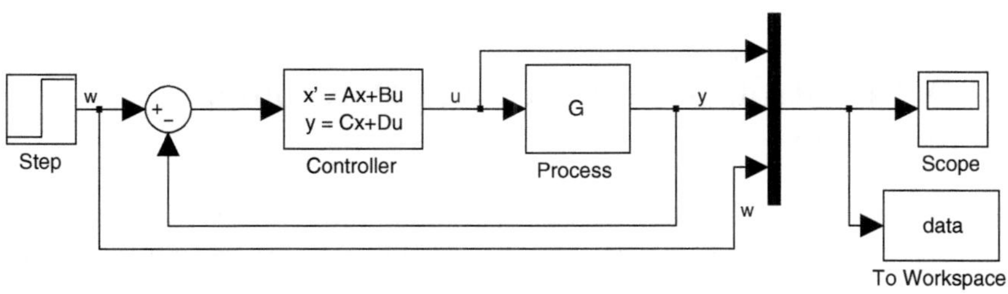

Fig. 8.31. Scheme `optlq22s.mdl` in Simulink for multivariable continuous-time LQ control

Note 8.6. Program 8.3 describes MIMO LQ control design. However, it is written to be used in PA, LQG, or H_2 control.

If pole assignment is desired, we comment the line in the program that computes spectral factorisation of the matrix $\boldsymbol{F}_R$

```
Fr = spf(Ar'*Ar+Brs'*Q*Brs)
```

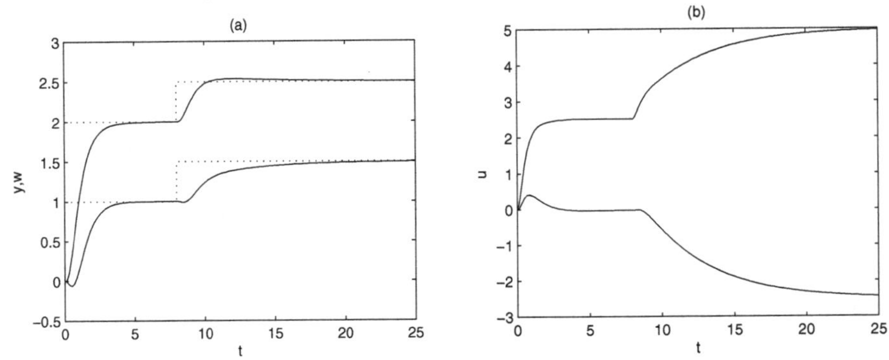

Fig. 8.32. Trajectories of (a) output, and setpoint variables, (b) input variables of the multivariable continuous-time process controlled with LQ controller

and substitute it with a corresponding matrix of appropriate dimensions that places the feedback poles.

To calculate LQG controller (see Section 8.9.2), we uncomment lines with the second spectral factorisation equation

```
V=10*eye(4)
O1 = spf(A1*A1'+B1s*V*B1s')
```

H_2 controller can be implemented similarly if differences between right hand sides of Riccati equations (8.539), (8.576), (8.607), and (8.609) are taken into account.

8.9 LQG Control, State-Space and Polynomial Interpretation

Based on the state LQG control with the Kalman filter we will derive expressions for polynomial LQG systems for both singlevariable and multivariable systems.

8.9.1 Singlevariable Polynomial LQG Control Design

Consider a linear singlevariable controllable and observable system

$$\dot{\boldsymbol{x}}(t) = \boldsymbol{A}\boldsymbol{x}(t) + \boldsymbol{B}u(t) + \boldsymbol{\xi}_x(t) \tag{8.528}$$

$$y(t) = \boldsymbol{C}\boldsymbol{x}(t) + \xi(t) \tag{8.529}$$

where $\boldsymbol{\xi}_x(t)$ is a random vector process with normal distribution and

$$E\left\{\boldsymbol{\xi}_x(t)\right\} = \boldsymbol{0} \tag{8.530}$$

$$E\left\{\boldsymbol{\xi}_x(t)\boldsymbol{\xi}_x^T(\tau)\right\} = \boldsymbol{V}\delta\left(t-\tau\right), \quad \boldsymbol{V} \geq 0 \tag{8.531}$$

$\xi(t)$ a random vector process with normal distribution and with

$$E\{\xi(t)\} = 0 \tag{8.532}$$

$$E\{\xi(t)\xi(\tau)\} = \delta(t-\tau), \tag{8.533}$$

We assume that $\boldsymbol{\xi}_x(t)$ and $\xi(t)$ are uncorrelated in time.

Linear Quadratic Gaussian (LQG) problem consists in finding a control law ($u = \text{function}(\boldsymbol{x})$) such that asymptotic stability of the closed-loop system is guaranteed and the cost function I is minimised where

$$I = E\left\{ \lim_{\substack{t_0 \to -\infty \\ t_f \to \infty}} \frac{1}{t_f - t_0} \int_{t_0}^{t_f} \left(\boldsymbol{x}^T(t)\boldsymbol{Q}\boldsymbol{x}(t) + u^2(t)\right) \mathrm{d}t \right\} \tag{8.534}$$

and $\boldsymbol{Q}$ is a real symmetric positive semidefinite matrix.

The separation theorem says that it is possible to solve LQG problem by solving the deterministic LQ problem at first

$$\dot{\boldsymbol{x}}(t) = \boldsymbol{A}\boldsymbol{x}(t) + \boldsymbol{B}u(t) \tag{8.535}$$

with the cost function

$$I = \frac{1}{2}\int_0^\infty \left(\boldsymbol{x}^T(t)\boldsymbol{Q}\boldsymbol{x}(t) + u^2(t)\right) \mathrm{d}t \tag{8.536}$$

This results in a control law of the form

$$u(t) = -\boldsymbol{K}\boldsymbol{x}(t) \tag{8.537}$$

where

$$\boldsymbol{K} = \boldsymbol{B}^T\boldsymbol{P} \tag{8.538}$$

and $\boldsymbol{P}$ is a symmetric positive semidefinite solution ($\boldsymbol{P}^T = \boldsymbol{P} \geq \boldsymbol{0}$) of the algebraic Riccati equation

$$\boldsymbol{P}\boldsymbol{A} + \boldsymbol{A}^T\boldsymbol{P} - \boldsymbol{P}\boldsymbol{B}\boldsymbol{B}^T\boldsymbol{P} = -\boldsymbol{Q} \tag{8.539}$$

In the second step, the state $\hat{\boldsymbol{x}}$ of the system (8.528), (8.529) is estimated using the Kalman filter

$$\dot{\hat{\boldsymbol{x}}}(t) = \boldsymbol{A}\hat{\boldsymbol{x}}(t) + \boldsymbol{B}u(t) + \boldsymbol{L}\left(y(t) - \boldsymbol{C}\hat{\boldsymbol{x}}(t)\right) \tag{8.540}$$

where

$$\boldsymbol{L} = \boldsymbol{N}\boldsymbol{C}^T \tag{8.541}$$

and $\boldsymbol{N}$ is a symmetric positive semidefinite solution ($\boldsymbol{N}^T = \boldsymbol{N} \geq \boldsymbol{0}$) of the algebraic Riccati equation

$$AN + NA^T - NC^TCN = -V \tag{8.542}$$

Finally, the control law is implemented as

$$u(t) = -K\hat{x}(t) \tag{8.543}$$

The LQG control is illustrated in Fig. 8.33.

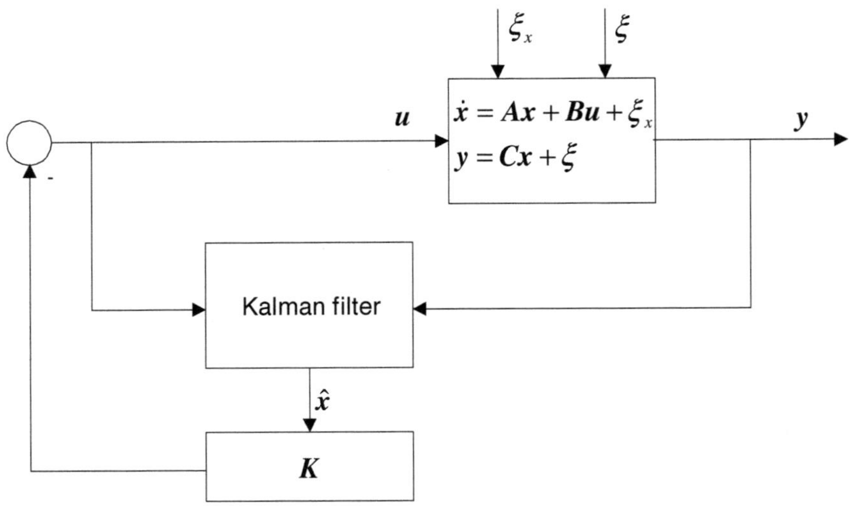

Fig. 8.33. Block diagram of LQG control

Theorem 8.7 (LQG control). *Consider a controllable and observable singlevariable system*

$$\dot{x}(t) = Ax(t) + Bu(t) + \xi_x(t) \tag{8.544}$$
$$y(t) = Cx(t) + \xi(t) \tag{8.545}$$

where

$$(sI - A)^{-1} B = \frac{B_{Rs}(s)}{a(s)} \tag{8.546}$$

$a(s)$, $B_{Rs}(s)$ *are right coprime for a reachable system,*

$$C(sI - A)^{-1} = \frac{B_{Ls}(s)}{a(s)} \tag{8.547}$$

$a(s)$, $B_{Ls}(s)$ *are left coprime for an observable system,*

$$C(sI - A)^{-1} B = \frac{b(s)}{a(s)} \tag{8.548}$$

where $a(s)$ is characteristic polynomial of the matrix $\boldsymbol{A}$ and polynomials $a(s)$, $b(s)$ are coprime.

The control law based on the state estimate from the Kalman filter and state feedback (8.543) is defined as

$$u = -\frac{q(s)}{p(s)}y \tag{8.549}$$

and polynomials $p(s)$, $q(s)$ are solution of the Diophantine equation

$$a(s)p(s) + b(s)q(s) = o(s)f(s) \tag{8.550}$$

Polynomial $p(s)$ is of degree n and polynomial $q(s)$ is of degree $n-1$. Stable monic polynomial $f(s)$ is defined from

$$a(-s)a(s) + \boldsymbol{B}_{Rs}^T(-s)\boldsymbol{Q}\boldsymbol{B}_{Rs}(s) = f(-s)f(s) \tag{8.551}$$

Stable monic polynomial $o(s)$ is defined from

$$a(-s)a(s) + \boldsymbol{B}_{Ls}(s)\boldsymbol{V}\boldsymbol{B}_{Ls}^T(-s) = o(s)o(-s) \tag{8.552}$$

Proof. We will first derive the equation for the polynomial $o(s)$ and the gain matrix $\boldsymbol{L}$. Equation (8.547) can be written as

$$\boldsymbol{B}_{Ls}(s)\left(s\boldsymbol{I} - \boldsymbol{A}\right) = a(s)\boldsymbol{C} \tag{8.553}$$

Adding $\boldsymbol{B}_{Ls}(s)\boldsymbol{L}\boldsymbol{C}$ to either side of the equation gives

$$\boldsymbol{B}_{Ls}(s)\left(s\boldsymbol{I} - \boldsymbol{A}\right) + \boldsymbol{B}_{Ls}(s)\boldsymbol{L}\boldsymbol{C} = a(s)\boldsymbol{C} + \boldsymbol{B}_{Ls}(s)\boldsymbol{L}\boldsymbol{C} \tag{8.554}$$

$$\boldsymbol{B}_{Ls}(s)\left(s\boldsymbol{I} - (\boldsymbol{A} - \boldsymbol{L}\boldsymbol{C})\right) = \left(a(s) + \boldsymbol{B}_{Ls}(s)\boldsymbol{L}\right)\boldsymbol{C} \tag{8.555}$$

$$\frac{\boldsymbol{B}_{Ls}(s)}{a(s) + \boldsymbol{B}_{Ls}(s)\boldsymbol{L}} = \boldsymbol{C}\frac{\operatorname{adj}\left(s\boldsymbol{I} - (\boldsymbol{A} - \boldsymbol{L}\boldsymbol{C})\right)}{\det\left(s\boldsymbol{I} - (\boldsymbol{A} - \boldsymbol{L}\boldsymbol{C})\right)} \tag{8.556}$$

From (8.556) follows that the Kalman gain filter $\boldsymbol{L}$ is a solution of

$$a(s) + \boldsymbol{B}_{Ls}(s)\boldsymbol{L} = o(s) \tag{8.557}$$

where $o(s)$ is the characteristic polynomial of the estimation matrix $\boldsymbol{A} - \boldsymbol{L}\boldsymbol{C}$. Stability of the estimate leads to the conclusion that $a(s) + \boldsymbol{B}_{Ls}(s)\boldsymbol{L}$ is a stable polynomial.

The polynomial $o(s)$ can be found as solution of the spectral factorisation equation (8.552). If $o(s)$ exists then it is unique.

Matrix $\boldsymbol{L} = \boldsymbol{N}\boldsymbol{C}^T$ from (8.557) is in LQG control involved in the Riccati equation

$$\boldsymbol{A}\boldsymbol{N} + \boldsymbol{N}\boldsymbol{A}^T - \boldsymbol{N}\boldsymbol{C}^T\boldsymbol{C}\boldsymbol{N} = -\boldsymbol{V} + s\boldsymbol{N} - s\boldsymbol{N} \tag{8.558}$$

This can be manipulated as follows

$$\boldsymbol{N}\left(-s\boldsymbol{I}-\boldsymbol{A}^T\right)+(s\boldsymbol{I}-\boldsymbol{A})\,\boldsymbol{N}=\boldsymbol{V}-\boldsymbol{N}\boldsymbol{C}^T\boldsymbol{C}\boldsymbol{N} \tag{8.559}$$

$$\begin{aligned}&\boldsymbol{C}\,(s\boldsymbol{I}-\boldsymbol{A})^{-1}\boldsymbol{N}\left(-s\boldsymbol{I}-\boldsymbol{A}^T\right)\left(-s\boldsymbol{I}-\boldsymbol{A}^T\right)^{-1}\boldsymbol{C}^T\\&+\boldsymbol{C}\,(s\boldsymbol{I}-\boldsymbol{A})^{-1}(s\boldsymbol{I}-\boldsymbol{A})\,\boldsymbol{N}\left(-s\boldsymbol{I}-\boldsymbol{A}^T\right)^{-1}\boldsymbol{C}^T\\&\qquad=\boldsymbol{C}\,(s\boldsymbol{I}-\boldsymbol{A})^{-1}\left(\boldsymbol{V}-\boldsymbol{N}\boldsymbol{C}^T\boldsymbol{C}\boldsymbol{N}\right)\left(-s\boldsymbol{I}-\boldsymbol{A}^T\right)^{-1}\boldsymbol{C}^T\end{aligned} \tag{8.560}$$

$$\begin{aligned}&\boldsymbol{C}\,(s\boldsymbol{I}-\boldsymbol{A})^{-1}\boldsymbol{N}\boldsymbol{C}^T+\boldsymbol{C}\boldsymbol{N}\left(-s\boldsymbol{I}-\boldsymbol{A}^T\right)^{-1}\boldsymbol{C}^T\\&\qquad=\boldsymbol{C}\,(s\boldsymbol{I}-\boldsymbol{A})^{-1}\left(\boldsymbol{V}-\boldsymbol{N}\boldsymbol{C}^T\boldsymbol{C}\boldsymbol{N}\right)\left(-s\boldsymbol{I}-\boldsymbol{A}^T\right)^{-1}\boldsymbol{C}^T\end{aligned} \tag{8.561}$$

From (8.547) follows for equation (8.561)

$$\begin{aligned}&\boldsymbol{B}_{Ls}(s)\boldsymbol{N}\boldsymbol{C}^T a(-s)+\boldsymbol{C}\boldsymbol{N}\boldsymbol{B}_{Ls}^T(-s)a(s)\\&\qquad=\boldsymbol{B}_{Ls}(s)\left(\boldsymbol{V}-\boldsymbol{N}\boldsymbol{C}^T\boldsymbol{C}\boldsymbol{N}\right)\boldsymbol{B}_{Ls}^T(-s)\end{aligned} \tag{8.562}$$

$$\boldsymbol{B}_{Ls}(s)\boldsymbol{L}a(-s)+\boldsymbol{L}\boldsymbol{B}_{Ls}^T(-s)a(s)=\boldsymbol{B}_{Ls}(s)\left(\boldsymbol{V}-\boldsymbol{L}\boldsymbol{L}^T\right)\boldsymbol{B}_{Ls}^T(-s) \tag{8.563}$$

$$\begin{aligned}&a(-s)a(s)+\boldsymbol{B}_{Ls}(s)\boldsymbol{L}a(-s)+\boldsymbol{L}^{-T}\boldsymbol{B}_{Ls}^T(-s)a(s)+\boldsymbol{B}_{Ls}(s)\,\boldsymbol{L}\boldsymbol{L}^T\boldsymbol{B}_{Ls}^T(-s)\\&\qquad=a(-s)a(s)+\boldsymbol{B}_{Ls}(s)\boldsymbol{V}\boldsymbol{B}_{Ls}^T(-s)\end{aligned} \tag{8.564}$$

$$\left(a(s)+\boldsymbol{B}_{Ls}(s)\boldsymbol{L}\right)\left(a(-s)+\boldsymbol{L}^T\boldsymbol{B}_{Ls}^T(-s)\right)=a(-s)a(s)+\boldsymbol{B}_{Ls}(s)\boldsymbol{V}\boldsymbol{B}_{Ls}^T(-s) \tag{8.565}$$

Considering equation (8.557), equation (8.565) is equal to (8.552). This concludes the first part of the proof.

Proof of relations (8.551) and (8.550) is the same as the proof of LQ control with observer. □

8.9.2 Multivariable Polynomial LQG Control Design

Consider a multivariable linear controllable and observable system

$$\dot{\boldsymbol{x}}(t)=\boldsymbol{A}\boldsymbol{x}(t)+\boldsymbol{B}\boldsymbol{u}(t)+\boldsymbol{\xi}_x(t) \tag{8.566}$$

$$\boldsymbol{y}(t)=\boldsymbol{C}\boldsymbol{x}(t)+\boldsymbol{\xi}(t) \tag{8.567}$$

where $\boldsymbol{\xi}_x(t)$ and $\boldsymbol{\xi}(t)$ are random processes uncorrelated in time with normal distribution, zero mean values, and covariances $\boldsymbol{V}$, $\boldsymbol{I}$.

The LQG problem consists in finding a control law ($u=\text{function}(\boldsymbol{x})$) such that the asymptotic stability of the closed-loop system is guaranteed and a the cost function I is minimised where

$$I = E\left\{ \lim_{\substack{t_0 \to -\infty \\ t_f \to \infty}} \frac{1}{t_f - t_0} \int_{t_0}^{t_f} \left(\boldsymbol{x}^T(t)\boldsymbol{Q}\boldsymbol{x}(t) + \boldsymbol{u}^T(t)\boldsymbol{u}(t) \right) \mathrm{d}t \right\} \tag{8.568}$$

As in the previous section, we first use the separation theorem and solve deterministic LQ problem for the system

$$\dot{\boldsymbol{x}}(t) = \boldsymbol{A}\boldsymbol{x}(t) + \boldsymbol{B}\boldsymbol{u}(t) \tag{8.569}$$

and the cost function

$$I = \frac{1}{2}\int_0^{\infty} \left(\boldsymbol{x}^T(t)\boldsymbol{Q}\boldsymbol{x}(t) + \boldsymbol{u}^T(t)\boldsymbol{u}(t) \right) \mathrm{d}t \tag{8.570}$$

that results in the control law of the form

$$\boldsymbol{u}(t) = -\boldsymbol{K}\boldsymbol{x}(t) \tag{8.571}$$

where

$$\boldsymbol{K} = \boldsymbol{B}^T\boldsymbol{P} \tag{8.572}$$

and $\boldsymbol{P}$ is a symmetric positive semidefinite solution of the algebraic Riccati equation

$$\boldsymbol{P}\boldsymbol{A} + \boldsymbol{A}^T\boldsymbol{P} - \boldsymbol{P}\boldsymbol{B}\boldsymbol{B}^T\boldsymbol{P} = -\boldsymbol{Q} \tag{8.573}$$

Then, in the next step we estimate the state $\hat{\boldsymbol{x}}$ for the system (8.566), (8.567) using the Kalman filter

$$\dot{\hat{\boldsymbol{x}}}(t) = \boldsymbol{A}\hat{\boldsymbol{x}}(t) + \boldsymbol{B}\boldsymbol{u}(t) + \boldsymbol{L}\left(\boldsymbol{y}(t) - \boldsymbol{C}\hat{\boldsymbol{x}}(t)\right) \tag{8.574}$$

where

$$\boldsymbol{L} = \boldsymbol{N}\boldsymbol{C}^T \tag{8.575}$$

and $\boldsymbol{N}$ is a symmetric positive semidefinite solution of the algebraic Riccati equation

$$\boldsymbol{A}\boldsymbol{N} + \boldsymbol{N}\boldsymbol{A}^T - \boldsymbol{N}\boldsymbol{C}^T\boldsymbol{C}\boldsymbol{N} = -\boldsymbol{V} \tag{8.576}$$

The LQG control law is implemented as

$$\boldsymbol{u}(t) = -\boldsymbol{K}\hat{\boldsymbol{x}}(t) \tag{8.577}$$

As (8.398) holds, matrix $\boldsymbol{K}$ can be determined from (8.522), (8.523), and (8.527). From (8.399) follows

$$\boldsymbol{L} = \boldsymbol{Y}_R \boldsymbol{X}_R^{-1} \tag{8.578}$$

where $\boldsymbol{X}_R$, $\boldsymbol{Y}_R$ are solution of the Diophantine equation

$$\boldsymbol{A}_L(s)\boldsymbol{X}_R + \boldsymbol{B}_{Ls}(s)\boldsymbol{Y}_R = \boldsymbol{O}_L(s) \tag{8.579}$$

$\boldsymbol{O}_L(s)$ is a polynomial matrix with $\det \boldsymbol{O}_L(s) \neq 0$ and is given from spectral factorisation as follows.

Adding $s\boldsymbol{N}$ to either side of (8.576) gives

$$(s\boldsymbol{I} - \boldsymbol{A})\,\boldsymbol{N} + \boldsymbol{N}\left(-s\boldsymbol{I} - \boldsymbol{A}^T\right) = \boldsymbol{V} - \boldsymbol{N}\boldsymbol{C}^T\boldsymbol{C}\boldsymbol{N} \tag{8.580}$$

Multiplying from left by $\boldsymbol{C}(s\boldsymbol{I} - \boldsymbol{A})^{-1}$ and from right by $(-s\boldsymbol{I} - \boldsymbol{A}^T)^{-1}\boldsymbol{C}^T$ and using (8.575), (8.399) yields

$$\begin{aligned}\boldsymbol{A}_L(s)\boldsymbol{L}^T\boldsymbol{B}_{Ls}^T(-s) + \boldsymbol{B}_{Ls}(s)\boldsymbol{L}\boldsymbol{A}_L^T(-s) + \boldsymbol{B}_{Ls}(s)\boldsymbol{L}\boldsymbol{L}^T\boldsymbol{B}_{Ls}(-s)&\\ = \boldsymbol{B}_{Ls}(s)\boldsymbol{V}\boldsymbol{B}_{Ls}^T(-s)&\end{aligned} \tag{8.581}$$

Adding $\boldsymbol{A}_L(s)\boldsymbol{A}_L^T(-s)$ to either side of this equation gives

$$\begin{aligned}(\boldsymbol{A}_L(s) + \boldsymbol{B}_{Ls}(s)\boldsymbol{L})(\boldsymbol{A}_L^T(-s) + \boldsymbol{L}^T\boldsymbol{B}_{Ls}^T(-s))&\\ = \boldsymbol{A}_L(s)\boldsymbol{A}_L^T(-s) + \boldsymbol{B}_{Ls}(s)\boldsymbol{V}\boldsymbol{B}_{Ls}^T(-s)&\end{aligned} \tag{8.582}$$

From (8.410) follows that the matrix $\boldsymbol{O}_L(s)$, needed in LQG control for determination of $\boldsymbol{L}$ from (8.579) can be found from the spectral factorisation equation

$$\boldsymbol{O}_L(s)\boldsymbol{O}_L^T(-s) = \boldsymbol{A}_L(s)\boldsymbol{A}_L^T(-s) + \boldsymbol{B}_{Ls}(s)\boldsymbol{V}\boldsymbol{B}_{Ls}^T(-s) \tag{8.583}$$

The rest of the multivariable LQG design is the same as the multivariable pole placement design.

8.10 H_2 Optimal Control

Robustness of the LQG control can be improved by changing the design to feedback controller with integral action in the sense of the H_2 norm minimisation. This design is more general than the LQG control because it removes some of its assumptions on properties of random sources as normal distribution that does not hold in general. However, it can be shown that LQG control is a special case of more general H_2 problem.

The LQG cost function is defined as

$$I = E\left\{\boldsymbol{x}^T(t)\boldsymbol{Q}\boldsymbol{x}(t) + \boldsymbol{u}^T(t)\boldsymbol{u}(t)\right\} \tag{8.584}$$

H_2 optimisation removes stochastic aspects from the controller design. To show this, we will transform the LQG control into a standard control configuration.

Define a new vector input $\boldsymbol{v}$ as

$$\begin{pmatrix} \boldsymbol{\xi}_x(t) \\ \boldsymbol{\xi}(t) \end{pmatrix} = \begin{pmatrix} \boldsymbol{B}_1 \\ \boldsymbol{D}_{21} \end{pmatrix} \boldsymbol{v}(t) \tag{8.585}$$

where $\boldsymbol{\xi}_x(t)$ and $\boldsymbol{\xi}(t)$ are defined by (8.220), (8.226). The new output vector is similarly defined as

$$\boldsymbol{z}(t) = \begin{pmatrix} \boldsymbol{C}_1 & \boldsymbol{D}_{12} \end{pmatrix} \begin{pmatrix} \boldsymbol{x}(t) \\ \boldsymbol{u}(t) \end{pmatrix} \tag{8.586}$$

Equations of the controlled system can thus be transformed as

$$\dot{\boldsymbol{x}}(t) = \boldsymbol{A}\boldsymbol{x}(t) + \boldsymbol{B}_1\boldsymbol{v}(t) + \boldsymbol{B}\boldsymbol{u}(t) \tag{8.587}$$
$$\boldsymbol{z}(t) = \boldsymbol{C}_1\boldsymbol{x}(t) + \boldsymbol{D}_{11}\boldsymbol{v}(t) + \boldsymbol{D}_{12}\boldsymbol{u}(t) \tag{8.588}$$
$$\boldsymbol{y}(t) = \boldsymbol{C}\boldsymbol{x}(t) + \boldsymbol{D}_{21}\boldsymbol{v}(t) + \boldsymbol{D}_{22}\boldsymbol{u}(t) \tag{8.589}$$

This standard control configuration is shown in Fig. 8.34. $\boldsymbol{G}(s)$ represents the generalised controlled process. Elements of the input vector $\boldsymbol{v}$ can in general be disturbances, measurement noise or reference inputs. Vector $\boldsymbol{z}$ represents the error of control, $\boldsymbol{y}$ is the vector of measured outputs, and $\boldsymbol{u}$ is the manipulated variable vector.

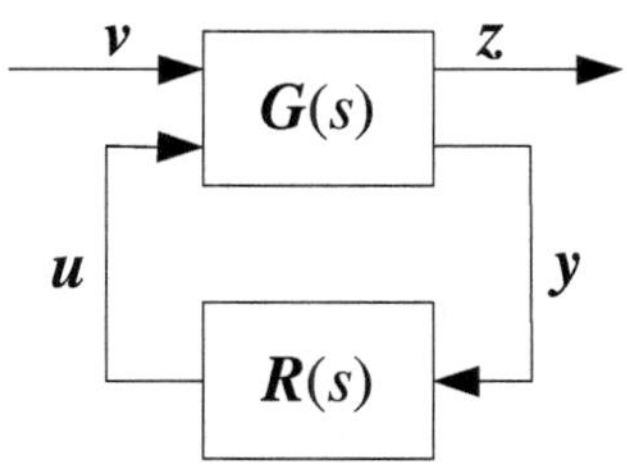

Fig. 8.34. Standard control configuration

Transfer function matrix of the generalised process is given as

$$\boldsymbol{G}(s) = \begin{pmatrix} \boldsymbol{G}_{11}(s) & \boldsymbol{G}_{12}(s) \\ \boldsymbol{G}_{21}(s) & \boldsymbol{G}_{22}(s) \end{pmatrix} \tag{8.590}$$

where

$$\boldsymbol{G}_{11}(s) = \boldsymbol{C}_1 \left(s\boldsymbol{I} - \boldsymbol{A}\right)^{-1} \boldsymbol{B}_1 + \boldsymbol{D}_{11} \tag{8.591}$$
$$\boldsymbol{G}_{12}(s) = \boldsymbol{C}_1 \left(s\boldsymbol{I} - \boldsymbol{A}\right)^{-1} \boldsymbol{B} + \boldsymbol{D}_{12} \tag{8.592}$$
$$\boldsymbol{G}_{21}(s) = \boldsymbol{C} \left(s\boldsymbol{I} - \boldsymbol{A}\right)^{-1} \boldsymbol{B}_1 + \boldsymbol{D}_{21} \tag{8.593}$$
$$\boldsymbol{G}_{22}(s) = \boldsymbol{C} \left(s\boldsymbol{I} - \boldsymbol{A}\right)^{-1} \boldsymbol{B} + \boldsymbol{D}_{22} \tag{8.594}$$

Following assumptions concerning $\boldsymbol{G}(s)$ are made.

A1: Pair $(\boldsymbol{A}, \boldsymbol{B}_1)$ or $(\boldsymbol{A}, \boldsymbol{B})$ is stabilisable, i. e. there is such a matrix $\boldsymbol{K}_1$ or $\boldsymbol{K}_2$ that matrix $(\boldsymbol{A}+\boldsymbol{B}_1\boldsymbol{K}_1)$ or $(\boldsymbol{A}+\boldsymbol{B}\boldsymbol{K}_2)$ is square and stable.
A2: Pair $(\boldsymbol{A}, \boldsymbol{C}_1)$ or $(\boldsymbol{A}, \boldsymbol{C})$ is detectable, i. e. there is such a matrix $\boldsymbol{L}_1$ or $\boldsymbol{L}_2$ that matrix $(\boldsymbol{A}+\boldsymbol{L}_1\boldsymbol{C}_1)$ or $(\boldsymbol{A}+\boldsymbol{L}_2\boldsymbol{C})$ is square and stable.
A3: Following relations hold:

$$\boldsymbol{D}_{11} = \boldsymbol{0}, \qquad \boldsymbol{D}_{22} = \boldsymbol{0} \tag{8.595}$$

$$\boldsymbol{D}_{12}^T\boldsymbol{C}_1 = \boldsymbol{0}, \qquad \boldsymbol{D}_{12}^T\boldsymbol{D}_{12} = \boldsymbol{I} \tag{8.596}$$

$$\boldsymbol{B}_1\boldsymbol{D}_{21}^T = \boldsymbol{0}, \qquad \boldsymbol{D}_{21}\boldsymbol{D}_{21}^T = \boldsymbol{I} \tag{8.597}$$

Assumptions A1 and A2 are needed to guarantee the existence of a stabilising controller. The third assumption is needed to guarantee the existence of suitable solutions of Riccati equations for the H_2 norm minimisation.

If equations (8.586), (8.595), (8.596), and (8.597), as well as

$$\boldsymbol{C}_1^T\boldsymbol{C}_1 = \boldsymbol{Q} \tag{8.598}$$

hold then LQG cost function can be written as

$$I = E\left\{\boldsymbol{z}^T(t)\boldsymbol{z}(t)\right\} \tag{8.599}$$

From Fig. 8.34 follows

$$\boldsymbol{z} = \boldsymbol{F}(s)\boldsymbol{v} \tag{8.600}$$

where $\boldsymbol{F}(s)$ is the closed-loop transfer function matrix

$$\boldsymbol{F}(s) = \boldsymbol{G}_{11}(s) + \boldsymbol{G}_{12}(s)\left(\boldsymbol{I} - \boldsymbol{R}(s)\boldsymbol{G}_{22}(s)\right)^{-1}\boldsymbol{R}(s)\boldsymbol{G}_{21}(s) \tag{8.601}$$

If $\boldsymbol{v}$ is white noise with the covariance matrix equal to $\boldsymbol{I}$ and if the closed-loop system is stable then

$$E\left\{\boldsymbol{z}^T(t)\boldsymbol{z}(t)\right\} = \frac{1}{2\pi}\,\mathrm{tr}\int_{-\infty}^{\infty}\boldsymbol{F}^T(-\mathrm{j}\omega)\,\boldsymbol{F}(\mathrm{j}\omega)\,\mathrm{d}\omega \tag{8.602}$$

$$E\left\{\boldsymbol{z}^T(t)\boldsymbol{z}(t)\right\} = \frac{1}{2\pi\mathrm{j}}\,\mathrm{tr}\int\boldsymbol{F}^T(-s)\,\boldsymbol{F}(s)\mathrm{d}s \tag{8.603}$$

H_2 – norm of a stable transfer function matrix $\boldsymbol{F}(s)$ is given as

$$\|\boldsymbol{F}\|_2 = \left(\frac{1}{2\pi}\,\mathrm{tr}\int_{-\infty}^{\infty}\boldsymbol{F}^T(-\mathrm{j}\omega)\,\boldsymbol{F}(\mathrm{j}\omega)\,\mathrm{d}\omega\right)^{\frac{1}{2}} \tag{8.604}$$

From this follows that LQG optimisation is equivalent to H_2 norm minimisation of the transfer function matrix $\boldsymbol{F}$.

If assumptions A1., A2., A3. hold, the H_2 optimal controller exists and is unique for the controlled system (8.590) and is defined as

$$R(s) = -K(sI - A + LC + BK)^{-1} L \tag{8.605}$$

where

$$K = B^T P \tag{8.606}$$

and P is a symmetric positive definite solution of the algebraic Riccati equation

$$PA + A^T P - PBB^T P = -C_1^T C_1 \tag{8.607}$$

and

$$L = NC^T \tag{8.608}$$

where N is a symmetric positive semidefinite solution of the algebraic Riccati equation

$$AN + NA^T - NC^T CN = -B_1 B_1^T \tag{8.609}$$

The transfer function matrix of the H_2 optimal controller $R(s)$ can also be determined by Diophantine equations. The derivation of the H_2 controller is the same as in the case of the pole placement, LQ or LQG controller.

If the controlled system is controllable and observable and if (8.398) holds then K can be determined from (8.522), (8.523) and $F_R(s)$ from

$$F_R^T(-s) F_R(s) = A_R^T(-s) A_R(s) + B_{Rs}^T(-s) C_1^T C_1 B_{Rs}(s) \tag{8.610}$$

If (8.399) holds then L can be determined from (8.578), (8.579) and $O_L(s)$ from

$$O_L(s) O_L^T(-s) = A_L(s) A_L^T(-s) + B_{Ls}(s) B_1 B_1^T B_{Ls}^T(-s) \tag{8.611}$$

Elements of matrices C_1, B_1 are design tunable parameters.

8.11 References

The bibiliography dealing with dynamic optimal control is very rich. Here we give only an overview of works that inspired development in this domain.

Controller design based on minimisation of a quadratic cost function and on state-space model was formulated and solved by Kalman in the early 1960s:

R. E. Kalman. Contribution to the theory of optimal control. *Boletin de la Sociedad Matematica Mexicana*, 5:102–119, 1960b.

R. E. Kalman. When is a linear system optimal? *Trans. ASME, Series D, Journal of Basic Engn.*, pages 51–60, 1964.

The problem formulated by Kalman was published and treated in many publications. Among them the most important are:

M. Athans and P. L. Falb. *Optimal Control.* McGraw-Hill, New York, 1966.

B. D. O. Anderson and J. B. Moore. *Linear Optimal Control.* Prentice Hall, Englewood Cliffs, New Jersey, 1971.

H. Kwakernaak and R. Sivan. *Linear Optimal Control Systems.* Wiley, New York, 1972.

If constraints on control variable are considered, the fundamental contribution to optimal control design was done by Pontryagin. We cite only one publication that inspired many others.

L. S. Pontryagin, V. G. Boltyanskii, R. V. Gamkrelidze, and E. F. Mishchenko. *The Mathematical Theory of Optimal Processes.* Wiley, New York, 1962.

Optimal control as solved by Kalman requires knowledge about state variables that all of which are usually not measurable. This issue was resolved using observers for deterministic systems and optimal state estimation for stochastic systems. The observer design was published in:

D. G. Luenberger. An introduction to observers. *IEEE Trans. AC*, 16: 596 – 602, 1971.

and the optimal state estimate in:

R. E. Kalman and R. S. Bucy. New results in linear filtering and prediction theory. *J. Basic Engr.*, 83:95 – 100, 1961.

Optimal feedback design can also be solved by polynomial approach. The Kalman approach leads to the solution of the Riccati equations. The basic idea of the input-output approach is to transform the design problem to the solution of Diophantine equations. The first reference in this domain is

L. N. Volgin. *The Fundamentals of the Theory of Controllers.* Soviet Radio, Moskva, 1962. (in Russian).

Polynomial approach can also be used for state-space systems. This can be found in:

V. Kučera. New results in state estimation and regulation. *Automatica*, 17: 745–748, 1981.

V. Kučera. A bridge between state-space and transfer-function methods. *Annual Reviews in Control*, 23:177–184, 1999.

and:

W. A. Wolovich. *Linear Multivariable Systems.* Springer Verlag, New York, 1974.

U. Shaked. A general transfer function approach to linear stationary filtering and steady-state optimal control problems. *Int. J. Control*, 24:741–770, 1976.

Traditional LQG optimal control is only a special case of a general H_2 optimal control. The standard H_2 problem of optimal control includes stabilisation of a linear system and its transfer function matrix attains a minimium norm in Hardy space H_2. The references dealing with H_2 control are for example:

J. C. Doyle, K. Glover, P. P. Khargonekar, and B. Francis. State-space solutions to the standard H_2 and H_∞ control problems. *IEEE Trans. Automatic Control*, 34:831–847, 1989.

K. Zhou, J. C. Doyle, and K. Glover. *Robust and Optimal Control.* Prentice Hall, Englewood Cliffs, New Jersey, 1995.

H. Kwakernaak. H_2-optimization – Theory and applications to robust control design. In *IFAC Symposium Robust Control Design*, Prague, 2000.

V. Kučera. A tutorial on H_2 control theory: The continuous-time case. In M. J. Grimble and V. Kučera, editors, *Polynomial Methods for Control System Design*, pages 1–56. Springer Verlag, London, 1996.

G. Meinsma. The standard H_2 problem. In *IFAC Symposium Robust Control Design*, Prague, 2000.

V. Kučera and D. Henrion. H_2 optimal control via pole placement. In *IFAC Symposium Robust Control Design*, Prague, 2000.

In

J. Mikleš, Š. Kožka, and Čirka L'. PID controller and LQ control design. In *IFAC Workshop on Digital Control*, pages 315–319, Terassa, 2000.

J. Mikleš. *Encyclopedia of Life Support Systems (EOLSS)*, chapter E6-43-34 Automation and Control in Process Industries, edited by H. Unbehauen. EOLSS Publishers, Oxford, UK, 2003. [http://www.eolss.net].

it is shown how the LQ, LQG, and H_2 control can be used for design and tuning of industrial controllers.

An alternative to variational approach in optimal control – *dynamic programming* was published for example in:

R. Bellman. *Dynamic Programming.* Princeton University Press, Princeton, New York, 1957.

and in many other sources.

Parametrisation of all stabilising controllers was published by Kučera and Youla:

V. Kučera. Stability of discrete linear feedback systems. In *Proc. 6th IFAC World Congress*, volume 1, Boston, 1975. Paper 44.1.

D. C. Youla, J. J. Bongiorno, and H.A. Jabr. Modern Wiener-Hopf design of optimal controller, Part I: The single-input case. *IEEE Trans. Automatic Control*, 21:3–14, 1976a.

D. C. Youla, J. J. Bongiorno, and H.A. Jabr. Modern Wiener-Hopf design of optimal controller, Part II: The multivariable case. *IEEE Trans. Automatic Control*, 21:319–338, 1976b.

V. Kučera. Diophantine equations in control – a survey. *Automatica*, 29: 1361–1375, 1993.

The dual parametrisation was formulated and published in:

F. R. Hansen and G. F. Franklin. On a fractional representation approach to closed-loop experiment design. In *Proc. ACC'88*, pages 1319–1320, Atlanta, GA, 1988.

Examples of use of the Youla-Kučera parametrisation for various control problems and of the simultaneous dual parametrisation are given in the works of the authors

M. Fikar and V. Kučera. On minimum finite length control problem. *Int. J. Control*, 73(2):152 – 158, 2000.

M. Fikar and H. Unbehauen. Youla-Kučera design of decoupled control systems. *Int. J. Control*, 75(3):213 – 218, 2002.

Ľ. Čirka, M. Fikar, and J. Mikleš. A deterministic LQ tracking problem: Parametrisation of the controller and the plant. *Journal of Electrical Engineering*, 53(5–6):126 – 131, 2002.

Some parts of the book use the Polynomial Toolbox for MATLAB. Authors of the book have cooperated within the European project Europoly and have used the newest version of the toolbox:

Polyx. *The Polynomial Toolbox for MATLAB*. PolyX, Ltd., Prague, 1999. http://www.polyx.com.

8.12 Exercises

Exercise 8.1:
Consider a controlled object with the transfer function of the form

$$G(s) = \frac{b_1 s + b_0}{s^2 + a_1 s + a_0}$$

where $b_0 = 2$, $b_1 = 0.1$, $b_0 = 4$, $a_1 = 3$.
Find the transfer fuction of a feedback LQ controller with integrator based on state observation. Choose $\boldsymbol{Q} = \boldsymbol{I}$ in the cost function (8.488).

Exercise 8.2:
Consider a controlled process with the transfer fuction

$$G(s) = \frac{b_1 s + b_0}{s^2 + a_1 s + a_0}$$

where $b_0 = 2$, $b_1 = 0.1$, $b_0 = 4$, $a_1 = 3$.
Find:

1. A feedback LQ controller with integrator based on state observation. Choose $\boldsymbol{Q} = \boldsymbol{I}$ in the cost function (8.488). Use the Polynomial toolbox for MATLAB.
2. Implement the closed-loop system in Simulink.

Exercise 8.3:
Consider a controlled process with the transfer fuction

$$G(s) = \frac{b_1 s + b_0}{s^2 + a_1 s + a_0}$$

where $b_0 = 2$, $b_1 = 0.1$, $b_0 = 4$, $a_1 = 3$.
Find:

1. State-space representation of the controlled system.
2. Use the Polynomial toolbox for MATLAB and design a H_2 controller.
3. Implement the closed-loop system in Simulink and test it for various choices of design parameters.

Exercise 8.4:
Consider a controlled process with the transfer fuction

$$G(s) = \frac{b_1 s + b_0}{s^2 + a_1 s + a_0}$$

where $b_0 = 2$, $b_1 = 0.1$, $b_0 = 4$, $a_1 = 3$.
Find:

1. State-space representation of the controlled system.
2. Use the Polynomial toolbox for MATLAB and design a H_2 controller with integrator.
3. Implement the closed-loop system in Simulink and test it for various choices of design parameters.

9

Predictive Control

This chapter deals with a popular control design method – predictive control. It explains its principles, features, and relations to other control methods. Compared to other parts of the book where continuous-time processes are discussed, predictive control is mainly based on discrete-time or sampled models. Therefore, derivations and expressions are presented mainly in the discrete-time domain.

9.1 Introduction

Model Based Predictive Control (MBPC) or only Predictive Control is a broad variety of control methods that comprise certain common ideas:

- a process model that is explicitly used to predict the process output for a fixed number of steps into future,
- a known future reference trajectory,
- calculation of a future control sequence minimising a certain objective function (usually quadratic, that involves future process output errors and control increments),
- receding strategy: at each sampling period only the first control signal of the sequence calculated is applied to a process controlled.

Among many useful features of MBPC, there is one that has created extensive industrial interest: the process constraints can easily be incorporated into the method at the design stage.

MBPC algorithms are reported to be very versatile and robust in process control applications. They usually outperform PID controllers and are applicable to non-minimum phase, open-loop unstable, time delay, and multivariable processes.

The principle of MBPC is shown in Fig. 9.1 and is as follows:

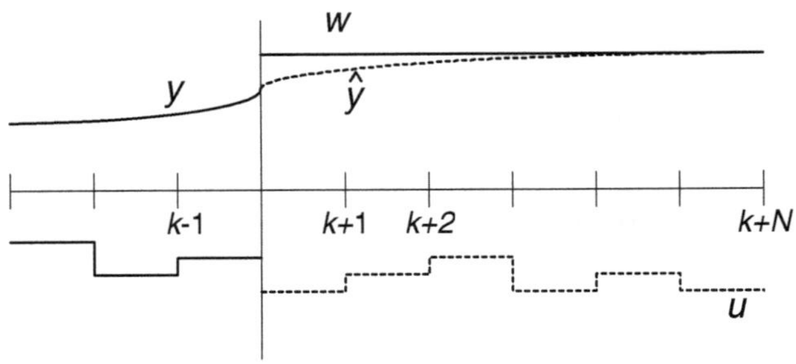

Fig. 9.1. Principle of MBPC

1. The process model is used to predict the future outputs $\hat{y}$ over some horizon N. The predictions are calculated based on information up to time k and on the future control actions that are to be determined.
2. The future control trajectory is calculated as a solution of an optimisation problem consisting of a cost function and possibly some constraints. The cost function comprises future output predictions, future reference trajectory, and future control actions.
3. Although the whole future control trajectory was calculated in the previous step, only its first element $u(k)$ is actually applied to the process. At the next sampling time the procedure is repeated. This is known as the *Receding Horizon* concept.

9.2 Ingredients of MBPC

9.2.1 Models

MBPC enables to plug-in directly any type of the process model. Of course, linear models are most often used. This is caused by the possibility of an analytic solution for the future control trajectory in unconstrained case.

The model should capture the process dynamics and to permit theoretical analysis. The process model is required to calculate the predicted future output trajectory. Some of the models incorporate directly disturbance model, in others it is simply assumed that disturbance is constant.

Impulse Response

The theoretical impulse sequence is usually truncated for practical reasons. The output is related to the input by the equation

$$y(k) = \sum_{i=1}^{N} h_i u(k-i) = H(q^{-1})u(k) \tag{9.1}$$

where $H(q^{-1}) = h_1 q^{-1} + h_2 q^{-2} + \cdots h_N q^{-N}$ and q^{-1} is the backward shift operator defined as $y(k)q^{-1} = y(k-1)$.

The drawbacks of this model are:

- high value of N needed ≈ 50,
- only stable processes can be represented.

Step Response

The step response model is very similar to the FIR model with the same drawbacks. Again, truncated step response is used for stable systems

$$y(k) = \sum_{i=1}^{N} g_i \Delta u(k-i) = G(q^{-1})(1-q^{-1})u(k) \tag{9.2}$$

As the step and impulse responses are easily collected, the methods based on them gained large popularity in the industry. The step model is for example used in DMC.

Transfer Function

This model is used in GPC, EHAC, EPSAC, and others. The output is modelled by the equation

$$A(q^{-1})y(k) = B(q^{-1})u(k) \tag{9.3}$$

The advantage of this representation is that it is also valid for unstable models. On the other side, order of the A, B polynomials is needed.

State Space

The representation of the state-space model is as follows:

$$\boldsymbol{x}(k+1) = \boldsymbol{A}\boldsymbol{x}(k) + \boldsymbol{B}\boldsymbol{u}(k) \tag{9.4}$$

$$\boldsymbol{y}(k) = \boldsymbol{C}\boldsymbol{x}(k) \tag{9.5}$$

Its advantage is an uncomplicated way of dealing with multivariable processes. However, the state observer is often needed.

Others

As it was stated before, any other process model is acceptable. Continuous nonlinear models in the form of ordinary differential equations are often used. Their drawbacks are large simulation times. The area of dynamic optimisation usually covers them.

Recently, neural and fuzzy models have gained popularity. Two approaches have emerged. The model is either directly used, or it only generates some process characteristics: step or impulse responses.

Disturbances

The most general disturbance model is an ARMA process (c. f. page 223) given by

$$n(k) = \frac{C(q^{-1})}{D(q^{-1})}\xi(k) \tag{9.6}$$

with $\xi(k)$ being white noise. Within the MBPC framework, the $D = \Delta A$ polynomial includes the integrator $\Delta = 1 - q^{-1}$ to cover random-walk disturbances. Another pleasing property of the integrator is the set-point tracking (integral action).

When the process model of the form (9.3) is used, the overall model is then called CARIMA (or ARIMAX) and is given as

$$\Delta A y(k) = B\Delta u(k) + C\xi(k) \tag{9.7}$$

9.2.2 Cost Function

The standard cost function used in predictive control contains quadratic terms of (possibly filtered) control error and control increments on a finite horizon into the future

$$I = \sum_{i=N_1}^{N_2} [P\hat{y}(k+i) - w(k+i)]^2 + \lambda \sum_{i=1}^{N_u} [\Delta u(k+i-1)]^2 \tag{9.8}$$

where $\hat{y}(k+i)$ is the process output of i steps in the future predicted on the base of information available upon the time k, $w(k+i)$ is the sequence of the setpoint variable and $\Delta u(k+i-1)$ is the sequence of the future control increments that have to be calculated.

Implicit constraints on Δu are placed between N_u and N_2 as

$$\Delta u(k+i-1) = 0, \quad N_u < i \leq N_2 \tag{9.9}$$

The cost function parameters are following:

- Horizons N_1, N_2, and N_u are called minimum, maximum, and control horizon, respectively. The horizons N_1 and N_2 mark the future time interval where it is desirable to follow the reference trajectory. N_1 should be at least equal to $T_d + 1$ where T_d is the assumed value of the process time delay. Also, the non-minimum phase behaviour of the process can be eliminated from the cost by letting N_1 to be sufficiently large. The value of N_2 should cover the important part of the step response curve, it is usually chosen to be about the settling time of the plant T_{90} (see page 257). The use of the control horizon N_u reduces the computational load of the method.

- Reference trajectory $w(k+i)$ is assumed to be known beforehand. If it is not the case, several approaches are possible. The simplest way is to assume that the future reference is constant and equal to the desired setpoint w^∞. The preferred approach is to use smooth reference trajectory that begins from the actual output value and approaches asymptotically via the first order filter the desired setpoint w^∞. It is thus given as

$$w(k) = y(k) \tag{9.10}$$

$$w(k+i) = \alpha w(k+i-1) + (1-\alpha)w^\infty \tag{9.11}$$

The parameter α determines smoothness of the trajectory with $\alpha \to 0$ being the fastest and $\alpha \to 1$ being the slowest trajectory.
The same effect can be achieved with the use of the filter polynomial $P(z^{-1})$. The output y follows the model trajectory $\frac{1}{P}w$. The corresponding filter to the previous first order trajectory is given as

$$P(z^{-1}) = \frac{1-\alpha z^{-1}}{1-\alpha} \tag{9.12}$$

9.3 Derivation and Implementation of Predictive Control

In this part, Generalised Predictive Control (GPC) will be derived. The GPC method is in the principle applicable to both SISO and MIMO processes and is based on input-output models. We begin the derivation for SISO systems for simplicity and show in the actual implementation of the method, how to treat MIMO systems. Alternatively, the predictor equations will also be derived for state-space models.

9.3.1 Derivation of the Predictor

The first step in the development of MBPC is derivation of the optimal predictor. We start with the CARIMA model (9.7) of the form

$$A(q^{-1})y(k) = B(q^{-1})u(k-1) + \frac{C(q^{-1})}{\Delta}\xi(k) \tag{9.13}$$

Note that we use explicitly $u(k-1)$ and thus the polynomial B has a non-zero absolute coefficient. We use $u(k-1)$ because $u(k)$ will constitute one element of the optimised variables.

Now let us think about this equation j steps in the future. This is accomplished by multiplication of this equation by q^j and it yields

$$y(t+j) = \frac{B}{A}u(t+j-1) + \frac{C}{\Delta A}\xi(t+j) \tag{9.14}$$

The last term of this equation contains past and future values of ξ. We may separate them by performing long division on the term $C/(\Delta A)$ and by separating the first j terms (quotient) with positive powers of q. This yields

$$\frac{C(q^{-1})}{\Delta A(q^{-1})} = E_j(q^{-1}) + q^{-j}\frac{F_j(q^{-1})}{\Delta A(q^{-1})} \tag{9.15}$$

where the polynomial E_j has degree $j-1$. Inserting back into (9.14) gives

$$y(t+j) = \frac{B}{A}u(t+j-1) + E_j\xi(t+j) + \frac{F_j}{\Delta A}\xi(k) \tag{9.16}$$

The last term contains actual value of the disturbance $\xi(k)$. This can be calculated from (9.13) and inserted back into the last equation

$$\begin{aligned} y(t+j) &= \frac{B}{A}u(t+j-1) - \frac{F_j B}{\Delta AC}\Delta u(k-1) + \frac{F_j}{C}y(k) + E_j\xi(t+j) \\ &= \left[\frac{B}{\Delta A} - q^{-j}\frac{F_j B}{\Delta AC}\right]\Delta u(t+j-1) + \frac{F_j}{C}y(k) + E_j\xi(t+j) \\ &= \frac{B}{C}\left[\frac{C}{\Delta A} - q^{-j}\frac{F_j}{\Delta A}\right]\Delta u(t+j-1) + \frac{F_j}{C}y(k) + E_j\xi(t+j) \end{aligned} \tag{9.17}$$

and finally substituting (9.15) into the term containing $\Delta u(t+j-1)$ yields

$$y(t+j) = \frac{BE_j}{C}\Delta u(t+j-1) + \frac{F_j}{C}y(k) + E_j\xi(t+j) \tag{9.18}$$

Again, we separate unknown (future and present) control actions from the known (past) ones by the means of the polynomial division

$$\frac{B(q^{-1})E_j(q^{-1})}{C(q^{-1})} = G_j(q^{-1}) + q^{-j}\frac{\Gamma_j(q^{-1})}{C(q^{-1})} \tag{9.19}$$

This gives the final form for the future value of the system output

$$y(t+j) = G_j\Delta u(t+j-1) + \frac{\Gamma_j}{C}\Delta u(k-1) + \frac{F_j}{C}y(k) + E_j\xi(t+j) \tag{9.20}$$

It is obvious that the minimum variance prediction of $y(t+j)$ for given data up to time k is obtained by replacing the last term containing future disturbances by zero and yields

$$\hat{y}(t+j) = G_j\Delta u(t+j-1) + \frac{\Gamma_j}{C}\Delta u(k-1) + \frac{F_j}{C}y(k) \tag{9.21}$$

$$\hat{y}(t+j) = G_j\Delta u(t+j-1) + y_0(t+j) \tag{9.22}$$

Thus, to obtain the j step predictor, two polynomial divisions (or equivalently Diophantine equations) are to be solved

$$C = E_j\Delta A + q^{-j}F_j \tag{9.23}$$

$$BE_j = G_jC + q^{-j}\Gamma_j \tag{9.24}$$

To implement calculation of the predictor efficiently, it is necessary to understand correctly the rôle of the equation (9.22) and the terms involved in it.

Let us first assume that all future control increments are zero. Equation (9.22) gives

$$\hat{y}(t+j) = y_0(t+j) \tag{9.25}$$

Hence, the term y_0 can be determined by the *free response* of the system if the input remains to be constant at the last computed value $u(k-1)$.

Similarly, let us assume that the system is at the time k in the steady-state and we may assume without loss of generality that the steady-state is zero. This gives a zero free response $y_0(t+j)$. If at time k the system is subject to unit step in input, the system output is given from (9.22) as

$$\begin{aligned}\hat{y}(t+j) &= G_j(q^{-1})\Delta u(t+j-1)\\ &= g_{j0}\Delta u(t+j-1) + g_{j1}\Delta u(t+j-2) + \cdots + g_{j,j-1}\Delta u(k)\\ &= g_{j,j-1}\end{aligned} \tag{9.26}$$

Thus, the polynomial $G_j(q^{-1})$ contains the system step response coefficients. As an alternative way to show this consider (9.15) multiplied by B/C:

$$\begin{aligned}\frac{B}{\Delta A} &= \frac{BE_j}{C} + q^{-j}\frac{BF_j}{\Delta AC}\\ &= G_j + q^{-j}\frac{\Gamma_j}{C} + q^{-j}\frac{BF_j}{\Delta AC}\end{aligned} \tag{9.27}$$

which shows that G_j is the quotient of the division $B/(\Delta A)$.

9.3.2 Calculation of the Optimal Control

The GPC cost function is given by (9.8). Let us now assume for simplicity that $N_1 = 1, N_u = N_2, P = 1$. It follows that all output prediction up to time $t + N_2$ are needed. Let us stack individual output predictions, future control increments, future reference trajectory, and free responses into corresponding vectors

$$\hat{\boldsymbol{y}}^T = [\hat{y}(t+1), \hat{y}(t+2), \ldots, \hat{y}(t+N_2)] \tag{9.28}$$

$$\boldsymbol{y}_0^T = [y_0(t+1), y_0(t+2), \ldots, y_0(t+N_2)] \tag{9.29}$$

$$\tilde{\boldsymbol{u}}^T = [\Delta u(k), \Delta u(t+1), , \ldots, \Delta u(t+N_2-1)] \tag{9.30}$$

$$\boldsymbol{w}^T = [w(t+1), w(t+2), \ldots, w(t+N_2)] \tag{9.31}$$

$$\tag{9.32}$$

To vectorise the predictor (9.22), let us form a matrix containing step response coefficients given as

$$\boldsymbol{G} = \begin{pmatrix} g_0 & 0 & \ldots\ldots & 0\\ g_1 & g_0 & 0 \;\ldots & 0\\ \vdots & & \ddots\;\ddots & \vdots\\ \vdots & & g_0 & 0\\ g_{N_2-1} & \cdots & \cdots & g_0 \end{pmatrix} \tag{9.33}$$

If we take into effect the real value of N_1, then the first $N_1 - 1$ rows of the matrix $\boldsymbol{G}$ should be removed. Similarly, only the first N_u columns are retained. Thus, the real matrix $\boldsymbol{G}$ has the dimension $[N_2 - N_1 + 1 \times N_u]$.

Hence, the predictor in the vector notation can be written as

$$\hat{\boldsymbol{y}} = \boldsymbol{G}\tilde{\boldsymbol{u}} + \boldsymbol{y}_0 \tag{9.34}$$

and the cost function (9.8) as

$$\begin{aligned} I &= (\hat{\boldsymbol{y}} - \boldsymbol{w})^T(\hat{\boldsymbol{y}} - \boldsymbol{w}) + \lambda\tilde{\boldsymbol{u}}^T\tilde{\boldsymbol{u}} \\ &= (\boldsymbol{G}\tilde{\boldsymbol{u}} + \boldsymbol{y}_0 - \boldsymbol{w})^T(\boldsymbol{G}\tilde{\boldsymbol{u}} + \boldsymbol{y}_0 - \boldsymbol{w}) + \lambda\tilde{\boldsymbol{u}}^T\tilde{\boldsymbol{u}} \\ &= c_0 + 2\boldsymbol{g}^T\tilde{\boldsymbol{u}} + \tilde{\boldsymbol{u}}^T\boldsymbol{H}\tilde{\boldsymbol{u}} \end{aligned} \tag{9.35}$$

where the gradient $\boldsymbol{g}$ and Hessian $\boldsymbol{H}$ are defined as

$$\boldsymbol{g}^T = \boldsymbol{G}^T(\boldsymbol{y}_0 - \boldsymbol{w}) \tag{9.36}$$

$$\boldsymbol{H} = \boldsymbol{G}^T\boldsymbol{G} + \lambda\boldsymbol{I} \tag{9.37}$$

Minimisation of the cost function 9.35 now becomes a direct problem of linear algebra. The solution in the unconstrained case can be found by setting partial derivative of I with respect to $\tilde{\boldsymbol{u}}$ to zero and yields

$$\tilde{\boldsymbol{u}} = -\boldsymbol{H}^{-1}\boldsymbol{g} \tag{9.38}$$

This equation gives the whole trajectory of the future control increments and as such it is an open-loop strategy. To close the loop, only the first element of $\tilde{\boldsymbol{u}}$, e. g. $\Delta u(k)$ is applied to the system and the whole algorithm is recomputed at time $t + 1$. This strategy is called the *Receding Horizon Principle* and is one of the key issues in the MBPC concept.

If we denote the first row of the matrix $(\boldsymbol{G}^T\boldsymbol{G} + \lambda\boldsymbol{I})^{-1}\boldsymbol{G}^T$ by $\boldsymbol{K}$ then the actual control increment can be calculated as

$$\Delta u(k) = \boldsymbol{K}(\boldsymbol{w} - \boldsymbol{y}_0) \tag{9.39}$$

Hence, if there is no difference between the free response and the setpoint sequence in the future, the actual control increment will be zero. If there will be some differences in the future, the actual control increment will be proportional to them with the factor $\boldsymbol{K}$.

To summarise the procedure, it should be noticed, that only two plant characteristics are needed: free response $\boldsymbol{y}_0$ that is changing at each sampling time and step response $G(z^{-1})$ which is in the case of time invariant system needed only once. Moreover, also the Hessian matrix $\boldsymbol{H}$ that should be inverted, contains only information from the step response and can also be calculated beforehand. The calculation of the actual control increment is thus dependent only on weighted sum of past inputs and outputs contained in $\boldsymbol{y}_0$ and forms therefore a linear control law.

9.3.3 Closed-Loop Relations

It follows from (9.39) that the control is a linear combination of the setpoint and free response. As the control law and the controlled process are linear, it is possible to derive the characteristic equation and the poles of the closed-loop system.

The equation (9.39) can be written in the time domain

$$\Delta u(k) = \sum_{j=N_1}^{N_2} k_j \left(w(t+j) - y_0(t+j)\right) \tag{9.40}$$

Using the free response from (9.22) yields

$$\Delta u(k) = \sum_{j=N_1}^{N_2} k_j \left(w(t+j) - \frac{\Gamma_j}{C} q^{-1}\Delta u(k) - \frac{F_j}{C} y(k)\right) \tag{9.41}$$

$$C\Delta u(k) = C\sum_{j=N_1}^{N_2} k_j w(t+j) - \sum_{j=N_1}^{N_2} k_j \Gamma_j q^{-1}\Delta u(k) - \sum_{j=N_1}^{N_2} k_j F_j y(k) \tag{9.42}$$

$$\left(C + q^{-1}\sum_{j=N_1}^{N_2} k_j \Gamma_j\right)\Delta u(k) = C\sum_{j=N_1}^{N_2} k_j w(t+j) - \sum_{j=N_1}^{N_2} k_j F_j y(k) \tag{9.43}$$

Let us assume for simplicity that the future setpoint change is constant, i. e. $w(t+j) = w(k)$. The control law constitutes a two degree-of-freedom (2DoF) controller

$$Q_c \Delta u(k) = R_c w(k) - P_c y(k) \tag{9.44}$$

where

$$Q_c = \frac{C + q^{-1}\sum_{j=N_1}^{N_2} k_j \Gamma_j}{\sum_{j=N_1}^{N_2} k_j} \tag{9.45}$$

$$R_c = C \tag{9.46}$$

$$P_c = \frac{\sum_{j=N_1}^{N_2} k_j F_j}{\sum_{j=N_1}^{N_2} k_j} \tag{9.47}$$

To derive the characteristic equation of the closed-loop systems, the control law is inserted into the CARIMA model (9.13)

$$A\Delta y(k) = Bq^{-1}\left(\frac{R_c}{Q_c}w(k) - \frac{P_c}{Q_c}y(k)\right) + C\xi(k) \tag{9.48}$$

$$y(z) = \frac{BR_c z^{-1}}{Q_c A\Delta + P_c B z^{-1}} w(z) + \frac{CQ_c}{Q_c A\Delta + P_c B z^{-1}} \xi(z) \tag{9.49}$$

The denominator polynomial is the characteristic equation. Its further manipulation gives

$$Q_c A\Delta + P_c B z^{-1} = \frac{CA\Delta + z^{-1}A\Delta \sum_{j=N_1}^{N_2} k_j \Gamma_j + Bz^{-1}\sum_{j=N_1}^{N_2} k_j F_j}{\sum_{j=N_1}^{N_2} k_j} \tag{9.50}$$

$$= \frac{CA\Delta + z^{-1}\sum_{j=N_1}^{N_2} k_j\left(A\Delta\Gamma_j + BF_j\right)}{\sum_{j=N_1}^{N_2} k_j} \tag{9.51}$$

$$= C\frac{A\Delta + \sum_{j=N_1}^{N_2} k_j z^{j-1}(B - A\Delta G_j)}{\sum_{j=N_1}^{N_2} k_j} \tag{9.52}$$

$$= CM \tag{9.53}$$

We can see that the characteristic polynomial is composed of two polynomials. The first determines the behaviour of the closed-loop system with respect to disturbances and corresponds to the observer poles. The second polynomial depends in a rather complicated way on the controlled process and the GPC parameters. It determines the closed-loop stability as it can also be seen from the output equation

$$y(z) = \frac{Bz^{-1}}{M}w(z) + \frac{R_c}{M}\xi(z) \tag{9.54}$$

where the polynomial C has been cancelled in both transfer functions.

9.3.4 Derivation of the Predictor from State-Space Models

To show that it is possible to use any linear model, we will derive here the predictor equations based in the state-space model.

Let us consider the standard state-space model of the form (9.4), (9.5). Like in the previous case when CARIMA model was used, it is necessary to include an integrator to the process. This can be accomplished by changing the input to the process from $\boldsymbol{u}(k)$ to its increment $\Delta\boldsymbol{u}(k)$

$$\boldsymbol{u}(k) = \Delta\boldsymbol{u}(k) + \boldsymbol{u}(k-1) \tag{9.55}$$

There are more possibilities how to incorporate an integrator in the process model. Among them, one way is to define a new state vector

$$\bar{\boldsymbol{x}}(k) = \begin{pmatrix} \boldsymbol{x}(k) \\ \boldsymbol{u}(k-1) \end{pmatrix} \tag{9.56}$$

From the state-space model (9.4), (9.5) then follows

$$\bar{\boldsymbol{x}}(t+1) = \begin{pmatrix} \boldsymbol{A} & \boldsymbol{B} \\ \boldsymbol{0} & \boldsymbol{I} \end{pmatrix} \bar{\boldsymbol{x}}(k) + \begin{pmatrix} \boldsymbol{B} \\ \boldsymbol{I} \end{pmatrix} \Delta\boldsymbol{u}(k) = \bar{\boldsymbol{A}}\bar{\boldsymbol{x}}(k) + \bar{\boldsymbol{B}}\Delta\boldsymbol{u}(k) \tag{9.57}$$

$$\boldsymbol{y}(k) = \begin{pmatrix} \boldsymbol{C} & \boldsymbol{0} \end{pmatrix} \bar{\boldsymbol{x}}(k) = \bar{\boldsymbol{C}}\bar{\boldsymbol{x}}(k) \tag{9.58}$$

We can see that formally both state-space models are equivalent.

The derivation of the predictor is simpler than in the input-output case. The predictor vector is given as

$$\hat{\boldsymbol{y}}(t+1) = \bar{\boldsymbol{C}}\bar{\boldsymbol{x}}(t+1) = \bar{\boldsymbol{C}}\left(\bar{\boldsymbol{A}}\bar{\boldsymbol{x}}(k) + \bar{\boldsymbol{B}}\Delta\boldsymbol{u}(k)\right) \tag{9.59}$$

$$\hat{\boldsymbol{y}}(t+2) = \bar{\boldsymbol{C}}\bar{\boldsymbol{x}}(t+2) = \bar{\boldsymbol{C}}\left(\bar{\boldsymbol{A}}\bar{\boldsymbol{x}}(t+1) + \bar{\boldsymbol{B}}\Delta\boldsymbol{u}(t+1)\right) \tag{9.60}$$

$$\vdots$$

$$\hat{\boldsymbol{y}}(t+N_2) = \bar{\boldsymbol{C}}\bar{\boldsymbol{x}}(t+N_2) = \bar{\boldsymbol{C}}\left(\bar{\boldsymbol{A}}\bar{\boldsymbol{x}}(t+N_2-1) + \bar{\boldsymbol{B}}\Delta\boldsymbol{u}(t+N_2-1)\right) \tag{9.61}$$

where the state predictions are of the form

$$\bar{\boldsymbol{x}}(t+1) = \bar{\boldsymbol{A}}\bar{\boldsymbol{x}}(k) + \bar{\boldsymbol{B}}\Delta\boldsymbol{u}(k) \tag{9.62}$$

$$\begin{aligned}\bar{\boldsymbol{x}}(t+2) &= \bar{\boldsymbol{A}}\bar{\boldsymbol{x}}(t+1) + \bar{\boldsymbol{B}}\Delta\boldsymbol{u}(t+1)\\ &= \bar{\boldsymbol{A}}^2\bar{\boldsymbol{x}}(k) + \bar{\boldsymbol{A}}\bar{\boldsymbol{B}}\Delta\boldsymbol{u}(k) + \bar{\boldsymbol{B}}\Delta\boldsymbol{u}(t+1)\end{aligned} \tag{9.63}$$

$$\vdots$$

$$\bar{\boldsymbol{x}}(t+N_2) = \bar{\boldsymbol{A}}^{N_2}\bar{\boldsymbol{x}}(k) + \bar{\boldsymbol{A}}^{N_2-1}\bar{\boldsymbol{B}}\Delta\boldsymbol{u}(k) + \cdots + \bar{\boldsymbol{B}}\Delta\boldsymbol{u}(t+N_2-1) \tag{9.64}$$

Joining all predictions to form the vector equation (9.34) yields matrices and vectors

$$\boldsymbol{G} = \begin{pmatrix} \bar{\boldsymbol{C}}\bar{\boldsymbol{B}} & \boldsymbol{0} & \dots & \dots & \boldsymbol{0} \\ \bar{\boldsymbol{C}}\bar{\boldsymbol{A}}\bar{\boldsymbol{B}} & \bar{\boldsymbol{C}}\bar{\boldsymbol{B}} & \boldsymbol{0} & \dots & \boldsymbol{0} \\ \vdots & & \ddots & \ddots & \vdots \\ \vdots & & & \bar{\boldsymbol{C}}\bar{\boldsymbol{B}} & \boldsymbol{0} \\ \bar{\boldsymbol{C}}\bar{\boldsymbol{A}}^{N_2-1}\bar{\boldsymbol{B}} & \dots & & \dots & \bar{\boldsymbol{C}}\bar{\boldsymbol{B}} \end{pmatrix} \tag{9.65}$$

and

$$\boldsymbol{y}_0 = \begin{pmatrix} \bar{\boldsymbol{C}}\bar{\boldsymbol{A}} \\ \bar{\boldsymbol{C}}\bar{\boldsymbol{A}}^2 \\ \vdots \\ \bar{\boldsymbol{C}}\bar{\boldsymbol{A}}^{N_2} \end{pmatrix} \bar{\boldsymbol{x}}(k) \tag{9.66}$$

Example 9.1:

To explain the principle let us consider a simple SISO process with the numerator and denominator polynomials of the form

$$B(z^{-1}) = 0.4 + 0.1z^{-1}, \qquad A(z^{-1}) = 1 - 0.5z^{-1}$$

This corresponds to the process transfer functions of the form

$$\frac{0.4z^{-1} + 0.1z^{-2}}{1 - 0.5z^{-1}}$$

Let us assume that $C(z^{-1}) = 1$. The CARIMA description of the system is of the form

$$\Delta A(q^{-1})y(k) = B(q^{-1})\Delta u(k-1) + \xi(k)$$

or

$$y(k) = 1.5y(k-1) - 0.5y(k-2) + 0.4\Delta u(k-1) + 0.1\Delta u(k-2) + \xi(k)$$

Let us now assume the cost function (9.8) with the parameters $N_1 = 1$, $N_2 = 3$, $N_u = 2$.
The predictions of future outputs can be obtained if $\xi(t+i) = 0$ and are as follows:

$$\begin{aligned}\hat{y}(t+1) &= 1.5y(k) - 0.5y(k-1) + 0.4\Delta u(k) + 0.1\Delta u(k-1)\\ \hat{y}(t+2) &= 1.5y(t+1) - 0.5y(k) + 0.4\Delta u(t+1) + 0.1\Delta u(k)\\ \hat{y}(t+3) &= 1.5y(t+2) - 0.5y(t+1) + 0.4\Delta u(t+2) + 0.1\Delta u(t+1)\end{aligned}$$

According to the assumptions, the term $\Delta u(t+2)$ is equal to zero. The higher output predictions contain the lower output predictions that can be back substituted and yield

$$\begin{aligned}\hat{y}(t+1) &= 1.5y(k) - 0.5y(k-1) + 0.4\Delta u(k) + 0.1\Delta u(k-1)\\ \hat{y}(t+2) &= 1.75y(k) - 0.75y(k-1)\\ &\quad + 0.4\Delta u(t+1) + 0.7\Delta u(k) + 0.15\Delta u(k-1)\\ \hat{y}(t+3) &= 1.875y(k) - 0.875y(k-1)\\ &\quad + 0.7\Delta u(t+1) + 0.85\Delta u(k) + 0.175\Delta u(k-1)\end{aligned}$$

Stacking all predictions into a vector and separating the terms unknown at time k from the known ones gives

$$\begin{pmatrix}\hat{y}(t+1)\\ \hat{y}(t+2)\\ \hat{y}(t+3)\end{pmatrix} = \begin{pmatrix}0.4 & 0\\ 0.7 & 0.4\\ 0.85 & 0.7\end{pmatrix}\begin{pmatrix}\Delta u(k)\\ \Delta u(t+1)\end{pmatrix} + \begin{pmatrix}1.5y(k) - 0.5y(k-1) + 0.1\Delta u(k-1)\\ 1.75y(k) - 0.75y(k-1) + 0.15\Delta u(k-1)\\ 1.875y(k) - 0.875y(k-1) + 0.175\Delta u(k-1)\end{pmatrix}$$

An alternative to obtain the matrix $\boldsymbol{G}$ would be to perform the long division

$$\frac{B}{\Delta A} = \frac{0.4 + 0.1z^{-1}}{1 - 1.5z^{-1} + 0.5z^{-2}} = 0.4 + 0.7z^{-1} + 0.85z^{-2} + \cdots$$

Now let us assume that the weighting coefficient λ is equal to zero. Inversion of the Hessian matrix gives

$$\boldsymbol{H}^{-1} = \begin{pmatrix}5.1383 & -6.9170\\ -6.9170 & 10.8498\end{pmatrix}$$

Finally, multiplication with $\boldsymbol{g}$ yields the closed-loop expression for the element

$$\begin{aligned}\Delta u(k) &= -3.6461y(k) + 1.2351y(k-1) - 0.2470\Delta u(k-1)\\ &\quad + 2.0553w(t+1) + 0.8300w(t+2) - 0.4743w(t+3)\end{aligned}$$

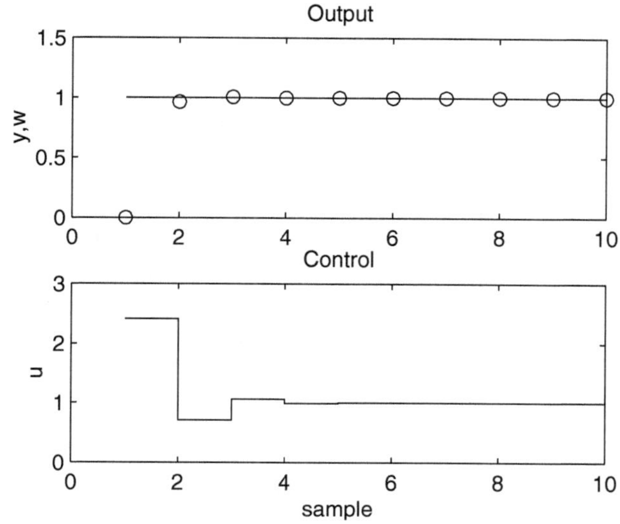

Fig. 9.2. Closed-loop response of the controlled system

Simulation results show the behaviour of the closed-loop system in the Fig. 9.2.

Let us also derive the predictor equations in the state-space form. The corresponding state-space model can for example be given as

$$\boldsymbol{x}(t+1) = \begin{pmatrix} 0 & 1 \\ 0 & 0.5 \end{pmatrix} \boldsymbol{x}(k) + \begin{pmatrix} 0 \\ 1 \end{pmatrix} u(k)$$

$$y(k) = \begin{pmatrix} 0.1 & 0.4 \end{pmatrix} \boldsymbol{x}(k)$$

The new state-space description with integrator has three states

$$\bar{\boldsymbol{x}}(t+1) = \begin{pmatrix} 0 & 1 & 0 \\ 0 & 0.5 & 1 \\ 0 & 0 & 1 \end{pmatrix} \bar{\boldsymbol{x}}(k) + \begin{pmatrix} 0 \\ 1 \\ 1 \end{pmatrix} \Delta u(k) = \bar{\boldsymbol{A}}\bar{\boldsymbol{x}}(k) + \bar{\boldsymbol{B}}\Delta u(k)$$

$$y(k) = \begin{pmatrix} 0.1 & 0.4 & 0 \end{pmatrix} \bar{\boldsymbol{x}}(k) = \bar{\boldsymbol{C}}\bar{\boldsymbol{x}}(k)$$

The state predictions of three steps to the future are then given as

$$\begin{aligned} \bar{\boldsymbol{x}}(t+1) &= \bar{\boldsymbol{A}}\bar{\boldsymbol{x}}(k) + \bar{\boldsymbol{B}}\Delta u(k) \\ \bar{\boldsymbol{x}}(t+2) &= \bar{\boldsymbol{A}}\bar{\boldsymbol{x}}(t+1) + \bar{\boldsymbol{B}}\Delta u(t+1) \\ &= \bar{\boldsymbol{A}}^2\bar{\boldsymbol{x}}(k) + \bar{\boldsymbol{A}}\bar{\boldsymbol{B}}\Delta u(k) + \bar{\boldsymbol{B}}\Delta u(t+1) \\ \bar{\boldsymbol{x}}(t+3) &= \bar{\boldsymbol{A}}\bar{\boldsymbol{x}}(t+2) \\ &= \bar{\boldsymbol{A}}^3\bar{\boldsymbol{x}}(k) + \bar{\boldsymbol{A}}^2\bar{\boldsymbol{B}}\Delta u(k) + \bar{\boldsymbol{A}}\bar{\boldsymbol{B}}\Delta u(t+1) \end{aligned}$$

The output predictions are then given as

$$\begin{aligned} \hat{y}(t+1) &= \bar{\boldsymbol{C}}\bar{\boldsymbol{A}}\bar{\boldsymbol{x}}(k) + \bar{\boldsymbol{C}}\bar{\boldsymbol{B}}\Delta u(k) \\ \hat{y}(t+2) &= \bar{\boldsymbol{C}}\bar{\boldsymbol{A}}^2\bar{\boldsymbol{x}}(k) + \bar{\boldsymbol{C}}\bar{\boldsymbol{A}}\bar{\boldsymbol{B}}\Delta u(k) + \bar{\boldsymbol{C}}\bar{\boldsymbol{B}}\Delta u(t+1) \\ \hat{y}(t+3) &= \bar{\boldsymbol{C}}\bar{\boldsymbol{A}}^3\bar{\boldsymbol{x}}(k) + \bar{\boldsymbol{C}}\bar{\boldsymbol{A}}^2\bar{\boldsymbol{B}}\Delta u(k) + \bar{\boldsymbol{C}}\bar{\boldsymbol{A}}\bar{\boldsymbol{B}}\Delta u(t+1) \end{aligned}$$

or in the vector form

$$\begin{pmatrix} \hat{y}(t+1) \\ \hat{y}(t+2) \\ \hat{y}(t+3) \end{pmatrix} = \begin{pmatrix} \bar{\boldsymbol{C}}\bar{\boldsymbol{B}} & 0 \\ \bar{\boldsymbol{C}}\bar{\boldsymbol{A}}\bar{\boldsymbol{B}} & \bar{\boldsymbol{C}}\bar{\boldsymbol{B}} \\ \bar{\boldsymbol{C}}\bar{\boldsymbol{A}}^2\bar{\boldsymbol{B}} & \bar{\boldsymbol{C}}\bar{\boldsymbol{A}}\bar{\boldsymbol{B}} \end{pmatrix} \begin{pmatrix} \Delta u(k) \\ \Delta u(t+1) \end{pmatrix} + \begin{pmatrix} \bar{\boldsymbol{C}}\bar{\boldsymbol{A}} \\ \bar{\boldsymbol{C}}\bar{\boldsymbol{A}}^2 \\ \bar{\boldsymbol{C}}\bar{\boldsymbol{A}}^3 \end{pmatrix} \bar{\boldsymbol{x}}(k)$$

When the actual values are substituted for the system matrices and states, it yields

$$\begin{pmatrix} \hat{y}(t+1) \\ \hat{y}(t+2) \\ \hat{y}(t+3) \end{pmatrix} = \begin{pmatrix} 0.4 & 0 \\ 0.7 & 0.4 \\ 0.85 & 0.7 \end{pmatrix} \begin{pmatrix} \Delta u(k) \\ \Delta u(t+1) \end{pmatrix} + \begin{pmatrix} 0.3x_2(k) + 0.4u(k-1) \\ 0.15x_2(k) + 0.7u(k-1) \\ 0.075x_2(k) + 0.85u(k-1) \end{pmatrix}$$

We can note that the matrix $\boldsymbol{G}$ is the same in both cases. The second part depends on the actual state definitions. However, the predictive controller will be the same.

9.3.5 Multivariable Input-Output Case

In the same line of thought as in the SISO case, the multivariable GPC algorithm can de derived via Diophantine equations. From the practical point of view, the multivariable controller can come from the prediction equation (9.22) that holds exactly as before, only the vector and matrix elements instead of being scalar are vectors and matrices. If m-input and n-output system is considered, then the matrix $\boldsymbol{G}$ is of the form

$$\boldsymbol{G} = \begin{pmatrix} \boldsymbol{G}_0 & 0 & \dots\dots & 0 \\ \boldsymbol{G}_1 & \boldsymbol{G}_0 & 0 \;\dots & 0 \\ \vdots & & \ddots \;\; \ddots & \vdots \\ \vdots & & \boldsymbol{G}_0 & 0 \\ \boldsymbol{G}_{N_2-1} & \dots & \dots & \boldsymbol{G}_0 \end{pmatrix} \tag{9.67}$$

with $\boldsymbol{G}_i$ being matrices of dimensions $[n \times m]$ and the overall $\boldsymbol{G}$ has dimensions $[n(N_2 - N_1 + 1) \times mN_u]$.

The vectors $\boldsymbol{y}_{0i}$ and matrices $\boldsymbol{G}_i$ can be obtained analogously as in the singlevariable case from free and step responses of the system.

The drawbacks of the proposed multivariable derivation are increased dimensions of the matrices involved in matrix inversion routines. The multivariable formulation can be broken into series of SISO GPC calculations if the system denominator matrix $A(z^{-1})$ is diagonal.

9.3.6 Implementation

As it was stated before, for the actual implementation of the linear GPC algorithm only two process characteristics are needed: step and free responses. The step response can be obtained directly from the process by performing a small step change in one of the manipulated inputs at a time if the process was before in a steady state. The magnitude of the step change is important if the process is non-linear. If the process steady-state gain is approximately known then the magnitude of the input step change should be chosen such that it produces step response approaching the desired setpoint value. This strategy may be applied if a non-linear model is used for predictions.

If the polynomials A, B are estimated on-line by means of a RLS algorithm, the step response is obtained from the polynomial division

$$\frac{B}{A\Delta} = g_0 + g_1 z^{-1} + \cdots + g_{N_1} z^{-N_1} + \cdots + g_{N_2} z^{-N_2} + \cdots \tag{9.68}$$

The matrix $\boldsymbol{G}$ can be formed from the coefficients of the step response and it is given by the last $N_2 - N_1 + 1$ rows and the first N_u columns of the Toeplitz matrix (9.33).

The free response is calculated as the process response from the actual initial conditions if the input is fixed to $u(k-1)$. At time k is should hold

$$y(k) = y_0(k) \tag{9.69}$$

However, the assumption about random-walk disturbance usually results in non-zero disturbance at time k and holds

$$y(k) = y_0(k) + d(k) \tag{9.70}$$

This disturbance is assumed to be constant in all future predictions. The disturbance and the free response are thus calculated as

$$d = \sum_{i=0}^{\deg(A)} a_i y_f(k-i) - \sum_{i=0}^{\deg(B)} b_i u_f(k-i-1) \tag{9.71}$$

$$y_0(t+j) = y_f(t+j) \tag{9.72}$$

$$= d - \sum_{i=1}^{\deg(A)} a_i y_f(k-i+j) + \sum_{i=0}^{\deg(B)} b_i \bar{u}(k-i+j-1) \tag{9.73}$$

where

$$\bar{u}(k-i+j) = \begin{cases} u_f(k-1) & j \geq i \\ u_f(k-i+j) & \text{inak} \end{cases} \tag{9.74}$$

and

$$y_f = \frac{y}{C}, \quad u_f = \frac{u}{C} \tag{9.75}$$

Note 9.1. If polynomial P is assumed to be non-unity, then the above is valid if the system denominator A is changed for PA. To ensure the offset-free setpoint following, the polynomial P should be specified subject to condition $P(1) = 1$.

Note 9.2. The polynomial C is normally not estimated on-line, but used as a user-design parameter. It can be shown from relation between state-space and input-oputput approaches that it acts as an observer polynomial and is used for disturbance rejection.

9.3.7 Relation to Other Approaches

One of the features of GPC approach is its generality. With different values of its parameters it can be reduced to some well-known controllers:

Mean level control

$$N_1 = 1,\ N_2 \to \infty,\ N_u = 1,\ P = 1,\ \lambda = 0$$

Exact model following

$$N_1 = 1,\ N_2 = T_d + 1,\ N_u = T_d + 1,\ P = 1,\ \lambda = 0$$

or

$$N_1 = 1,\ N_2 > T_d,\ N_u = N_2 - T_d,\ P = 1,\ \lambda = 0$$

Dead-beat control

$$N_1 \geq \deg(B) + 1,\ N_2 \geq N_u + N_1 - 1,\ N_u \geq \deg(A) + 1,\ P = 1,\ \lambda = 0$$

Pole placement Dead-beat $+P$. Poles are placed at zeros of P.

$$N_1 \geq \deg(B) + 1,\ N_2 \geq N_u + N_1 - 1,\ N_u \geq \deg(A) + 1,\ P \neq 1,\ \lambda = 0$$

9.3.8 Continuous-Time Approaches

Predictive control has been developed in discrete-time domain. The discrete formulation allows for an easy prediction generation, because time response of discrete systems can be obtained from polynomial division of the system numerator and denominator.

The analogical formulation for continuous-time systems is by no means so simple. The principial polynomial equations remain the same, however, their interpretation is different. For the signal predictions, the Taylor expansion is used. However, the main advantage of MBPC – constraints handling, is very difficult with this approach.

There are some approaches that are more realistic while still allowing constraints. The main principle is to approximate the future control and output signals as a linear combination of selected continuous-time base functions – for example splines. The real optimised variables become parameters of the splines. It can be shown that both input and output signal approximations are affine functions of these parameters and thus constrained continuous-time predictive control can be solved as a Quadratic Programming task.

The advantage of this approach is a truly continuous-control where the choice of the sampling time is not so crucial as in the discrete time case. For the disadvantages, we mention a larger number of user parameters and not very clear stability properties.

9.4 Constrained Control

The GPC algorithm derived in the preceding section did not consider the presence of constraints. This is not very realistic, as in practice some kind of constraints is usually present in the process control. Most often, inputs are constrained to be between some minimal and maximal values (flows cannot be negative, valves can be opened at 100% maximally) or input rate changes are limited. Usually, there also exist some recommended values of process outputs; these are often formulated as *soft* constraints as opposed to *hard* input constraints.

The ability to handle constraints is one of the key properties of MBPC and also caused its spread, use, and popularity in industry. Nowadays, most of the industrial processes run at the constraints, if not, the process is unnecessarily overdesigned.

One might argue that input constraints can be respected if the calculated control by some control method is subsequently clipped to be within limits. There are at least two reasons not to do so:

- There is a loss of anticipating action. As the control is on its limit, it cannot influence the process in a suitable way (one degree of freedom is lost). The process may go totally unstable, out of limits of safety, or to an emergency mode. This usually causes heavy economic losses connected with emergency stop and start-up procedures.
- If multivariable control is considered, certain influence between the input vector elements has to be respected. Clipping one input element may cause entirely different transient responses. This phenomenon is called *directionality* of a multivariable plant.

The cost function used in GPC is quadratic and of the form (9.35). If we assume only constraints that are linear with respect to the optimised vector $\tilde{\boldsymbol{u}}$ then the resulting optimisation problem may be cast as the *Quadratic Programming* problem which is known to be convex and for which efficient programming codes exist. The general constrained GPC formulation is thus given as

$$\min_{\tilde{\boldsymbol{u}}} 2\boldsymbol{g}^T\tilde{\boldsymbol{u}} + \tilde{\boldsymbol{u}}^T\boldsymbol{H}\tilde{\boldsymbol{u}} \quad \text{subject to: } \boldsymbol{A}\tilde{\boldsymbol{u}} \geq \boldsymbol{b} \tag{9.76}$$

Several types of constraints may be written in this general form:

Input rate limits $\Delta u_{min} \leq \Delta u \leq \Delta u_{max}$:

$$\boldsymbol{A} = \begin{pmatrix} \boldsymbol{I} \\ -\boldsymbol{I} \end{pmatrix}, \quad \boldsymbol{b} = \begin{pmatrix} \mathbf{1}\Delta u_{min} \\ -\mathbf{1}\Delta u_{max} \end{pmatrix}$$

where $\mathbf{1}$ is a vector whose entries are ones.

Input amplitude limits $u_{min} \leq u \leq u_{max}$:

$$\boldsymbol{A} = \begin{pmatrix} \boldsymbol{L} \\ -\boldsymbol{L} \end{pmatrix}, \quad \boldsymbol{b} = \begin{pmatrix} \mathbf{1}u_{min} - \mathbf{1}u(t-1) \\ -\mathbf{1}u_{max} + \mathbf{1}u(t-1) \end{pmatrix}$$

where $\boldsymbol{L}$ is a lower triagonal matrix whose entries are ones.

Output constraints $y_{min} \leq y \leq y_{max}$:

$$\boldsymbol{A} = \begin{pmatrix} \boldsymbol{G} \\ -\boldsymbol{G} \end{pmatrix}, \quad \boldsymbol{b} = \begin{pmatrix} \boldsymbol{y}_{min} - \boldsymbol{y}_0 \\ -\boldsymbol{y}_{max} + \boldsymbol{y}_0 \end{pmatrix}$$

Several other types of output constraints can be handled similarly: overshoot, undershoot, monotony, etc.

Although input constraints can always be met, presence of output constraints can cause infeasibility of the Quadratic Programming. Therefore from practical point of view, hard output constraints should be changed to soft constraints where amount ε of constraint violation is penalised. In such a case the output constraints are of the form

$$-\varepsilon + y_{min} \leq y \leq y_{max} - \varepsilon, \quad \varepsilon > 0 \tag{9.77}$$

and the cost function (9.35) is of the form

$$I = 2\boldsymbol{g}^T\tilde{\boldsymbol{u}} + \tilde{\boldsymbol{u}}^T\boldsymbol{H}\tilde{\boldsymbol{u}} + \boldsymbol{\varepsilon}^T\bar{\boldsymbol{H}}\boldsymbol{\varepsilon} \tag{9.78}$$

and the variables $\boldsymbol{\varepsilon}$ are added to optimised variables.

9.5 Stability Results

Any predictive method minimising the finite horizon cost function may become unstable in some cases. This can easily be imagined if the system controlled contains right half plane zeros, the output horizon is equal to one, and control penalisation is equal to zero. Inevitably, the predictive controller that minimises only the output error, is able to set it to zero at each sampling time. The price however is, that the control signals are increasing in magnitude and the system will be unstable.

This is one of the main issues against MBPC. Although the methods may work well in practice, some systems exists in theory for which the methods are highly sensitive. Even more significant is, that there is no clear theory which predicts the closed-loop behaviour for arbitrary horizons and control penalisations.

Therefore, two main streams toward stability have been developed. In the first case, some combinations of GPC parameters have been proven to be stabilising. The second line of research has been devoted to methods that overcome the basic GPC drawbacks.

9.5.1 Stability Results in GPC

Theorem 9.3. *The closed-loop system is stable if the system is stabilisable and detectable and if:*

- $N_2 \to \infty$, $N_u = N_2$, *and* $\lambda > 0$ *or*
- $N_2 \to \infty$, $N_u \to \infty$, $N_u \leq N_2 - n + 1$, *and* $\lambda = 0$ *where* n *is the system state dimension.*

Theorem 9.4. *For open-loop stable processes the closed-loop system is stable and the control tends to a mean level law for* $N_u = 1$ *and* $\lambda = 0$ *as* $N_2 \to \infty$.

Theorem 9.5. *The closed-loop system is equivalent to a stable state dead-beat controller if*

1. *the system is observable and controllable and*
2. $N_1 = n$, $N_2 \geq 2n - 1$, $N_u = n$, *and* $\lambda = 0$ *where* n *is the system state dimension.*

9.5.2 Terminal Constraints

The first approach that forces MBPC methods to be stable is based on the state *terminal constraints.* Roughly speaking, the system is stable if it is subject to the moving-terminal constraint on final states

$$\boldsymbol{x}(k + N_2) = \boldsymbol{0} \tag{9.79}$$

Several different algorithm have emerged that are based on this result.

CRHPC

CRHPC (Constrained Receding Horizon Predictive Control), and **SIORHC** (Stable Input Output Receding Horizon Control) were developed independently, but are in fact equivalent. The idea behind these techniques is an equivalent of the state terminal constraint within input/output system description. Hence, these methods optimise the usual quadratic function over finite horizons subject to condition that the output exactly matches a reference value over a future constraint range (after $k + N_2$). Some degrees of freedom force the output to stay at setpoints while the remaining degrees of freedom are available to minimise the cost function. The output constraint description is

$$y(k + N_2 + i) = w(k + N_2), \quad i = 1, \ldots, n \tag{9.80}$$

and n is the dimension of the system state vector.

Although the output constraints are added to the original formulation, the solution in the unconstrained case can still be found analytically.

Theorem 9.6 (SIORHC). *Let the system polynomials ΔA and B be coprime. If $\lambda > 0$ and $\deg(B) \leq \deg(A) + 1$ then provided that*

$$N_2 \geq n = \max(\deg(A) + 1, \deg(B))$$

then

- *the SIORHC control law is unique;*
- *SIORHC stabilises the plant, and, irrespective of $\deg(A), \deg(B), \lambda$, for*

  $$N_2 = n$$

  yields a state dead-beat closed-loop system;
- *whenever stabilising, SIORHC yields asymptotic rejection of constant disturbances and offset free closed-loop system.*

Derivation

As usual in GPC, consider the predicted output to be of the form

$$\hat{\boldsymbol{y}} = \boldsymbol{G}\tilde{\boldsymbol{u}} + \boldsymbol{y}_0 \tag{9.81}$$

where all vectors are stacked from $k + 1$ to $k + N_2$. After this time, up to $k + N_2 + n$, output predictions are constrained to be equal to the setpoint and future control increments to zero

$$\bar{\boldsymbol{w}} = \bar{\boldsymbol{G}}\tilde{\boldsymbol{u}} + \bar{\boldsymbol{y}}_0 \tag{9.82}$$

where the vector $\bar{\boldsymbol{y}}_0$ denotes the free response between $k + N_2 + 1$ and $k + N_2 + n$ and $\bar{\boldsymbol{w}}$ is vector of $w(k + N_2)$ of the corresponding dimension.

The cost function is as usual of the form

$$I = (\hat{\boldsymbol{y}} - \boldsymbol{w})^T(\hat{\boldsymbol{y}} - \boldsymbol{w}) + \lambda\tilde{\boldsymbol{u}}^T\tilde{\boldsymbol{u}} \tag{9.83}$$

subject to the constraint (9.82). This can be solved analytically by the Lagrange multipliers.

Let us denote $\bar{\boldsymbol{h}} = \bar{\boldsymbol{w}} - \bar{\boldsymbol{y}}_0$ and $\boldsymbol{h} = \boldsymbol{w} - \boldsymbol{y}_0$. Hence

$$I = (\boldsymbol{G}\tilde{\boldsymbol{u}} - \boldsymbol{h})^T(\boldsymbol{G}\tilde{\boldsymbol{u}} - \boldsymbol{h}) + \lambda\tilde{\boldsymbol{u}}^T\tilde{\boldsymbol{u}} + \boldsymbol{x}^T(\bar{\boldsymbol{G}}\tilde{\boldsymbol{u}} - \bar{\boldsymbol{h}}) \tag{9.84}$$

Partial derivatives of I with respect to $\tilde{\boldsymbol{u}}, \boldsymbol{x}$ ($\boldsymbol{x}$ are the Lagrange multipliers) are zero. We obtain a system linear equations as follows

$$\begin{pmatrix} 2(\boldsymbol{G}^T\boldsymbol{G}+\lambda\boldsymbol{I}) & \bar{\boldsymbol{G}}^T \\ \bar{\boldsymbol{G}} & \boldsymbol{0} \end{pmatrix} \begin{pmatrix} \tilde{\boldsymbol{u}} \\ \boldsymbol{x} \end{pmatrix} = \begin{pmatrix} 2\boldsymbol{G}^T\boldsymbol{h} \\ \bar{\boldsymbol{h}} \end{pmatrix} \tag{9.85}$$

The block matrix inversion formula states

$$\begin{pmatrix} A^{-1} & D \\ C & B \end{pmatrix}^{-1} = \begin{pmatrix} A + AD\Delta CA & -AD\Delta \\ -\Delta CA & -\Delta \end{pmatrix}, \quad \Delta^{-1} = B - CAD \tag{9.86}$$

Therefore, the future control increment vector is given as

$$\tilde{\boldsymbol{u}} = \tilde{\boldsymbol{G}}(\boldsymbol{I} - \bar{\boldsymbol{G}}^T\boldsymbol{Q}\bar{\boldsymbol{G}}\tilde{\boldsymbol{G}})\boldsymbol{G}^T\boldsymbol{h} + \tilde{\boldsymbol{G}}\bar{\boldsymbol{G}}^T\boldsymbol{Q}\bar{\boldsymbol{h}} \tag{9.87}$$

where

$$\tilde{\boldsymbol{G}} = \left(\boldsymbol{G}^T\boldsymbol{G} + \lambda\boldsymbol{I}\right)^{-1} \tag{9.88}$$

$$\boldsymbol{Q} = \left(\bar{\boldsymbol{G}}\tilde{\boldsymbol{G}}\bar{\boldsymbol{G}}^T\right)^{-1} \tag{9.89}$$

SGPC

Another method that can be shown to be equivalent to the preceding methods is **SGPC**. Its advantages include more efficient computational implementation and better numerical robustness.

In this approach GPC is invoked after application of the stabilising feedback control law

$$Y(q^{-1})\Delta u(t) = c(t) - X(q^{-1})y(t) \tag{9.90}$$

where the polynomials X, Y are calculated as the dead-beat controller from the Diophantine equation

$$\Delta AY + BX = 1 \tag{9.91}$$

and where $c(t)$ denoted the reference signal for the closed-loop system that forms the vector of optimised variables. The dead-beat controller results in the control and output predictions of the form

$$y(t) = B(q^{-1})c(t) \tag{9.92}$$

$$u(t) = A(q^{-1})c(t) \tag{9.93}$$

These can be simulated forward in time to give the vectors of future output and control predictions and treated in the same way as in GPC.

Hence, this methods optimises future reference trajectory subject to (terminal) constraint that this trajectory should be equal to the desired setpoint after some horizon.

It can be shown that SGPC is equivalent to the pole-placement method with the controller polynomials Y_r, X_r given from the Diophantine equation

$$\Delta AY_r + BX_r = P_r \tag{9.94}$$

and its stability depends on the roots of the P_r polynomial.

Theorem 9.7 (SGPC). *For $N_2 \geq \deg(A) + 1 + N_c$, where N_c is the number of reference points optimised, is SGPC stable for any N_c.*

YKPC

The last approach within finite horizon formulation is the predictive control algorithm based on the Youla Kučera parametrisation of all stabilising controllers YKPC.

As in the previous approach, in the first step a nominal controller with pole-placement technique is calculated and a nominal controller is given as a solution of 2 Diophantine equations

$$P_0 \Delta A + Q_0 B = M \tag{9.95}$$

$$\Delta S_0 + B R_0 = M \tag{9.96}$$

where M is the desired closed-loop polynomial and the two degree of freedom controller with integral action is defined as $Q/\Delta P$ (feedback) and $R/\Delta P$ (feedforward).

The minimum degree controller P_0, Q_0, R_0 only serves as a basis to find an expression for the set of all stabilizing controllers. Among these controllers the one is chosen that minimises the GPC cost function.

The expression of such controllers (Youla-Kučera parametrisation) is as follows:

Theorem 9.8 (YK Controllers). *A controller $(P(z), Q(z), R(z))$ gives rise to the closed-loop denominator matrix $M(z)$ if and only if it can be expressed as*

$$P = P_0 + ZB \tag{9.97}$$

$$Q = Q_0 - ZA\Delta \tag{9.98}$$

$$R = R_0 + \Delta X \tag{9.99}$$

X, Z are assumed to be polynomials for simplicity. Their coefficients form the vector of the optimised parameters.

Stability is proved as in the previous approaches via terminal constraint. It is interesting to note, that in this approach the state terminal constraint does not have to be specified and is implicitly assured.

Theorem 9.9 (Choice of horizons). *Let $n = \max(\deg(X), \deg(Z))$, $N_1 = 1$ and let the horizons be equal or greater than*

$$N_2 = \deg(B) + n \tag{9.100}$$

$$N_u = \deg(A\Delta) + n \tag{9.101}$$

Further assume that the sequences w, d (reference, disturbance) are bounded. Then the unconstrained YKPC is uniformly asymptotically stable.

This controller is time-varying in spite of the fact that the system is assumed to be time-invariant.

9.5.3 Infinite Horizons

Another line of research has been focused on reformulation of the basic GPC method when N_2, N_u are infinity. Of course, if such a method can be implemented, stability problems disappear. However, a number of the optimised parameters (future control moves) is also infinity and the original problem is untractable. Therefore, several suboptimal algorithms have emerged. The basic principle of all of them is to leave $N_2 = \infty$ but to play with N_u or with its equivalents.

Rawlings a Muske developed a method in state-space formulation where the number of control moves N_u is finite. The feedback gain is calculated via the recursive ARE.

Theorem 9.10 (Stable plants). *For stable system matrix* $\boldsymbol{A}$ *and* $N_u \geq 1$ *is the receding horizon controller stabilising.*

Theorem 9.11 (Unstable plants). *For stabilisable plant* $(\boldsymbol{A}, \boldsymbol{B})$ *with* r *unstable modes and* $N_u \geq r$ *is the receding horizon controller stabilising.*

Constrained control can also be dealt with in this approach. The requirement added to the previous theorems is that the initial state at time k is feasible (within constraints).

The SGPC and YKPC methods can be modified to use both input and output horizons infinite. The SGPC approach utilises the finite reference sequence as the vector of optimised variables. The solution is found via Lyapunov equation.

The YKPC method utilises coefficients of the Youla-Kučera polynomials as the optimised variables. It is shown that in the unconstrained case the optimal predictive controller coincides with the nominal pole-placement controller whose poles are calculated via spectral factorisation equation – hence it is the standard LQ controller. If the constraints are active, piece-wise linear controller results.

9.5.4 Finite Terminal Penalty

The third approach to MBPC stability is to adopt a finite input and state horizon with a finite terminal weighting matrix. This is equivalent to the condition that the terminal state has to be within some neighbourhood of the origin. Compared to the previous approaches when it had to be exactly zero, here the state can be in such a neighbourhood of the origin that is asymptotically stable.

With a state-space formulation

$$\boldsymbol{x}(k+1) = \boldsymbol{A}(k)\boldsymbol{x}(k) + \boldsymbol{B}(k)\boldsymbol{u}(k) \tag{9.102}$$

is the cost given as

$$I = \boldsymbol{x}^T(k+N)\boldsymbol{P}(k+N)\boldsymbol{x}(k+N) + \sum_{i=0}^{N}\Big(\boldsymbol{x}^T(k+i)\boldsymbol{Q}(k+i)\boldsymbol{x}(k+i) + \boldsymbol{u}^T(k+i)\boldsymbol{R}(k+i)\boldsymbol{u}(k+i)\Big) \quad (9.103)$$

If $\boldsymbol{P}(k+N) \to \infty$ then the terminal state constraint approach results. However, also some "smaller" terminal penalty matrix $\boldsymbol{P}$ can give a stabilising receding horizon control. Equivalently, in terms of state constraints, we will speak about stabilising state terminal sets.

Theorem 9.12. *Assume that the terminal weighting matrix $\boldsymbol{P}(k+1)$ satisfies the following matrix difference inequality for some matrix $\boldsymbol{H}(k)$*

$$\boldsymbol{P}(k) \geq \boldsymbol{F}^T(k)\boldsymbol{P}(k+1)\boldsymbol{F}(k) + \boldsymbol{Q}(k) + \boldsymbol{H}^T(k)\boldsymbol{R}(k+1)\boldsymbol{H}(k), \ \forall k \in [N, \infty) \quad (9.104)$$

where

$$\boldsymbol{F}(k) = \boldsymbol{A}(k) + \boldsymbol{B}(k)\boldsymbol{H}(k) \quad (9.105)$$

Further suppose that $\boldsymbol{Q}(k)$ is positive definite. Then the receding horizon control which stems from the optimisation problem minimising the performance index I asymptotically stabilises the system. In addition, if $\boldsymbol{A}(k), \boldsymbol{Q}(k)$, and $\boldsymbol{P}(k)$ are bounded above $\forall k \geq 0$, then the receding horizon control exponentially stabilises the system.

9.6 Explicit Predictive Control

In receding horizon control, a finite-time optimal control problem is solved at each time step to obtain the optimal input sequence. Subsequently, only the first element of that sequence is applied to the system. At the next time step, the state is measured and the procedure is repeated from the beginning. The input sequence can be computed by solving an optimisation problem on-line at each time step. Hence the complexity of the underlying optimisation problem limits the minimal admissible sampling time of the plant.

Alternatively, it is possible to solve the optimisation problem off-line as a *multi-parametric* program. Then, the on-line effort reduces to finding the correct feedback law entry in a lookup table, allowing MPC to be applied to systems with very low sampling times.

9.6.1 Quadratic Programming Definition

Consider optimal control problems for a discrete-time linear, time-invariant (LTI) system

$$\boldsymbol{x}(k+1) = \boldsymbol{A}\boldsymbol{x}(k) + \boldsymbol{B}\boldsymbol{u}(k) \tag{9.106}$$

with $\boldsymbol{A}[n \times n]$ and $\boldsymbol{B}[m \times n]$.

Now consider the constrained finite-time optimal control problem

$$\begin{aligned}\min I(\boldsymbol{x}(k)) = \boldsymbol{x}^T(k+N)\boldsymbol{P}(k+N)\boldsymbol{x}(k+N)\\ + \sum_{i=0}^{N}\Big(\boldsymbol{x}^T(k+i)\boldsymbol{Q}(k+i)\boldsymbol{x}(k+i)\\ + \boldsymbol{u}^T(k+i)\boldsymbol{R}(k+i)\boldsymbol{u}(k+i)\Big)\end{aligned} \tag{9.107a}$$

$$\text{subj. to } \boldsymbol{x}(k+i) \in \mathbb{X},\ \boldsymbol{u}(k+i) \in \mathbb{U}, \qquad i \in \{0, \ldots, N-1\}, \tag{9.107b}$$

$$\boldsymbol{x}(k+N) \in \mathcal{T}_{\text{set}}, \tag{9.107c}$$

$$\boldsymbol{x}(k+1) = \boldsymbol{A}\boldsymbol{x}(k) + \boldsymbol{B}\boldsymbol{u}(k), \quad \boldsymbol{x}(k) = \boldsymbol{x}_0 \tag{9.107d}$$

where we assume that states and control are subject to constraints (9.107b). The terminal set constraint (9.107c) is an additional constraint which is often added to obtain certain properties (i.e. stability and constraint satisfaction).

As future state predictions are constrained by (9.106), we can recursively substitute for them yielding

$$\boldsymbol{x}(k+i) = \boldsymbol{A}^i\boldsymbol{x}(k) + \sum_{j=0}^{i-1}\boldsymbol{A}^j\boldsymbol{B}\boldsymbol{u}(k+i-1-j) \tag{9.108}$$

Thus, optimal solution of problem (9.107) can be reformulated as

$$I^*(\boldsymbol{x}_0) = \boldsymbol{x}_0^T\boldsymbol{Y}\boldsymbol{x}_0 + \min_{\boldsymbol{U}_N}\frac{1}{2}\left\{\boldsymbol{U}_N^T\boldsymbol{H}\boldsymbol{U}_N + \boldsymbol{x}_0^T\boldsymbol{F}\boldsymbol{U}_N\right\} \tag{9.109a}$$

$$\text{subj. to } \boldsymbol{G}\boldsymbol{U}_N \le \boldsymbol{W} + \boldsymbol{E}\boldsymbol{x}_0 \tag{9.109b}$$

where the column vector $\boldsymbol{U}_N = [\boldsymbol{u}_0^T, \ldots, \boldsymbol{u}_{N-1}^T]^T$ is the optimisation vector and $\boldsymbol{H}$, $\boldsymbol{F}$, $\boldsymbol{Y}$, $\boldsymbol{G}$, $\boldsymbol{W}$, $\boldsymbol{E}$ can easily be obtained from the original formulation.

The reformulated problem (9.109) is a standard quadratic programming formulation that was also obtained in previous sections. Taking any initial value $\boldsymbol{x}_0$, optimal future control trajectory $\boldsymbol{U}_N(x)$ can be found from which only the first element is used to close the loop.

9.6.2 Explicit Solution

In explicit solution of (9.109) we use the so-called multi-parametric programming approach to optimisation. In multi-parametric programming, the objective is to obtain the optimiser $\boldsymbol{U}_N$ for a whole range of parameters $\boldsymbol{x}_0$, i.e. to obtain $\boldsymbol{U}_N(\boldsymbol{x})$ as an explicit function of the parameter $\boldsymbol{x}$. The term *multi* is used to emphasise that the parameter $\boldsymbol{x}$ (in our case the actual state vector $\boldsymbol{x}_0$) is a vector and not a scalar. If the objective function is quadratic in the optimisation variable $\boldsymbol{U}_N$, the terminology *multi-parametric Quadratic Program* (mp-QP) is used.

In this formulation, it is useful to define

$$\boldsymbol{z} = \boldsymbol{U}_N + \boldsymbol{H}^{-1}\boldsymbol{F}^T\boldsymbol{x}_0 \tag{9.110}$$

and to transform the formulation to problem, where the state vector $\boldsymbol{x}_0$ apears only in constraints

$$I^*(\boldsymbol{x}_0) = \min_{\boldsymbol{z}} \frac{1}{2}\left\{\boldsymbol{z}^T\boldsymbol{H}\boldsymbol{z}\right\} \tag{9.111a}$$

$$\text{subj. to } \boldsymbol{G}\boldsymbol{z} \leq \boldsymbol{W} + \boldsymbol{S}\boldsymbol{x}_0 \tag{9.111b}$$

where $\boldsymbol{S} = \boldsymbol{E} + \boldsymbol{G}\boldsymbol{H}^{-1}\boldsymbol{F}^T$.

An mp-QP computation scheme consists of the following three steps:

1. Active Constraint Identification: A feasible parameter $\hat{\boldsymbol{x}}$ is determined and the associated QP (9.111) is solved. This will yield the optimiser $\boldsymbol{z}$ and active constraints $\mathcal{A}(\hat{\boldsymbol{x}})$ defined as inequalities that are active at solution, i.e.

$$\mathcal{A}(\hat{\boldsymbol{x}}) = \{i \in \mathcal{J} \mid \boldsymbol{G}_{(i)}\boldsymbol{z} = \boldsymbol{W}_{(i)} + \boldsymbol{S}_{(i)}\hat{\boldsymbol{x}}\},\ \mathcal{J} = \{1, 2, \ldots, q\}, \tag{9.112}$$

   where $\boldsymbol{G}_{(i)}$, $\boldsymbol{W}_{(i)}$, and $\boldsymbol{S}_{(i)}$ denote the i-th row of the matrices $\boldsymbol{G}$, $\boldsymbol{W}$, and $\boldsymbol{S}$, respectively, and q denotes the number of constraints. The rows indexed by the active constraints $\mathcal{A}(\hat{\boldsymbol{x}})$ are extracted from the constraint matrices $\boldsymbol{G}, \boldsymbol{W}$ and $\boldsymbol{S}$ in (9.111) to form the matrices $\boldsymbol{G}_{\mathcal{A}}, \boldsymbol{W}_{\mathcal{A}}$ and $\boldsymbol{S}_{\mathcal{A}}$.
2. Region Computation: Next, it is possible to use the Karush-Kuhn-Tucker (KKT) conditions to obtain an explicit representation of the optimiser $\boldsymbol{U}_N(\boldsymbol{x})$ which is valid in some neighborhood of $\hat{\boldsymbol{x}}$. These are for our problem defined as

$$\boldsymbol{H}\boldsymbol{z} + \boldsymbol{G}^T\boldsymbol{\lambda} = \boldsymbol{0} \tag{9.113a}$$

$$\boldsymbol{\lambda}^T(\boldsymbol{G}\boldsymbol{z} - \boldsymbol{W} - \boldsymbol{S}\hat{\boldsymbol{x}}) = \boldsymbol{0} \tag{9.113b}$$

$$\boldsymbol{\lambda} \geq \boldsymbol{0} \tag{9.113c}$$

$$\boldsymbol{G}\boldsymbol{z} \leq \boldsymbol{W} + \boldsymbol{S}\hat{\boldsymbol{x}} \tag{9.113d}$$

   Optimised variable $\boldsymbol{z}$ can be solved from (9.113a)

$$\boldsymbol{z} = -\boldsymbol{H}^{-1}\boldsymbol{G}^T\boldsymbol{\lambda} \tag{9.114}$$

   Condition (9.113b) can be separated into active and inactive constraints. For inactive constraints holds $\boldsymbol{\lambda}_{\mathcal{I}} = 0$. For active constraints are the corresponding Lagrange multipliers $\boldsymbol{\lambda}_{\mathcal{A}}$ positive and inequality constraints are changed to equalities. Substituting for $\boldsymbol{z}$ from (9.114) into equality constraints gives

$$-\boldsymbol{G}_{\mathcal{A}}\boldsymbol{H}^{-1}\boldsymbol{G}_{\mathcal{A}}^T\boldsymbol{\lambda}_{\mathcal{A}} + \boldsymbol{W}_{\mathcal{A}} + \boldsymbol{S}_{\mathcal{A}}\hat{\boldsymbol{x}} = \boldsymbol{0} \tag{9.115}$$

and yields expressions for active Lagrange multipliers

$$\boldsymbol{\lambda}_{\mathcal{A}} = -(\boldsymbol{G}_{\mathcal{A}}\boldsymbol{H}^{-1}\boldsymbol{G}_{\mathcal{A}}^T)^{-1}(\boldsymbol{W}_{\mathcal{A}} + \boldsymbol{S}_{\mathcal{A}}\hat{\boldsymbol{x}}) \tag{9.116}$$

The optimal value of optimiser $\boldsymbol{z}$ and optimal control trajectory $\boldsymbol{U}_N$ are thus given as affine functions of $\hat{\boldsymbol{x}}$

$$\boldsymbol{z} = -\boldsymbol{H}^{-1}\boldsymbol{G}_{\mathcal{A}}^T(\boldsymbol{G}_{\mathcal{A}}\boldsymbol{H}^{-1}\boldsymbol{G}_{\mathcal{A}}^T)^{-1}(\boldsymbol{W}_{\mathcal{A}} + \boldsymbol{S}_{\mathcal{A}}\hat{\boldsymbol{x}}) \tag{9.117}$$

$$\begin{aligned}\boldsymbol{U}_N &= \boldsymbol{z} - \boldsymbol{H}^{-1}\boldsymbol{F}^T\hat{\boldsymbol{x}} \\ &= -\boldsymbol{H}^{-1}\boldsymbol{G}_{\mathcal{A}}^T(\boldsymbol{G}_{\mathcal{A}}\boldsymbol{H}^{-1}\boldsymbol{G}_{\mathcal{A}}^T)^{-1}(\boldsymbol{W}_{\mathcal{A}} + \boldsymbol{S}_{\mathcal{A}}\hat{\boldsymbol{x}}) - \boldsymbol{H}^{-1}\boldsymbol{F}^T\hat{\boldsymbol{x}} \\ &= \boldsymbol{F}_r\hat{\boldsymbol{x}} + \boldsymbol{G}_r\end{aligned} \tag{9.118}$$

where

$$\boldsymbol{F}_r = \boldsymbol{H}^{-1}\boldsymbol{G}_{\mathcal{A}}^T(\boldsymbol{G}_{\mathcal{A}}\boldsymbol{H}^{-1}\boldsymbol{G}_{\mathcal{A}}^T)^{-1}\boldsymbol{S}_{\mathcal{A}} - \boldsymbol{H}^{-1}\boldsymbol{F}^T \tag{9.119}$$

$$\boldsymbol{G}_r = \boldsymbol{H}^{-1}\boldsymbol{G}_{\mathcal{A}}^T(\boldsymbol{G}_{\mathcal{A}}\boldsymbol{H}^{-1}\boldsymbol{G}_{\mathcal{A}}^T)^{-1}\boldsymbol{W}_{\mathcal{A}} \tag{9.120}$$

In a next step, the set of states is determined where the optimiser $\boldsymbol{U}_N(\boldsymbol{x})$ satisfies the the same active constraints and is optimal. Such a region is characterised by two inequalities (9.113c), (9.113d) and is written compactly as $\boldsymbol{H}_r\boldsymbol{x} \leq \boldsymbol{K}_r$ where

$$\boldsymbol{H}_r = \begin{bmatrix}\boldsymbol{G}(\boldsymbol{F}_r + \boldsymbol{H}^{-1}\boldsymbol{F}^T) - \boldsymbol{S} \\ (\boldsymbol{G}_{\mathcal{A}}\boldsymbol{H}^{-1}\boldsymbol{G}_{\mathcal{A}}^T)^{-1}\boldsymbol{S}_{\mathcal{A}}\end{bmatrix} \tag{9.121}$$

$$\boldsymbol{K}_r = \begin{bmatrix}\boldsymbol{W} - \boldsymbol{G}\boldsymbol{G}_r \\ -(\boldsymbol{G}_{\mathcal{A}}\boldsymbol{H}^{-1}\boldsymbol{G}_{\mathcal{A}}^T)^{-1}\boldsymbol{W}_{\mathcal{A}}\end{bmatrix} \tag{9.122}$$

3. State Space Exploration: Once the controller region is computed, the algorithm proceeds iteratively in neighbouring regions until the entire feasible state space is covered with controller regions.

After completing the algorithm, the explicit model predictive controller consists of definitions of multiple state regions with different affine control laws. Its actual implementation reduces to search for an active region of states and calculation of the corresponding control.

9.6.3 Multi-Parametric Toolbox

The Multi-Parametric Toolbox (MPT) is a free, open-source, MATLAB-based toolbox for design, analysis and deployment of optimal controllers for constrained linear, nonlinear and hybrid systems. The toolbox offers a broad spectrum of algorithms developed by the international community and compiled in a user friendly and accessible format: starting from different performance objectives (linear, quadratic, minimum time) to the handling of systems with persistent additive and polytopic uncertainties. Users can add custom constraints, such as polytopic, contraction, or collision avoidance constraints, or create custom objective functions. Resulting optimal control laws can either be embedded into applications in the form of C code, or deployed to target platforms using the Real Time Workshop.

Although the primal focus of MPT is on design of optimal controllers based on multi-parametric programming, it can be used to formulate and solve online MPC problems as well. The toolbox contains efficient algorithms to solve linear and quadratic problems parametrically and features a rich library for performing operations on convex polytopes and non-convex unions thereof.

WWW **Example 9.2:**

To illustrate the capabilities of the Multi-Parametric Toolbox, we consider a double integrator whose transfer function representation is given by

$$P(s) = \frac{1}{s^2}$$

With a sampling time $T_s = 1$ second, the corresponding state-space form can be written as

$$\begin{aligned} \boldsymbol{x}(k+1) &= \begin{pmatrix} 1 & 1 \\ 0 & 1 \end{pmatrix} \boldsymbol{x}(k) + \begin{pmatrix} 1 \\ 0.5 \end{pmatrix} u(k) \\ y(k) &= x_2(k) \end{aligned}$$

We want to design an explicit optimal state-feedback law which minimizes the performance index (9.107a) with $N = 5$, $\boldsymbol{P} = \boldsymbol{0}$, $\boldsymbol{Q} = \boldsymbol{I}$, and $R = 1$. System states and the control signals are subject to constraints $\boldsymbol{x}(k) \in [-1, 1] \times [-1, 1]$ and $u(k) \in [-1, 1]$, respectively. In order to solve the problem (9.107) explicitly, one has first to describe the dynamical model of the plant:

```
model.A = [1 1; 0 1];
model.B = [1; 0.5];
model.C = [0 1];
model.D = 0;
```

along with the system constraints:

```
model.umin = -1;
model.umax = 1;
model.xmin = [-1; -1];
model.xmax = [1; 1];
```

Next, parameters of the performance index have to be specified:

```
cost.N = 5;
cost.Q = [1 0; 0 1];
cost.R = 1;
cost.P_N = 0;
cost.norm = 2;
```

Finally, the explicit optimal state-feedback control law can be calculated by executing

```
controller = mpt_control(model, cost)
```

The obtained explicit control law is defined by

$$u = \begin{cases} \begin{pmatrix}-0.52 & -0.94\end{pmatrix} \boldsymbol{x}, & \text{if } \begin{pmatrix} 0.48 & 0.88 \\ -0.48 & -0.88 \\ -1 & 0 \\ 0 & -1 \\ 1 & 0 \\ 0 & 1 \end{pmatrix} \boldsymbol{x} \le \begin{pmatrix} 0.93 \\ 0.93 \\ 1 \\ 1 \\ 1 \\ 1 \end{pmatrix} \\ & \text{(Region \#1)} \\ -1, & \text{if } \begin{pmatrix} 1 & 0 \\ 0 & 1 \\ -0.48 & -0.88 \end{pmatrix} \boldsymbol{x} \le \begin{pmatrix} 1 \\ 1 \\ -0.93 \end{pmatrix} \\ & \text{(Region \#2)} \\ 1, & \text{if } \begin{pmatrix} -1 & 0 \\ 0 & -1 \\ 0.48 & 0.88 \end{pmatrix} \boldsymbol{x} \le \begin{pmatrix} 1 \\ 1 \\ -0.93 \end{pmatrix} \\ & \text{(Region \#3)} \end{cases}$$

and it can be plotted using the command

```
plot(controller)
```

which will show the regions of the state-space over which the optimal control law is defined, as illustrated in Figure 9.3.

9.7 Tuning

Let us recall the GPC parameters: horizons N_1, N_2, N_u, control weighting λ, and output weighting polynomial P. The effects of a change of the parameters are strongly coupled and the strategies dealing with adjustment of GPC parameters usually adjust only one parameter while all others are at some default values.

9.7.1 Tuning based on the First Order Model

The tuning strategy based on the analysis of the first order system with time delay is given as follows: The strategy is given as follows:

1. Approximate the process dynamics with the first order plus dead time model:

$$G(s) = \frac{Z}{Ts+1} \mathrm{e}^{-T_d s} \tag{9.123}$$

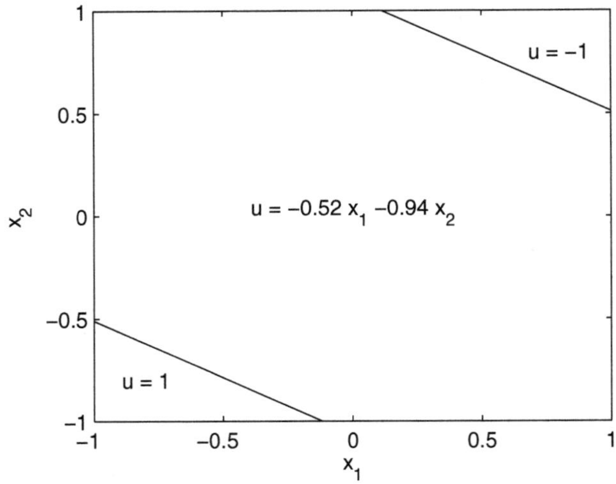

Fig. 9.3. Controller regions and the optimal control for the double integrator example

2. If the sampling time has not yet been specified, select it as the larger value that satisfies

$$T_s \le 0.1T, \qquad T_s \le 0.5T_d \tag{9.124}$$

3. Calculate the discrete dead time $\bar{T}_d$ (rounded to the next integer)

$$\bar{T}_d = T_d/T_s + 1 \tag{9.125}$$

4. Set $N_1 = 1$, and

$$N_2 = 5T/T_s + \bar{T}_d \tag{9.126}$$

5. Select the control horizon N_u (usually between 1–6) and calculate the control weighting λ as

$$f = \begin{cases} 0 & N_u = 1 \\ \frac{N_u}{500}\left(\frac{3.5T}{T_s} + 2 - \frac{N_u-1}{2}\right) & N_u > 1 \end{cases} \tag{9.127}$$

$$\lambda = fZ^2 \tag{9.128}$$

9.7.2 Multivariable Tuning based on the First Order Model

This tuning strategy is a generalisation of the previous approach to multivariable systems with R outputs and S inputs. Again, it is based on the analysis of the first order system with time delay.

1. Approximate the process dynamics of all controller output-process variable pairs with first order plus dead time models:

$$\frac{y_r(s)}{u_s(s)} = \frac{Z_{rs}}{T_{rs}s+1}\mathrm{e}^{-T_{d,rs}s} \tag{9.129}$$

2. Select the sampling time as close as possible to:

$$T^s_{rs} = \max(0.1T_{rs}, 0.5T_{d,rs}), \quad T_s = \min(T^s_{rs}) \tag{9.130}$$

3. Set $N_1 = 1$, compute the prediction horizon N_2:

$$N_2 = \max\left(\frac{5T_{rs}}{T_s} + k_{rs}\right), \quad k_{rs} = \frac{T_{d,rs}}{T_s} + 1 \tag{9.131}$$

4. Select a control horizon N_u, equal to 63.2% of the settling time of the slowest sub-process in the multivariable system:

$$N_u = \max\left(\frac{T_{rs}}{T_s} + k_{rs}\right) \tag{9.132}$$

5. Select the controlled variable weights γ_r, to scale process variable measurements to similar magnitudes
6. Compute the control weightings λ_s as

$$\lambda_s = \frac{M}{500}\sum_{r=1}^{R}\left[\gamma_r Z^2_{rs}\left\{N_2 - k_{rs} - \frac{3}{2}\frac{T_{rs}}{T_s} + 2 - \frac{M-1}{2}\right\}\right] \tag{9.133}$$

Fine tuning of the method is performed by increasing the corresponding γ_r of the process variable for which tighter control is desired and increasing the corresponding λ_s of the manipulated variable for which less aggressive moves are desired.

9.7.3 Output Horizon Tuning

This tuning strategy assumes the active tuning parameter to be the output horizon N_2 with all other fixed at the values

$$N_1 = 1, \; N_u = 1, \; P = 1, \; \lambda = 0 \tag{9.134}$$

It is well known that if $N_2 \to \infty$, mean-level controller results. This controller is rather conservative as its speed is the same as step responses.

The other limit for N_2 is the value of the process dead time. If $N_2 = T_d + 1$, where T_d represents process dead time, then we have the Minimum Variance (MV) controller known to be unstable for non-minimum phase plants.

The practical range for N_2 can be specified as

$$T_d + 1 < N_2 \leq T_{90}/T_s \tag{9.135}$$

where T_{90} is the time when process reaches about 90% of its final value after input step change and T_s is the sampling time.

If the process is uncertain, it is better to start with a larger value of N_2. The minimum value of N_2 for non-minimum phase plants should be such that $\sum_i g_i$ has the same sign as the process gain.

9.7.4 λ Tuning

In this case the active tuning parameter is the penalisation of control moves λ. All other parameters are fixed as

$$N_1 = \deg(B) + 1, \; N_u = \deg(A) + 1, \; N_2 \geq N_u + N_1 - 1 \approx t_r/T_s, \; P = 1 \tag{9.136}$$

With λ equal to zero, dead-beat controller results. This is too rapid in the majority of cases. Hence, with increasing value of λ the controller is made more conservative. It might be shown that the closed-loop poles converge to the open loop poles if $\lambda \to \infty$. Thus λ tuning is not recommended for unstable plants.

It has been found that to desensitise the closed-loop system to changes in process dynamics, the actual λ should be proportional to $B(1)^2$:

$$\lambda = \lambda_0 B(1)^2 \tag{9.137}$$

with λ_0 being a constant.

To determine a starting value of λ, the following relation can be used:

$$\lambda = \frac{m \; \mathrm{tr}(G^T G)}{N_u} \tag{9.138}$$

and m is a factor of detuning the controller relative to the dead-beat control. The control increments are approximately reduced by a factor $m+1$ compared to that of the dead-beat strategy.

From this starting value of λ, an initial guess for λ_0 can be determined from (9.137).

9.7.5 Tuning based on Model Following

As it has been shown before, the P polynomial can be used to generate reference trajectory w/P. GPC can be set up to follow this trajectory exactly and so to place the closed-loop poles at the process zeros. In order to have a more practical controller, the model following can be detuned. This may be accomplished by either increasing N_2 or λ.

The fixed parameters are as follows:

$$N_1 = 1, \; N_u = \deg(A) + 1, \; N_2 \geq N_u + T_d \approx t_r/T_s, \; \lambda = 0 \tag{9.139}$$

Most often, the models $M = 1/P$ are of the first and the second order. If the first order closed-loop model is assumed to be of the form

$$M(s) = \frac{1}{Ts+1} \tag{9.140}$$

then its discrete equivalent is

$$M(z^{-1}) = \frac{(1-p_1)z^{-1}}{1-p_1 z^{-1}} \tag{9.141}$$

where $p_1 = \exp(-T_s/T)$. The polynomial P can thus be chosen as (cf. equation (9.12)) and $P(1)$ is equal to 1 to ensure the offset-free behaviour. This model is applicable mainly for simpler plants as the first order trajectory may sometimes generate excessive control actions.

The second order model can be of the form

$$M(s) = \frac{1}{T^2 s^2 + 2T\zeta s + 1} \tag{9.142}$$

Its discrete time equivalent is

$$M(z^{-1}) = \frac{n_1 z^{-1} + n_2 z^{-2}}{1 + p_1 z^{-1} + p_2 z^{-2}} \tag{9.143}$$

where

$$p_1 = -2\exp\left(\frac{-\zeta T_s}{T}\right)\cos\left[\left(\frac{T_s}{T}\right)\sqrt{1-\zeta^2}\right] \tag{9.144}$$

$$p_2 = \exp\left(\frac{-2\zeta T_s}{T}\right) \tag{9.145}$$

Ignoring the numerator dynamics, the polynomial P may be specified as

$$P(z^{-1}) = \frac{1 + p_1 z^{-1} + p_2 z^{-2}}{1 + p_1 + p_2} \tag{9.146}$$

The dominant time constant of the closed-loop system is approximately $2T$ and the fractional overshoot is solely a function of the damping factor ζ:

$$o_v = \exp\left[\frac{-\pi\zeta}{\sqrt{1-\zeta^2}}\right] \tag{9.147}$$

and thus the user can then specify the desired rise time and overshoot and translate these settings into an appropriate P polynomial.

9.7.6 The C polynomial

The CARIMA model includes knowledge about the disturbance properties in the polynomial C. This can be estimated on-line using a suitable recursive identification algorithm. However, this is rather difficult in practice, because the convergence of the C polynomial coefficients is rather slow.

Therefore, a more realistic approach is to set C by user directly. The value that has been suggested as a default in the literature is of the form

$$C = (1 - 0.8z^{-1})^2 \tag{9.148}$$

Another possibility that follows from the optimal LQ theory is to calculate it as a stable polynomial from spectral factorisation of the denominator polynomial A as

$$C^*C = A^*A \tag{9.149}$$

9.8 Examples

In this section some examples of the GPC control algorithm are shown. The first example shows some effects of the tuning parameters that have been described in the previous sections.

The bioreactor control example demonstrates the possibility of using a non-linear model for predictions. Here, an artificial neural network model is used. Comparison with an adaptive control based on a linear model shows some drawbacks of adaptive methods applied to non-linear processes.

Finally, the pH control example shows a real-time control problem. It is demonstrated that GPC is able to control such a strong non-linear process.

9.8.1 A Linear Process

Let us consider a linear continuous-time system with transfer function

$$G(s) = \frac{1}{(s+1)^2} \tag{9.150}$$

that is discretised with the sampling time $T_s = 1$.

Two simulation runs were performed. In the first, mean level settings were given. The GPC parameters were $N_1 = 1$, $N_u = 1$, $\lambda = 0$, and N_2 varied between $2 - 20$. The results are shown in Fig. 9.4 and illustrate that with the increasing N_2 the control actions are smoother, more conservative and converge to input step change.

In the second simulation λ was given as the varying parameter. The settings of other parameters were $N_1 = 3$, $N_u = 3$, $N_2 = 5$ which for $\lambda = 0$ gives the dead-beat controller. Increasing λ gives more weight to control increments and slows down the controller. The results are shown in Fig. 9.5.

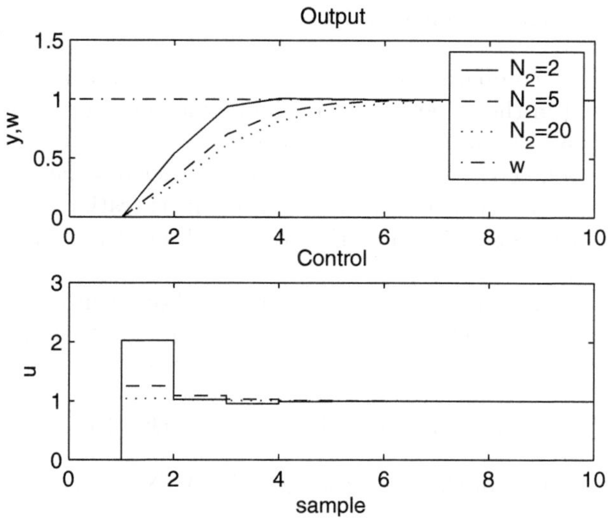

Fig. 9.4. Increasing value of N_2 leads towards mean-level control

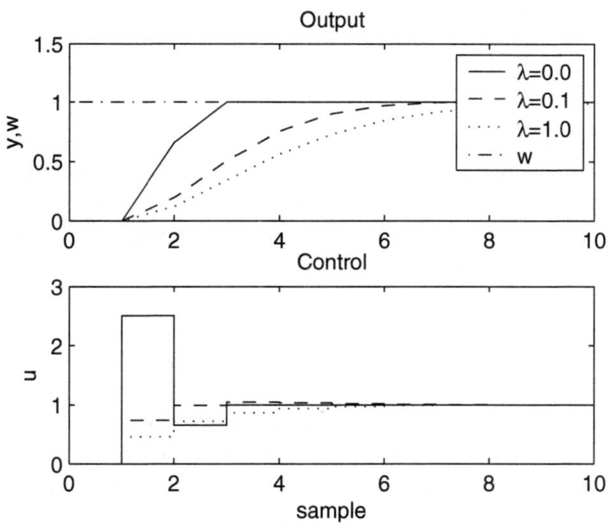

Fig. 9.5. Increasing value of λ - detuned dead-beat control

9.8.2 Neural Network based GPC

In this example we compare an adaptive GPC based on linear model (AGPC) and implemented in the same way as in the previous example to GPC based on non-linear neural network model (GPCNN).

The process studied was the bioprocess that describes the growth of *Saccharomyces cerevisiae* on glucose. The oxygen concentration c_o and the dilution rate D_g have been selected as the controlled and the manipulated variables, respectively.

A feedforward ANN plant model with the third order input dynamics and one hidden layer was used. This means six neurons in the input layer with signals

$$y(k-1), y(k-2), y(k-3), u(k-1), u(k-2), u(k-3) \tag{9.151}$$

For the calculation of the step response, the ANN inputs are

$$y(k-1), y(k-1), y(k-1), u_n, u(k-1), u(k-1) \tag{9.152}$$

where the step change magnitude u_n was specified as

$$u_n = u(k-1) + \frac{w - y(k-1)}{Z} \tag{9.153}$$

and the process static gain Z was determined from the step response estimated in the previous sampling period. The static gain Z was initially set equal to 1. To take into account the fact that the initial conditions are not equal to zero and the step input is not of unit value, the ANN approximation of the step response is subsequently normalised.

For the free response the ANN inputs are

$$u(k-1), u(k-2), u(k-3), y(k-1), y(k-2), y(k-3) \tag{9.154}$$

and it is assumed that the input is constant in the future.

The sampling period was set equal to 0.5 h. The training and validation data sets (800 input-output pair samples) were obtained using a pseudo random binary sequence input. The conjugate gradients algorithm was used as a learning method and a genetic algorithm was used for the initialization of the ANN weights.

For the AGPC, a third order discrete model has been considered for process modelling. The model parameters have been estimated using the parameters estimation algorithm LDDIF.

The GPC parameters were $N_1 = 1, N_2 = 14, N_u = 4, \lambda = 0.1$. The obtained profiles of the process output controlled by the AGPC (dashed line) and the GPCNN (solid line) are shown in Fig. 9.6. Figure 9.7 shows the profiles of the control actions generated by AGPC (dashed line) and GPCNN (solid line), respectively. As it can be seen from these figures, both algorithms

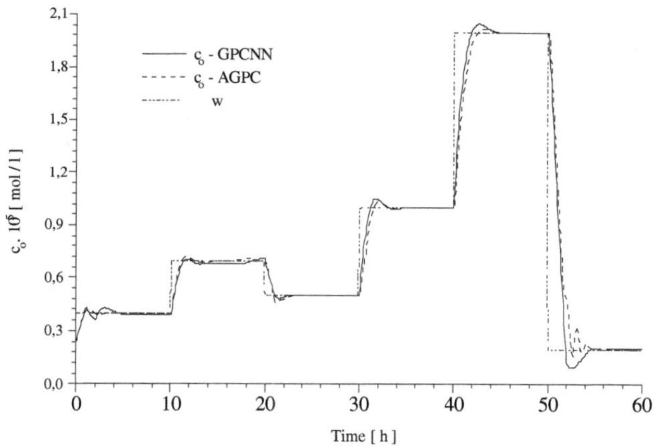

Fig. 9.6. Trajectories of the oxygen concentration (the process output)

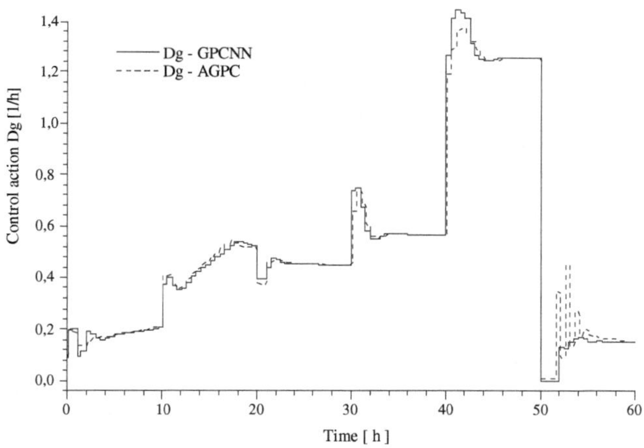

Fig. 9.7. Trajectories of the dilution rate (the manipulated variable)

achieve good results. When a large change of the setpoint occurs (see Fig. 9.7, $t = 50\,\mathrm{h}$), GPC based on linear model leads to a generation of a bad transient behaviour. Unlike AGPC, GPC based on ANN generates a smooth control action which leads to a good control behaviour. This behaviour was expected as the AGPC is based on linear model. Due to the nonlinear characteristic of the bioprocess, a large change of the setpoint or some disturbance can bring the process into other operating points with different dynamical properties.

9.8.3 pH Control

The experimental pH control has been studied at the University of Dortmund.

The neutralisation plant to be controlled consists of a laboratory-scale continuous stirred tank with two inlets and one outlet (see Fig. 9.8)), in which acetic acid is neutralized with sodium hydroxide.

The hold-up of the tank is 5.57 l, the concentrations of the acid and the sodium hydroxide solution streams are approximately 0.01 mol/l. The acid flow rate F_A is fixed at 0.33 l min^{-1}, whereas the NaOH flow F_B is manipulated by the controller. In order to obtain the necessary precision of the flow rates diaphragm pumps were chosen. All control actions are performed by a PC-based control system. The flow F_B is controlled by the modulation of an impulse frequency f, which leads to a quantisation of the control amplitude because the frequency can assume only certain discrete values.

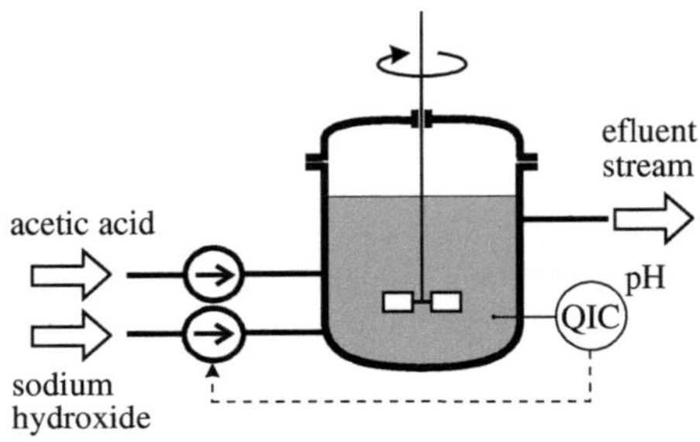

Fig. 9.8. Neutralisation plant

In the tank, the following reaction takes place:

$$NaOH + CH_3COOH \rightleftharpoons CH_3COONa + H_2O \tag{9.155}$$

Due to the incomplete dissociation of the acetic acid in water and the equilibrium reaction with sodium hydroxide the system behaves like a buffer solution between pH 4 and 6.5. Consequently, the process gain varies extremely over the range of pH-values that can be controlled.

The controlled variable pH and the control variable F_B have been scaled for control and identification purposes as

$$y = \frac{\text{pH} - 7}{7} \qquad u = \frac{F_B - F_B^s}{F_B^s} \tag{9.156}$$

where F_B^s denotes the approximate steady state value of F_B corresponding to pH = 7.

The parameters for the GPC controller were chosen as $N = 50, N_u = 4, \lambda = 1, \alpha = 0.3$. Several model orders have been tried, the best results have been obtained with the third order model. The sampling time was set at 5s.

The tuning of the predictive algorithm was performed at pH = 9 with the requirement, that the deviations from the steady-state have to be within

±0.1pH. It was observed, that the small values of N_u led to rather active control actions and the final value $N_u = 4$ was chosen as the result of trade-off between the performance and the complexity of the calculations. The parameter λ influenced the penalisation of the future control increments. Very large values caused limit cycles of pH as the control was unable to compensate satisfactorily the disturbances, therefore the smallest possible value was chosen.

For the final tuning of the algorithm, the P polynomial was used. Only the first order polynomial of the form $P(z^{-1}) = (1 - \alpha z^{-1})/(1 - \alpha)$ was assumed. The smaller values resulted in increased steady-state deviations and the larger values in very slow and oscillatory response to setpoint changes. The value $\alpha = 0.3$ was chosen as a compromise.

The experimental results of the adaptive GPC controller were compared with a carefully tuned PI controller. All experiments were carried out with the same pattern of setpoint changes. At first the reactor was stabilised at pH = 7 and then controlled to pH = 9, 7, 8.3 (Fig. 9.9). Finally, the disturbance rejection performance has been studied. As a disturbance, a 20% decrease of the acid flow was performed at $t = 0$ and held constant afterwards (see Fig. 9.10).

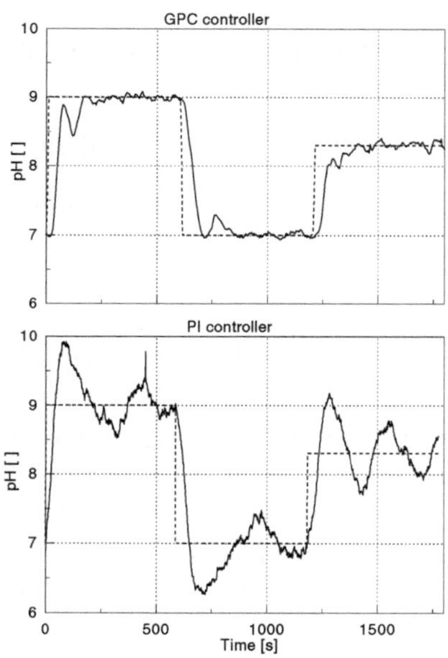

Fig. 9.9. Setpoint tracking

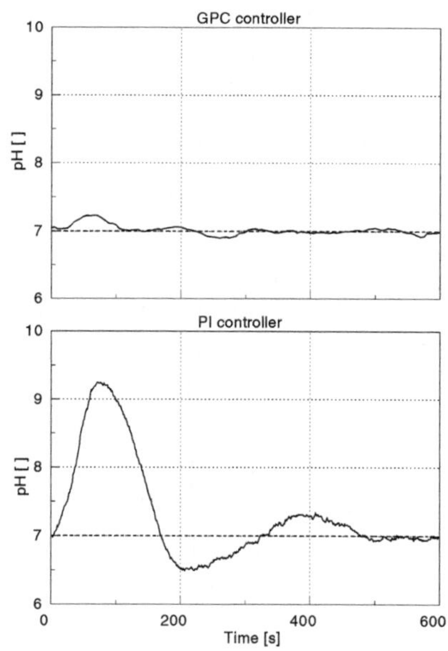

Fig. 9.10. Disturbance rejection

The experiments have confirmed, that the adaptive GPC method is able to control the strongly nonlinear plant and that it behaves much better compared to the linear PI control. However the tuning of the controller parameters must have been done with some care and only the parameters which slowed down the closed-loop system substantially, gave good results. This is because a neutralisation reactor control is known to be not very succesful with linear controllers.

9.9 References

There are several books dealing with predictive control:

D. M. Prett and C. E. García. *Fundamental Process Control.* Butterworths, 1988.

R. R. Bitmead, M. Gevers, and V. Wertz. *Adaptive Optimal Control.* Prentice Hall, Englewood Cliffs, New Jersey, 1990.

E. F. Camacho and C. Bordons. *Model Predictive Control.* Springer-Verlag, London, 1999.

B. Kouvaritakis and M. Cannon, editors. *Non-linear Predictive Control: Theory and Practice.* The Institution of Engineering and Technology, London, 2001.

J. M. Maciejowski. *Predictive Control with Constraints.* Prentice Hall, London, 2002.

J. A. Rossiter. *Model-Based Predictive Control: A Practical Approach.* CRC Press Inc., 2003.

Some selected journal papers dealing with predictive control and stability:

D. W. Clarke, C. Mohtadi, and P. S. Tuffs. Generalized predictive control - part I. The basic algorithm. *Automatica*, 23(2):137 – 148, 1987.

D. W. Clarke and R. Scattolini. Constrained receding-horizon predictive control. *IEE Proc. D*, 138(4):347 – 354, 1991.

E. Mosca and J. Zhang. Stable redesign of predictive control. *Automatica*, 28: 1229 – 1233, 1992.

B. Kouvaritakis, J. A. Rossiter, and A. O. T. Chang. Stable generalised predictive control: an algorithm with guaranteed stability. *IEE Proc. D*, 139(4):349 – 362, 1992.

M. Fikar and S. Engell. Receding horizon predictive control based upon Youla-Kučera parametrization. *European Journal of Control*, 3(4):304 – 316, 1997.

M. Fikar, S. Engell, and P. Dostál. Design of predictive LQ controller. *Kybernetika*, 35(4):459–472, 1999.

H. Chen and F. Allgöwer. A quasi-infinite horizon nonlinear model predictive control scheme with guaranteed stability. *Automatica*, 34(10):1205 – 1217, 1998.

W. H. Kwon and A. E. Pearson. On feedback stabilization of time-varying discrete linear systems. *IEEE Tr. Aut. Control*, 23(3):479 – 481, 1978.

J.-W. Lee, W. H. Kwon, and J. Choi. On stability of constrained receding horizon control with finite terminal weighting matrix. In *CD-ROM Proceedings of ECC'97, Bruxelles, Paper No. 93*, 1997.

J. A. Rossiter, J. R. Gossner, and B. Kouvaritakis. Infinite horizon stable predictive control. *IEEE Trans. Automatic Control*, 41(10):1522 – 1527, 1996.

J. B. Rawlings and K. R. Muske. The stability of constrained receding horizon control. *IEEE Trans. Automatic Control*, 38(10):1512 – 1516, 1993.

A. Zheng and M. Morari. Stability of model predictive control with mixed constraints. *IEEE Trans. Automatic Control*, 40(10):1818 – 1823, 1995.

P. Zelinka, B. Rohal'-Ilkiv, and A.G. Kuznetsov. Experimental verification of stabilising predictive control. *Control Engineering Practice*, 7(5):601–610, 1999.

Tuning of the MBPC parameters can be found for example in these papers:

A. R. McIntosh, S. L. Shah, and D. G. Fisher. Analysis and tuning of adaptive generalized predictive control. *Can. J. Chem. Eng.*, 69:97 – 110, 1991.

R. Shridhar and D. J. Cooper. A tuning strategy for unconstrained SISO model predictive control. *Ind. Eng. Chem. Res.*, 36(3):729 – 746, 1997.

R. Shridhar and D. J. Cooper. A tuning strategy for unconstrained multivariable model predictive control. *Ind. Eng. Chem. Res.*, 37(10):4003 – 4016, 1998.

References to explicit predictive control and Multi-Parametric Toolbox that can efficiently solve a broad range of explicit and on-line formulations:

A. Bemporad, M. Morari, V. Dua, and E. N. Pistikopoulos. The explicit linear quadratic regulator for constrained systems. *Automatica*, 38(1): 3–20, 2002.

P. Grieder, M. Kvasnica, M. Baotić, and M. Morari. Stabilizing low complexity feedback control of constrained piecewise affine systems. *Automatica*, 41 (10):1683–1694, 2005.

M. Kvasnica, P. Grieder, and M. Baotić. Multi-Parametric Toolbox (MPT), 2004. Available from `http://control.ee.ethz.ch/~mpt/`.

V. Sartori, E. Zanderigo, M. Kvasnica, and M. Morari. Optimal Infusion of Intravenous Morphine and Ketamine – A Mixed-Integer Linear Programming Application. In *IEEE Engineering in Medicine and Biology Society*, Shanghai, China, 2005.

P. Tøndel, T.A. Johansen, and A. Bemporad. An algorithm for multiparametric quadratic programming and explicit MPC solutions. *Automatica*, 39(3):489–497, 2003.

Simulation and experimental studies from section 9.8 have been published in more detail in:

A. Rusnák, K. Fikar, M. Najim, and A. Mészaros. Generalized predictive control based on neural networks. *Neural Processing Letters*, 4(2):107 – 112, 1996.

M. Fikar and A. Draeger. Adaptive predictive control of a neutralization reactor. In J. Mikleš, editor, *Preprints of 10th Conf. Process Control'95, June 4 –7, Tatranské Matliare, Slovakia*, volume 1, pages 153–156, 1995.

9.10 Exercises

Exercise 9.1:
Derive the control law from Example 9.1 on page 413 for the state-space model and show that it is equivalent to the input-output model.

Exercise 9.2:
One possible way of including an integrator to a state-space model has been demonstrated in section 9.3.4. Another possibility is to consider the new state vector of the form

$$\bar{\boldsymbol{x}}(k) = \begin{pmatrix} \Delta \boldsymbol{x}(k) \\ y(k) \end{pmatrix} \tag{9.157}$$

Derive the modified state-space description in this case.

Exercise 9.3:
Unstable processes can pose problems when the predictor is derived in section 9.3.4. Calculation of the terms $\boldsymbol{A}^i$ for unstable matrix $\boldsymbol{A}$ can in this case be ill-conditioned as higher powers of $\boldsymbol{A}^i$ do not converge to zero. One possible remedy is to stabilise the controlled process using the state feedback control law of the form

$$\boldsymbol{u}(k+i) = -\boldsymbol{K}\boldsymbol{x}(k+i) + \boldsymbol{c}_i \tag{9.158}$$

where $\boldsymbol{c}_i$ is a new optimised parameter and it is supposed that it is zero after $k + N_u$. Derive prediction equations and find the solution minimising the cost function without constraints.

Exercise 9.4:
Consider an unstable process with the transfer function of the form

$$G(z^{-1}) = \frac{1}{1 - 2z^{-1} + z^{-2}} \tag{9.159}$$

- Calculate the closed-loop poles for $N_1 = 1$, $\lambda = 0$, $N_u = 1$ a $N_2 = 1, 2, 3, \ldots, 10$. Find the smallest N_2 that guarantees stable closed-loop system.
- Choose suitable parameters of the predictive controller to realise the dead-beat control.

10

Adaptive Control

Dynamic and static properties of controlled processes often change in time. Process control is always based on a process model. If process characteristics are changing in time, it seems to be appropriate to use a time-variant process model – model that adapts to changes in the process.

There are many approaches and definitions of adaptive control.

Adaptive control is such a control that adapts to unknown and changing process behaviour.

Adaptive control can be in principle realised in several ways. One of them is the self-tuning control. Here, one starts with a control design for a known process. However, as process parameters are unknown, they are obtained using a recursive parameter identification algorithm assuming that the process model is linear.

There are several reasons for use of self-tuning controllers. The first are changes of operating points of technologic processes that are nonlinear. Linear models for each operating point are different and the controller should change for each of them. Another reason are changes in physical process properties caused by material ageing, wearing etc. The third reason is the presence and elimination of unmeasurable disturbances, perturbations, and unmodelled process dynamics.

The self-tuning controller structure consists of two loops. The inner loop contains a linear feedback controller with time-varying parameters. The outer loop contains identification algorithm and a box that adjusts the controller parameters based on the identification results. The linear controller design is separated from the identification part. It can be based on LQ, LQG or other control design. Parameter estimation is usually implemented using a least squares method. A self-tuning control system is a complex nonlinear system that converges under some conditions.

To start adaptive control, it is generally advisable to use a nominal stabilising controller until the identified process parameters converge. As it was shown in Section 6.3.1, if parameter estimation algorithm starts without any a priori knowledge about the process (covariance matrix contains large values on the main diagonal), convergence can be attained in at minimum so many steps as there is the number of parameters. After that it is possible to switch the controllers usually using some kind of bumpless transfer methods.

Self-tuning control can be used based on both discrete-time and continuous-time models.

10.1 Discrete-Time Self-Tuning Control

This section deals with an explicit self-tuning control based on discrete-time process models. The explicit self-tuning algorithm is implemented in three steps.

The first step is to estimate process model parameters of a given structure.

In the second step controller parameters are calculated given the process model parameters from the first step and a suitable design method.

Manipulated variable is calculated and realised in the third step based on the control law.

This procedure is repeated in each sampling instant.

The self-tuning control system with a discrete-time controller calculates identified parameters and a new controller in each sampling instant. The manipulated variable and identified parameters then remain constant until the new sampling time.

Block diagram of a self-tuning discrete-time control system of a singlevariable process is shown in Fig. 10.1.

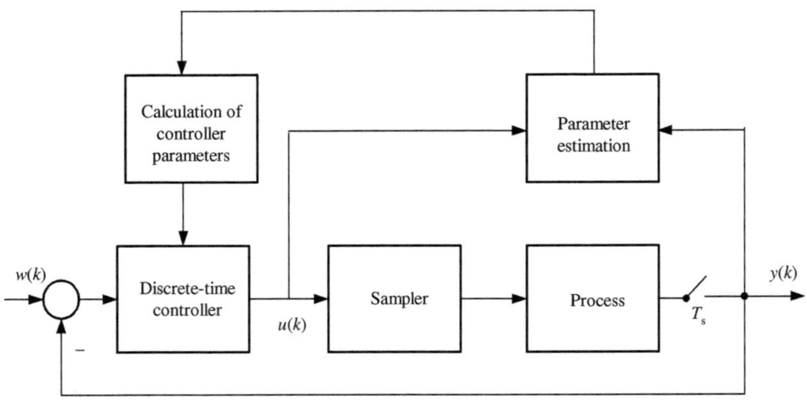

Fig. 10.1. Block diagram of a self-tuning discrete-time controller

10.2 Continuous-Time Self-Tuning Control

Block diagram of a self-tuning control with a continuous-time controller and singlevariable controlled process is shown in Fig. 10.2.

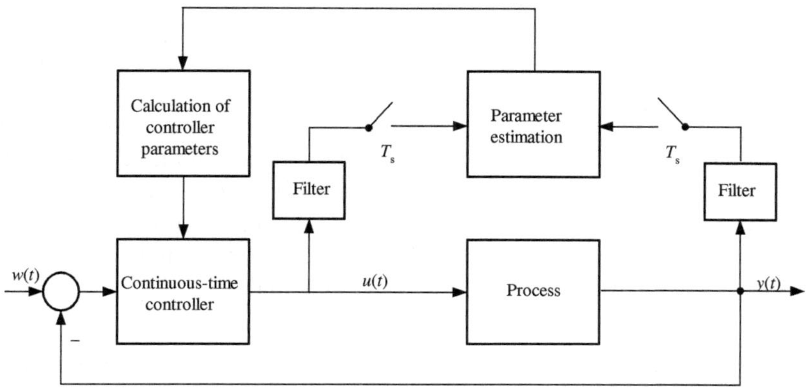

Fig. 10.2. Block diagram of a self-tuning continuous-time controller

The self-tuning continuous-time control uses a continuous-time controller but its parameters are changed in discrete time intervals based on the recursive identification of the continuous-time process. Inputs to identification are filtered inputs, outputs and their derivatives. Outputs of the identification block are piece-wise constant (with respect to time) parameters of the continuous-time process model. See Section 6.3.3 for more details.

The self-tuning continuous-time control system is implemented in three steps in the same way as its discrete-time counterpart.

10.3 Examples of Self-Tuning Control

The above described adaptive control algorithms will be tested on several simulation examples. Testing of algorithms is usually based on step changes of reference and disturbance variables, slowly varying parameters, etc for an unknown system after some adaptation phase.

In all studies the recursive parameter estimation LDDIF with directional forgetting is employed. Dead-beat, pole-placement, LQ, and predictive control algorithms are used. Simulations results show that adaptive control algorithms converge quickly.

Simulation studies in MATLAB/Simulink environment use the Polynomial Toolbox for MATLAB and Simulink identification blockset IDTOOL. Although other choices are possible, this corresponds to the authors and their colleagues experience.

One of the fundamental problems of the feedback linear control in general and of ST control in particular is the issue of system structure selection. This in linear case means determination of degrees of numerator and denominator. If the model order is chosen too low, control performance will suffer. On the other side, if the system degree is too high, its transfer function may contain similar factors in numerator and denominator. As we have seen, the Diophantine equation can be solved only if both polynomials are coprime. Therefore, numerical problems and ill-conditioning can be observed in this case.

Also the choice of the sampling period is very important in the ST control.

If not otherwise explicitly stated, all MATLAB simulations used default values of the library IDTOOL. This includes LDDIF algorithm with increasing exponential forgetting strategy (6.98) with $\lambda_0 = 0.985$, minimum trace of the covariance matrix $\delta = 0.01$, zero initial parameter estimates, and initial covariance matrix with values of 10^6 on the diagonal.

10.3.1 Discrete-Time Adaptive Dead-Beat Control of a Second Order System

www

Consider a continuous-time controlled system with the transfer function

$$G(s) = \frac{b_{s1}s + a_{s0}}{a_{s2}s^2 + a_{s1}s + 1} \tag{10.1}$$

This system will be controlled by an adaptive feedback controller according to Simulink scheme `ad031s.mdl` shown in Fig. 10.3. Initialisation of this diagram is performed with the script `ad031init.m` that sets the process and discrete integrator blocks. This file can be loaded automatically when the diagram is opened using the MATLAB command (the scheme has to be already opened)

```
set_param('ad031s', 'PreLoadFcn', 'ad031init');
```

It is necessary to save the scheme afterwards. The first parameter of this function determines the name of the Simulink scheme and the third determines the name of the script to be executed.

Program 10.1 (Initialisation of the scheme in Fig. 10.3 – `ad031init.m`**)**

```
z = 2; t1 = 1; t2 = 2; ts = 1;
G = tf(z, conv([t1 1],[t2 1]));
zinteg = tf([1 0],[1 -1], ts);
```

The corresponding discrete-time transfer function is in the fractional representation given as

$$G(z^{-1}) = \frac{b_1 z^{-1} + b_2 z^{-2}}{1 + a_1 z^{-1} + a_2 z^{-2}} \tag{10.2}$$

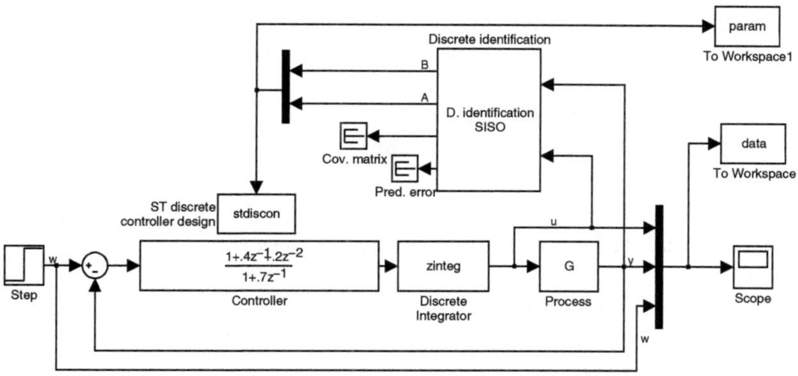

Fig. 10.3. Scheme `ad031s.mdl` in Simulink for dead-beat tracking of a continuous-time second order system with a discrete-time adaptive controller

Parameters a_1, a_2, b_1, b_2 are estimated in each sampling instant. The controller without integral part is given by polynomials

$$Y(z^{-1}) = y_0 + y_1 z^{-1} + y_2 z^{-2}, \quad X(z^{-1}) = x_0 + x_1 z^{-1} \tag{10.3}$$

that are solution of the Diophantine equation

$$(1 + a_1 z^{-1} + a_2 z^{-2})(1 - z^{-1})X(z^{-1}) + (b_1 z^{-1} + b_2 z^{-2})Y(z^{-1}) = 1 \tag{10.4}$$

The general transfer function of the controller is then of the form

$$R(z^{-1}) = \frac{Y(z^{-1}) + (1 - z^{-1})A(z^{-1})T(z^{-1})}{X(z^{-1}) - B(z^{-1})T(z^{-1})} \tag{10.5}$$

If $T(z^{-1}) = 0$ then the controller defined as

$$R(z^{-1}) = \frac{y_0 + y_1 z^{-1} + y_2 z^{-2}}{x_0 + x_1 z^{-1}} \tag{10.6}$$

guarantees the finite-time response to the step change of reference and disturbance variables in a minimum number of control steps.

The controller parameters are determined in the each sampling instant from (10.4) based on estimated parameters a_1, a_2, b_1, b_2. From the actual control law is calculated the control action. This algorithm is repeated in each sampling instant.

Parameter estimation is for the case shown in Fig. 10.3 implemented using the discrete-time SISO identification block from the library IDTOOL and the controller parameters are calculated in the S-function `stdiscon`.

We can notice in the Simulink scheme that in contrast to Fig. 10.1 it contains only one feedback loop consisting of the controlled process and the controller. The second loop is closed using the information about the actual controller parameter values from the block `stdiscon` into the block

Controller. To achieve this, MATLAB function set_param is used within stdiscon. The implementation of the function stdiscon is as follows.

Program 10.2 (S-function stdiscon.m**)**

```
function [sys,x0,str,ts] = stdiscon(t,x,u,flag,tsamp)
  switch flag,
  case 0,
    [sys,x0,str,ts]=mdlInitializeSizes(tsamp);
  case 2,
    sys=mdlUpdate(t,x,u);
  case 3,
    sys=mdlOutputs(t,x,u);
  case 9,
    sys=mdlTerminate(t,x,u);
  otherwise
    error(['Unhandled flag = ',num2str(flag)]);
end

function [sys,x0,str,ts]=mdlInitializeSizes(tsamp)
  sizes = simsizes;
  sizes.NumContStates  = 0;
  sizes.NumDiscStates  = 0;
  sizes.NumOutputs     = 0;
  sizes.NumInputs      = 5;
  sizes.DirFeedthrough = 1;
  sizes.NumSampleTimes = 1;
  sys = simsizes(sizes);
  x0  = [];
  str = [];
  ts  = [tsamp 0];
  % Controller initialisation
  Denominator = '[1 .7]';
  Numerator = '[1 .4 .2]';
  set_param('ad031s/Controller', 'Numerator', Numerator, ...
                                 'Denominator', Denominator);

function sys=mdlOutputs(t,x,u)
  if t>4  % begin adaptive control
    b = [1 zi zi^2] * [0; u(1:2)];
    a = [1 zi zi^2] * u(3:5);
    a = a * (1-zi); % add integrator
    [x0,y0]=axbyc(a,b,1);
    tt = 0;
    xx = x0 - b * tt;
    yy = y0 + a * tt;
```

```
    Denominator = sprintf('[%f %f]',xx{0},xx{1});
    Numerator = sprintf('[%f %f %f]',yy{0},yy{1}, yy{2});
    set_param('ad031s/Controller', 'Numerator', Numerator, ...
                                   'Denominator', Denominator);
  end
  sys=[];

function sys=mdlUpdate(t,x,u)
  sys = [];

function sys=mdlTerminate(t,x,u)
  sys = [];
```

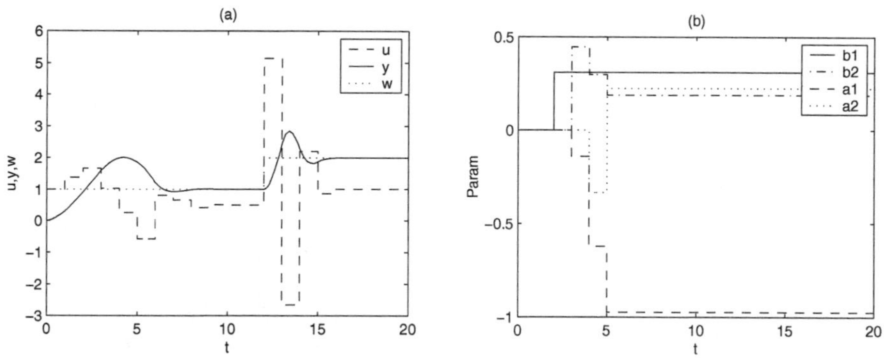

Fig. 10.4. Trajectories of (a) input, output, and setpoint variables, (b) identified parameters of the discrete-time process model controlled with a dead-beat controller

Fig. 10.4a shows trajectories of input and output variables of the controlled process and Fig. 10.4b gives the trajectories of the estimated parameters.

The nominal controller was chosen as

$$R(z^{-1}) = \frac{1 + 0.4z^{-1} + 0.2z^{-2}}{1 + 0.7z^{-1}} \tag{10.7}$$

The adaptive closed-loop system was controlled in the first four sampling periods with this nominal controller. Afterwards, the controller calculated in the block `stdiscon` was used. The controlled process used in the simulation was the same as its model.

10.3.2 Continuous-Time Adaptive LQ Control of a Second Order System

Consider a continuous-time controlled system with the transfer function WWW

$$G(s) = \frac{b(s)}{a(s)} = \frac{b_1 s + b_0}{a_2 s^2 + a_1 s + a_0} \tag{10.8}$$

This system will be controlled by an adaptive feedback controller according to the Simulink scheme `ad033s.mdl` shown in Fig. 10.5. Initialisation of this diagram is performed with the script `ad033init.m` that sets the process, identification and controller update sampling time. This file can be loaded automatically in the same way as in the previous example by the command

```
set_param('ad031s', 'PreLoadFcn', 'ad031init');
```

Program 10.3 (Initialisation of the scheme in Fig. 10.5 – file) ad033init.m

```
ts = .5;
G = tf([1 1.5],[1 3 2]);
```

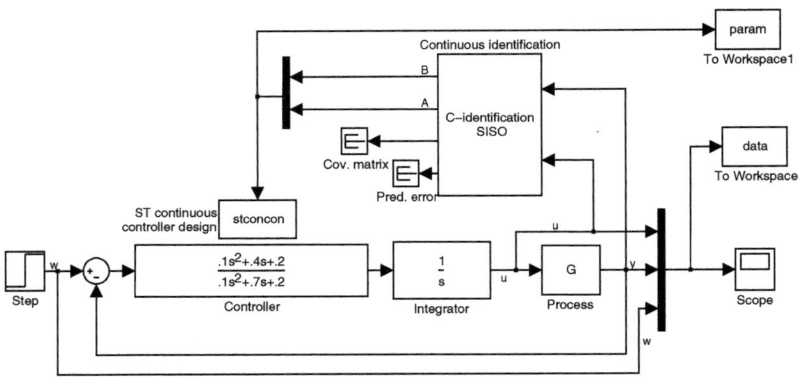

Fig. 10.5. Scheme `ad0313.mdl` in Simulink for LQ tracking of a continuous-time second order system with a continuous-time adaptive controller

Parameters a_0, a_1, a_2, b_1, b_2 are estimated in the each sampling time. The feedback LQ control law based on state observation is given as

$$u = -\frac{q(s)}{sp(s)} e \tag{10.9}$$

where the polynomials

$$q(s) = q_2 s^2 + q_1 s + q_0, \quad p(s) = p_2 s^2 + p_1 s + p_0 \tag{10.10}$$

are solution of the Diophantine equation

$$a(s)sp(s) + b(s)q(s) = o(s)f(s) \tag{10.11}$$

Polynomials

$$o(s) = s^2 + o_1 s + o_0, \quad f(s) = s^3 + f_2 s^2 + f_1 s + f_0 \tag{10.12}$$

are stable and monic and $f(s)$ is defined by the spectral factorisation equation

$$\tilde{a}(-s)\tilde{a}(s) + \tilde{\boldsymbol{B}}_{Rs}(-s)\tilde{\boldsymbol{Q}}_n\tilde{\boldsymbol{B}}_{Rs}(s) = f(-s)f(s) \tag{10.13}$$

where

$$\tilde{a}(s) = sa(s), \qquad \tilde{\boldsymbol{B}}_{Rs}(s) = \tilde{a}(s)\left(s\boldsymbol{I} - \boldsymbol{A}_n\right)^{-1}\boldsymbol{B}_n \tag{10.14}$$

$$\boldsymbol{A}_n = \begin{pmatrix} \boldsymbol{0} & \boldsymbol{I} \\ \boldsymbol{0} & \boldsymbol{A} \end{pmatrix}, \qquad \boldsymbol{B}_n = \begin{pmatrix} \boldsymbol{0} \\ \boldsymbol{B} \end{pmatrix} \tag{10.15}$$

$$\boldsymbol{A} = \begin{pmatrix} 0 & 1 \\ -a_0 & -a_1 \end{pmatrix}, \qquad \boldsymbol{B} = \begin{pmatrix} 0 \\ 1 \end{pmatrix} \tag{10.16}$$

Weighting matrix $\tilde{\boldsymbol{Q}}_n$ of the form

$$\tilde{\boldsymbol{Q}}_n = \begin{pmatrix} \boldsymbol{Q} & \boldsymbol{0} \\ \boldsymbol{0} & \boldsymbol{0} \end{pmatrix} \tag{10.17}$$

follows from the cost function to be minimised

$$I = \frac{1}{2}\int_0^\infty \left(\boldsymbol{x}^T\boldsymbol{Q}\boldsymbol{x} + \dot{u}^2\right) \mathrm{d}t \tag{10.18}$$

where

$$\boldsymbol{Q} = \begin{pmatrix} Q_1 & 0 \\ 0 & Q_1 \end{pmatrix} \tag{10.19}$$

Coefficients of the control law are determined in each sampling instant from equation (10.11) based on estimated parameters a_0, a_1, a_2, b_0, b_1. Finally, the control action is realised.

Parameter estimation is for the case shown in Fig. 10.5 implemented using the continuous-time SISO identification block from the library IDTOOL and the controller parameters are calculated in the S-function `stconcon`.

MATLAB function `set_param` is used to transfer the calculated controller parameters into the block `Controller` and thus to close the adaptive control loop. The implementation is as follows.

Program 10.4 (S-funkcia `stconcon.m`)

```
function [sys,x0,str,ts] = stconcon(t,x,u,flag,tsamp)
  switch flag,
  case 0,
    [sys,x0,str,ts]=mdlInitializeSizes(tsamp);
  case 2,
    sys=mdlUpdate(t,x,u);
  case 3,
```

```
    sys=mdlOutputs(t,x,u);
  case 9,
    sys=mdlTerminate(t,x,u);
  otherwise
    error(['Unhandled flag = ',num2str(flag)]);
end

function [sys,x0,str,ts]=mdlInitializeSizes(tsamp)
  sizes = simsizes;
  sizes.NumContStates  = 0;
  sizes.NumDiscStates  = 0;
  sizes.NumOutputs     = 0;
  sizes.NumInputs      = 5;
  sizes.DirFeedthrough = 1;
  sizes.NumSampleTimes = 1;
  sys = simsizes(sizes);
  x0  = [];
  str = [];
  ts  = [tsamp 0];
  % Controller initialisation
  Denominator = '[.1 .7 .2]';
  Numerator = '[.1 .4 .2]';
  set_param('ad033s/Controller', 'Numerator', Numerator, ...
                                 'Denominator', Denominator);

function sys=mdlOutputs(t,x,u)
  if t>4 % begin adaptive control
    b1 = u(1); b0 = u(2); b = b1*s + b0;
    a2 = u(3); a1 = u(4); a0 = u(5); a = a2*s^2 + a1*s + a0;

    An=[0 0 1 0;0 0 0 1;0 0 0 1;0 0 -a0 -a1];
    Bn=[0;0;0;1];
    Cn=[b0 b1 0 0];
    [BRs,AR]=ss2rmf(An,Bn,eye(4));
    BR=Cn*BRs;

    Q1=1; Q2=1;
    Qn=[Q1 0 0 0;0 Q2 0 0;0 0 0 0;0 0 0 0];
    [Dc,J]=spf(AR'*AR+BRs'*Qn*BRs);
    Df=s^2+1.5*s+3;
    [x0,y0]=axbyc(AR,BR,Dc*Df);
    tt = 0;
    xx = x0 - b * tt;
    yy = y0 + a * tt;
```

```
    Denominator = sprintf('[%f %f %f]',xx{2},xx{1},xx{0});
    Numerator = sprintf('[%f %f %f]',yy{2},yy{1}, yy{0});
    set_param('ad033s/Controller', 'Numerator', Numerator, ...
                                  'Denominator', Denominator);
  end
  sys=[];

function sys=mdlUpdate(t,x,u)
  sys = [];

function sys=mdlTerminate(t,x,u)
  sys = [];
```

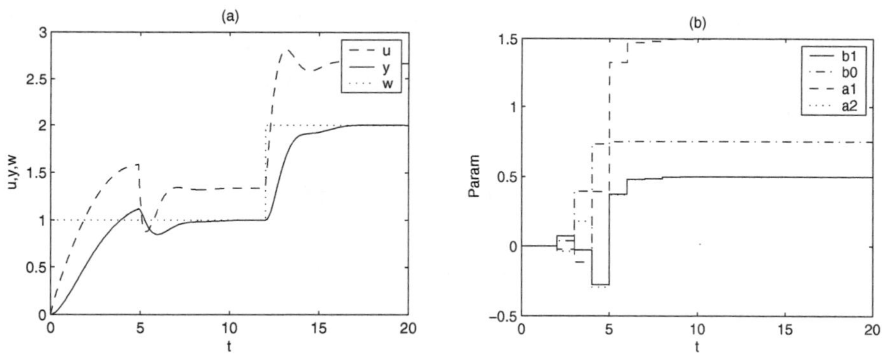

Fig. 10.6. Trajectories of (a) input, output, and setpoint variables, (b) identified parameters of the continuous-time process model controlled with LQ controller

Fig. 10.6a shows trajectories of input and output variables of the controlled process and Fig. 10.6b gives the trajectories of the estimated parameters. IDTOOL normalises the estimated transfer function, thus $a_0 = 1$. LQ weighting coefficients were set as $\varphi = 1$, $\psi = 10$.

The nominal controller and the observer polynomial were set as

$$R(s) = \frac{0.1s^2 + 0.4s + 0.2}{0.1s^2 + 0.7s + 0.2}, \quad o(s) = s^2 + 1.5s + 3 \tag{10.20}$$

The adaptive closed-loop system was controlled in the first eight sampling periods ($T_s = 0.5$) with this nominal controller. Afterwards, the controller calculated in the block `stconcon` was used. The choice of the sampling period follows from the dynamical properties of the process. The controlled process used in the simulation was the same as its model.

10.3.3 Continuous-Time Adaptive MIMO Pole Placement Control

Consider a continuous-time multivariable controlled system with the transfer
www
function matrix

$$G(s) = A_L^{-1}(s)B_L(s) \tag{10.21}$$

where

$$A_L(s) = \begin{pmatrix} 1 + a_1 s & a_2 s \\ a_3 s & 1 + a_4 s \end{pmatrix}, \quad B_L(s) = \begin{pmatrix} b_1 & b_2 \\ b_3 & b_4 \end{pmatrix} \tag{10.22}$$

and $a_1 a_4 \neq a_2 a_3$.

Nonadaptive LQ control of this system was illustrated in Example 8.11 on page 386.

This system will be controlled by an adaptive feedback controller according to the Simulink scheme `ad035s.mdl` shown in Fig. 10.7. Initialisation of this diagram is performed with the script `ad035init.m` that sets the process, integrator, identification and controller update sampling time. This file can be loaded automatically in the same way as in the previous example by the command

```
set_param('ad035s', 'PreLoadFcn', 'ad035init');
```

Program 10.5 (Initialisation of the scheme in Fig. 10.7 – file) ad035init.m

```
ts = 0.5;
al = [1+.3*s, .5*s; .1*s 1+0.7*s];
bl = [.2 .4;.6 .8];
bl = pol(bl);
[a,b,c,d] = lmf2ss(bl,al);
G = ss(a,b,c,d);
```

Parameters of the left matrix fraction a_i, b_i are estimated in each sampling instant.

To calculate the controller, it is necessary to determine the right matrix fraction of the controlled process

$$A_L^{-1}(s)B_L(s) = B_R(s)A_R^{-1}(s) \tag{10.23}$$

The feedback controller

$$R(s) = P_L^{-1}(s)Q(s) = \begin{pmatrix} p_1 s & p_2 s \\ p_3 s & p_4 s \end{pmatrix}^{-1} \begin{pmatrix} q_1 + q_2 s & q_3 + q_4 s \\ q_5 s + q_6 s & q_7 s + q_8 s \end{pmatrix} \tag{10.24}$$

is then obtained by solving the matrix Diophantine equation

$$P_L(s)A_R(s) + Q_L(s)B_R(s) = \tilde{O}_L(s)F_R(s) \tag{10.25}$$

where the matrix $\tilde{O}_L(s)F_R(s)$ determines poles of the closed-loop system.

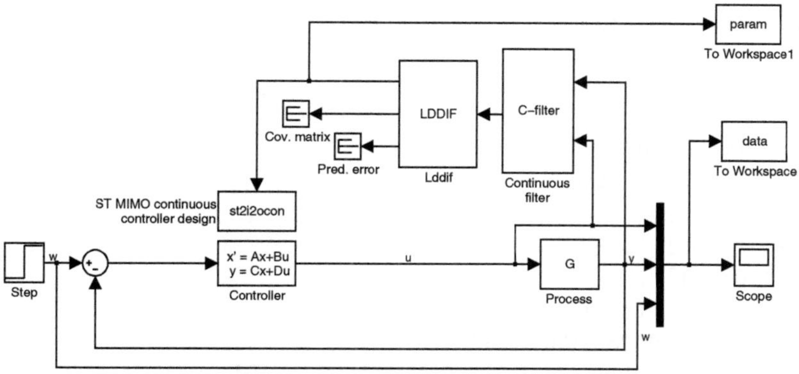

Fig. 10.7. Scheme `ad035s.mdl` in Simulink for tracking of a multivariable system with a continuous-time adaptive controller

Parameter estimation is for the case shown in Fig. 10.7 implemented using the blocks from the library IDTOOL and the controller parameters are calculated in the S-function `st2i2ocon`.

MATLAB function `set_param` is used to transfer the calculated controller parameters into the block `Controller` and thus to close the adaptive control loop. The implementation is as follows.

Program 10.6 (S-funkcia `st2i2ocon.m`)

```
function [sys,x0,str,ts] = st2i2ocon(t,x,u,flag,tsamp)
  switch flag,
  case 0,
    [sys,x0,str,ts]=mdlInitializeSizes(tsamp);
  case 2,
    sys=mdlUpdate(t,x,u);
  case 3,
    sys=mdlOutputs(t,x,u);
  case 9,
    sys=mdlTerminate(t,x,u);
  otherwise
    error(['Unhandled flag = ',num2str(flag)]);
end

function [sys,x0,str,ts]=mdlInitializeSizes(tsamp)
  sizes = simsizes;
  sizes.NumContStates  = 0;
  sizes.NumDiscStates  = 0;
  sizes.NumOutputs     = 0;
  sizes.NumInputs      = 8;
  sizes.DirFeedthrough = 1;
```

```
  sizes.NumSampleTimes = 1;
  sys = simsizes(sizes);
  x0  = [];
  str = [];
  ts  = [tsamp 0];
  % Controller initialisation
  A='[0 0; 0 0]';
  B='[1 0; 0 1]';
  C='[1 0;0 1]';
  D='[.1 .1 ; .1 .1]';
  set_param('ad035s/Controller', ...
                'A', A, 'B', B,  'C', C,  'D', D);

function sys=mdlOutputs(t,x,u)
  if t>4 % begin adaptive control
    al = [1+u(1)*s, u(2)*s; u(5)*s, 1+u(6)*s];
    bl = pol([u(3), u(4); u(7), u(8)]);
    [br,ar] = lmf2rmf(bl,al);
    ar = ar * s; % add integrator
    % closed-loop poles
    clpoles = [(0.7*s+1)*(1.5*s+1),0;0,(0.7*s+1)*(1.5*s+1)];
    [pl,ql]=xaybc(ar, br, clpoles);
    [ac,bc,cc,dc] = lmf2ss(ql,pl*s);
    A = sprintf('[%f %f; %f %f]',ac');
    B = sprintf('[%f %f; %f %f]',bc');
    C = sprintf('[%f %f; %f %f]',cc');
    D = sprintf('[%f %f; %f %f]',dc');
    set_param('ad035s/Controller', ...
                'A', A, 'B', B,  'C', C,  'D', D);
  end
  sys=[];

function sys=mdlUpdate(t,x,u)
  sys = [];

function sys=mdlTerminate(t,x,u)
  sys = [];
```

Fig. 10.6a,b shows trajectories of input and output variables of the controlled process and Fig. 10.6c gives the trajectories of the estimated parameters.

The nominal controller and the closed-loop poles were set as

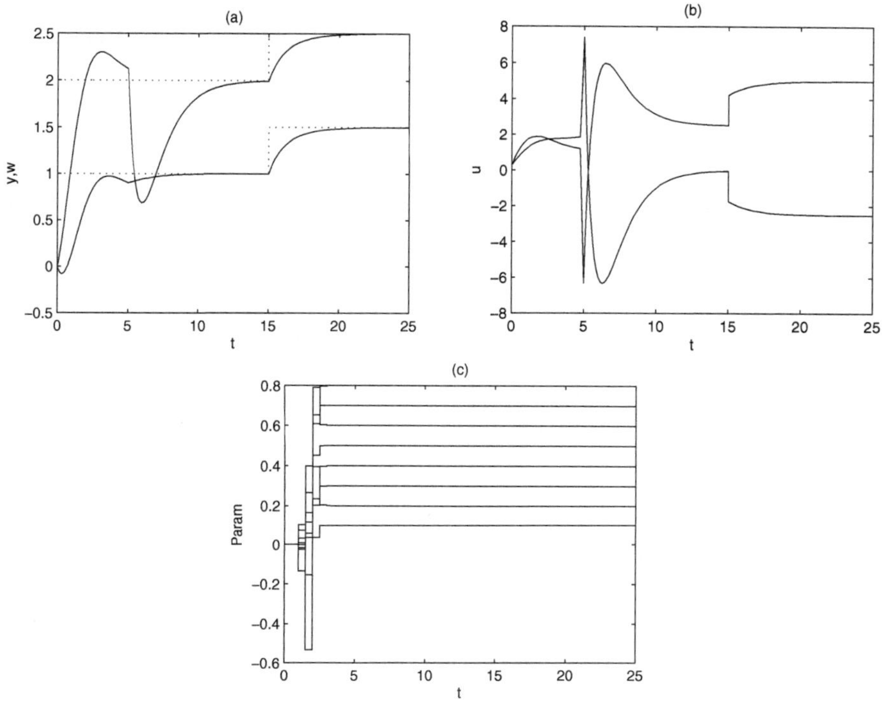

Fig. 10.8. Trajectories of (a) output, and setpoint variables, (b) input variables, (c) identified parameters of the continuous-time process controlled with a pole placement controller

$$\boldsymbol{R}(s) = \begin{pmatrix} s & 0 \\ 0 & s \end{pmatrix}^{-1} \begin{pmatrix} 1+0.1s & 0.1s \\ 0.1s & 1+0.1s \end{pmatrix} \tag{10.26}$$

$$\tilde{\boldsymbol{O}}_L(s)\boldsymbol{F}_R(s) = (0.7s+1)(0.5s+1)\boldsymbol{I} \tag{10.27}$$

The adaptive closed-loop system was controlled in the first eight sampling periods ($T_s = 0.5$) with this nominal controller. Afterwards, the controller calculated in the block `st2i2ocon` was used. The controlled process used in the simulation was the same as its model.

10.3.4 Adaptive Control of a Tubular Reactor

This example dealing with the control of a tubular chemical reactor describes the adaptive implementation of GPC. A linear model is estimated on-line with a recursive least squares algorithm and successively controlled.

We consider an ideal plug-flow tubular chemical reactor with an exothermic consecutive reaction $A \to B \to C$ in the liquid phase and with countercurrent cooling. It is assumed that A is the educt, B is the desired product and C the unwanted by-product of the reaction. Such reactors are central components of many plants in the chemical industry and exhibit highly nonlinear dynamics.

Mathematical model of this reactor is given as

$$\frac{\partial c_A}{\partial t} = -v\frac{\partial c_A}{\partial z} - k_1 c_A \tag{10.28}$$

$$\frac{\partial c_B}{\partial t} = -v\frac{\partial c_B}{\partial z} k_1 c_A - k_1 c_B \tag{10.29}$$

$$\frac{\partial T}{\partial t} = -v\frac{\partial T}{\partial z} + \frac{q_r}{\rho c_p} - \frac{4U_1}{d_1 \rho c_p}(T - T_w) \tag{10.30}$$

$$\frac{\partial T_w}{\partial t} = \frac{4}{(d_2^2 - d_1^2)\rho_w c_{pw}}[d_1 U_1 (T - T_w) + d_2 U_2 (T_c - T_w)] \tag{10.31}$$

$$\frac{\partial T_c}{\partial t} = v\frac{\partial T_c}{\partial z} + \frac{4 n_1 d_2 U_2}{(d_3^2 - n_1 d_2^2)\rho_c c_{pc}}(T_w - T_c) \tag{10.32}$$

with initial conditions

$$\begin{aligned} c_A(z,0) &= c_A^s(z),\ T(z,0) = T^s(z),\ \ T_c(z,0) = T_c^s(z) \\ c_B(z,0) &= c_B^s(z),\ T_w(z,0) = T_w^s(z) \end{aligned} \tag{10.33}$$

and with boundary conditions

$$\begin{aligned} c_A(0,t) &= c_{A0}(t),\ T(0,t) = T_0(t) \\ c_B(0,t) &= c_{B0}(t),\ T_c(L,t) = T_{cL}(t). \end{aligned} \tag{10.34}$$

Here t is time, z space variable along the reactor, c are concentrations, T are temperatures, v are fluid velocities, d are diameters, ρ are densities, c_p are specific heat capacities, and U are heat transfer coefficients. The subscripts are $(.)_w$ for metal wall of tubes, $(.)_c$ for coolant, and $(.)^s$ for steady-state values. The reaction rates k and the heat of reactions are expressed as

$$k_j = k_{0j} \exp(-E_j/RT)\ , \quad j = 1, 2 \tag{10.35}$$

$$q_r = h_1 k_1 c_A + h_2 k_2 c_B \tag{10.36}$$

where k_0 are exponential factors, E are activation energies and h are reaction enthalpies.

Assuming the reactant temperature measurement along the reactor at points z_j, the mean temperature profile can be expressed as

$$T_m(t) = \frac{1}{n}\sum_{j=1}^{n} T(z_j, t) \tag{10.37}$$

where n is the number of measured temperatures along the reactor.

The input variable, the value of q_c, has been assumed to be constrained in the interval

$$0.2 \le q_c \le 0.35 \tag{10.38}$$

For the control purposes both the manipulated input and the controlled output were defined as scaled deviations from their steady-state values

$$u(t) = \frac{q_c(t) - q_c^s}{q_c^s}, \quad y(t) = \frac{T_m(t) - T_m^s}{T_m^s}. \tag{10.39}$$

This scaling helps to obtain variables with approximately the same magnitude and reduces the possibility of ill-conditioned control problem and round-off errors.

The sampling time was chosen $T_s = 3s$ and the reactor was on-line identified as SISO discrete system with $\deg(A) = 2,\ \deg(B) = 3$ of the form

$$y(k) = -a_1 y(k-1) - a_2 y(k-2) + b_1 u(k-1) + b_2 u(k-2) + b_3 u(k-3) + d_c + \xi \tag{10.40}$$

The estimation method used is the recursive least-squares algorithm LDDIF with exponential and directional forgetting. The value of exponential forgetting was set at 0.8 and the minimum of the covariance matrix was constrained to $0.01\boldsymbol{I}$. The purpose of these settings was to improve tracking properties of the estimation algorithm.

The result of the first simulation is shown in Fig. 10.9. It shows comparison of two GPC settings: mean-level (ml) and dead-beat (db) control.

The upper graph represents behaviour of the controlled variable T_m together with its reference value and the lower graph manipulated variable q_c.

The values of GPC tuning parameters $[N_1, N_2, N_u, \lambda]$ were $[1, 15, 1, 10^{-1}]$ for mean-level and $[3, 7, 3, 10^{-5}]$ for dead-beat, respectively. These values correspond to the slow open loop response (ml) and the fastest dead-beat response. The polynomials P, C were set to 1 as the effect of disturbances is very small. One can notice that the dead-beat control strategy actively uses constraints on manipulated variable defined by Eq. (10.38).

The purpose of the second simulation was to investigate the behaviour of GPC with respect to unmeasured disturbances. The output variable was corrupted by measurement noise with variance 0.1K. The inlet concentration c_{A0} of the component A varied in steps and was given as

t	0	100	300	500
$c_{A0} - c_{A0}^s$	0	0.1	0	0.1

Due to the presence of disturbances, the design polynomials P, C were used. The polynomial C attenuates effects of measurement noise and the polynomial P shapes responses of the closed-loop system subject to load disturbances in c_{A0} and also to reference step changes. The degrees of the polynomials were chosen 1 and their values as

$$P = 0.6 - \frac{2}{3} z^{-1}, \quad C = 1 - 0.8 z^{-1}. \tag{10.41}$$

The GPC controller was implemented with the mean-level strategy and had the values of the tuning parameters given as $[1, 15, 1, 10^{-1}]$.

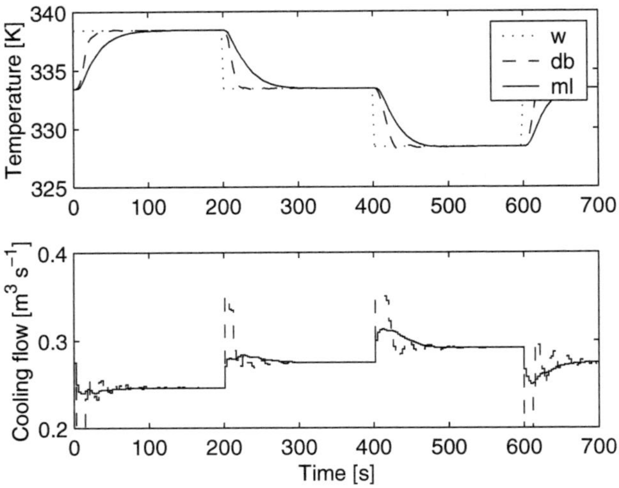

Fig. 10.9. Comparison of the dead-beat (db) and mean-level (ml) control strategy

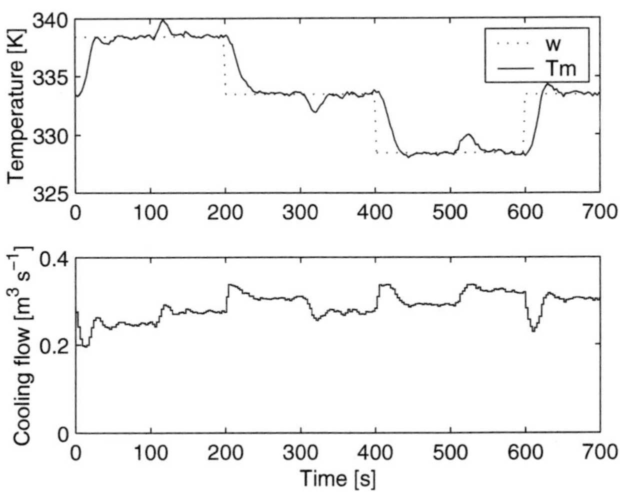

Fig. 10.10. Effect of step disturbances in inlet concentration

The result of the simulation is shown in Fig. 10.10. One can notice that the behaviour of GPC controller is very good and no abrupt control actions can be observed. Moreover, the controlled variable quickly tracks the reference temperature and the effects of load changes in c_{A0} are suppressed very well.

10.4 References

Publication activity in the field of adaptive control has been very rich in the last decades. We will cite at this place only some of the works that have had the largest impact and have shown the principal directions.

V. Peterka. Adaptive digital regulation of noisy systems. In *Proc. 2nd IFAC Symposium on Identification and Process Parameter Estimation*, Prague, 1970.

K. J. Åström and B. Wittenmark. On self-tuning regulators. *Automatica*, 9: 185–189, 1973.

K. J. Åström. Theory and application of adaptive control – A survey. *Automatica*, 19:471–486, 1983.

K. J. Åström and B. Wittenmark. *Adaptive Control.* Addison-Wesley, Massachusetts, 1989.

Part of this chapter is based on the authors' publications:

J. Mikleš. A multivariable self-tuning controller based on pole-placement design. *Automatica*, 26(2):293 – 302, 1990.

J. Mikleš, H. Unbehauen, and V. Keuchel. Entwurf eines adaptiven kontinuierlichen Reglers für Mehrgrößenregelstrecken. *Automatisierungstechnik*, 40 (9):333–342, 1992.

M. Fikar, P. Dostál, and J. Mikleš. Adaptive predictive control of tubular chemical reactor. *Petroleum and Coal*, 38(3):51 – 57, 1996.

References

K. J. Åström. *Introduction to Stochastic Control Theory.* Academic Press, New York, 1970.

K. J. Åström. Theory and application of adaptive control – A survey. *Automatica*, 19:471–486, 1983.

K. J. Åström and T. J. Hägglund. *PID Controllers: Theory, Design, and Tuning.* Instrument Society of America, Research Triangle Park, 2 edition, 1995.

K. J. Åström and B. Wittenmark. On self-tuning regulators. *Automatica*, 9: 185–189, 1973.

K. J. Åström and B. Wittenmark. *Computer Controlled Systems.* Prentice Hall, 1984.

K. J. Åström and B. Wittenmark. *Adaptive Control.* Addison-Wesley, Massachusetts, 1989.

B. D. O. Anderson and J. B. Moore. *Linear Optimal Control.* Prentice Hall, Englewood Cliffs, New Jersey, 1971.

M. Athans and P. L. Falb. *Optimal Control.* McGraw-Hill, New York, 1966.

R. Bellman. *Dynamic Programming.* Princeton University Press, Princeton, New York, 1957.

A. Bemporad, M. Morari, V. Dua, and E. N. Pistikopoulos. The explicit linear quadratic regulator for constrained systems. *Automatica*, 38(1):3–20, 2002.

V. A. Besekerskij and E. P. Popov. *Theory of Automatic Control Systems.* Nauka, Moskva, 1975. (in Russian).

J. Bhat, M. Chidambaram, and K. P. Madhavan. Robust control of a batch-fed fermentor. *Journal of Process Control*, 1:146 – 151, 1991.

G. J. Bierman. *Factorization Methods for Discrete Sequential Estimation.* Academic Press, New York, 1977.

R. R. Bitmead, M. Gevers, and V. Wertz. *Adaptive Optimal Control.* Prentice Hall, Englewood Cliffs, New Jersey, 1990.

S. Bittanti, P. Bolzern, and M. Campi. Exponential convergence of a modified directional forgetting identification algorithm. *Systems & Control Letters*, 14:131–137, 1990.

V. Bobál, J. Böhm, and J. Fessl. *Digital Self-tuning Controllers.* Springer Verlag, London, 2005.

E. F. Camacho and C. Bordons. *Model Predictive Control.* Springer-Verlag, London, 1999.

B. K. Čemodanov et al. *Mathematical Foundations of Automatic Control I.* Vyššaja škola, Moskva, 1977. (in Russian).

C. T. Chen. *Linear System Theory and Design.* Holt, Rinehart and Winston, New York, 1984.

H. Chen and F. Allgöwer. A quasi-infinite horizon nonlinear model predictive control scheme with guaranteed stability. *Automatica*, 34(10):1205 – 1217, 1998.

D. Chmúrny, J. Mikleš, P. Dostál, and J. Dvoran. *Modelling and Control of Processes and Systems in Chemical Technology.* Alfa, Bratislava, 1985. (in Slovak).

L'. Čirka, M. Fikar, and J. Mikleš. A deterministic LQ tracking problem: Parametrisation of the controller and the plant. *Journal of Electrical Engineering*, 53(5–6):126 – 131, 2002.

L'. Čirka and M. Fikar. Identification tool for Simulink. Technical report KAMF9803, Department of Process Control, FCT STU, Bratislava, Slovakia, 1998.

D. W. Clarke and R. Scattolini. Constrained receding-horizon predictive control. *IEE Proc. D*, 138(4):347 – 354, 1991.

D. W. Clarke, C. Mohtadi, and P. S. Tuffs. Generalized predictive control - part I. The basic algorithm. *Automatica*, 23(2):137 – 148, 1987.

J.-P. Corriou. *Process Control. Theory and Applications.* Springer Verlag, 2004.

D. R. Coughanouwr and L. B. Koppel. *Process System Analysis and Control.* McGraw-Hill, New York, 1965.

Y. Z. Cypkin. Theory of discontinuous control. *Automat. i Telemech.*, 3,5,5, 1949, 1949, 1950.

Y. Z. Cypkin. *Foundations of Theory of Automatic Systems.* Nauka, Moskva, 1977. (in Russian).

Y. Z. Cypkin. *Grundlagen der Theorie automatischer Systeme.* VEB Verlag Technik, Berlin, 1981.

B. Dahhou, M. Lakrori, I. Queinnec, E. Ferret, and A. Cherny. Control of continuous fermentation process. *Journal of Process Control*, 2:103 – 111, 1992.

J. Danko, M. Ondrovičová, and A. Mészáros. Experimental model of a hydraulic system. *Automatizace*, (12):478 – 481, 1995. (in Slovak).

C. A. Desoer and M. Vidyasagar. *Feedback Systems: Input-Output Properties.* Academic Press, New York, 1975.

P. Dostál, J. Mikleš, and A. Mészáros. *Theory of Automatic Control. Exercises II.* ES SVŠT, Bratislava, 1983. (in Slovak).

J. M. Douglas. *Process Dynamics and Control.* Prentice Hall, Inc., New Jersey, 1972.

J. C. Doyle, K. Glover, P. P. Khargonekar, and B. Francis. State-space solutions to the standard H_2 and H_∞ control problems. *IEEE Trans. Automatic Control*, 34:831–847, 1989.

P. Eykhoff. *Trends and Progress in System Identification.* Pergamon Press, Oxford, 1981.

A. A. Feldbaum and A. G. Butkovskij. *Methods in Theory of Automatic Control.* Nauka, Moskva, 1971. (in Russian).

M. Fikar and A. Draeger. Adaptive predictive control of a neutralization reactor. In J. Mikleš, editor, *Preprints of 10th Conf. Process Control'95, June 4 –7, Tatranské Matliare, Slovakia*, volume 1, pages 153–156, 1995.

M. Fikar and S. Engell. Receding horizon predictive control based upon Youla-Kučera parametrization. *European Journal of Control*, 3(4):304 – 316, 1997.

M. Fikar and V. Kučera. On minimum finite length control problem. *Int. J. Control*, 73(2):152 – 158, 2000.

M. Fikar and J. Mikleš. *System Identification.* STU Press, 1999. (in Slovak).

M. Fikar and H. Unbehauen. Youla-Kučera design of decoupled control systems. *Int. J. Control*, 75(3):213 – 218, 2002.

M. Fikar, P. Dostál, and J. Mikleš. Adaptive predictive control of tubular chemical reactor. *Petroleum and Coal*, 38(3):51 – 57, 1996.

M. Fikar, S. Engell, and P. Dostál. Design of predictive LQ controller. *Kybernetika*, 35(4):459–472, 1999.

T. R. Fortescue, L. S. Kershenbaum, and B. E. Ydsie. Implementation of self tuning regulators with variable forgetting factors. *Automatica*, 17:831–835, 1981.

C. J. Friedly. *Dynamic Behavior of Processes.* Prentice Hall, Inc., New Jersey, 1972.

K. F. Gauss. *Theoria motus corporum celestium.* English translation: Theory of the Motion of the Heavenly Bodies, Dover, New York, 1963, 1809.

H. P. Geering. *Mess- und Regelungstechnik.* Springer Verlag, Berlin, 1990.

E. D. Gilles. *Systeme mit verteilten Parametern, Einführung in die Regelungstheorie.* Oldenbourg Verlag, München, 1973.

G. C. Goodwin and K. S. Sin. *Adaptive Filtering Prediction and Control.* Prentice Hall, Inc., Englewood Cliffs, New Jersey, 1984.

A. Grace, A. J. Laub, J. N. Little, and C. Thompson. *Control System Toolbox for Use with MATLAB.* The Mathworks, Inc., Natich, MA, 1990.

P. Grieder, M. Kvasnica, M. Baotić, and M. Morari. Stabilizing low complexity feedback control of constrained piecewise affine systems. *Automatica*, 41 (10):1683–1694, 2005.

F. R. Hansen and G. F. Franklin. On a fractional representation approach to closed-loop experiment design. In *Proc. ACC'88*, pages 1319–1320, Atlanta, GA, 1988.

B. Hanuš. *Introduction to linear discrete control*, volume I. ES VSŠT, Liberec, 1972. (in Czech).

M. Huba. Constrained pole assignment control. In L. Menini, L. Zaccarian, and C. T. Abdallah, editors, *Current Trends in Nonlinear Systems and Control*, pages 395 – 398. Birkhäuser, Boston, 2006a.

M. Huba. *Theory of Automatic Control 3. Constrained PID Control.* STU Press, Bratislava, 2006b. (in Slovak).

P. Hudzovič. *Identifcation and Modelling.* ES SVŠT, Bratislava, 1986. (in Slovak).

A. Hurwitz. Über die Bedinungen, unter welchen eine Gleichung nur Wurzeln mit negativen reellen Teilen besitzt. *Math. Annalen*, 46:273 – 284, 1895.

J. Ingham, I. J. Dunn, E. Henzle, and J. E. Přenosil. *Chemical Engineering Dynamics.* VCH Verlagsgesselschaft, Weinheim, 1994.

R. Isermann. *Digitale Regelsysteme.* Springer Verlag, Berlin, 1977.

V. A. Ivanov, V. S. Medvedev, B. K. Čemodanov, and A. S. Juščenko. *Mathematical Foundations of Theory of Automatic Control. Vol II.* Vyššaja škola, Moskva, 1977. (in Russian).

M. Jamshidi, M. Tarokh, and B. Shafai. *Computer-aided Analysis and Design of Linear Control Systems.* Prentice Hall, Englewood Cliffs, New Jersey, 1992.

J. Ježek. Symmetric matrix polynomial equations. *Kybernetika*, 22:19 – 30, 1986.

J. Ježek and V. Kučera. Efficient algorithm for matrix spectral factorization. *Automatica*, 21:663 – 669, 1985.

E. I. Jury. *Digital Control Design.* GIFML, Moskva, 1963. (in Russian).

V. V. Kafarov, V. L Perov, and B. P. Mešalkin. *Principles of Mathematical Modelling of Systems in Chemical Technology.* Chimija, Moskva, 1974. (in Russian).

T. Kailaith. *Linear Systems.* Prentice Hall, Englewood Cliffs, New Jersey, 1980.

P. L. Kalman, R. E. Falb and M. Arib. *Examples of Mathematical Systems Theory.* Mir, Moskva, 1971. (in Russian).

R. E. Kalman. On the general theory of control systems. In *Proc. First IFAC Congress, Moscow, Butterworths*, volume 1, pages 481 – 492, 1960a.

R. E. Kalman. Contribution to the theory of optimal control. *Boletin de la Sociedad Matematica Mexicana*, 5:102–119, 1960b.

R. E. Kalman. When is a linear system optimal? *Trans. ASME, Series D, Journal of Basic Engn.*, pages 51–60, 1964.

R. E. Kalman and J. E. Bertram. Control system analysis and design via the second method of Lyapunov. *J. Basic Engineering*, 82:371 – 399, 1960.

R. E. Kalman and R. S. Bucy. New results in linear filtering and prediction theory. *J. Basic Engr.*, 83:95 – 100, 1961.

R. E. Kalman, Y. C. Ho, and K. S. Narendra. Controllability of linear dynamical systems in contributions to differential equations. *Interscience Publishers*, V1(4):189 – 213, 1963.

L. B. Koppel. *Introduction to Control Theory with Application to Process Control.* Prentice Hall, Englewood Cliffs, New Jersey, 1968.

B. Kouvaritakis and M. Cannon, editors. *Non-linear Predictive Control: Theory and Practice.* The Institution of Engineering and Technology, London, 2001.

B. Kouvaritakis, J. A. Rossiter, and A. O. T. Chang. Stable generalised predictive control: an algorithm with guaranteed stability. *IEE Proc. D*, 139(4):349 – 362, 1992.

S. Kubík, Z. Kotek, V. Strejc, and J. Štecha. *Theory of Automatic Control I.* SNTL/ALFA, Praha, 1982. (in Czech).

V. Kučera. *Discrete Linear Control: The Polynomial Equation Approach.* Wiley, Chichester, 1979.

V. Kučera. Stability of discrete linear feedback systems. In *Proc. 6th IFAC World Congress*, volume 1, Boston, 1975. Paper 44.1.

V. Kučera. New results in state estimation and regulation. *Automatica*, 17: 745–748, 1981.

V. Kučera. Diophantine equations in control – a survey. *Automatica*, 29: 1361–1375, 1993.

V. Kučera. A tutorial on H_2 control theory: The continuous-time case. In M. J. Grimble and V. Kučera, editors, *Polynomial Methods for Control System Design*, pages 1–56. Springer Verlag, London, 1996.

V. Kučera. A bridge between state-space and transfer-function methods. *Annual Reviews in Control*, 23:177–184, 1999.

V. Kučera and D. Henrion. H_2 optimal control via pole placement. In *IFAC Symposium Robust Control Design*, Prague, 2000.

R. Kulhavý and M. Kárný. Tracking of slowly varying parameters by directional forgetting. In *Proc. 9th IFAC World Congr., Budapest*, pages 79–83, 1984.

M. Kvasnica, P. Grieder, and M. Baotić. Multi-Parametric Toolbox (MPT), 2004. Available from `http://control.ee.ethz.ch/~mpt/`.

H. Kwakernaak. H_2-optimization – Theory and applications to robust control design. In *IFAC Symposium Robust Control Design*, Prague, 2000.

H. Kwakernaak and R. Sivan. *Linear Optimal Control Systems.* Wiley, New York, 1972.

W. H. Kwon and A. E. Pearson. On feedback stabilization of time-varying discrete linear systems. *IEEE Tr. Aut. Control*, 23(3):479 – 481, 1978.

J.-W. Lee, W. H. Kwon, and J. Choi. On stability of constrained receding horizon control with finite terminal weighting matrix. In *CD-ROM Proceedings of ECC'97, Bruxelles, Paper No. 93*, 1997.

L. N. Lipatov. *Typical Processes of Chemical Technology as Objects for Control.* Chimija, Moskva, 1973. (in Russian).

L. Ljung. *System Identification: Theory for the User.* MIT Press, Cambridge, Mass., 1987.

L. Ljung and T. Söderström. *Theory and Practice of Recursive Identification.* MIT Press, Cambridge, Mass., 1983.

D. G. Luenberger. An introduction to observers. *IEEE Trans. AC*, 16: 596 – 602, 1971.

J. Lunze. *Regelungstechnik 2.* Springer Verlag, Berlin, 1997.

W. L. Luyben. *Process Modelling, Simulation and Control for Chemical Engineers.* McGraw Hill, Singapore, 2 edition, 1990.

A. M. Lyapunov. *General Problem of Stability of Motion.* Fizmatgiz, Moskva, 1959. (in Russian).

J. M. Maciejowski. *Predictive Control with Constraints.* Prentice Hall, London, 2002.

P. S. Maybeck. *Stochastic Models, Estimation, and Control*, volume 1. Academic Press, New York, 1979.

A. R. McIntosh, S. L. Shah, and D. G. Fisher. Analysis and tuning of adaptive generalized predictive control. *Can. J. Chem. Eng.*, 69:97 – 110, 1991.

G. Meinsma. The standard H_2 problem. In *IFAC Symposium Robust Control Design*, Prague, 2000.

A. Mészáros and J. Mikleš. On observability and controllability of a tubular chemical reactor. *Chem. prumysl*, 33(2):57 – 60, 1983. (in Slovak).

J. Mikleš. *Encyclopedia of Life Support Systems (EOLSS)*, chapter E6-43-34 Automation and Control in Process Industries, edited by H. Unbehauen. EOLSS Publishers, Oxford, UK, 2003. [http://www.eolss.net].

J. Mikleš. *Foundations of Technical Cybernetics.* ES SVŠT, Bratislava, 1973. (in Slovak).

J. Mikleš. *Theory of Automatic Control of Processes in Chemical Technology, Part I.* ES SVŠT, Bratislava, 1978. (in Slovak).

J. Mikleš. *Theory of Automatic Control of Processes in Chemical Technology, Part II.* ES SVŠT, Bratislava, 1980. (in Slovak).

J. Mikleš. A multivariable self-tuning controller based on pole-placement design. *Automatica*, 26(2):293 – 302, 1990.

J. Mikleš and V. Hutla. *Theory of Automatic Control.* Alfa, Bratislava, 1986. (in Slovak).

J. Mikleš, H. Unbehauen, and V. Keuchel. Entwurf eines adaptiven kontinuierlichen Reglers für Mehrgrößenregelstrecken. *Automatisierungstechnik*, 40 (9):333–342, 1992.

J. Mikleš, P. Dostál, and A. Mészáros. *Control of Processes in Chemical Technology.* STU, Bratislava, 1994. (in Slovak).

J. Mikleš, Š. Kožka, and Čirka Ľ. PID controller and LQ control design. In *IFAC Workshop on Digital Control*, pages 315–319, Terassa, 2000.

E. Mosca and J. Zhang. Stable redesign of predictive control. *Automatica*, 28: 1229 – 1233, 1992.

H. Nyquist. Regeneration theory. *Bell Syst. techn. J.*, 11:126 – 147, 1932.

K. Ogata. *Discrete-time control systems.* Prentice Hall, Englewood Cliffs, New Jersey, 1987.

J. Olehla, M. Olehla, et al. *BASIC for Microcomputers.* NADAS, Praha, 1998. (in Czech).

W. Oppelt. Vergleichende Betrachtung verschiedener Regelaufgaben hinsichtlich der geeigneten Regelgesetzmäßigkeit. *Luftfahrtforschung*, 16: 447 – 472, 1939.

W. Oppelt. *Kleines Handbuch technischer Regelvorgänge.* Verlag Chemie, Weinheim, 1972.

J. Penny and G. Lindfield. *Numerical Methods Using MATLAB.* ELLIS HORWOOD, London, 1995.

V. Peterka. Adaptive digital regulation of noisy systems. In *Proc. 2nd IFAC Symposium on Identification and Process Parameter Estimation*, Prague, 1970.

Polyx. *The Polynomial Toolbox for MATLAB.* PolyX, Ltd., Prague, 1999. http://www.polyx.com.

L. S. Pontryagin, V. G. Boltyanskii, R. V. Gamkrelidze, and E. F. Mishchenko. *The Mathematical Theory of Optimal Processes.* Wiley, New York, 1962.

W. H. Press, S. A. Teukolsky, W. T. Vetterling, and B. P. Flannery. *Numerical Recipes in C. The Art of Scientific Computing.* Cambridge University Press, 1992.

D. M. Prett and C. E. García. *Fundamental Process Control.* Butterworths, 1988.

J. R. Ragazzini and L. A. Zadeh. The analysis of sampled-data control systems. *AIEE Trans.*, 75:141 – 151, 1952.

M. Rauseier, P. Agrawal, and D. A. Melichamp. Non-linear adaptive control of fermentation processes utilizing a priori modelling knowledge. *Journal of Process Control*, 2:129 – 138, 1992.

J. B. Rawlings and K. R. Muske. The stability of constrained receding horizon control. *IEEE Trans. Automatic Control*, 38(10):1512 – 1516, 1993.

W. H. Ray. *Advanced Process Control.* McGraw-Hill, New York, 1981.

K. Reinisch. *Kybernetische Grundlagen und Beschreibung kontinuericher Systeme.* VEB Verlag Technik, Berlin, 1974.

B. Rohál'-Il'kiv. *System Identification.* SjF SVŠT, Bratislava, 1987. (in Slovak).

K. Rörentrop. *Entwicklung der modernen Regelungstechnik.* Oldenbourg-Verlag, München, 1971.

H. H. Rosenbrock. *State-space and Multivariable Theory.* Nelson, London, 1970.

J. A. Rossiter. *Model-Based Predictive Control: A Practical Approach.* CRC Press Inc., 2003.

J. A. Rossiter, J. R. Gossner, and B. Kouvaritakis. Infinite horizon stable predictive control. *IEEE Trans. Automatic Control*, 41(10):1522 – 1527, 1996.

E. J. Routh. *A Treatise on the Stability of a Given State of Motion.* Mac Millan, London, 1877.

A. Rusnák, K. Fikar, M. Najim, and A. Mészaros. Generalized predictive control based on neural networks. *Neural Processing Letters*, 4(2):107 – 112, 1996.

V. Sartori, E. Zanderigo, M. Kvasnica, and M. Morari. Optimal Infusion of Intravenous Morphine and Ketamine – A Mixed-Integer Linear

Programming Application. In *IEEE Engineering in Medicine and Biology Society*, Shanghai, China, 2005.

D. E. Seborg, T. F. Edgar, and D. A. Mellichamp. *Process Dynamics and Control.* Wiley, New York, 1989.

J. A. Seinfeld and L. Lapidus. *Mathematical Methods in Chemical Engineering, Vol. 3, Process Modeling, Estimation, and Identification.* Prentice Hall, Inc., New Jersey, 1980.

U. Shaked. A general transfer function approach to linear stationary filtering and steady-state optimal control problems. *Int. J. Control*, 24:741–770, 1976.

F. G. Shinskey. *Process Control Systems.* McGraw-Hill, New York, 1979.

R. Shridhar and D. J. Cooper. A tuning strategy for unconstrained SISO model predictive control. *Ind. Eng. Chem. Res.*, 36(3):729 – 746, 1997.

R. Shridhar and D. J. Cooper. A tuning strategy for unconstrained multivariable model predictive control. *Ind. Eng. Chem. Res.*, 37(10):4003 – 4016, 1998.

T. Söderström and P. Stoica. *System Identification.* Prentice Hall, Inc., 1989.

G. Stephanopoulos. *Chemical Process Control, An Introduction to Theory and Practice.* Prentice Hall, Englewood Cliffs, New Jersey, 1984.

A. Stodola. Über die Regulierung von Turbinen. *Schweizer Bauzeitung*, 22,23: 27 – 30, 17 – 18, 1893, 1894.

V. Strejc. Approximation aperiodischer Übertragungscharakteristiken. *Regelungstechnik*, 7:124 – 128, 1959.

V. Strejc. *State-space Theory of Linear Control.* Academia, Praha, 1978. (in Czech).

V. Strejc and kol. *Control Design with Digital Computers.* NČSAV, 1965. (in Czech).

K. K. Tan, Q.-G. Wang, C. C. Hang, and T. J. Hägglund. *Advances in PID Control.* Springer Verlag, London, 1999.

P. Tøndel, T.A. Johansen, and A. Bemporad. An algorithm for multiparametric quadratic programming and explicit MPC solutions. *Automatica*, 39 (3):489–497, 2003.

H. Unbehauen. *Regelungstechnik I.* Vieweg, Braunschweig/Wiesbaden, 1986.

H. Unbehauen. *Regelungstechnik II.* Vieweg, Braunschweig/Wiesbaden, 1987.

H. Unbehauen. *Regelungstechnik III.* Vieweg, Braunschweig/Wiesbaden, 1988.

H. Unbehauen and G. P. Rao. *Identification of Continuous Systems.* North-Holland, Amsterdam, 1987.

J. Čermák, V. Peterka, and J. Závorka. *Dynamics of Controlled Systems in Thermal Energetics and Chemistry.* Academia, Praha, 1968. (in Czech).

L. N. Volgin. *The Fundamentals of the Theory of Controllers.* Soviet Radio, Moskva, 1962. (in Russian).

A. A. Voronov. *Stability, Controllability, Observability.* Nauka, Moskva, 1979. (in Russian).

M. Šalamon. *Control Theory.* ES SVŠT, Bratislava, 1979. (in Slovak).

T. W. Weber. *An Introduction to Process Dynamics and Control.* Wiley, New York, 1973.

B. Wittenmark, J. K. Åström, and S. B. Jørgensen. *Process Control.* Lund Institute of Technology, Technical University of Denmark, Lyngby, 1992.

W. A. Wolovich. *Linear Multivariable Systems.* Springer Verlag, New York, 1974.

D. C. Youla, J. J. Bongiorno, and H.A. Jabr. Modern Wiener-Hopf design of optimal controller, Part I: The single-input case. *IEEE Trans. Automatic Control*, 21:3–14, 1976a.

D. C. Youla, J. J. Bongiorno, and H.A. Jabr. Modern Wiener-Hopf design of optimal controller, Part II: The multivariable case. *IEEE Trans. Automatic Control*, 21:319–338, 1976b.

C.-C. Yu. *Autotuning of PID Controllers. Relay Feedback Approach.* Springer Verlag, London, 1999.

L. A. Zadeh and C. A. Desoer. *Linear System Theory - the State-space Approach.* McGraw-Hill, New York, 1963.

P. Zelinka, B. Rohal'-Ilkiv, and A.G. Kuznetsov. Experimental verification of stabilising predictive control. *Control Engineering Practice*, 7(5):601–610, 1999.

A. Zheng and M. Morari. Stability of model predictive control with mixed constraints. *IEEE Trans. Automatic Control*, 40(10):1818 – 1823, 1995.

K. Zhou, J. C. Doyle, and K. Glover. *Robust and Optimal Control.* Prentice Hall, Englewood Cliffs, New Jersey, 1995.

J. G. Ziegler and N. B. Nichols. Optimum settings for automatic controllers. *Trans. ACME*, 64(8):759–768, 1942.

Index

adaptive control, 445
amplitude, 148
analysis
  frequency, 145
ARMA model, 224
ARMAX model, 224
ARX model, 224
Åström-Hägglund methods, 290

bandwidth, 261
BASIC, 130
Bezout identity, 111
block diagram, 8
Bode diagram, 152
Broida method of identification, 232
bumpless transfer, 279

C language, 131
canonical decomposition, 83
canonical form
  controllable, 104
  observable, 107
characteristics
  frequency, 145
  statistical, 162
closed-loop system, 253
column
  distillation, 28
  packed absorption, 26
connection
  feedback, 98
  parallel, 97
  serial, 97
control
  dead-beat, 376
  direct digital, 189, 205
  error, 5
  feedback, 6
  feedforward, 9
  integral action, 321
  law, 5
  optimal, 297
    feedback, 305
  predictive, 403
  problem, 253
  problem definition, 255
  proportional, 5
  standard configuration, 394
  steady state error, 7
control device, 5
control error
  steady-state, 212, 259
control performance, 257
  frequency indices, 260
  integral criteria, 259
  time domain, 257
control quality, 212
controllability, 78
  complete, 79
controllable canonical form, 210
controlled object, 3
controller
  H_2, 394
  derivative, 272
  integral, 270
  integral action, 355

LQ, 380
polynomial, 380
state-space, 380
LQG, 388
multivariable, 392
PID, 266
polynomial design, 350
proportional, 5, 266
gain, 5
stabilising, 368
state-space, 370
converter
analog-to-digital, 189
digital-to-analog, 189
correlation coefficient, 168
cost
IAE, 260
ISE, 260
covariance, 168
covariance matrix, 169
cross-power spectral density, 175
cutoff rate, 261

damping ratio, 258
dead-beat control, 376
derivation operator, 85
Diophantine equation, 351
solvability, 352
directional forgetting, 242
dominant poles, 263
dominant time constant, 258
dual parametrisation, 372
dynamic programming, 322
continuous-time systems, 322
discrete-time systems, 329

equation
algebraic, 34
differential, 34
set, 34
solution, 62
Diophantine, 351
Hamilton-Jacobi-Bellman, 326
Lyapunov, 77
output, 36
Riccati, 309
state, 34, 36
state balance, 14
state space, 66
balance, 66
solution, 67
Euler identity, 149
Euler method, 126
expected value, 171
exponential forgetting, 241

feedback
negative, 6
Fourier
transform, 102
Fourier expansion, 149
Fourier law, 25
fraction
partial, 61
polynomial
matrix, 107
fractional representation, 366
frequency
transfer function matrix, 102
frequency response
magnitude, 151, 152
phase, 151
phase angle, 152
frequency transfer function, 150
frequency transfer function matrix, 151
function
auto-correlation, 171
auto-covariance, 171
correlation, 172
cross-correlation, 174
cumulative distribution, 164
dirac delta, 60
distribution, 164
exponential, 54
Lyapunov, 75
negative definite, 76
positive definite, 75
positive semidefinite, 76
probability density, 167
ramp, 54
step, 53
unit, 53
transfer, 85
trigonometric, 55
unit impulse, 60
weighting, 117

gain, 120
gain crossover frequency, 261
gain margin, 260

Hamilton-Jacobi-Bellman
equation, 326
Hamiltonian
function, 301
heat exchanger
jacketed, 19

identifiability, 239
identification
continuous-time transfer
function, 245
discrete-time transfer
function, 233
parameter convergence, 239
step responses, 225
Broida method, 232
first order system, 226
higher order system, 230
second order system, 227
Strejc method, 230
identification methods, 222
impossible event, 162
integrator windup, 277
inverse Z transform, 197
partial fraction expansion, 199
polynomial division, 198

Kalman filter, 337

Laplace
transform, 51, 52
convolution, 58
derivatives, 57
integral, 57
inverse, 52, 61
properties, 57
time delay, 59
unit impulse, 60
least squares method, 233
recursive, 235
linearisation, 39
method, 39
LQ control, 380
LQG control, 388, 390
Luenberger observer, 335

magnitude frequency spectrum, 191
MATLAB, 130
matrix
auto-covariance, 172
eigenvalues, 70
exponential, 69
frequency transfer function, 102
fundamental, 69
impulse responses, 100
observability, 82
polynomial, 107
common divisor, 110
coprime, 110
degree, 108
determinant degree, 110
division algorithm, 111
greatest common divisor, 110
irreducibility, 110
left fraction, 110
rank, 109
relatively prime, 111
right fraction, 110
spectral factorisation, 112
stable, 109
unimodular, 109
rational
proper, 102
strictly proper, 102
state transition, 69
system, 70, 77
transfer function, 100, 101
matrix inversion lemma, 236
maximum overshoot, 257
method
Euler, 126
Runge-Kutta, 127
mixing
ideal, 14
model
batch, 14
continuous, 14
discrete-time, 189
controllability, 212
examples, 214
input-output, 204, 208
observability, 212
stability, 211
state-space, 207
empirical, 14

empirical-theoretical, 14
input output, 84
MIMO, 100
realisation, 101
linear, 39
mathematical, 13
theoretical, 14
moment
m-th, 167
central, 167
second, 167

Naslin method, 282
norm
vector, 73
normal distribution, 168
Nyquist diagram, 152

observability, 81
matrix, 82
observable canonical form, 210
observer, 335
optimal control, 297
optimal feedback, 331
optimal tracking, 317
Output Error model, 225

parametrisation
discrete-time systems, 375
persistent excitation, 239
phase margin, 261
physical realisability, 85
PID controller, 266
digital implementation, 279
integrator windup, 277
tuning, 281
Åström-Hägglund methods, 290
experimental methods, 287
Naslin method, 282
pole placement, 284
standard forms, 281
Strejc method, 283
Ziegler-Nichols methods, 288
PID controller structures, 274
pole placement, 284, 317, 350
polynomial, 351
pole placement design, 345
poles, 101, 102
polynomial
characteristic, 77
polynomial design, 350
multivariable systems, 361
polynomial matrix
fraction, 107
power spectral density, 175
predictive control, 403
closed-loop, 411
constraints, 419
cost function, 406
derivation, 407
implementation, 416
model, 404
stability, 420
state-space, 412
tuning, 431
principle of optimality, 323
principle of minimum, 297
probability, 162
conditional, 162
density, 171
joint density, 168
law of distribution, 164
theory, 162
probability density, 167
process, 3
analysis, 51
automatic control, 4
control, 4
feedback, 5
distributed parameters, 14
dynamical model, 13
dynamical properties, 4
ergodic, 173
general model, 33
heat transfer, 19
liquid storage, 16
interacting tank, 17
single tank, 16
lumped parameters, 14
mass transfer, 26
mathematical model, 13
model
input-output, 51
non-stationary, 173
random, 170
response, 5, 6
space discretisation, 35
state space model, 101

stationary, 173
stochastic, 170
expected value, 171
realisation, 170
variance, 171
vector mean value, 172
process identification, 221
process models, 223
stochastic, 223

random event, 162
reachability, 79
complete, 79
reactor
biochemical, 31, 33
chemical, 31
CSTR, 31
realisation
minimum, 101
recursive least squares method, 235
modifications, 241
resonance peak, 261
resonant frequency, 261
Riccati equation, 309
rise time, 258
Runge-Kutta method, 127

sampler, 189
sampling period, 189
self-tuning control, 445
continuous-time, 447
discrete-time, 446
separation principle, 341
servo problem, 318, 320
settling time, 258
Simulink, 130
stability, 72
asymptotic, 73, 77
linear system, 77
asymptotic in large, 73, 77
BIBO, 103
external, 103
Lyapunov, 73, 76
system
continuous, 72
state
concept, 45, 67
observability, 81, 82
complete, 81
reconstructibility, 81
complete, 81
steady, 3
deviation, 39
unsteady, 5
state estimation, 335
state plane, 75
state space, 75
representation, 66
state trajectory, 75
steady state, 3
steady-state behaviour, 256
step function, 119
step response, 119
step responses
identification, 225
stochastic process models, 223
Strejc method of controller tuning, 283
Strejc method of identification, 230
sure event, 162
system
abstract, 45
autonomous, 72
causal, 118
continuous, 45
stability, 72
deterministic
MIMO, 45
SISO, 45
dynamical, 46
equilibrium state, 72
forced, 72
free, 69, 72
linear, 45
nonautonomous, 72
order, 101
physical, 45
poles, 101, 207, 263
dominant, 263
representation
input-output, 51
state space, 51
response, 117, 119
stochastic, 46
unit impulse response, 117
zeros, 101, 207

Taylor series, 41
Taylor theorem, 39

theorem
  existence of optimal control, 327
  LQ control with observer, 382
  pole placement, 345
time constant, 6, 120
time delay, 59
tracking problem, 318
transfer function, 85
  algebra, 96
  discrete-time, 200, 202
  frequency, 145
  matrix, 100, 210
  parametrisation, 371
transform
  Fourier, 150
transformation
  canonical, 70

value
  expected, 164
  random, 162
variable
  disturbance, 4, 35
  expected value, 164
  input, 3
  manipulated, 4, 35
    deviation, 40
  output, 3, 35
  random, 162, 163
    continuous, 166
    discrete, 163
    standard deviation, 168
  setpoint, 4
  state, 14, 35
    deviation, 40
  variance, 165
variance, 171
variational calculus, 299

white noise, 176

Youla-Kučera parametrisation, 365

zeros, 101
Ziegler-Nichols methods, 288
Z transform, 195
  exponential function, 197
  inverse, 197
  linearity, 197
  table, 198
  unit step function, 196

MIX
Papier aus verantwortungsvollen Quellen
Paper from responsible sources
FSC® C105338
FSC
www.fsc.org

If you have any concerns about our products,
you can contact us on
ProductSafety@springernature.com

In case Publisher is established outside the EU,
the EU authorized representative is:
Springer Nature Customer Service Center GmbH
Europaplatz 3, 69115 Heidelberg, Germany

Printed by Libri Plureos GmbH
in Hamburg, Germany